U0358693

全国科学技术名词审定委员会

公　布

科学技术名词·工程技术卷（全藏版）

38

铁 道 科 技 名 词

CHINESE TERMS IN RAILWAY SCIENCE AND TECHNOLOGY

铁道科学名词审定委员会

国家自然科学基金资助项目

科学出版社

北京

内 容 简 介

本书是全国科学技术名词审定委员会审定公布的第一批铁道科技名词。包括通类、铁道工程建设、铁道工务、铁道牵引动力、铁道车辆、铁道运输管理与运输经济、铁道通信、铁道信号等八类，共 6 891 条。部分名词有简明定义性注释。书末附有英汉和汉英两种索引，以利读者检索。本书是科研、教学、生产、经营以及新闻出版等部门使用的铁道科技规范名词。

图书在版编目(CIP)数据

科学技术名词. 工程技术卷：全藏版 / 全国科学技术名词审定委员会审定.
—北京：科学出版社，2016.01
ISBN 978-7-03-046873-4

I. ①科…　II. ①全…　III. ①科学技术–名词术语　②工程技术–名词术语
IV. ①N-61 ②TB-61

中国版本图书馆 CIP 数据核字(2015)第 307218 号

责任编辑：李玉英 / 责任校对：陈玉凤
责任印制：张　伟 / 封面设计：铭轩堂

科学出版社 出版
北京东黄城根北街 16 号
邮政编码：100717
http://www.sciencep.com
北京厚诚则铭印刷科技有限公司印刷
科学出版社发行　各地新华书店经销
*
2016 年 1 月第　一　版　　开本：787×1092 1/16
2016 年 1 月第一次印刷　　印张：28 3/4
字数：834 000
定价：7800.00 元(全 44 册)
(如有印装质量问题，我社负责调换)

全国自然科学名词审定委员会
第三届委员会委员名单

特邀顾问：　吴阶平　　　钱伟长　　　朱光亚

主　　任：　卢嘉锡

副 主 任：　路甬祥　　　章　综　　　林　泉　　　左铁镛　　　马　阳

　　　　　　孙　枢　　　许嘉璐　　　于永湛　　　丁其东　　　汪继祥

　　　　　　潘书祥

委　　员 （以下按姓氏笔画为序）：

马大猷	王　夔	王大珩	王之烈	王亚辉
王树岐	王绵之	玉窝骧	方鹤春	卢良恕
叶笃正	吉木彦	师昌绪	朱照宣	仲增墉
华茂昆	刘天泉	刘瑞玉	米吉提·扎克尔	
祁国荣	孙家栋	孙儒泳	李正理	李廷杰
李行健	李　竞	李星学	李焯芬	肖培根
杨　凯	吴凤鸣	吴传钧	吴希曾	吴钟灵
吴鸿适	沈国舫	宋大祥	张　伟	张光斗
张钦楠	陆建勋	陆燕荪	陈运泰	陈芳允
范维唐	周　昌	周明煜	周定国	罗钰如
季文美	郑光迪	赵凯华	侯祥麟	姚世全
姚贤良	姚福生	夏　铸	顾红雅	钱临照
徐　僖	徐士珩	徐乾清	翁心植	席泽宗
谈家桢	黄昭厚	康景利	章　申	梁晓天
董　琨	韩济生	程光胜	程裕淇	鲁绍曾
曾呈奎	蓝　天	褚善元	管连荣	薛永兴

铁道科学名词审定委员会委员名单

顾　　问：谭葆宪　　屠由瑞　曹建猷

主　　任：沈之介

副主任：程庆国　　袁国铎　　梁家瑞

委　　员（按姓氏笔画为序）：

万复光	马时亮	王克让	王泳琨	王效良
卢祖文	叶培礼	吕孝	朱思本	刘怿
刘友梅	纪宴宁	苏民	杜庆萱	李雨生
李承恕	李绍光	杨国桢	杨照久	吴凤
张阳生	陈朝贵	胡安洲	费名申	秦淞君
顾发祥	钱仲侯	徐威	高家驹	黄问盈
商福崑	韶能仁			

秘　　书：韶能仁（兼）

序

科技名词术语是科学概念的语言符号。人类在推动科学技术向前发展的历史长河中，同时产生和发展了各种科技名词术语，作为思想和认识交流的工具，进而推动科学技术的发展。

我国是一个历史悠久的文明古国，在科技史上谱写过光辉篇章。中国科技名词术语，以汉语为主导，经过了几千年的演化和发展，在语言形式和结构上体现了我国语言文字的特点和规律，简明扼要，蓄意深切。我国古代的科学著作，如已被译为英、德、法、俄、日等文字的《本草纲目》、《天工开物》等，包含大量科技名词术语。从元、明以后，开始翻译西方科技著作，创译了大批科技名词术语，为传播科学知识，发展我国的科学技术起到了积极作用。

统一科技名词术语是一个国家发展科学技术所必须具备的基础条件之一。世界经济发达国家都十分关心和重视科技名词术语的统一。我国早在1909年就成立了科技名词编订馆，后又于1919年中国科学社成立了科学名词审定委员会，1928年大学院成立了译名统一委员会。1932年成立了国立编译馆，在当时教育部主持下先后拟订和审查了各学科的名词草案。

新中国成立后，国家决定在政务院文化教育委员会下，设立学术名词统一工作委员会，郭沫若任主任委员。委员会分设自然科学、社会科学、医药卫生、艺术科学和时事名词五大组，聘任了各专业著名科学家、专家，审定和出版了一批科学名词，为新中国成立后的科学技术的交流和发展起到了重要作用。后来，由于历史的原因，这一重要工作陷于停顿。

当今，世界科学技术迅速发展，新学科、新概念、新理论、新方法不断涌现，相应地出现了大批新的科技名词术语。统一科技名词术语，对科学知识的传播，新学科的开拓，新理论的建立，国内外科技交流，学科和行业之间的沟通，科技成果的推广、应用和生产技术的发展，科技图书文献的编纂、出版和检索，科技情报的传递等方面，都是不可缺少的。特别是计算机技术的推广使用，对统一科技名词术语提出了更紧迫的要求。

为适应这种新形势的需要，经国务院批准，1985年4月正式成立了全国自然科学名词审定委员会。委员会的任务是确定工作方针，拟定科技名词术语审定工作计划、实施方案和步骤,组织审定自然科学各学科名词术语,并予以公布。根据国务院授权,委员会审定公布的名词术语,科研、教学、生产、经营以及新闻出版等各部门,均应遵照使用。

全国自然科学名词审定委员会由中国科学院、国家科学技术委员会、国家教育委员会、中国科学技术协会、国家技术监督局、国家新闻出版署、国家自然科学基金委员会分别委派了正、副主任担任领导工作。在中国科协各专业学会密切配合下,逐步建立各专业审定分委员会,并已建立起一支由各学科著名专家、学者组成的近千人的审定队伍,负责审定本学科的名词术语。我国的名词审定工作进入了一个新的阶段。

这次名词术语审定工作是对科学概念进行汉语订名,同时附以相应的英文名称,既有我国语言特色,又方便国内外科技交流。通过实践,初步摸索了具有我国特色的科技名词术语审定的原则与方法,以及名词术语的学科分类、相关概念等问题,并开始探讨当代术语学的理论和方法,以期逐步建立起符合我国语言规律的自然科学名词术语体系。

统一我国的科技名词术语,是一项繁重的任务,它既是一项专业性很强的学术性工作,又涉及到亿万人使用习惯的问题。审定工作中我们要认真处理好科学性、系统性和通俗性之间的关系;主科与副科间的关系;学科间交叉名词术语的协调一致;专家集中审定与广泛听取意见等问题。

汉语是世界五分之一人口使用的语言,也是联合国的工作语言之一。除我国外,世界上还有一些国家和地区使用汉语,或使用与汉语关系密切的语言。做好我国的科技名词术语统一工作,为今后对外科技交流创造了更好的条件,使我炎黄子孙,在世界科技进步中发挥更大的作用,作出重要的贡献。

统一我国科技名词术语需要较长的时间和过程,随着科学技术的不断发展,科技名词术语的审定工作,需要不断地发展、补充和完善。我们将本着实事求是的原则,严谨的科学态度作好审定工作,成熟一批公布一批,提供各界使用。我们特别希望得到科技界、教育界、经济界、文化界、新闻出版界等各方面同志的关心、支持和帮助,共同为早日实现我国科技名词术语的统一和规范化而努力。

全国自然科学名词审定委员会主任

钱 三 强

1990 年 2 月

前　　言

多年来,铁道界的专家们在名词标准化方面进行了大量的工作。但过去的标准,大都是根据某一专业或某项工作当时的需要制订的,彼此间缺乏足够的联系,在铁道内更缺乏整体的、系统的安排。1986 年以后,全国性的科技名词标准化工作相继开展,作为铁道科技基础或支撑的许多学科,也纷纷进行工作,这就为此项工作提供了新的基础,因而铁道界对科技名词进行全面和系统的编纂和修订势在必然。

根据全国科学技术名词审定委员会(以下简称全国名词委,原名为全国自然科学名词审定委员会)的统一规划,铁道科学名词审定委员会(以下简称铁道名词委)于 1991 年 1 月成立。铁道名词委下成立了工程建设、工务、牵引动力、车辆、运输管理与运输经济、通信信号六个专业编写组,广泛参阅了五、六十种相关的国标、行标和大量国内外其他资料,于 1991 年秋天提出草稿。经初步整理形成了拥有 7 000 多词条的初稿,并提交第二次全体委员会议(1992 年 1 月)进行深入细致的审议,审定出 6 810 条的"征求意见稿",发出约250 份向全路广泛征求意见。收回意见和建议约 180 份。经各专业组反复筛选评议后,编纂成拥有词条总数为 7 118 条的"送审稿"提交给第三次全体委员会议(1994 年 1 月)审议。审议后编纂成上报国家的词条为 6 891 条的"报批稿"。全国名词委接稿后还特聘谭葆宪、曹建猷、沈之介、程庆国、张锡第等五位专家对报批稿进行了复审,尔后予以批准公布。现在出版的铁道科技名词拥有词 6 891 条,内分"通类"、"铁道工程建设"、"铁道工务"、"铁道牵引动力"、"铁道车辆"、"铁道运输管理与运输经济"、"铁道通信"、"铁道信号"八大部份。

对铁道而言,有关的科技名词数量大、实用性强。选词的深度和广度,一直是我们工作注意的焦点。作为基础文献的"铁道科技名词",必须既有科学性又有实用性。过细,对词条的要求将庞杂无章;过粗,实用中将难具参考价值。因此,在编纂中我们严格遵循词的单义性、科学性、系统性、简明性和约定俗成五项原则,精选一些基础词、重要词和常用词,并注意各词在实际工作中的应用情况。例如:"铁道"和"铁路"原为二个词条,经反复考虑,现统称"铁道",注明又称"铁路";在采用"内燃机车"或"柴油机车"词条问题上,意见纷纷,现按科学性、实用性和约定俗成相结合原则,采用"内燃机车",并注明曾用名"柴油机车";又如"大陆桥"一词,在我国已约定俗成,采用后注明又称"洲际铁路"。关于"重量"与"质量"、"压力"与"压强"、"速度"与"速率"等词,虽然从科学性看,后者为宜,但考虑到铁道多年实际情况,仍尊重工程界习惯而采用"重量"、"压力"、"速度"。至于汉文词条的英文译名,各方关心之情更切。有的词条收集到的译名多达四、五条。因篇幅所限,有的不得不割爱,现规定每一汉文词条的英文译名不超过三个,一般为一个或两个。

从 1991 到 1994 的三年时间里,铁道名词委的委员们为高质量完成这项国家任务作出了无私的奉献。在全国名词委和铁道部的大力支持下,我们铁道名词委的工作还得到了全路许多单位、专家、学者和职工的热情关怀和多方帮助。现在出版的"铁道科技名词"尽管还有不完善之处,但它凝聚了全路科技工作者的心血和智慧,也是他们对推动我国科技基础事业所作的具体奉献。对于这项工作,铁道部给予了大量资助,作为挂靠单位的铁道学会一直在人力和工作上热情支持,铁道科学研究院标准所不仅在人力物力上大力支持,且自始至终承担了大量汇总、整理工作。此外,还有不少专家和同志,如梁秉智、吴钰、李志炘、刘彦邦、张锡第、陈崎生等,在编纂过程中协助铁道名词委做了许多工作。在此,对所有关心和支持本书编审出版的单位、领导、专家、学者和同志,表示衷心的感谢和诚挚的敬意。希望各界继续提出宝贵意见,以便今后进一步修订。

铁道科学名词审定委员会

1996 年 6 月

编 排 说 明

一、本批公布的是铁道科技的基本名词。

二、全书按学科分为通类、铁道工程建设、铁道工务、铁道牵引动力、铁道车辆、铁道运输管理与运输经济、铁道通信、铁道信号等八类。

三、正文中的汉文名词按学科的相关概念排列，并附有与该词对应的同一概念英文名。

四、一个汉文名词如对应几个英文同义词时，一般只配最常用的一个或两个英文，并用"，"分开。

五、凡英文词的首字母大、小写均可时，一律小写。

六、对某些新词及概念易混淆的词给出简明的定义性注释。

七、主要异名列在注释栏内。"又称"为不推荐用名；"曾用名"为被淘汰的旧名。

八、名词中 〔 〕部分的内容表示可以省略。

九、书末所附的英汉索引，按英文字母顺序排列；汉英索引，按汉语拼音顺序排列。所示号码为该词在正文中的序号。索引中带"＊"者为在注释栏内的条目。

目　　录

01. 通 类

序 码	汉 文 名	英 文 名	注 释
01.0001	铁道	railway, railroad	又称"铁路"。
01.0002	铁路线	railway line, railroad line	
01.0003	铁路网	railway network, railroad network	
01.0004	铁道科学	railway science	
01.0005	铁路技术	railway technology	
01.0006	铁路等级	railway classification	
01.0007	国有铁路	national railway, state railway	
01.0008	地方铁路	local railway, regional railway	
01.0009	私有铁路	private railway	
01.0010	合资铁路	joint investment railway, jointly owned railway	
01.0011	准轨铁路	standard-gage railway	
01.0012	窄轨铁路	narrow-gage railway	
01.0013	米轨铁路	meter-gage railway	
01.0014	宽轨铁路	broad-gage railway	
01.0015	单线铁路	single track railway	
01.0016	双线铁路	double track railway	
01.0017	多线铁路	multiple track railway	
01.0018	重载铁路	heavy haul railway	
01.0019	高速铁路	high speed railway	
01.0020	电气化铁路	electrified railway, electric railway	又称"电力铁道"。
01.0021	干线铁路	main line railway, trunk railway	
01.0022	市郊铁路	suburban railway	
01.0023	地下铁道	subway, metro, underground railway	简称"地铁"。
01.0024	工业企业铁路	industry railway	
01.0025	矿山铁路	mine railway	
01.0026	轻轨铁路	light railway, light rail	
01.0027	高架铁路	elevated railway	
01.0028	单轨铁路	monorail, monorail railway	又称"独轨铁路"。
01.0029	磁浮铁路	magnetic levitation railway, maglev	

序　码	汉 文 名	英 文 名	注　释
01.0030	森林铁路	forest railway	
01.0031	山区铁路	mountain railway	
01.0032	既有铁路	existing railway	
01.0033	新建铁路	newly-built railway	
01.0034	改建铁路	reconstructed railway	
01.0035	运营铁路	railway in operation, operating railway	
01.0036	专用铁路	special purpose railway	
01.0037	干线	trunk line, main line	
01.0038	支线	branch line	
01.0039	铁路专用线	railway special line	
01.0040	货运专线	railway line for freight traffic, freight special line, freight traffic only line	
01.0041	客运专线	railway line for passenger traffic, passenger special line, passenger traffic only line	
01.0042	客货运混合线路	railway line for mixed passenger and freight traffic	
01.0043	铁路运营长度	operating length of railway, operating distance, revenue length	又称"运营里程"。
01.0044	列车运行图	train diagram	
01.0045	铁路建筑长度	construction length of railway	
01.0046	区间	section	
01.0047	区段	district	
01.0048	轨距	rail gage, rail gauge	
01.0049	轮重	wheel load	
01.0050	轴重	axle load	
01.0051	最大轴重	maximum allowable axle load	车轴允许负担的最大重量与轮对自重之总和。
01.0052	限制轴重	axle load limited	
01.0053	限界	clearance, gauge	
01.0054	限界图	clearance diagram	
01.0055	铁路建筑限界	railway construction clearance, structure clearance for railway, railway structure gauge	

序 码	汉 文 名	英 文 名	注 释
01.0056	基本建筑限界	fundamental construction clearance, fundamental structure gauge	
01.0057	桥梁建筑限界	bridge construction clearance, bridge structure gauge	
01.0058	隧道建筑限界	tunnel construction clearance, tunnel structure gauge	
01.0059	铁路机车车辆限界	rolling stock clearance [for railway], vehicle gauge	
01.0060	机车车辆上部限界	clearance limit for upper part of rolling stock	
01.0061	机车车辆下部限界	clearance limit for lower part of rolling stock	
01.0062	装载限界	loading clearance limit, loading gauge	
01.0063	阔大货物限界	clearance limit for freight with exceptional dimension, clearance limit for oversize commodities	
01.0064	接触网限界	clearance limit for overhead contact wire, clearance limit for overhead catenary system, overhead catenary system gauge	
01.0065	列车与线路相互作用	track-train interaction	
01.0066	轮轨关系	wheel-rail relation, wheel-rail interaction	
01.0067	粘着系数	adhesion coefficient	
01.0068	车轮滑行	wheel sliding, wheel skid	制动使车辆轮对与轨面间产生滑动摩擦。
01.0069	车轮空转	wheel slipping	
01.0070	牵引种类	kinds of traction, category of traction	
01.0071	牵引方式	mode of traction	
01.0072	牵引定数	tonnage rating, tonnage of traction	
01.0073	装载系数	loading coefficient	
01.0074	速度	speed	
01.0075	持续速度	continuous speed	
01.0076	限制速度	limited speed, speed restriction	

序 码	汉 文 名	英 文 名	注 释
01.0077	均衡速度	balancing speed	
01.0078	构造速度	construction speed, design speed	
01.0079	最高速度	maximum speed	
01.0080	临界速度	critical speed	
01.0081	重载列车	heavy haul train	
01.0082	高速列车	high speed train	
01.0083	超长超重列车	exceptionally long and heavy train	
01.0084	列车正面冲突	train collision	
01.0085	列车尾追	train tail collision	
01.0086	列车尾部防护	train rear end protection	
01.0087	伸缩运动	fore and aft motion	
01.0088	蛇行运动	hunting, nosing	
01.0089	列车压缩	train running in	
01.0090	列车拉伸	train running out	
01.0091	列车分离	train separation	
01.0092	列车颠覆	train overturning	
01.0093	列车动力学	train dynamics	
01.0094	列车空气动力学	train aerodynamics	
01.0095	机车车辆振动	vibration of rolling stock	
01.0096	纵向振动	longitudinal vibration	
01.0097	横向振动	lateral vibration	
01.0098	垂向振动	vertical vibration	
01.0099	摆滚振动	rock-roll vibration	
01.0100	浮沉振动	bouncing [vibration]	
01.0101	侧滚振动	rolling [vibration]	
01.0102	侧摆振动	swaying [vibration]	
01.0103	点头振动	pitching [vibration], nodding [vibration]	
01.0104	摇头振动	yawing [vibration], hunting [vibration]	
01.0105	机车车辆共振	resonance of rolling stock	
01.0106	机车车辆冲击	impact of rolling stock	
01.0107	纵向冲击	longitudinal impact	
01.0108	横向冲击	lateral impact	
01.0109	垂向冲击	vertical impact	
01.0110	货运站综合作业自动化	automation of synthetic operations at freight station	

序　码	汉文名	英文名	注　释
01.0111	行车指挥自动化	automation of traffic control	
01.0112	编组场综合作业自动化	automation of synthetic operations in marshalling yard	
01.0113	铁路运营信息系统	railway operation information system	
01.0114	铁路数据交换系统	railway data exchange system	
01.0115	运营系统模拟	simulation of operation system	
01.0116	铁路法	Railway Law	
01.0117	铁道法规	Railway Act	
01.0118	铁路条例	railway code	
01.0119	铁路技术管理规程	regulations of railway technical operation	
01.0120	综合运输	comprehensive transport, multimode transport, intermode transport	
01.0121	国际铁路联运	international railway through traffic	
01.0122	大陆桥	transcontinental railway, intercontinental railway, land-railway	又称"洲际铁路"。
01.0123	国际联运协定	agreement of international through traffic	
01.0124	国际联运议定书	protocol of international through traffic	
01.0125	国际铁路联运公约	convention of international railway through traffic	

02. 铁道工程建设

序　码	汉文名	英文名	注　释
02.0001	铁路新线建设	newly-built railway construction	
02.0002	铁路技术改造	technical reform of railway, technical renovation of railway, betterment and improvement of railway	

序　码	汉　文　名	英　文　名	注　　释
02.0003	铁路主要技术条件	main technical standard of railway, main technical requirement of railway	
02.0004	单位工程	unit project	
02.0005	分部工程	part project	
02.0006	分项工程	item project	
02.0007	预可行性研究	pre-feasibility study	
02.0008	项目建议书	proposed task of project	
02.0009	可行性研究	feasibility study	
02.0010	设计阶段	design phase, design stage	
02.0011	三阶段设计	three-step design, three-phase design	
02.0012	两阶段设计	two-step design, two-phase design	
02.0013	一阶段设计	one-step design, one-phase design	
02.0014	初步设计	preliminary design	
02.0015	技术设计	technical design	
02.0016	扩大初步设计	enlarged preliminary design, expanded preliminary design	
02.0017	施工图设计	construction detail design, working-drawing design	
02.0018	变更设计	altered design	
02.0019	设计概算	approximate estimate of design, budgetary estimate of design	
02.0020	个别概算	individual approximate estimate	
02.0021	综合概算	comprehensive approximate estimate	
02.0022	总概算	sum of approximate estimate, total estimate, summary estimate	
02.0023	修正总概算	amended sum of approximate estimate, revised general estimate	
02.0024	调整总概算	adjusted sum of approximate estimate	
02.0025	投资检算	checking of investment	
02.0026	预算定额	rating of budget, rating form for budget	
02.0027	概算定额	rating of approximate estimate, rating form for estimate	

序 码	汉 文 名	英 文 名	注 释
02.0028	投资估算	investment estimate	
02.0029	估算指标	index of estimate	
02.0030	机械台班定额	rating per machine per team, rating per machine-team	
02.0031	工程直接费	direct expense of project, direct cost of project	
02.0032	工程间接费	indirect expense of project, indirect cost of project	
02.0033	工程预备费	reserve fund of project	
02.0034	设计鉴定	certification of design, appraisal of design	
02.0035	竣工决算	final accounts of completed project	
02.0036	铁路用地	right-of-way	
02.0037	铁路勘测	railway reconnaissance	
02.0038	调查测绘	survey and drawing of investigation, investigation survey, investigation surveying and sketching	
02.0039	地形调查	topographic survey	
02.0040	地貌调查	topographic feature survey, geomorphologic survey	
02.0041	地质调查	geologic survey	
02.0042	经济调查	economic investigation, economic survey	
02.0043	水文地质调查	hydrogeologic survey	
02.0044	土石成分调查	survey of soil and rock composition	
02.0045	土石物理力学性质	physical and mechanical properties of soil and rock	
02.0046	土石分类	classification of soil and rock	
02.0047	地基承载力	bearing capacity of foundation, bearing capacity of ground, bearing capacity of subgrade	
02.0048	隧道围岩分级	classification of tunnel surrounding rock	
02.0049	地质图测绘	survey and drawing of geological map, surveying and sketching of geological map	

序 码	汉 文 名	英 文 名	注 释
02.0050	勘探	exploration, prospecting	
02.0051	挖探	excavation prospecting	
02.0052	钻探	boring [prospecting], exploration drilling	
02.0053	物探	geophysical prospecting	
02.0054	室内测试	indoor test, laboratory test	
02.0055	原位测试	*in situ* test	
02.0056	静力触探	static sounding, static probing, cone penetration test	
02.0057	动力触探试验	dynamic penetration test	
02.0058	标准贯入试验	standard penetration test	
02.0059	区域地质	regional geology	
02.0060	工程地质	engineering geology	
02.0061	不良地质	unfavorable geology	
02.0062	特殊地质	special geology	
02.0063	工程地质条件	engineering geologic requirement, engineering geologic condition	
02.0064	气象资料	meteorological data	
02.0065	冻结深度	freezing depth	
02.0066	地震基本烈度	basic intensity of earthquake, seismic basic intensity	
02.0067	工程地质图	engineering geological map	
02.0068	地层柱状图	column diagram of stratum, graphic logs of strata, drill log of stratum	
02.0069	洪水调查	flood survey	
02.0070	河道调查	river course survey	
02.0071	冰凌调查	ice floe survey, frazil ice survey	
02.0072	汇水区流域特征调查	survey of catchment basin characteristics	
02.0073	水文断面	hydrologic sectional drawing, hydrologic section, hydrologic cross-section	
02.0074	主河槽	main river channel	
02.0075	设计流速	design current velocity	
02.0076	设计高程	design elevation	又称"设计标高"。
02.0077	河流比降	slope of river, comparable horizon	

序　码	汉　文　名	英　文　名	注　　释
		of river	
02.0078	历史洪水位	historic flood level	
02.0079	最高水位	highest water level, HWL	
02.0080	通航水位	navigation water level, NWL	
02.0081	桥涵水文	hydrology of bridge and culvert	
02.0082	水力半径	hydraulic radius	
02.0083	桥前壅水高度	backwater height in front of bridge, top water level in front of bridge	
02.0084	桥渡勘测设计	survey and design of bridge crossing	
02.0085	水面坡度	slope of water surface	
02.0086	水文测量	hydrological survey	
02.0087	泥石流流域	catchment basin of debris flow	
02.0088	分水岭	watershed, dividing ridge	
02.0089	汇水面积	catchment area, water collecting area, drainage area	
02.0090	洪水频率	flood frequency	
02.0091	设计流量	design discharge	
02.0092	设计水位	design water level	
02.0093	施工水位	construction level, construction water level, working water level	
02.0094	设计洪水过程线	designed flood hydrograph	
02.0095	容许冲刷	allowable scour	
02.0096	一般冲刷	general scour	
02.0097	局部冲刷	local scour, partial scour	
02.0098	铁路测量	railway survey	
02.0099	线路踏勘	route reconnaissance	又称"草测"。
02.0100	初测	preliminary survey	
02.0101	定测	location survey, alignment, final location survey	
02.0102	导线测量	traversing, traverse survey	
02.0103	光电导线	photoelectric traverse	
02.0104	地形测量	topographical survey	
02.0105	横断面测量	cross leveling, cross-section survey, cross-section leveling	
02.0106	线路测量	route survey, profile survey, lon-	

序　码	汉　文　名	英　文　名	注　释
		gitudinal survey	
02.0107	既有线测量	survey of existing railway	又称"旧线测量"。
02.0108	线路复测	repetition survey of existing railway, resurvey of existing railway	
02.0109	测量精度	survey precision, precision of survey	
02.0110	均方[误]差	mean square error	又称"中误差"。
02.0111	最大误差	maximum error, limiting error	又称"极限误差"。
02.0112	中线测量	center line survey	
02.0113	中线桩	center line stake	
02.0114	加桩	additional stake, plus stake	
02.0115	外移桩	shift out stake, stake outward, offset stake	
02.0116	水准点高程测量	benchmark leveling	又称"基平"。
02.0117	中桩高程测量	center stake leveling	又称"中平"。
02.0118	曲线控制点	curve control point	
02.0119	放线	setting-out of route, lay out of route	
02.0120	交点	intersection point	
02.0121	副交点	auxiliary intersection point	
02.0122	转向角	deflection angle	
02.0123	分转向角	auxiliary deflection angle	
02.0124	坐标方位角	plane-coordinate azimuth	
02.0125	象限角	quadrantal angle	
02.0126	经纬距	plane rectangular coordinate	
02.0127	断链	broken chain	
02.0128	投影断链	projection of broken chain	
02.0129	断高	broken height	
02.0130	铁路航空摄影测量	railway aerial photogrammetry	简称"铁路航测"。
02.0131	铁路航空勘测	railway aerial surveying	
02.0132	航带设计	flight strip design, design of flight strip	
02.0133	铁路工程地质遥感	remote sensing of railway engineering geology	
02.0134	测段	segment of survey	

序 码	汉 文 名	英 文 名	注 释
02.0135	航测选线	aerial surveying alignment	
02.0136	航测外控点	field control point of aerophoto-grammetry	
02.0137	全球定位系统	global positioning system, GPS	
02.0138	镶辑复照图	index of photography	又称"像片索引图"。
02.0139	三角测量	trigonometric survey, triangula-tion	
02.0140	精密导线测量	precise traverse survey, accurate traverse survey	
02.0141	三角高程测量	trigonometric leveling	
02.0142	隧道洞外控制测量	outside tunnel control survey	
02.0143	隧道洞内控制测量	intunnel control survey, through survey	又称"贯通测量"。
02.0144	隧道洞口投点	horizontal point of tunnel portal, geodetic control point of portal location of adit	
02.0145	桥轴线测量	survey of bridge axis	
02.0146	铁路选线	railway location, approximate rail-way location, location of rail-way route selection	
02.0147	平原地区选线	location in plain region, plain loca-tion	
02.0148	越岭选线	location of mountain line, location of line in mountain region, loca-tion over mountain	
02.0149	山区河谷选线	mountain and valley region loca-tion, location of line in moun-tain and valley region	
02.0150	丘陵地段选线	hilly land location, location of line on hilly land	
02.0151	工程地质选线	engineering geologic location of line	
02.0152	线间距[离]	distance between centers of tracks, midway between tracks	
02.0153	车站分布	distribution of stations	
02.0154	方案比选	scheme comparison, route alterna-	

序 码	汉 文 名	英 文 名	注 释
		tive	
02.0155	投资回收期	repayment period of capital cost	
02.0156	纸上定线	paper location of line	
02.0157	缓坡地段	section of easy grade, section of gentle slope	
02.0158	紧坡地段	section of sufficient grade	
02.0159	非紧坡地段	section of unsufficient grade, section of insufficient grade	
02.0160	导向线	leading line, alignment guiding line	
02.0161	拔起高度	height of lifting, lifting height, ascent of elevation	又称"克服高度"。
02.0162	横断面选线	cross-section method of railway location, location with cross-section method, cross-section method for location of line	
02.0163	展线	extension of line, development of line, line development	
02.0164	展线系数	coefficient of extension line, coefficient of developed line	
02.0165	套线	overlapping line	
02.0166	线路平面图	track plan, line plan	
02.0167	线路纵断面图	track profile, line profile	
02.0168	站坪长度	length of station site	
02.0169	站坪坡度	grade of station site	
02.0170	控制区间	control section, controlling section	
02.0171	最小曲线半径	minimum radius of curve	
02.0172	圆曲线	circular curve	
02.0173	单曲线	simple curve	
02.0174	缓和曲线	transition curve, easement curve, spiral transition curve	
02.0175	缓和曲线半径变更率	rate of easement curvature, rate of transition curve	
02.0176	复曲线	compound curve	
02.0177	同向曲线	curves of same sense, adjacent curves in one direction	
02.0178	反向曲线	reverse curve, curve of opposite	

序 码	汉 文 名	英 文 名	注 释
		sense	
02.0179	夹直线	intermediate straight line, tangent between curves	
02.0180	坡度	grade, gradient, slope	
02.0181	人字坡	double spur grade	
02.0182	限制坡度	ruling grade, limiting grade	
02.0183	加力牵引坡度	pusher grade, assisting grade	
02.0184	最大坡度	maximum grade	
02.0185	临界坡度	critical grade	
02.0186	长大坡道	long steep grade, long heavy grade	
02.0187	动力坡度	momentum grade	
02.0188	均衡坡度	balanced grade	
02.0189	有害地段	harmful district	
02.0190	无害地段	harmless district	
02.0191	变坡点	point of gradient change, break in grade	
02.0192	坡段	grade section	
02.0193	坡段长度	length of grade section	
02.0194	坡度[代数]差	algebraic difference between adjacent gradients	
02.0195	竖曲线	vertical curve	
02.0196	分坡平段	level stretch between opposite sign gradient	
02.0197	缓和坡度	slight grade, flat grade, easy grade	
02.0198	起动缓坡	flat gradient for starting	
02.0199	加速缓坡	easy gradient for acceleration, accelerating grade	
02.0200	坡度折减	compensation of gradient, gradient compensation, grade compensation	
02.0201	曲线折减	compensation of curve, curve compensation	
02.0202	隧道坡度折减	compensation of gradient in tunnel, compensation grade in tunnel	
02.0203	绕行地段	detouring section, round section	

序　码	汉　文　名	英　文　名	注　释
02.0204	换侧	change side of double line	又称"换边"。
02.0205	容许应力设计法	allowable stress design method	
02.0206	破损阶段设计法	plastic stage design method	
02.0207	极限状态设计法	limit state design method	
02.0208	概率极限状态设计法	probabilitic limit state design method	又称"可靠度设计法"。
02.0209	地震系数法	seismic coefficient method	
02.0210	路基	subgrade, road bed, formation subgrade	
02.0211	岩石路基	rock subgrade	
02.0212	渗水土路基	permeable soil subgrade, pervious embankment	
02.0213	非渗水土路基	non-permeable soil subgrade, impervious embankment	
02.0214	特殊土路基	subgrade of special soil	
02.0215	软土地区路基	subgrade in soft soil zone, subgrade in soft clay region	
02.0216	泥沼地区路基	subgrade in bog [soil] zone, subgrade in morass region, subgrade in swampland	
02.0217	膨胀土地区路基	subgrade in swelling soil zone, subgrade in expansive soil region	又称"裂土地区路基"。
02.0218	盐渍土地区路基	subgrade in salty soil zone, subgrade in saline soil region	
02.0219	多年冻土路基	subgrade in permafrost soil zone	
02.0220	特殊条件下的路基	subgrade under special condition	
02.0221	河滩路堤	embankment on plain river beach	
02.0222	滨河路堤	embankment on river bank	
02.0223	水库路基	subgrade in reservoir, embankment crossing reservoir	
02.0224	崩塌地段路基	subgrade in rock fall district, subgrade in collapse zone	
02.0225	岩堆地段路基	subgrade in rock deposit zone, subgrade in talus zone, subgrade in scree zone	
02.0226	滑坡地段路基	subgrade in slide	

序 码	汉 文 名	英 文 名	注 释
02.0227	喀斯特地段路基	subgrade in karst zone	又称"岩溶地段路基"。
02.0228	洞穴地段路基	subgrade in cavity zone, subgrade in cavern zone	
02.0229	风沙地段路基	subgrade in windy and sandy zone, subgrade in desert	
02.0230	雪害地段路基	subgrade in snow damage zone, subgrade in snow disaster zone	
02.0231	泥石流地段路基	subgrade in debris flow zone	
02.0232	路基横断面	subgrade cross-section	
02.0233	路基面	subgrade surface, formation	
02.0234	路基面宽度	width of the subgrade surface, formation width	
02.0235	路拱	road crown, [subgrade] crown	
02.0236	路肩	[road] shoulder, subgrade shoulder	
02.0237	路肩高程	formation level, shoulder level	
02.0238	路堤	embankment, fill	
02.0239	路堑	cut, road cutting	
02.0240	半堤半堑	part-cut and part-fill section, cut and fill section	
02.0241	基床	subgrade bed, formation	
02.0242	基床表层	surface layer of subgrade bed, formation top layer, surface layer of subgrade	
02.0243	基床底层	bottom layer of subgrade, formation base layer, bottom layer of subgrade bed	
02.0244	一般路基	general subgrade, ordinary subgrade	
02.0245	最小填筑高度	minimum fill height of subgrade, minimum height of fill	
02.0246	临界高度	critical height	
02.0247	基底	foundation base, base	
02.0248	路堤边坡	side slope of embankment, fill slope	
02.0249	坡脚	toe of side slope	

序　码	汉　文　名	英　文　名	注　释
02.0250	护道	berm	
02.0251	取土坑	borrow pit	
02.0252	路堤填料	embankment fill material, embankment filler, filling material of embankment	
02.0253	填料分类	classification of filling material	
02.0254	岩块填料	rock block filler, rock filler, rock fill	
02.0255	粗粒土填料	coarse-grained soil filler, coarse-grained soil fill	
02.0256	细粒土填料	fine-grained soil filler, fine-grained soil fill	
02.0257	压实标准	compacting criteria	
02.0258	相对密度	relative density	
02.0259	压实系数	compacting factor, compacting coefficient	
02.0260	最佳含水量	optimum moisture content, best moisture content	
02.0261	最佳密度	optimum density, best density	
02.0262	核子密度湿度测定	determination of nuclear density-moisture	
02.0263	路基承载板测定	determination of bearing slab of subgrade, bearing slab method for subgrade testing, bearing plate test on subgrade	
02.0264	预留沉落量	reserve settlement, settlement allowance	
02.0265	反压护道	berm with superloading, berm for back pressure, counter swelling berm	
02.0266	石灰砂桩	lime sand pile	
02.0267	换土	change soil, soil replacement	
02.0268	爆破排淤	blasting discharging sedimentation, silt arresting by explosion, discharge of sedimentation by blasting	
02.0269	抛石挤淤	throwing stones to packing sedi-	

序　码	汉　文　名	英　文　名	注　　释
		mentation, packing sedimentation by throwing stones, packing up sedimentation by dumping stones	
02.0270	路堑石方爆破	rock cutting blasting, rock blasting in cut	
02.0271	定向爆破	directional blasting	
02.0272	浅孔爆破	shallow hole blasting	
02.0273	深孔爆破	deep hole blasting	
02.0274	洞室药包爆破	chamber explosive package blasting, chamher blasting	
02.0275	扬弃爆破	abandoned blasting, abandonment blasting	
02.0276	抛掷爆破	pin-point blasting	
02.0277	松动爆破	blasting for loosening rock	
02.0278	药壶爆破	pot hole blasting	
02.0279	裸露药包爆破	adobe blasting, contact blasting	
02.0280	路堑边坡	cutting slope, side slope of cut	
02.0281	堑顶	top of cutting slope, top of cutting	
02.0282	路堑平台	platform of cutting, berm in cutting	
02.0283	弃土堆	waste bank, bankette, spoil bank	
02.0284	挡土墙	retaining wall	
02.0285	重力式挡土墙	gravity retaining wall	
02.0286	衡重式挡土墙	balance weight retaining wall, gravity retaining wall with relieving platform, balanced type retaining wall	
02.0287	锚定板挡土墙	anchored retaining wall by tie rods, anchored bulkhead retaining wall, anchored plate retaining wall	
02.0288	加筋土挡土墙	reinforced earth retaining wall, reinforced soil retaining wall	
02.0289	锚杆挡墙	anchored bolt retaining wall, anchored retaining wall by tie rods	
02.0290	管柱挡墙	cylindrical shaft retaining wall	

序　码	汉　文　名	英　文　名	注　　释
02.0291	沉井挡墙	caisson retaining wall	
02.0292	抗滑桩	anti-slide pile, counter-sliding pile	
02.0293	护墙	guard wall	
02.0294	护坡	slope protection, revetment, pit-ching	
02.0295	排水沟	weep drain, drainage ditch, drain ditch	
02.0296	边沟	side ditch	又称"侧沟"。
02.0297	天沟	gutter, overhead ditch, intercepting ditch	
02.0298	吊沟	suspended ditch	
02.0299	跌水	hydraulic drop	
02.0300	截水沟	intercepting ditch, catch-drain	
02.0301	急流槽	chute	
02.0302	排水槽	drainage channel	
02.0303	渗水暗沟	blind drain	
02.0304	渗水隧洞	leak tunnel, permeable tunnel, drainage tunnel	
02.0305	渗井	leaching well, seepage well	
02.0306	渗管	leaky pipe	
02.0307	平孔排水	horizontal hole drainage	
02.0308	反滤层	reverse filtration layer, inverted filter, protective filter	
02.0309	检查井	inspection well, manhole	
02.0310	砂井	sand drain	又称"排水砂井"。
02.0311	隔断层	insulating course, insulating layer	
02.0312	透水路堤	pervious embankment, permeable embankment	
02.0313	浸水路堤	immerseable embankment	
02.0314	排水砂垫层	sand filled drainage layer, drainage sand blanket	
02.0315	坡面防护	slope protection	
02.0316	护岸	revetment, shore protection	
02.0317	导流堤	diversion dike	
02.0318	拦石墙	stone cut off wall, stone falling wall, buttress wall for intercepting falling rocks	

序 码	汉 文 名	英 文 名	注 释
02.0319	落石槽	stone falling channel, trough for catching falling rocks	
02.0320	柴排	firewood raft, mattress, willow fascine	
02.0321	固沙造林	stabilization for sands by afforestation	
02.0322	挡风墙	wind-break wall	
02.0323	防风栅栏	wind break fence	
02.0324	砂土液化	sand liquefaction	
02.0325	中-活载	CR-live loading, China railway standard live loading	
02.0326	桥梁标准活载	standard live load for bridge	
02.0327	桥梁荷载谱	bridge load spectrum	
02.0328	换算均布活载	equivalent uniform live load	
02.0329	设计荷载	design load	
02.0330	主力	principal load	
02.0331	恒载	dead load	
02.0332	土压力	earth pressure	
02.0333	静水压力	hydrostatic pressure	
02.0334	浮力	buoyancy	
02.0335	列车活载	live load of train	
02.0336	列车离心力	centrifugal force of train	
02.0337	列车冲击力	impact force of train	又称"冲击荷载"。
02.0338	冲击系数	coefficient of impact	
02.0339	人行道荷载	sidewalk loading	
02.0340	附加力	subsidiary load, secondary load	
02.0341	列车制动力	braking force of train	
02.0342	列车牵引力	tractive force of train	
02.0343	风[荷]载	wind load	
02.0344	列车横向摇摆力	lateral swaying force of train	
02.0345	流水压力	pressure of water flow	
02.0346	冰压力	ice pressure	
02.0347	冻胀力	frost heaving force	
02.0348	特殊荷载	particular load	
02.0349	船只或排筏的撞击力	collision force of ship or raft	
02.0350	地震力	seismic force	

序 码	汉 文 名	英 文 名	注 释
02.0351	地震烈度	earthquake intensity	
02.0352	地震震级	earthquake magnitude	
02.0353	施工荷载	constructional loading	
02.0354	荷载组合	loading combination	
02.0355	铁路桥	railway bridge	
02.0356	公铁两用桥	combined bridge, combined highway and railway bridge, combined rail-cum-road bridge	
02.0357	跨线桥	overpass bridge, grade separation bridge, flyover	又称"立交桥"。
02.0358	高架桥	viaduct	
02.0359	旱桥	dry bridge	
02.0360	人行桥	foot bridge, pedestrian bridge	
02.0361	圬工桥	masonry bridge	
02.0362	钢桥	steel bridge	
02.0363	铆接钢桥	riveted steel bridge	
02.0364	栓焊钢桥	bolted and welded steel bridge	
02.0365	全焊钢桥	all welded steel bridge	
02.0366	摩擦结合式高强度螺栓	high strength friction grip bolt	
02.0367	扭剪式高强度螺栓	torshear type high strength bolt	
02.0368	螺栓示功扳手	bolt wrench with indicator	
02.0369	混凝土桥	concrete bridge	
02.0370	钢筋混凝土桥	reinforced concrete bridge	
02.0371	预应力混凝土桥	prestressed concrete bridge	
02.0372	先张法预应力梁	pretensioned prestressed concrete girder	
02.0373	后张法预应力梁	post-tentioned prestressed concrete girder	
02.0374	部分预应力混凝土桥	partially prestressed concrete bridge	
02.0375	结合梁桥	composite beam bridge	
02.0376	低高度梁	shallow girder	
02.0377	无碴无枕梁	girder without ballast and sleeper	
02.0378	型钢混凝土梁	girder with rolled steel section encased in concrete, skeleton rein-	又称"劲性骨架混凝土梁"。

序　码	汉 文 名	英 文 名	注　释
		forced concrete girder	
02.0379	简支梁桥	simply supported beam bridge	
02.0380	连续梁桥	continuous beam bridge	
02.0381	悬臂梁桥	cantilever beam bridge	
02.0382	板桥	slab bridge	
02.0383	空心板桥	hollow slab bridge	
02.0384	板梁	plate girder	
02.0385	工形梁	I-beam	
02.0386	箱形梁	box girder	
02.0387	槽型梁	trough girder	
02.0388	桁架	truss	
02.0389	拆装式桁架	demountable truss	
02.0390	刚架桥	rigid frame bridge	又称"刚构桥"。
02.0391	斜腿刚架桥	strutted beam bridge, slant-legged rigid frame bridge	又称"斜腿刚构桥"。
02.0392	悬板桥	stressed ribbon bridge	又称"悬带桥"。
02.0393	悬索桥	suspension bridge	又称"吊桥"。
02.0394	斜拉桥	cable-stayed bridge	又称"斜张桥"。
02.0395	浮桥	pontoon bridge, floating bridge, bateau bridge	
02.0396	拱桥	arch bridge	
02.0397	固端拱	fixed-end arch	又称"无铰拱"。
02.0398	双铰拱	two-hinged arch	
02.0399	三铰拱	three-hinged arch	
02.0400	实腹拱	spandrel-filled arch, solid-spandrel arch	
02.0401	空腹拱	open-spandrel arch	
02.0402	双曲拱	two-way curved arch, cross-curved arch	
02.0403	系杆拱	tied arch	又称"柔性系杆刚性拱"。
02.0404	朗格尔式桥	Langer bridge, flexible arch bridge with rigid tie	又称"刚性系杆柔性拱桥"。
02.0405	洛泽式桥	Lohse bridge, rigid arch bridge with rigid tie and vertical suspenders	又称"直悬杆式刚性拱刚性梁桥"。
02.0406	尼尔森体系桥	Nielsen system bridge	

序　码	汉　文　名	英　文　名	注　释
02.0407	尼尔森式洛泽梁桥	Nielsen type Lohse bridge, rigid arch bridge with rigid tie and inclined suspenders	又称"斜悬杆式刚性拱刚性梁桥"。
02.0408	活动桥	movable bridge	
02.0409	竖旋桥	bascule bridge	
02.0410	平旋桥	swing bridge	
02.0411	升降桥	lift bridge	
02.0412	正交桥	right bridge	
02.0413	斜交桥	skew bridge	
02.0414	曲线桥	curved bridge	
02.0415	曲梁	curved beam	
02.0416	特大桥	super major bridge	
02.0417	大桥	major bridge	
02.0418	中桥	medium bridge	
02.0419	小桥	minor bridge	
02.0420	单线桥	single track bridge	
02.0421	双线桥	double track bridge	
02.0422	多线桥	multi-track bridge	
02.0423	正桥	main bridge	又称"主桥"。
02.0424	引桥	approach spans	
02.0425	上承式桥	deck bridge	
02.0426	半穿式桥	half through bridge, midheight deck bridge	又称"中承式桥"。
02.0427	下承式桥	through bridge	
02.0428	双层桥	double-deck bridge	
02.0429	永久性桥	permanent bridge	
02.0430	临时性桥	temporary bridge	又称"便桥"。
02.0431	跨径	span	又称"跨度"。
02.0432	净跨	clear span	
02.0433	桥梁全长	overall length of bridge	
02.0434	桥下净空	underneath clearance	
02.0435	主梁中心距	center to center distance between main girder	
02.0436	节间长度	panel length	
02.0437	梁高	depth of girder	
02.0438	拱度	camber	
02.0439	挠度	deflection	

序　码	汉 文 名	英 文 名	注　释
02.0440	节间	panel	
02.0441	锚跨	anchor span	又称"锚孔"。
02.0442	悬跨	suspended span	又称"吊孔"。
02.0443	桥梁上部结构	superstructure	
02.0444	腹板	web plate	
02.0445	翼缘	flange	
02.0446	翼缘板	flange plate	
02.0447	弦杆	chord member	
02.0448	腹杆	web member	
02.0449	斜杆	diagonal member	
02.0450	竖杆	vertical member	
02.0451	吊杆	suspender, hanger	
02.0452	加劲杆	stiffener	
02.0453	节点	panel point	
02.0454	节点板	gusset plate	
02.0455	拼接板	splice plate	
02.0456	缀条	lacing bar	
02.0457	缀板	stay plate, tie plate	
02.0458	侧向水平联结系	lateral bracing	
02.0459	横联	sway bracing	
02.0460	制动撑架	braking bracing	
02.0461	桥门架	portal frame	
02.0462	纵梁	stringer	
02.0463	横梁	floor beam, transverse beam	
02.0464	桥面系	floor system	
02.0465	端横梁	end floor beam	
02.0466	起重横梁	jacking floor beam	
02.0467	梁端缓冲梁	auxiliary girder for controlling angle change	
02.0468	应变时效	strain ageing	
02.0469	碳当量	carbon equivalent	
02.0470	钢丝	steel wire	
02.0471	钢丝束	bundled steel wires	
02.0472	钢绞线	steel strand	
02.0473	钢筋	reinforcement, steel bar	
02.0474	箍筋	stirrup	
02.0475	纵向钢筋	longitudinal reinforcement	

序　码	汉　文　名	英　文　名	注　释
02.0476	弯起钢筋	bent-up bar	
02.0477	架立钢筋	erection bar	
02.0478	构造钢筋	constructional reinforcement	
02.0479	预应力筋	tendon	
02.0480	套管	sheath	
02.0481	梁腋	haunch	
02.0482	[桥涵]拱圈	arch ring	
02.0483	拱肋	arch rib	
02.0484	拱顶	arch crown	
02.0485	拱矢	rise of arch	
02.0486	起拱点	springing	
02.0487	拱腹	soffit	
02.0488	拱腹线	intrados	
02.0489	拱背线	extrados	
02.0490	桥塔	bridge tower, pylon	
02.0491	索平面	cable plane	
02.0492	缆索	cable	
02.0493	斜缆	stay cable, inclined cable	
02.0494	吊缆	suspension cable	
02.0495	索鞍	cable saddle	
02.0496	索夹	cable band, cable clamp	
02.0497	锚座	socket	
02.0498	锚碇	anchorage	
02.0499	明桥面	open deck, ballastless deck, open floor	
02.0500	桥梁道碴槽	ballast trough	
02.0501	道碴桥面	ballasted deck, ballasted floor	
02.0502	桥梁护轨	guard rail of bridge	
02.0503	桥梁护木	guard timber of bridge	
02.0504	桥枕	bridge tie, bridge sleeper	
02.0505	桥上人行道	sidewalk on bridge	
02.0506	步行板	foot plank	
02.0507	避车台	refuge platform	
02.0508	伸缩缝	expansion joint	
02.0509	正交异性板	orthotropic plate	
02.0510	栏杆	railing, handrail, handrailing	
02.0511	泄水孔	drainage opening	

序　码	汉　文　名	英　文　名	注　释
02.0512	直结轨道	track fastened directly to steel girders	
02.0513	抗剪连接件	shear connector	又称"抗剪结合件"。
02.0514	支座	bearing	
02.0515	固定支座	fixed bearing	
02.0516	活动支座	expansion bearing, movable bearing	
02.0517	平板支座	plate bearing	
02.0518	摇轴支座	rocker bearing	
02.0519	滚轴支座	roller bearing	
02.0520	球面支座	spherical bearing	
02.0521	板式橡胶支座	laminated rubber bearing	
02.0522	盆式橡胶支座	pot rubber bearing	
02.0523	聚四氟乙烯支座	poly-tetrafluoroethylene bearing, PTFE bearing	
02.0524	涡流激振	vortex-excited oscillation	
02.0525	驰振	galloping	
02.0526	颤振	flutter	
02.0527	扰流板	spoiler	
02.0528	风嘴	wind fairing	
02.0529	桥梁自振周期	natural vibration period of bridge	
02.0530	浮运架桥法	bridge erection by floating	
02.0531	架桥机架设法	erection by bridge girder erecting equipment	
02.0532	顶推式架设法	erection by incremental launching method	
02.0533	拖拉架设法	launching method	
02.0534	鹰架式架设法	erection with scaffolding	
02.0535	悬臂架设法	cantilever erection, erection by protrusion	
02.0536	悬臂灌注法	cast-in-place cantilever construction, free cantilever segmental concreting with suspended formwork	
02.0537	悬臂拼装法	cantilevered assembling construction, free cantilever erection with segments of precast conc-	

序　码	汉　文　名	英　文　名	注　释
		rete	
02.0538	预制混凝土构件	precast concrete units, precast concrete members	
02.0539	活动模架逐跨施工法	segmental span-by-span construction using form traveller	
02.0540	桥梁合龙	closure	
02.0541	就地灌筑法	cast-in-place method, cast-*in-situ* method	
02.0542	活动吊篮	travelling cradle	
02.0543	顶进法	jack-in method	
02.0544	旋转法施工	erection by swing method	又称"转体施工"。
02.0545	液压式张拉千斤顶	hydraulic tensioning jack	
02.0546	桥梁下部结构	substructure	
02.0547	桥台	abutment	
02.0548	重力式桥台	gravity abutment	
02.0549	埋置式桥台	buried abutment	
02.0550	锚定板式桥台	anchor slab abutment	
02.0551	U 形桥台	U-shaped abutment	
02.0552	耳墙式桥台	abutment with cantilevered retaining wall	
02.0553	台身	abutment body	
02.0554	前墙	front wall	
02.0555	台帽	abutment coping	
02.0556	翼墙	wing wall	
02.0557	锥体护坡	quadrant revetment, truncated cone banking	
02.0558	台后填方	filling behind abutment	
02.0559	桥墩	pier	
02.0560	空心桥墩	hollow pier	
02.0561	实体桥墩	solid pier	
02.0562	重力式桥墩	gravity pier	
02.0563	柔性[桥]墩	flexible pier	
02.0564	拼装式桥墩	assembly pier, pier constructed with precast units	
02.0565	制动墩	braking pier	
02.0566	柱式桥墩	column pier	

序码	汉文名	英文名	注释
02.0567	V形桥墩	V-shaped pier	
02.0568	圆端形桥墩	round-ended pier	
02.0569	圆形桥墩	circular pier	
02.0570	矩形桥墩	rectangular pier	
02.0571	排架式桥墩	pile bent pier	
02.0572	墩身	pier body, pier shaft	
02.0573	墩帽	pier coping	
02.0574	围栏	railing around coping of pier or abutment	
02.0575	承台	bearing platform	
02.0576	破冰体	ice apron, ice-breaking cutwater, ice guard	
02.0577	地基	foundation, foundation soil, sub-grade	
02.0578	加固地基	improved foundation, improved ground	
02.0579	天然地基	natural foundation, natural ground	
02.0580	桥梁基础	bridge foundation	
02.0581	扩大基础	spread foundation	
02.0582	明挖基础	open-cut foundation, open excavation foundation	
02.0583	沉井基础	open caisson foundation	
02.0584	浮式沉井基础	floating caisson foundation	
02.0585	沉井刃脚	cutting edge of open caisson	
02.0586	围堰	cofferdam	
02.0587	双壁钢围堰钻孔基础	double wall steel cofferdam bored foundation	
02.0588	预制钢壳钻孔基础	prefabricated steel shell bored foundation	
02.0589	泥浆套沉井法	slurry jacket method for sinking caisson	
02.0590	空气幕沉井法	air curtain method for sinking caisson	
02.0591	沉箱基础	pneumatic caisson foundation	
02.0592	管柱基础	tubular column foundation	
02.0593	桩基础	pile foundation	
02.0594	预制桩	precast pile	

序 码	汉 文 名	英 文 名	注 释
02.0595	就地灌注桩	cast-in-place concrete pile, cast-in-situ concrete pile	
02.0596	螺旋喷射桩	auger injected pile	
02.0597	摩擦桩	friction pile	
02.0598	支承桩	bearing pile	
02.0599	钻孔桩	bored pile	
02.0600	挖孔桩	dug pile	
02.0601	钢桩	steel pile	
02.0602	钢管桩	steel pipe pile	
02.0603	钢板桩	steel sheet pile	
02.0604	板桩	sheet pile	
02.0605	木桩	timber pile	
02.0606	钢筋混凝土桩	reinforced concrete pile	
02.0607	砂桩	sand pile	
02.0608	挤密砂桩	sand compaction pile	
02.0609	流砂	quick sand, drift sand	
02.0610	送桩	pile follower	
02.0611	试桩	test pile	
02.0612	斜桩	batter pile, raking pile, spur pile	
02.0613	护筒	pile casing	
02.0614	重锤夯实法	heavy tamping method	
02.0615	灰土换填夯实法	method of lime-soil replacement and tamping	
02.0616	灌筑水下混凝土	underwater concreting, concreting with tremie method	
02.0617	导流建筑物	regulating structure	
02.0618	丁坝	spur dike	又称"挑水坝"。
02.0619	顺坝	longitudinal dam	
02.0620	河床铺砌	river bed paving	
02.0621	码头	wharf	
02.0622	排架	bent	
02.0623	脚手架	scaffold	
02.0624	悬空脚手架	hanging stage, hanging scaffold	
02.0625	铁路涵洞	railway culvert	
02.0626	涵洞孔径	aperture of culvert	
02.0627	管涵	pipe culvert	
02.0628	箱涵	box culvert	

序　码	汉　文　名	英　文　名	注　释
02.0629	拱涵	arch culvert	
02.0630	盖板涵	slab culvert	
02.0631	无压力涵洞	inlet unsubmerged culvert	
02.0632	压力式涵洞	outlet submerged culvert	
02.0633	半压力式涵洞	inlet submerged culvert	
02.0634	明渠	open channel, open ditch, open drain	
02.0635	倒虹吸管	inverted siphon	
02.0636	潮汐河流	tidal river	
02.0637	淤积	silting, siltation	
02.0638	流冰	ice drift	
02.0639	铁路轮渡	railway car ferries	
02.0640	轮渡站	ferry station	
02.0641	轮渡栈桥	ferry trestle bridge	
02.0642	渡轮	ferry boat	
02.0643	轮渡引线	ferry slip	又称"轮渡斜引道"。
02.0644	铁路隧道	railway tunnel	
02.0645	山岭隧道	mountain tunnel	
02.0646	越岭隧道	over mountain line tunnel	
02.0647	水下隧道	subaqueous tunnel, underwater tunnel	又称"水底隧道"。
02.0648	地铁隧道	subway tunnel, underground railway tunnel, metro tunnel	
02.0649	浅埋隧道	shallow tunnel, shallow-depth tunnel, shallow burying tunnel	
02.0650	深埋隧道	deep tunnel, deep-depth tunnel, deep burying tunnel	
02.0651	单线隧道	single track tunnel	
02.0652	双线隧道	double track tunnel	
02.0653	多线隧道	multiple track tunnel	
02.0654	车站隧道	station tunnel	
02.0655	地铁车站	subway station, metro station	
02.0656	特长隧道	super long tunnel	
02.0657	长隧道	long tunnel	
02.0658	中长隧道	medium tunnel	
02.0659	短隧道	short tunnel	
02.0660	隧道群	tunnel group	

序　码	汉　文　名	英　文　名	注　释
02.0661	地铁工程	subway engineering, metro engineering	
02.0662	洞口	tunnel adit, tunnel opening	
02.0663	隧道进口	tunnel entrance	
02.0664	隧道出口	tunnel exit	
02.0665	仰坡	front slope	又称"正面坡"。
02.0666	洞门	tunnel portal	
02.0667	洞门框	tunnel portal frame	
02.0668	端墙式洞门	end wall tunnel portal	
02.0669	柱式洞门	post tunnel portal	
02.0670	翼墙式洞门	wing wall tunnel portal	
02.0671	耳墙式洞门	ear wall tunnel portal	
02.0672	台阶式洞门	bench tunnel portal	
02.0673	正洞门	orthonormal tunnel portal, straight tunnel portal	
02.0674	斜洞门	skew tunnel portal	
02.0675	明洞门	open-cut-tunnel portal, gallery portal	
02.0676	衬砌	lining	
02.0677	[隧道]拱圈	arch	
02.0678	边墙	side wall	
02.0679	仰拱	invert, inverted arch	
02.0680	底板	floor	
02.0681	整体式衬砌	integral lining	
02.0682	装配式衬砌	precast lining, prefabricated lining	
02.0683	模筑衬砌	moulded lining	
02.0684	洞口段衬砌	lining of tunnel portal section	
02.0685	偏压衬砌	unsymmetrically loading lining eccentrically compressed lining	
02.0686	组合衬砌	composite lining	又称"复合衬砌"。
02.0687	初期支护	primary support	
02.0688	二次衬砌	secondary lining	
02.0689	隔离层	isolation layer	
02.0690	喷锚衬砌	shotcrete bolt lining	
02.0691	下锚段衬砌	anchor-section lining	又称"接触网锚段衬砌"。
02.0692	挤压混凝土衬砌	extruding concrete tunnel lining	

序 码	汉 文 名	英 文 名	注 释
02.0693	隔热层	thermal insulation layer	
02.0694	明洞	open-cut tunnel, tunnel without cover, gallery	
02.0695	拱形明洞	arch open cut tunnel, arch tunnel without cover, arch gallery	
02.0696	棚洞	shed tunnel, shed gallery	
02.0697	路堑式明洞	cut-type open cut tunnel, cut-type tunnel without cover, cut-type gallery	
02.0698	半路堑式明洞	part cut-type open cut tunnel, part cut-type tunnel without cover, part cut-type gallery	
02.0699	抗滑明洞	anti-skid-type open cut tunnel, anti-skid-type tunnel without cover, anti-skid-type gallery	
02.0700	盖板式棚洞	slab shed tunnel, slad shed gallery	
02.0701	刚架式棚洞	framed shed tunnel, framed shed gallery	
02.0702	悬臂式棚洞	cantilever shed tunnel, cantilever shed gallery	
02.0703	隧道专家系统	expert system of tunnel	
02.0704	围岩	surrounding rock	
02.0705	围岩压力	pressure of surrounding rock	
02.0706	地层压力	ground pressure, stratum pressure	
02.0707	松弛压力	relaxation pressure	
02.0708	形变压力	deformation pressure	
02.0709	围岩自承能力	self-supporting capacity of surrounding rock	
02.0710	坑道自稳时间	self-stabilization time of tunnel	
02.0711	弹性抗力	elastic resistance	
02.0712	灌浆压力	grouting pressure	
02.0713	落石冲击力	impact force of falling stone	
02.0714	隧道埋置深度	buried depth of tunnel	简称"埋深"。
02.0715	地温梯度	geothermal gradient	
02.0716	地应力	crustal stress	
02.0717	预留变形量	deformation allowance	
02.0718	暗挖法	subsurface excavation method	

序　码	汉　文　名	英　文　名	注　释
02.0719	明挖法	open-cut method	
02.0720	矿山法	mining method, mine tunnelling method	
02.0721	钻[眼]爆[破]法	drilling and blasting method	
02.0722	新奥法	New Austrian Tunnelling Method, NATM	
02.0723	喷锚构筑法	shotcrete-bolt construction method	
02.0724	掘进机法	tunnel boring machine method	
02.0725	盾构法	shield method	
02.0726	沉管法	immersed tunnelling method	
02.0727	地下连续墙法	underground diaphragm wall method, underground wall method, diaphragm wall method	
02.0728	全断面开挖法	full-face tunnelling method	
02.0729	分部开挖法	partial excavation method	
02.0730	台阶法	benching tunnelling method	
02.0731	正台阶法	positive benching tunnelling method	
02.0732	反台阶法	negative benching tunnelling method	
02.0733	导坑	heading	
02.0734	上下导坑法	top and bottom heading method	
02.0735	上导坑法	top-heading method	
02.0736	漏斗棚架法	hopper-shed support tunnelling method	
02.0737	蘑菇形开挖法	mushroom-type tunnelling method	
02.0738	侧壁导坑法	side heading method	
02.0739	单侧导坑法	single side heading method	
02.0740	双侧导坑法	twin-side heading method	
02.0741	品字形导坑法	top and twin-side bottom heading method	
02.0742	眼镜式开挖法	spectacles type tunnelling method	
02.0743	超前导坑	advance heading	
02.0744	扩大	enlargement	
02.0745	拉槽	pull trough, trench excavated	
02.0746	马口	excavation of side wall at intervals	
02.0747	控制爆破	controlled blasting	

序　码	汉　文　名	英　文　名	注　释
02.0748	光面爆破	smooth blasting	
02.0749	预裂爆破	presplit blasting	
02.0750	先墙后拱法	side wall first lining method	
02.0751	先拱后墙法	archfirst lining method, flying arch method	
02.0752	开挖工作面	excavated surface, excavated work face	
02.0753	超挖	overbreak	
02.0754	欠挖	underbreak	
02.0755	找顶	top cleaning	
02.0756	出碴	mucking and removing	
02.0757	有轨运输	track transportation	
02.0758	无轨运输	trackless transportation	
02.0759	拱圈封顶	closing the top of lining	
02.0760	墙顶封口	seal, seal at the top of wall	
02.0761	喷射混凝土回弹	rebound of shotcrete	
02.0762	隧道贯通	tunnel holing-through	
02.0763	隧道贯通误差	tunnel through error	
02.0764	隧道贯通面	tunnel through plane	
02.0765	临时支护	temporary support	
02.0766	构件支撑	member support	
02.0767	钢拱支撑	steel arch support	
02.0768	喷射混凝土支护	shotcrete support	
02.0769	锚杆支护	anchor bolt support	
02.0770	喷锚支护	shotcrete and rock bolt support	又称"锚喷支护"。
02.0771	系统锚杆	system anchor bolt	
02.0772	超前支护	advance support	
02.0773	超前锚杆	advance anchor bolt	
02.0774	插板支护	inserting plate support, forepoling	
02.0775	小导管预注浆	pre-grouting with small duct	
02.0776	管棚支护	pipe-shed support, pipe roofing support	
02.0777	格构拱	trellis arch, lattice arch	又称"钢花拱"、"格栅拱"。
02.0778	喷射混凝土	shotcrete, spray concrete	
02.0779	干喷混凝土	dry shotcreting	
02.0780	湿喷混凝土	wet shotcreting	

序 码	汉 文 名	英 文 名	注 释
02.0781	潮喷混凝土	half wet shotcreting	又称"半湿喷混凝土"。
02.0782	喷射钢纤维混凝土	steel fiber shotcrete, steel fiber reinforced shotcrete	
02.0783	防水混凝土	waterproof concrete	
02.0784	预切槽	precutting trough	
02.0785	辅助坑道	service gallery	
02.0786	竖井	shaft	
02.0787	斜井	inclined shaft	
02.0788	横洞	transverse gallery	
02.0789	平行导坑	parallel heading	
02.0790	横通道	transverse passage-way	
02.0791	调查坑道	investigation gallery, survey tunnel	
02.0792	竖井联系测量	shaft connection survey	
02.0793	隧道防水	waterproofing of tunnel	
02.0794	防水等级	classification of waterproof	
02.0795	防水层	waterproof layer	
02.0796	防水板	waterproof board, waterproof sheet	
02.0797	防水涂层	waterproofing coating	
02.0798	隧道压浆	pressure grouting of tunnel	
02.0799	止水带	water stop tie	
02.0800	盲沟	blind ditch	
02.0801	泄水洞	drain cavern, drain tunnel	
02.0802	隧道运营通风	permanent ventilation of tunnel	
02.0803	自然通风	natural ventilation	
02.0804	机械通风	mechanical ventilation	
02.0805	风道式通风	ventilation by air passage	
02.0806	隧道通风帘幕	ventilation curtain	
02.0807	列车活塞作用	piston action of train	
02.0808	隧道射流式通风	tunnel efflux ventilation, tunnel injector type ventilation	
02.0809	隧道通风试验	tunnel ventilation test	
02.0810	隧道施工通风	construction ventilation of tunnel	
02.0811	风管式通风	ventilation by pipes	
02.0812	巷道式通风	ventilation by ducts	

序 码	汉 文 名	英 文 名	注 释
02.0813	隧道照明	tunnel illumination, tunnel lighting	
02.0814	隧道防灾设施	tunnel anti-disaster equipment	
02.0815	避车洞	refuge hole, refuge recess, refuge niche	
02.0816	隧道火灾	tunnel fire hazard	
02.0817	隧道报警装置	tunnel warning equipment	
02.0818	隧道消防系统	tunnel fire-fighting system	
02.0819	隧道瓦斯爆炸	tunnel gas explosion	
02.0820	隧道监控量测	tunnel monitoring measurement	
02.0821	隧道地表沉陷	tunnel ground subsidence	
02.0822	隧道拱顶下沉	tunnel arch top settlement	
02.0823	隧道底鼓	tunnel floor heave	
02.0824	隧道周边位移	tunnel perimeter deflection	
02.0825	隧道地中位移	tunnel surrounding mass deflection	
02.0826	隧道改建	tunnel reconstruction	
02.0827	隧道落底	under cut of tunnel	
02.0828	隧道挑顶	top picking of tunnel	
02.0829	隧道套拱	cover arch of tunnel	
02.0830	施工调查	construction investigation	
02.0831	施工准备	construction preparation	
02.0832	征地	expropriation	
02.0833	拆迁	removing	
02.0834	租地	rented land	
02.0835	交接桩	delivery-receiving stake	
02.0836	施工复测	construction repetition, repetive survey, construction repetition survey	
02.0837	测量放样	staking out in survey	
02.0838	土[石]方调配	cut-fill transition	
02.0839	土[石]方体积图	volume diagram of earth-rock work	又称"土积图"。
02.0840	隐蔽工程	hidden project, hidden construction work	
02.0841	废弃工程	abandoned project, abandoned construction work	
02.0842	工程运输	engineering transportation	

序　码	汉　文　名	英　文　名	注　释
02.0843	冬季施工	cold weather construction, winter season construction	
02.0844	雨季施工	raining season construction, rainy season construction	
02.0845	工程招标	calling for tenders of project, calling for tending of project	
02.0846	工程投标	bidding for project	
02.0847	工程报价	project quoted price	
02.0848	工程发包	contracting out of project	
02.0849	工程承包	contracting of project	
02.0850	工程监理	supervision of construction, supervision of project	
02.0851	总预算	total budget	
02.0852	单项预算	individual budget	
02.0853	建筑业总产值	total output of building industry	
02.0854	指导性施工组织设计	design for guiding construction organization, guiding design for construction scheme	
02.0855	实施性施工组织设计	design for practical construction organization, practical design for construction scheme, operative construction organization design	
02.0856	临时工程	temporary project	
02.0857	大型临时工程	large-scale temporary project	
02.0858	小型临时工程	small-scale temporary project	
02.0859	过渡工程	transition project	
02.0860	控制工程	dominant project	
02.0861	关键工程	key project	
02.0862	辅助设施	auxiliary facilities	
02.0863	通车期限	time limit for opening to traffic	
02.0864	施工总工期	total time of construction, total construction time	
02.0865	施工组织方案	construction scheme	
02.0866	开工报告	report on starting of construction work, commencement report of construction work, construction	

序　码	汉　文　名	英　文　名	注　释
		starting report	
02.0867	竣工报告	completion report of construction work	
02.0868	施工计划管理	planned management of construction	
02.0869	网络计划技术	network planning technique	
02.0870	施工工艺流程	construction technology process, construction process	
02.0871	工法制度	construction method system	
02.0872	施工产值	construction output value	
02.0873	施工利润	construction profit	
02.0874	施工机械利用率	utilization ratio of construction machinery	
02.0875	施工机械完好率	ratio of construction machinery in good condition	
02.0876	施工形象进度	construction figure progress, figurative progress of construction work	
02.0877	挖掘机	excavating machine, excavator	
02.0878	单斗挖土机	power shovel	
02.0879	索铲挖土机	dragline	
02.0880	反铲挖土机	backhoe	
02.0881	蛤壳式抓斗	clamshell bucket	
02.0882	推土机	bulldozer	
02.0883	铲运机	scoper, scraper, carrying scraper	
02.0884	装载机	loader	
02.0885	带式输送机	belt conveyer	
02.0886	平地机	grader	
02.0887	振动辗压机	vibration compactor, vibration roller	
02.0888	压路机	roller	
02.0889	潜孔钻机	diving drill	
02.0890	凿岩机	air hammer drill, rock drill	
02.0891	碎石机	crusher	
02.0892	挖沟机	trench cutting machine, trenching machine, trencher	
02.0893	塔式起重机	tower crane, column crane	

序　码	汉　文　名	英　文　名	注　　释
02.0894	汽车式起重机	automobile crane, autocrane, truck crane	
02.0895	轮胎式起重机	rubber tired crane, rough-terrain crane	
02.0896	履带式起重机	crawler crane, caterpiller crane	
02.0897	轨道起重机	track crane, rail crane	
02.0898	浮式起重机	floating crane	又称"浮吊"。
02.0899	缆索起重机	cable crane	
02.0900	空中索道	aerial ropeway	
02.0901	电动卷扬机	electric winch	
02.0902	电动吊车	electric hoist	又称"电动葫芦"。
02.0903	千斤顶	jack	
02.0904	打桩机	pile driver	
02.0905	蒸汽打桩机	steam pile driver, steam pile hammer	
02.0906	柴油打桩机	diesel pile driver	
02.0907	液压打桩机	hydraulic pile driver	
02.0908	振动打桩机	vibrating pile driver, vibratory driver	
02.0909	振动沉拔桩机	vibro-driver extractor	
02.0910	静压力拔桩机	static pressure pile extractor, static pressure pile drawing machine	
02.0911	履带式桩架	crawler pile frame	
02.0912	旋转式钻机	swiveling drill machine, rotary drill machine	
02.0913	潜水钻机	diving drill machine	
02.0914	套管钻机	drill machine with casing	
02.0915	冲击式钻机	impact-type drill machine, percussion type drill machine	
02.0916	离心水泵	centrifugal pump	
02.0917	真空水泵	vacuum pump	
02.0918	深井水泵	deep well pump	
02.0919	轴流水泵	axial flow pump	
02.0920	潜水泵	diving pump	
02.0921	往复式空气压缩机	reciprocating compressor	
02.0922	螺杆式空气压缩	screw compressor	

序 码	汉 文 名	英 文 名	注 释
	机		
02.0923	单转子滑片式空压机	single rotary compressor	
02.0924	水泥输送泵	cement pump	
02.0925	混凝土搅拌机	concrete mixer	
02.0926	鼓筒式混凝土搅拌机	drum concrete mixer	
02.0927	锥形反转出料混凝土搅拌机	tapered reverse tilting concrete mixer	
02.0928	锥形倾翻出料混凝土搅拌机	tapered tilting concrete mixer	
02.0929	强制式混凝土搅拌机	forced concrete mixer	
02.0930	混凝土搅拌楼	concrete mixing plant	
02.0931	简易混凝土搅拌站	simple concrete mixing plant	
02.0932	混凝土搅拌运输车	concrete mixing and transporting car, truck mixer, transit mixer	
02.0933	混凝土输送泵	concrete pump	
02.0934	混凝土泵车	concrete pump truck	
02.0935	混凝土吊斗	concrete lifting bucket	
02.0936	钢筋调直机	bar straightener	
02.0937	钢筋切断机	bar cutter	
02.0938	钢筋弯曲机	bar bender	
02.0939	钢筋冷拉机	bar cold-drawing machine	
02.0940	插入式混凝土振捣器	immersion type vibrator for concrete	
02.0941	附着式混凝土振捣器	attached type vibrator for concrete, attached type vibrator	
02.0942	平板式混凝土振捣器	plate vibrator	
02.0943	混凝土振动台	concrete vibrating stand	
02.0944	长线张拉台座	stretching bed for longline production	
02.0945	斜撑桅杆式架梁起重机	cross stay derrick girder erecting machine	
02.0946	全回转式架梁起	full circle girder erecting crane	

序 码	汉 文 名	英 文 名	注 释
	重机		
02.0947	双悬臂式架桥机	double cantilever girder-erecting machine	
02.0948	单梁式架桥机	single beam girder-erecting machine	
02.0949	双梁式架桥机	double beam girder-erecting machine	
02.0950	组拼式架桥机	assembly type girder-erecting machine	
02.0951	造桥机	bridge fabrication machine	
02.0952	工程船舶	engineering ship, engineering vessel	
02.0953	舟桥	bateau bridge	
02.0954	浮箱	floating box, pontoon	
02.0955	万能杆件	fabricated universal steel members	
02.0956	模板	form	
02.0957	组合钢模板	combined steel formwork	
02.0958	滑动模板	slip form	
02.0959	爬升模板	climbing shuttering	
02.0960	砌拱支架	soffit scaffolding	又称"碗扣式脚手架"。
02.0961	风镐	air pick, pneumatic pick	
02.0962	凿岩台车	rock drilling jumbo	又称"钻孔台车"。
02.0963	立爪式装岩机	vertical claw rock loader	
02.0964	蟹爪式装岩机	crab rock loader	
02.0965	梭式矿车	shuttle car	
02.0966	槽式列车	bunker train	
02.0967	电瓶车	storage battery car	
02.0968	斗车	bucket loader	
02.0969	大型矿车	large scale ore car	
02.0970	装岩机	rock loader	
02.0971	干喷混凝土机	dry shotcreting machine	
02.0972	湿喷混凝土机	wet shotcreting machine	
02.0973	半湿喷混凝土机	half wet shotcreting machine	
02.0974	喷射混凝土机械手	shotcrete manipulator	
02.0975	风动凿岩机	pneumatic rock drill	

序 码	汉 文 名	英 文 名	注 释
02.0976	手持式凿岩机	portable rock drill, jack hammer	
02.0977	通风机	ventilating set, ventilating machine, ventilator	
02.0978	注浆机	grouting machine, grouter	
02.0979	注浆泵	grouting pump, injection pump	
02.0980	隧道衬砌模板台车	working jumbo for tunnel lining, tunnelling shutter jumbo for tunnel lining	
02.0981	隧道掘进机	tunnel boring machine, TBM	
02.0982	盾构	shield	
02.0983	接触网架线车	installation vehicle for contact wire	
02.0984	接触网作业车	operation vehicle for contact wire	
02.0985	自卸汽车	dumping truck	
02.0986	斗式提升机	bucket elevator	
02.0987	安全技术措施	safety technical measures	
02.0988	安全教育	safety education	
02.0989	事故报告	accident report	
02.0990	灾害性地质	disaster geology	
02.0991	突泥	projecting mud soil	
02.0992	突水	gushing water	
02.0993	冒顶	roof fall	
02.0994	岩爆	rock burst	
02.0995	瓦斯突出	gas projection	
02.0996	瓦斯浓度	gas density, gas consistency	
02.0997	掏底开挖	cut the vertical earthwork bottom	俗称"挖神仙土"。
02.0998	施工防护	construction protection	
02.0999	线路封锁	track blockade, closure of track, traffic interruption	
02.1000	防护音响信号	protecting acoustic signal	
02.1001	覆盖防护	covered protection	
02.1002	近体防护	nearby protection	
02.1003	安全标志	safety mark, safety symbol, safety indicator	
02.1004	停车防护	stopping train protection, standing train protection	
02.1005	安全距离	safety distance	
02.1006	人员掩蔽所	hiding-place for personnel	

序　码	汉　文　名	英　文　名	注　释
02.1007	待避所	refuge place	
02.1008	安全地点	safety place	
02.1009	安全帽	safety cap, safety helmet	
02.1010	安全绳	safety rope, safety strap	
02.1011	安全网	safety net	
02.1012	保险带	safety belt	
02.1013	安全梯	emergency staircase, fire escape	
02.1014	防护栏杆	protection railing	
02.1015	安全用电	safety in utilizing electric energy	
02.1016	安全电压	safety voltage	
02.1017	消防设施	fire-fighting equipment	
02.1018	防火门	fire protection gate	
02.1019	消防器材	fire-fighting apparatus and materials	
02.1020	隔火带	fire protection strip	
02.1021	防火净距	fire protection distance	
02.1022	瓦斯治理	gas control	
02.1023	安全炸药	safety explosion, explosive charge, safety explosive	
02.1024	湿式凿岩	wet boring for rock, wet drilling for rock	
02.1025	防尘	dust prevention, dust control	
02.1026	防滑鞋	antiskid shoe	
02.1027	绝缘鞋	insulant shoe	
02.1028	防坠器	falling protector	
02.1029	阻车器	stop device	
02.1030	安全装置	safety device, safety equipment	
02.1031	救生设施	lifesaving appliance	
02.1032	防雷装置	lightning protection device	
02.1033	保护地线	protective earth wire	
02.1034	保护接地	protective grounding	
02.1035	触电保安器	electric shock protector	
02.1036	防爆设施	blasting protection facilities	
02.1037	质量管理	quality management	
02.1038	质量保证	quality assurance	
02.1039	质量控制	quality control	
02.1040	质量体系	quality system	

序 码	汉 文 名	英 文 名	注 释
02.1041	质量监督	quality superintendence, quality surveillance	
02.1042	工程质量检验	inspection of engineering quality	
02.1043	隐蔽工程检查	hidden project inspection	
02.1044	取样试验检查	sampling inspection	
02.1045	目测	visual measurement, visual observation	
02.1046	丈量	measure	
02.1047	仪器检查	inspect by instrument	
02.1048	全验	overall acceptance	
02.1049	抽验	selective acceptance	
02.1050	工程质量评定	evaluation of engineering quality	
02.1051	工程质量验收	acceptance of engineering quality	
02.1052	随工验收	acceptance following construction, follow-up acceptance	
02.1053	中间验收	intermediate acceptance	
02.1054	缺陷工程	defect project, drawback project	
02.1055	合格工程	qualified project	
02.1056	样板工程	sample project	
02.1057	优质工程	high grade project, high quality project	又称"优良工程"。
02.1058	施工验收	delivery-receiving acceptance	

03. 铁 道 工 务

序 码	汉 文 名	英 文 名	注 释
03.0001	工务段	track division, track district, track maintenance division	
03.0002	养路领工区	track subdivision, track maintenance subdivision	
03.0003	养路工区	track maintenance section, permanent way gang	
03.0004	养路领工员	track master, track supervisor	
03.0005	养路工长	track foreman	
03.0006	养路工	trackman, machine operator, track mechanics	

序 码	汉 文 名	英 文 名	注 释
03.0007	巡道工	track walker, track patrolling man	
03.0008	道口看守工	grade crossing watchman, level crossing watchman	
03.0009	桥隧巡守工	bridge and tunnel watchman, bridge and tunnel patrolling man	
03.0010	轨道	track	曾用名"线路上部建筑"。
03.0011	轨道类型	classification of track, track standard	
03.0012	轨道结构	track structure	
03.0013	有碴轨道	ballasted track	
03.0014	无碴轨道	ballastless track	
03.0015	线路	track, permanent way	
03.0016	有缝线路	jointed track	
03.0017	无缝线路	continuously welded rail track, jointless track	
03.0018	长轨线路	long welded rail track	
03.0019	轨排	track panel, track skeleton	
03.0020	轨节	rail link	
03.0021	长轨条	long rail string	
03.0022	轨道几何形位	track geometry	
03.0023	轨距加宽	gauge widening	
03.0024	螺旋曲线	spiral curve, clothoid curve	
03.0025	三次抛物线曲线	cubic parabola curve	
03.0026	曲线超高	superelevation, cant, elevation of curve	
03.0027	欠超高	deficient superelevation	
03.0028	过超高	surplus superelevation, excess elevation	
03.0029	反超高	reverse superelevation, counter superelevation, negative superelevation	
03.0030	未被平衡离心加速度	unbalanced centrifugal acceleration	
03.0031	曲线正矢	curve versine	
03.0032	轨道力学	track mechanics	
03.0033	轨道动力学	track dynamics	

序 码	汉 文 名	英 文 名	注 释
03.0034	轨道强度计算	track strength analysis	
03.0035	轨道几何状态恶化	track deterioration	
03.0036	轨道失效	track failure	
03.0037	轨道框架刚度	rigidity of track panel	
03.0038	道床系数	ballast coefficient, ballast modulus	
03.0039	钢轨基础模量	rail supporting modulus, track modulus	
03.0040	钢轨支点弹性模量	modulus of elasticity of rail support	
03.0041	轨道应力	track stresses	
03.0042	轨道稳定性	stability of track	
03.0043	接头阻力	joint resistance	
03.0044	道床阻力	ballast resistance	
03.0045	扣件扣压力	toe load of fastening	
03.0046	轮/轨接触应力	rail/wheel contact stress	
03.0047	钢轨	rail	
03.0048	轨头	rail head	
03.0049	轨腰	rail web	
03.0050	轨底	rail base, rail bottom	
03.0051	淬火轨	head hardened rail, quenched rail	
03.0052	合金轨	alloy steel rail	
03.0053	合金淬火轨	head hardened alloy steel rail, quenched alloy steel rail	
03.0054	耐磨轨	wear resistant rail	
03.0055	耐腐蚀轨	corrosion resistant rail	
03.0056	标准长度钢轨	standard length rail	
03.0057	短轨	short rail	
03.0058	缩短轨	standard shortened rail, fabricated short rail used on curves, standard curtailed rail	
03.0059	异形轨	compromise rail	
03.0060	护轨	guard rail, check rail	
03.0061	轨底坡	rail cant	
03.0062	钢轨工作边	gage line	
03.0063	轨缝	rail gap, joint gap	又称"接头缝"。
03.0064	构造轨缝	structural joint gap, maximum	

序　码	汉　文　名	英　文　名	注　释
		joint gap structurally obtainable	
03.0065	钢轨接头	rail joint	
03.0066	相对式接头	opposite joint, square joint	
03.0067	相错式接头	alternate joint, staggered joint, broken joint	又称"相互式接头"。
03.0068	垫接接头	supported joint	又称"承接接头"。
03.0069	悬接接头	suspended joint	
03.0070	绝缘接头	insulated joint	
03.0071	胶结绝缘接头	glued insulated joint	
03.0072	焊接接头	welded joint	
03.0073	冻结接头	frozen joint	
03.0074	异型接头	compromise joint	
03.0075	钢轨伸缩调节器	expansion rail joint, rail expansion device, switch expansion joint	又称"温度调节器"。
03.0076	接头联结零件	rail joint accessories, rail joint fastenings	
03.0077	接头夹板	joint bar, splice bar, fish plate	曾用名"鱼尾板"。
03.0078	平型双头夹板	flat joint bar	
03.0079	轨头底面接触夹板	head contact flat joint bar	又称"楔型接头夹板"。
03.0080	轨腹上圆弧接触夹板	head free flat joint bar	又称"铰型接头夹板"。
03.0081	带裙鱼尾板	aproned fish plate, fish plate with apron	
03.0082	异型接头夹板	compromise joint bar	
03.0083	接头螺栓	track bolt, fish bolt	曾用名"鱼尾螺栓"。
03.0084	钩螺栓	claw bolt, hook bolt, anchor bolt	
03.0085	扣件	rail fastening	又称"中间联结零件"。
03.0086	分开式扣件	separated rail fastening, indirect holding fastening	
03.0087	不分开式扣件	nonseparated rail fastening, direct holding fastening	
03.0088	半分开式扣件	semi-separated rail fastening, mixed holding fastening	
03.0089	弹性扣件	elastic rail fastening	
03.0090	道钉	track spike, rail spike, dog spike	又称"狗头钉","钩

序　码	汉文名	英文名	注　释
			头钉"。
03.0091	弹簧道钉	elastic rail spike	
03.0092	螺纹道钉	screw spike	
03.0093	螺栓螺纹钉	bolt-screw spike	
03.0094	硫磺锚固	sulphur cement mortar anchor, sulphur cement mortar anchorage	
03.0095	垫板	tie plate	
03.0096	橡胶垫板	rubber tie plate	又称"弹性垫板"。
03.0097	衬垫	pad	
03.0098	冻害垫板	frost heave board, track shim for frost heaving roadbed, frost shim	
03.0099	弹簧垫圈	spring washer	
03.0100	防爬器	anti-creeper, rail anchor	
03.0101	穿销防爬器	wedged rail anchor	
03.0102	弹簧防爬器	spring rail anchor	
03.0103	防爬支撑	anti-creep strut	
03.0104	轨撑	rail brace	
03.0105	轨距杆	gage tie bar, gage rod, gage tie	
03.0106	绝缘轨距杆	insulated gage rod	
03.0107	轨下基础	sub-rail foundation, sub-rail track bed	
03.0108	轨枕	tie, cross tie, sleeper	
03.0109	宽混凝土轨枕	broad concrete tie	曾用名"轨枕板"。
03.0110	纵向轨枕	longitudinal tie	
03.0111	板式轨道	slab-track	
03.0112	木枕	wooden tie	又称"枕木"。
03.0113	油枕	treated wooden tie	
03.0114	素枕	untreated wooden tie	
03.0115	钢枕	steel tie	
03.0116	混凝土枕	concrete tie	
03.0117	岔枕	switch tie, turnout tie	
03.0118	短枕	short tie, block tie	
03.0119	枕盒	crib	曾用名"枕木盒"
03.0120	道床	ballast bed	
03.0121	整体道床	solid bed, integrated ballast bed,	

序　码	汉 文 名	英 文 名	注　释
		monolithic concrete bed	
03.0122	沥青道床	asphalt cemented ballast bed	
03.0123	道碴层	ballast layer	曾用名"道碴床"。
03.0124	道碴	ballast	
03.0125	道碴级配	ballast grading	
03.0126	洛杉矶磨损试验	Los Angeles abrasion test	
03.0127	碎石道碴	stone ballast	
03.0128	矿渣道碴	slag ballast	
03.0129	卵石道碴	gravel ballast	
03.0130	砂道碴	sand ballast	
03.0131	底碴	subballast	
03.0132	道床厚度	thickness of ballast bed, depth of ballast	
03.0133	道床宽度	width of ballast bed	
03.0134	道床碴肩	shoulder of ballast bed	
03.0135	轮轨游间	clearance between wheel flange and gage line	
03.0136	曲线内接	inscribed to curves	
03.0137	自由内接	free inscribing	
03.0138	强制内接	compulsory inscribing	
03.0139	楔形内接	wedging inscribing	
03.0140	静力内接	static inscribing	
03.0141	动力内接	dynamic inscribing	
03.0142	轨道不平顺	track irregularity	
03.0143	静态不平顺	static track irregularity, irregularity without load	
03.0144	动态不平顺	dynamic track irregularity	
03.0145	轨道几何尺寸容许公差	track geometry tolerances	
03.0146	轨道变形	track deformation, track disorder, track distortion	
03.0147	轨道残余变形	track residual deformation, track permanent deformation	
03.0148	弹性挤开	gage elastically widened, elastic squeeze-out	
03.0149	轨道水平	track cross level	又称"左右水平"。
03.0150	轨道方向	track alignment	

序 码	汉 文 名	英 文 名	注 释
03.0151	轨道前后高低	longitudinal level of rail, track profile	简称"前后高低"。
03.0152	三角坑	twist, warp	又称"扭曲"。
03.0153	轨道明坑	visible pit of track, visible low spot of track, track depression	
03.0154	轨道暗坑	loose tie	又称"空吊板"。
03.0155	翻浆冒泥	mud-pumping	
03.0156	钢轨低接头	depressed joint, battered joint of rail	
03.0157	接头瞎缝	closed joint, tight joint	
03.0158	大轨缝	excessive joint gap, wide joint gap	
03.0159	线路爬行	track creeping	
03.0160	轨道鼓出	track buckling	俗称"跑道"。
03.0161	轨头总磨损	total wear of rail head	曾用名"轨头总磨耗"。
03.0162	轨头垂直磨损	vertical wear of rail head	曾用名"轨头垂直磨耗"。
03.0163	轨头侧面磨损	side wear of rail head	曾用名"轨头侧面磨耗"。
03.0164	轨头波形磨损	wave-type deformation of rail head	曾用名"轨头波浪形磨耗"。
03.0165	轨头波纹磨损	corrugation of rail head, rail corrugation	
03.0166	轨头短波浪磨损	short wave undulation of rail head	
03.0167	轨头长波浪磨损	long wave undulation of rail head	
03.0168	轨端马鞍形磨损	rail end batter, saddle wear of rail end	
03.0169	轨头肥边	flow of rail head, lipping of rail head	
03.0170	错牙接头	rail ends unevenness in line or surface	
03.0171	钢轨伤损	rail defects and failures	
03.0172	钢轨擦伤	engine burn, wheel burn	
03.0173	钢轨锈蚀	rail corrosion	
03.0174	白点	flake crack, shatter crack, small nucleus fissure	
03.0175	核伤	nucleus flaw, oval flaw	

序　码	汉　文　名	英　文　名	注　释
03.0176	轨头剥离	gage line shelly cracks	
03.0177	轨头掉块	spalling of rail head	
03.0178	轨头压溃	crushing of rail head	
03.0179	轨头垂直劈裂	vertical split of rail head	
03.0180	轨头微细裂纹	detail fracture of rail head	
03.0181	轨端崩裂	rail end breakage	
03.0182	钢轨打标记	branding and stamping of rails	
03.0183	轨端削角	rail end chamfering	又称"轨端倒棱"。
03.0184	螺栓孔削角	bolt hole chamfering	又称"螺栓孔倒棱"。
03.0185	螺栓孔加强	bolt hole cold-working strenthening	
03.0186	轨头水平劈裂	horizontal split of rail head	
03.0187	轨腰劈裂	piped rail, split of rail web	
03.0188	钢轨裂纹	rail cracks	
03.0189	轨头发裂	head checks, hair crack of rail head	
03.0190	轨底崩裂	burst of rail base, burst of rail bottom, broken rail base	
03.0191	钢轨折断	brittle fractures of rail, sudden rupture of rail	
03.0192	钢轨落锤试验	drop test of rail	
03.0193	螺孔裂纹	bolt hole crack	
03.0194	线路大修	major repair of track, overhaul of track, track renewal	
03.0195	线路中修	intermediate repair of track	
03.0196	线路维修	maintenance of track	
03.0197	找小坑	spot surfacing	
03.0198	全面起道捣固	out-of-face surfacing	
03.0199	起道	raising of track, track lifting	
03.0200	落道	under cutting of track, lowering of track	
03.0201	垫砂起道	measured shovel packing	
03.0202	改道	gage correction, gaging of track	
03.0203	拨道	track lining	
03.0204	整正水平	adjusting of cross level	
03.0205	整正曲线	curve adjusting, curve lining	
03.0206	绳正法整正曲线	string lining of curve	

序　码	汉　文　名	英　文　名	注　释
03.0207	绳度整正曲线计算器	string lining computer, string-line calculator	
03.0208	调整轨缝	adjusting of rail gaps, evenly distributing joint gaps	又称"均匀轨缝"。
03.0209	整正轨缝	dispersal of rail gaps, adjusting joint gaps up to standard	
03.0210	矫直钢轨	straightening of kinked rail	
03.0211	焊修钢轨	resurfacing of rail	又称"堆焊钢轨"。
03.0212	打磨钢轨	rail grinding	
03.0213	轨头整形	rail head reprofiling	
03.0214	轨头非对称断面打磨	asymmetrical rail head profile grinding	
03.0215	方正轨枕	tie respacing, squaring of ties	
03.0216	削平木枕	tie adzing	
03.0217	修补木枕	tie repairing	
03.0218	木枕防腐	preservation of wooden tie	
03.0219	木枕预钻孔	preboring of wooden tie, preboring of spike holes	
03.0220	木枕刻痕	incising of wooden tie	
03.0221	枕木塞	tie plug	
03.0222	木枕防裂装置	anti splitting device	
03.0223	捣固道床	ballast tamping	
03.0224	夯实道床	ballast ramming, ballast consolidating	
03.0225	清筛道床	ballast cleaning	
03.0226	整理道床	ballast trimming	
03.0227	锁定轨温	fastening-down temperature of rail	
03.0228	零应力轨温	stress free rail temperature	
03.0229	中和轨温	neutral temperature	
03.0230	放散温度力	destressing, stress liberation	
03.0231	温度力	temperature stress	
03.0232	温度力峰	temperature stress peak	
03.0233	轨道鼓出临界温度	critical temperature of track buckling	
03.0234	伸缩区	breathing zone	又称"呼吸区"。
03.0235	缓冲区	buffer zone, transition zone	
03.0236	固定区	nonbreathing zone, fixed zone,	

序　码	汉　文　名	英　文　名	注　释
		deformation-free zone	
03.0237	调节轨	buffer rail	
03.0238	气压焊	oxyacetylene pressure welding	
03.0239	电阻焊	flash butt welding	
03.0240	铝热焊	[alumino-]thermit welding	
03.0241	周转轨	inventory stock rails	
03.0242	备用轨	stock rails per kilometer of track, emergency rail stored along the way	
03.0243	再用轨	second hand rail, relaying rail	
03.0244	顺坡	run-off elevation	
03.0245	作业标志	working signal	
03.0246	停车牌	red board, stop indicator	
03.0247	慢行牌	yellow board, speed indicator	
03.0248	线路标志	road way signs, permanent way signs	
03.0249	公里标	kilometer post	
03.0250	曲线标	curve post	
03.0251	坡度标	grade post	
03.0252	桥梁标	bridge post	
03.0253	隧道标	tunnel post	
03.0254	管界标	section sign	
03.0255	钢轨位移观测桩	rail creep indication posts	
03.0256	鸣笛标	whistle board	
03.0257	线路中断	line interruption	
03.0258	施工封闭线路	line occupation for works	
03.0259	封闭线路作业时间	work occupation time, working time of closed section	又称"开天窗作业时间"。
03.0260	列车空隙作业时间	working time between trains	
03.0261	养路表报	maintenance-of-way report and forms, track work forms	
03.0262	工务设备台帐	technical record of track, bridge and other equipments	
03.0263	养路管理电脑系统	computer aided track maintenance and management system	
03.0264	线路平剖面图	track charts	

序　码	汉　文　名	英　文　名	注　释
03.0265	养路费用	maintenance of way expenditures	
03.0266	线路维修规则	rules of maintenance of way	
03.0267	轨道养护标准	standards of track maintenance	
03.0268	换算线路长度	equivalent track kilometerage	
03.0269	轨检车评分	track evaluation by recording car, evaluation by track inspection car	
03.0270	轨道质量指数	track quality index	
03.0271	平面交叉	level crossing	
03.0272	立体交叉	grade separation	
03.0273	道口	grade crossing, level crossing	
03.0274	道口铺面	grade crossing pavement, surface of grade crossing	
03.0275	道口平台	level stretch of grade crossing	
03.0276	道口警标	warning sign at grade crossing	
03.0277	道口栏木	cross barrier at grade crossing	
03.0278	道口栅栏	side barrier at grade crossing	
03.0279	道口护桩	protective stake at grade crossing	
03.0280	限界架	clearance limit frame	
03.0281	道岔	turnout, switches and crossings	
03.0282	单开道岔	simple turnout, lateral turnout	
03.0283	单式对称道岔	symmetrical double curve turnout, equilateral turnout	又称"双开道岔"。
03.0284	单式不对称道岔	unsymmetrical double curve turnout, unequilateral turnout	又称"不对称双开道岔"。
03.0285	单式同侧道岔	unsymmetrical double curve turnout in the same direction	
03.0286	三开道岔	symmetrical three throw turnout, three-way turnout	
03.0287	不对称三开道岔	unsymmetrical three-way turnout, unsymmetrical three throw turnout	
03.0288	左开道岔	left hand turnout	
03.0289	右开道岔	right hand turnout	
03.0290	交叉	crossing	
03.0291	菱形交叉	diamond crossing	
03.0292	直角交叉	rectangular crossing, square cross-	

序 码	汉 文 名	英 文 名	注 释
		ing	
03.0293	交分道岔	slip switch	
03.0294	单式交分道岔	single slip switches	
03.0295	复式交分道岔	double slip switches	
03.0296	渡线	crossover	
03.0297	交叉渡线	scissors crossing, double crossover	
03.0298	平行渡线	parallel crossover	
03.0299	梯线	ladder track	
03.0300	套线道岔	mixed gage turnout	
03.0301	道岔主线	main line of turnout, main track of turnout, turnout main	
03.0302	道岔侧线	branch line of turnout, branch track of turnout, turnout branch	
03.0303	导曲线	lead curve	
03.0304	导曲线半径	radius of lead curve	
03.0305	导曲线支距	offset of lead curve	
03.0306	道岔中心	center of turnout	
03.0307	辙叉	frog, crossing	
03.0308	辙叉角	frog angle	
03.0309	辙叉号数	frog number	
03.0310	道岔号数	turnout number	
03.0311	道岔始端	beginning of turnout	
03.0312	道岔终端	end of turnout	
03.0313	尖轨	switch rail, tongue rail, blade	
03.0314	直线尖轨	straight switch	
03.0315	曲线尖轨	curved switch	
03.0316	尖轨理论尖端	theoretical point of switch rail	
03.0317	尖轨尖端	actual point of switch rail	
03.0318	尖轨跟端	heel of switch rail	
03.0319	曲线出岔道岔	turnout from curved track	
03.0320	辙叉心轨理论尖端	theoretical point of frog	
03.0321	辙叉心轨尖端	actual point of frog	
03.0322	辙叉趾端	toe end of frog, frog toe	
03.0323	辙叉跟端	heel end of frog, frog heel	
03.0324	道岔基线	reference line of turnout	

序　码	汉　文　名	英　文　名	注　释
03.0325	道岔全长	total length of turnout	
03.0326	道岔理论长度	theoretical length of turnout	
03.0327	道岔实际长度	actual length of turnout	
03.0328	道岔理论导程	theoretical lead of turnout	
03.0329	道岔前部理论长度	front part theoretical length of turnout	
03.0330	道岔后部理论长度	rear part theoretical length of turnout	
03.0331	道岔后部实际长度	rear part actual length of turnout	
03.0332	尖轨长度	length of switch rail	
03.0333	辙叉趾长	toe length of frog	
03.0334	辙叉跟长	heel length of frog	
03.0335	辙叉趾宽	toe spread of frog	
03.0336	辙叉跟宽	heel spread of frog	
03.0337	尖轨动程	throw of switch	
03.0338	护轨与心轨的查照间隔	[guard rail] check gage	
03.0339	护背距离	guard rail face gage, back gage	
03.0340	辙叉有害空间	gap in the frog, open throat, unguarded flange-way	
03.0341	转辙角	switch angle	
03.0342	辙叉咽喉	throat of frog	
03.0343	轮缘槽	flange-way, flange clearance	
03.0344	转辙器	switch	
03.0345	可弯式尖轨转辙器	flexible switch	
03.0346	间隔铁式尖轨转辙器	loose heel switch	
03.0347	高锰钢整铸辙叉	solid manganese steel frog, cast manganese steel frog	
03.0348	钢轨组合辙叉	bolted rigid frog, assembled frog	
03.0349	可动心轨辙叉	movable-point frog	
03.0350	可动翼轨辙叉	movable-wing frog	
03.0351	钝角辙叉	obtuse frog	
03.0352	锐角辙叉	end frog, acute frog	
03.0353	曲线辙叉	curved frog	

序　码	汉　文　名	英　文　名	注　释
03.0354	基本轨	stock rail	
03.0355	密贴尖轨	closed switch rail, close contact between switch point and stock rail	
03.0356	淬火尖轨	surface-hardened switch rail, quenched switch rail	
03.0357	特种断面尖轨	special heavy section switch rail, tongue rail made of special section rail, full-web section switch rail	
03.0358	翼轨	wing rail	
03.0359	心轨	point rail, nose rail	
03.0360	道岔护轨	turnout guard rail	
03.0361	尖轨护轨	switch point guard rail	
03.0362	尖轨保护器	switch protector	
03.0363	道岔拉杆	switch rod, stretcher bar	
03.0364	道岔连接杆	connecting bar, following stretcher bar	
03.0365	间隔铁	filler, spacer block	
03.0366	尖轨补强板	reinforcing bar	
03.0367	滑床板	slide plate, switch plate	
03.0368	线路机械	permanent way machine	
03.0369	线路机具	permanent way tool	
03.0370	路基机械	subgrade machine, machine for roadway work	
03.0371	轨道机械	track machine	
03.0372	道碴机械	ballast machine	
03.0373	捣固机械	tamping machine	
03.0374	夯实机械	ballast consolidating machine	
03.0375	道碴清筛机械	ballast cleaning machine	
03.0376	轨行式装运机械	rail-mounted handling and transportation machine	
03.0377	轨道检测设备	track geometry measuring device	
03.0378	安全防护设备	safety protection equipment	
03.0379	线路清理机械	permanent way clearing machine	
03.0380	公铁两用线路机械	rail/road permanent way machine	

序 码	汉 文 名	英 文 名	注 释
03.0381	大型线路机械	heavy permanent way machine, large permanent way machine	又称"重型线路机械"。
03.0382	小型线路机械	light permanent way machine, small permanent way machine	又称"轻型线路机械"。
03.0383	拨道机	track lining machine	
03.0384	拨道器	track lining tool	
03.0385	起道机	track lifting machine	
03.0386	起道器	rail jack, track jack	
03.0387	起拨道机	track lifting and lining machine	
03.0388	起拨道器	track lifting and lining tool	
03.0389	钢轨钻孔机	rail drilling machine	
03.0390	钢轨钻孔器	rail drilling tool	
03.0391	磨轨机	rail grinding machine	
03.0392	磨轨车	rail grinding car	
03.0393	磨轨列车	rail grinding train	
03.0394	锯轨机	rail cutting machine, rail sawing machine	
03.0395	轨缝调整器	rail gap adjuster, rail puller	
03.0396	钢轨拉伸器	rail tensor	
03.0397	直轨器	rail straightening tool, rail straightener	
03.0398	平轨器	rail bending tool, rail bender	
03.0399	钢轨刨边机	rail-head edges planing machine	
03.0400	钢轨推凸机	rail weld seam shearing machine	
03.0401	钢轨推凸器	rail shearing device	
03.0402	机动螺钉－螺栓搬手	rail screw-bolt power wrench	
03.0403	地面钢轨涂油器	on-track rail lubricator	
03.0404	车载钢轨涂油器	on-board rail lubricator	
03.0405	焊轨机	rail welding machine	
03.0406	轨枕抽换机	tie replacing machine	
03.0407	木枕钻孔机	wooden tie drilling machine	
03.0408	打道钉机	spike driver	
03.0409	起道钉机	spike puller	
03.0410	混凝土枕螺栓钻取机	concrete tie dowel drilling and pulling machine	
03.0411	方枕器	tie respacer	

序 码	汉 文 名	英 文 名	注 释
03.0412	木枕削平机	wooden sleeper adzing machine	
03.0413	手提电动捣固机	portable electric tamper, hand electric tamper	又称"电镐"。
03.0414	手提风动捣固机	portable pneumatic tamper	
03.0415	手提内燃捣固机	portable gasoline-powered tamper	
03.0416	小型液压捣固机	light hydraulic tamping machine	
03.0417	液压捣固机	hydraulic tamping machine	又称"液压捣固车"。
03.0418	自动液压大型捣固车	auto-leveling-lifting-lining-tamping machine	又称"自动整平－起道－拨道－捣固车"。
03.0419	配碴整形机	ballast distributing and regulating machine	
03.0420	道碴犁	ballast plow	
03.0421	扒碴机	crib ballast removers	
03.0422	收碴机	ballast recollecting machine	
03.0423	枕间夯实机	crib consolidating machine	
03.0424	道床边坡夯实机	ballast shoulder consolidating machine	
03.0425	全断面道床夯实机	full section ballast consolidating machine	
03.0426	道床底碴夯实机	subballast consolidating machine	
03.0427	动力稳定机	dynamic track stabilizer	
03.0428	边坡清筛机	ballast shoulder cleaning machine	
03.0429	大型全断面清筛机	large ballast undercutting cleaners, on-track full section undercutting cleaners	
03.0430	中型清筛机	medium ballast undercutting cleaners	
03.0431	大揭盖清筛机	ballast cleaning machine with removed track panels	
03.0432	小型枕底清筛机	small ballast undercutting cleaners	
03.0433	轻型轨道车	light rail motor car, light motor trolley	
03.0434	重型轨道车	heavy rail motor car, heavy motor trolley	
03.0435	发电轨道车	power generating rail car	
03.0436	发电走行两用车	self-propelled power generating car	
03.0437	轨道平车	rail flat car	

序　码	汉　文　名	英　文　名	注　释
03.0438	长钢轨运输作业列车	long welded rail transporting and working train	
03.0439	铺轨机	track laying machine	
03.0440	轨行式起重机	crane on-track, rail-mounted crane	
03.0441	悬臂式铺轨排机	track panel laying machine with cantilever	
03.0442	悬臂式铺轨机	rail laying machine with cantilever	
03.0443	门式铺轨排机	track panel laying gantry crane	
03.0444	铺轨列车	track laying train	
03.0445	工程宿营车	work train with camp cars	
03.0446	轨道检查车	track recording car, track inspection car	
03.0447	轨道检查小车	track geometry measuring trolley	
03.0448	桥梁实验车	bridge test car	
03.0449	钢轨探伤车	rail flaw detection car	
03.0450	钢轨探伤仪	rail flaw detector	
03.0451	钢轨磨损检查车	rail profile measuring car	
03.0452	钢轨磨损检查仪	rail profile gauge	
03.0453	列车速度检测仪	train speed monitoring device	
03.0454	列车接近报警器	train approach warning device	
03.0455	限界检测车	clearance car, clearance inspection car	
03.0456	塌方落石报警器	land slide warning device	
03.0457	除雪机	snow removing machine, snow remover	
03.0458	道岔熔冰器	switch heater	
03.0459	除砂机	sand removing machine	
03.0460	开沟机	ditcher	
03.0461	开沟平路机	track shaving machine, ditching and grading machine	
03.0462	灌木切割机	brush cutting machine, brush cutter	
03.0463	除草机	weed cutting machine, weed cutter	
03.0464	喷洒除草机	weed killing machine, weed killer	
03.0465	轨距尺	track gage	又称"道尺"。
03.0466	万能道尺	universal rail gage	

序 码	汉 文 名	英 文 名	注 释
03.0467	捣镐	packer, tamping pick, beater	又称"道镐"。
03.0468	道钉锤	spike hammer	
03.0469	撬棍	lining bar, claw bar	
03.0470	轨道水平尺	track level	
03.0471	螺栓扳手	track wrenches	
03.0472	单轨小车	hand cart	
03.0473	地表排水	surface drainage	
03.0474	地下排水	subsurface drainage	
03.0475	过滤层	filter	
03.0476	路基松软	soft spots of road bed	
03.0477	滑坡	landslip, landslide	
03.0478	坍方	slide, slip	
03.0479	蠕变	creep, crawl	
03.0480	崩塌	collapse, toppling	
03.0481	冲刷	erosion, scouring	
03.0482	落石	rock fall	
03.0483	泥石流	earth flow, debris flow, mud and rock flow	
03.0484	冻害	frost heaving	
03.0485	雪崩	snow slip, snow slide, snow avalanche	
03.0486	道碴槽	ballast tub	
03.0487	道碴箱	ballast box	
03.0488	道碴袋	ballast pocket	
03.0489	道碴巢	ballast nest	
03.0490	路基下沉	subgrade settlement	
03.0491	煤矿沉陷	mining subsidence	
03.0492	路基挤起	subgrade bulge, subgrade squeeze-out	
03.0493	防洪	flood control	
03.0494	防砂	sand-drift control, sand protection	
03.0495	防雪	snow-drift control, snow protection	
03.0496	雪害	snow blockade, snow drift	
03.0497	水害	flood damage, washout	
03.0498	砂害	sand blockade, sand drift	
03.0499	隧道冰害	frost damage in tunnel	

序 码	汉 文 名	英 文 名	注 释
03.0500	防雪栅	snow fence, snow guard	
03.0501	防雪障	snow protection bank	
03.0502	防雪树篱	snow protection hedge	
03.0503	拉沟	slotting	
03.0504	刷坡	slope cutting	
03.0505	边坡植被防护	vegetation on slope	
03.0506	石笼	gabion	
03.0507	土工织物	geotextile	又称"土工布"。
03.0508	土工格栅	geogrid	又称"土工网格"。
03.0509	铺草皮	sodding	
03.0510	整治冻害轨道	treatment of frost heaving track	
03.0511	垫冻害垫板	track shimming	
03.0512	电化学加固土壤	electro-chemical treatment of soil	
03.0513	桥梁浅基	shallow foundation of bridge, un-safe depth foundation of bridge	
03.0514	限界检查	checking of clearance, clearance check measurement	
03.0515	河调－防护失修	disrepair of flow regulating and shore protecting structure, river bank protection out of repair	
03.0516	桥址水文观测	hydrologic observation of bridge site	
03.0517	桥梁检定试验	bridge rating test	
03.0518	容许强度检定	working stress rating	
03.0519	承载系数检定	load factor rating	
03.0520	自动力矩检定	autostress rating	
03.0521	桥梁检定承载系数	rated load-bearing coefficient for bridge, rated load-bearing coefficient for bridge as compared with standard live loading	
03.0522	特种车辆活载	special live load, live load of special type wagon	
03.0523	墩台基础挖验	excavating foundation for checking purpose, foundation examination by excavation	
03.0524	墩台基础钻探	drilling foundation for checking purpose, foundation examina-	

序　码	汉　文　名	英　文　名	注　释
		tion by drilling	
03.0525	圬工梁裂损	cracking of concrete and masonry beam	
03.0526	隧道漏水	tunnel leak	
03.0527	衬砌裂损	lining cracking	
03.0528	衬砌腐蚀	lining corrosion	
03.0529	衬砌变形	lining deformation	
03.0530	桥梁孔径不足	unsufficient span of bridge	
03.0531	桥梁上拱度	camber of bridge span	
03.0532	桥梁挠度	deflection of bridge span	
03.0533	静载试验	static test	
03.0534	动载试验	dynamic test	
03.0535	冲击因数	impact factor	
03.0536	应变仪	strain gage, strainometer	又称"应变计"。
03.0537	阻尼因数	damping factor	
03.0538	次应力	secondary stress	
03.0539	钢梁疲劳损伤	steel bridge fatigue damage	
03.0540	桥梁疲劳剩余寿命	fatigue residual life of bridge	
03.0541	桥梁疲劳剩余寿命评估	evaluating fatigue residual life of bridge	
03.0542	桥梁横向刚度	lateral rigidity of bridge	
03.0543	桥梁最大横向振幅	maximum lateral amplitude of bridge	
03.0544	桥梁自振频率	self-excited vibrational frequency of bridge span, natural frequency of bridge span	
03.0545	桥隧大修	major repair of bridge and tunnel, capital repair of bridge and tunnel	
03.0546	桥隧改造	upgrading of bridge and tunnel	
03.0547	钢梁油漆	protective coating of steel bridge	
03.0548	喷砂除锈	sand blasting	
03.0549	火焰除锈	flame cleaning	
03.0550	钢丝刷除锈	brush cleaning	
03.0551	底漆	prime coat	
03.0552	中间漆	intermediate coat	

序　码	汉　文　名	英　文　名	注　释
03.0553	面漆	top coat	
03.0554	工地漆	field coat	
03.0555	工厂漆	shop coat	
03.0556	干捣水泥砂浆	dry tamped cement mortar	
03.0557	混凝土裂纹	concrete cracks	
03.0558	钢梁腐蚀裂纹	corrosion cracking of steel bridge	
03.0559	钢梁应力腐蚀裂纹	stress corrosion cracking of steel bridge	
03.0560	隧道防火措施	measures against tunnel fire	
03.0561	隧道通风	tunnel ventilation	
03.0562	隧道防排水	tunnel water handling	
03.0563	喷射混凝土修理	shotcrete repair	
03.0564	桥隧养护	maintenance of bridge and tunnel	
03.0565	桥隧经常保养	regular maintenance of bridge and tunnel	
03.0566	桥隧综合维修	comprehensive maintenance of bridge and tunnel structure	
03.0567	桥隧病害整治	damage repair for bridge and tunnel, repair bridge and tunnel fault	
03.0568	桥涵扩孔	opening enlargement of bridge and culvert	
03.0569	桥梁浅基防护	unsafe depth foundation protection, bridge shallow foundation protection	
03.0570	墩周冲淤	scouring and depositing around pier	
03.0571	桥隧巡守	bridge and tunnel patrolling	
03.0572	限界改善	clearance improvement	
03.0573	墩台防撞	collision prevention around pier, pier protection against collision	
03.0574	钢梁加固	strengthening of steel bridge	
03.0575	临时便线	shoofly	
03.0576	防洪预抢工程	precautionary work against flood	
03.0577	水害断道	railroad break down due to flood, line blockade due to flood	
03.0578	水害抢修	rush repair of flood damage to	

序 码	汉 文 名	英 文 名	注 释
		open for traffic	
03.0579	抛石防护	riprap protection	
03.0580	临险抢护	emergency rush engineering, emergency repairs	
03.0581	水害复旧	restoration work for flood damage, restoration of flood damaged structures	
03.0582	日降雨量	daily precipitation	
03.0583	雨量计	rain gage	
03.0584	洪水标记	flood mark	
03.0585	河床冲刷	channel erosion	
03.0586	洪水淹没	flood inundation on tracks	
03.0587	防水林	forestation against flood	
03.0588	防撞破凌	breaking up ice run, breaking up ice floe prevent collision	
03.0589	房建维修	regular maintenance of buildings and structures	
03.0590	房建大修	major repair of buildings and structures	
03.0591	房建检修	inspection and repair of buildings and structures	

04. 铁道牵引动力

序 码	汉 文 名	英 文 名	注 释
04.0001	铁道牵引动力	railway traction power, railway motive power	
04.0002	机车	locomotive	
04.0003	机车种类	types of locomotive	
04.0004	机车比率	locomotive ratio	又称"机车比值"。
04.0005	蒸汽机车	steam locomotive	
04.0006	内燃机车	diesel locomotive	曾用名"柴油机车"。
04.0007	电力机车	electric locomotive	
04.0008	燃气轮机车	gasturbine locomotive	
04.0009	动车组	motor train unit, motor train set	
04.0010	动车	motor car	

序 码	汉 文 名	英 文 名	注 释
04.0011	拖车	trailer	
04.0012	铁路干线机车	railway trunk line locomotive	
04.0013	工矿机车	industrial and mining locomotive	
04.0014	客运机车	passenger locomotive	
04.0015	货运机车	freight locomotive, goods locomotive	
04.0016	调车机车	shunting locomotive, switcher	
04.0017	小运转机车	locomotive for district transfer, transfer locomotive train	
04.0018	路用机车	locomotive of service train, service locomotives	
04.0019	配属机车	allocated locomotive	
04.0020	非配属机车	un-allocated locomotive	
04.0021	支配机车	disposal locomotive	
04.0022	非支配机车	un-disposal locomotive	
04.0023	运用机车	locomotive in operation	
04.0024	检修机车	locomotive under repairing	
04.0025	备用机车	locomotive in reserve	
04.0026	封存机车	locomotive stored up	
04.0027	待修机车	locomotive waiting for repair	
04.0028	机车出租	leased locomotive	
04.0029	机车报废	locomotive retirement	
04.0030	机车储备	locomotive reservation, locomotive storage	
04.0031	机车整备	locomotive servicing, locomotive running preparation	
04.0032	机车整备能力	locomotive service capacity	
04.0033	机车技术规范	locomotive technical specification	
04.0034	轴列式	axle arrangement	简称"轴式"。
04.0035	转向架中心	bogie pivot center	
04.0036	转向架中心距离	distance between bogie pivot centers, bogie pivot pitch	
04.0037	机车全轴距	locomotive total wheel base	
04.0038	机车转向架轴距	locomotive wheel base of bogie	
04.0039	机车固定轴距	locomotive rigid wheel base	
04.0040	机车长度	locomotive overall length	
04.0041	机车宽度	locomotive width	

序 码	汉 文 名	英 文 名	注 释
04.0042	机车高度	locomotive height	
04.0043	机车计算重量	calculated weight of locomotive	
04.0044	机车整备重量	locomotive service weight	
04.0045	机车重量	locomotive weight	
04.0046	机车粘着重量	locomotive adhesive weight	
04.0047	轴重转移	axle load transfer	又称"轴载荷转移"。
04.0048	粘着重量利用系数	adhesive weight utility factor	
04.0049	机车重量分配	weight distribution of locomotive	
04.0050	轮对横动量	lateral play of wheel set	
04.0051	机车噪声	locomotive noise	
04.0052	热值	heat value	又称"发热量"。
04.0053	机车万吨公里能耗	energy consumption per 10 000 t·km of locomotive	
04.0054	机车用煤	coal for locomotive	
04.0055	机车用柴油	diesel oil for locomotive	
04.0056	机车用电	electricity for locomotive	
04.0057	机车用换算煤	converted coal for locomotive	
04.0058	标准煤	standard coal	
04.0059	机车用润滑剂	lubricant for locomotive	
04.0060	给水	water supply	
04.0061	水鹤	water crane	
04.0062	给水处理	water [supply] treatment	
04.0063	炉内软水	water softened in boiler	
04.0064	炉外软水	water softened out of boiler	
04.0065	软水与净水	water softened and purified	
04.0066	给水站	water supply station	又称"给水所"。
04.0067	机务段	locomotive depot	
04.0068	机车运用段	locomotive running depot	
04.0069	机车检修段	locomotive repair depot	
04.0070	机务折返段	locomotive turnaround depot	
04.0071	列车无线电调度系统	train radio dispatching system	
04.0072	机车监控记录装置	locomotive supervise and record apparatus	
04.0073	机车故障	locomotive failure	
04.0074	机车牵引特性	locomotive tractive characteristic	

序 码	汉 文 名	英 文 名	注 释
04.0075	机车效率	total locomotive efficiency	
04.0076	机车轮周效率	efficiency of locomotive at wheel rim	
04.0077	机车传动效率	transmission efficiency of locomotive	
04.0078	机车牵引特性曲线	locomotive tractive characteristic curve	
04.0079	机车轮周功率曲线	locomotive power curve at wheel rim	
04.0080	机车牵引力曲线	locomotive tractive effort curve	
04.0081	机车预期牵引特性曲线	predetermined tractive characteristic curve of locomotive	
04.0082	机车功率	locomotive power	
04.0083	标称功率	nominal power	又称"额定功率"。
04.0084	最大运用功率	maximum service output power	又称"装车功率"。
04.0085	轮周功率	output power at wheel rim	
04.0086	比功率	power/weight ratio	
04.0087	比重量	weight/power ratio	
04.0088	单位体积功率	specific volume power	
04.0089	机车牵引力	locomotive tractive effort	
04.0090	轮周牵引力	tractive effort at wheel rim	
04.0091	车钩牵引力	tractive effort at coupler, drawbar pull	
04.0092	起动牵引力	starting tractive effort	
04.0093	粘着牵引力	adhesive tractive effort	
04.0094	持续牵引力	continuous tractive effort	
04.0095	基本阻力	basic resistance	
04.0096	运行阻力	running resistance	
04.0097	惰行阻力	idle running resistance, coasting resistance	
04.0098	起动阻力	starting resistance	
04.0099	附加阻力	additional resistance	
04.0100	坡道阻力	gradient resistance	
04.0101	曲线阻力	curve resistance	
04.0102	空气阻力	air resistance	
04.0103	单位阻力	unit resistance, specific resistance	
04.0104	换算阻力	converted resistance	又称"加算阻力"。

序 码	汉 文 名	英 文 名	注 释
04.0105	速度控制系统	speed control system	
04.0106	加速	acceleration	
04.0107	减速	deceleration	
04.0108	恒速	constant speed	
04.0109	加速力	acceleration force	
04.0110	减速力	deceleration force	
04.0111	列车制动	train braking	
04.0112	制动方式	brake mode	
04.0113	空气制动	air brake	
04.0114	真空制动	vacuum brake	
04.0115	动力制动	dynamic brake	
04.0116	液力制动	hydraulic brake	
04.0117	电阻制动	rheostatic brake	
04.0118	再生制动	regenerative brake	
04.0119	电空制动	electropneumatic brake	
04.0120	蓄能制动	energy-storing brake	
04.0121	涡流制动	eddy current brake	
04.0122	磁轨制动	electromagnetic rail brake	
04.0123	踏面制动	tread brake	
04.0124	盘型制动	disc brake	
04.0125	机车制动周期	locomotive braking period	
04.0126	机车制动距离	locomotive braking distance	
04.0127	机车每轴闸瓦作用力	brake shoe force per axle of locomotive	
04.0128	制动	braking	
04.0129	常用制动	service braking, service application	
04.0130	最大常用制动	full service braking, full service application	又称"常用全制动"。
04.0131	阶段制动	graduated application	
04.0132	自然制动	unintended braking, undesired braking	
04.0133	紧急制动	emergency braking, emergency application	
04.0134	意外紧急制动	undesirable emergency braking, UDE	
04.0135	缓解	release	
04.0136	直接缓解	direct release	又称"一次缓解"。

序 码	汉 文 名	英 文 名	注 释
04.0137	阶段缓解	graduated release	
04.0138	自然缓解	unintended release, undesired release	
04.0139	漏泄	leakage	
04.0140	充风	charging	又称"充气"。
04.0141	充风位	charge position	又称"充气位"。
04.0142	过充风	overcharging	又称"过充气"。
04.0143	再充风	recharging	又称"再充气"。
04.0144	过量充风	overcharge	又称"过量充气"。
04.0145	阶段提升	graduated increasing	
04.0146	保持位	suppression, maintaining position, holding position	
04.0147	制动管减压量	brake pipe pressure reduction	
04.0148	过量减压	over reduction	
04.0149	局部减压	local reduction	
04.0150	分段减压	split reduction	
04.0151	最大常用减压	full service reduction	
04.0152	常用局减	quick service	
04.0153	紧急局减	quick action	
04.0154	前后风压差	false gradient	
04.0155	列车管压差	train pipe pressure gradient	
04.0156	保压停车	stopping at maintaining position	
04.0157	缓解停车	stopping at release	
04.0158	缓解波速	release propagation rate	
04.0159	制动波速	braking propagation rate	
04.0160	机车制动机	locomotive brake gear	
04.0161	空气压缩机	air compressor	俗称"风泵"。
04.0162	调压器	pressure regulator	
04.0163	给风阀	feed valve	
04.0164	减压阀	reducing valve	
04.0165	机车分配阀	locomotive distributing valve	
04.0166	切换阀	transfer valve	
04.0167	自动制动阀	automatic brake valve	俗称"大闸"。
04.0168	单独制动阀	independent brake valve	俗称"小闸"。
04.0169	滤尘止回阀	strainer check valve	
04.0170	机车紧急放风阀	locomotive emergency vent valve	
04.0171	高压保安阀	high pressure safety valve	

序 码	汉 文 名	英 文 名	注 释
04.0172	低压保安阀	low pressure safety valve	
04.0173	无载起动电空阀	no-load starting electro pneumatic valve	
04.0174	油水分离器	oil-water separator	
04.0175	总风缸	main air reservoir	
04.0176	空气干燥器	air dryer	
04.0177	撒砂装置	sanding device	
04.0178	砂箱	sand box	
04.0179	撒砂阀	sanding valve	
04.0180	撒砂器	sanding sprayer	
04.0181	紧急撒砂	emergency sanding	
04.0182	自动撒砂	automatic sanding	
04.0183	间隙效应	slack action	
04.0184	L/V 比值	L/V ratio	
04.0185	弓网关系	pantograph-contact line relation	
04.0186	机车牵引区段	locomotive tractive district	
04.0187	机车交路	locomotive routing	
04.0188	单肩回交路	single-arm routing	
04.0189	双肩回交路	double-arm routing	
04.0190	半循环交路	semi-loop routing	
04.0191	循环交路	loop routing	
04.0192	环形交路	circular routing	
04.0193	短交路	short routing	
04.0194	长交路	long routing	
04.0195	直通交路	through routing	
04.0196	机车乘务制度	locomotive crew working system	
04.0197	机车包乘制	system of assigning crew to designated locomotive	
04.0198	机车轮乘制	locomotive crew pooling system	
04.0199	机车随乘制	locomotive caboose crew system	
04.0200	机车乘务组	locomotive crew	
04.0201	司机	driver	
04.0202	副司机	assistant driver	
04.0203	指导司机	driver instructor	
04.0204	司炉	fireman	
04.0205	机车运用指标	index of locomotive operation	
04.0206	机车出入段作业	preparation of locomotive for leav-	

序 码	汉 文 名	英 文 名	注 释
		ing and arriving at depot	
04.0207	机车全周转	complete turnround of locomotive	
04.0208	机车在段停留时间	detention time of locomotive at depot	
04.0209	机车全周转距离	distance of one complete turnround of locomotive	
04.0210	机车全周转时间	period of one complete turnround of locomotive	
04.0211	机车走行公里	locomotive running kilometers	
04.0212	换算走行公里	converted running kilometers	
04.0213	沿线走行公里	running kilometers on the road	
04.0214	辅助走行公里	auxiliary running kilometers	
04.0215	本务走行公里	leading locomotive running kilometers	
04.0216	单机走行公里	light locomotive running kilometers	
04.0217	重联机车走行公里	multi-locomotive running kilometers	
04.0218	机车日车公里	average daily locomotive running kilometers	
04.0219	机车平均牵引总重	average gross weight hauled by locomotive	
04.0220	机车日产量	average daily output of locomotive	
04.0221	运用机车台数	number of locomotives in service	
04.0222	机车需要系数	coefficient of locomotive requirment	
04.0223	单机运行	light locomotive running	
04.0224	双机牵引	double locomotive traction	又称"双机重联牵引"。
04.0225	多机牵引	multi-locomotive traction	
04.0226	主机	leading locomotive	又称"本务机车"。
04.0227	辅机	assisting locomotive	牵引区段不摘挂机车。
04.0228	补机	banking locomotive	牵引区段摘挂机车。
04.0229	机车超重牵引	traction for train exceed mass norm	
04.0230	机车调度命令	locomotive dispatching order	

序　码	汉　文　名	英　文　名	注　　释
04.0231	机务段运行揭示	running service-bulletin of depot	
04.0232	司机运转报单	driver's service-report, driver's log	
04.0233	司机室	driver's cab	
04.0234	司机操纵台	driver's desk	
04.0235	司机模拟操纵装置	simulator for driver train-handling	
04.0236	优化操纵	optimum handling, optimum operation	
04.0237	机车自动操纵	automatic locomotive operation	
04.0238	机车保养	locomotive maintenance	
04.0239	机车检修	locomotive inspection and repair	
04.0240	机车检修修程	classification of locomotive repair	
04.0241	机车大修	locomotive overhaul [repair], locomotive general overhaul	
04.0242	架修	intermediate repair	
04.0243	定修	periodical repair, light repair	
04.0244	蒸汽机车洗修	steam locomotive boiler washout repair	
04.0245	机车厂修	locomotive repair in works	
04.0246	机车段修	locomotive repair in depot	
04.0247	机车定期修	locomotive periodical repair	即机车半年、一、二、四、五年修。
04.0248	日常检查	routine inspection	
04.0249	运行检查	running inspection	又称"行修"。
04.0250	机车临修	locomotive temporary repair	
04.0251	定期检修	repair based on time or running kilometers	
04.0252	状态检修	repair based on condition of component	
04.0253	换件大修	component exchange repair	
04.0254	配件互换修	repair with interchangeable component	
04.0255	预防维修制	preventive maintenance system	
04.0256	检修周期	period of inspection and repair	
04.0257	月检	monthly inspection	
04.0258	定检公里	running kilometers between prede-	

序　码	汉　文　名	英　文　名	注　释
		termined repairs	
04.0259	定检时间	time between predetermined repairs	
04.0260	检修范围	scope of repairing course, scope of repair	
04.0261	超范围修理	repair beyond the scope of repairing course	
04.0262	检修停时	standing time under repair	
04.0263	集中化修理	centralization of repair	又称"集中修"。
04.0264	检修基本技术条件	fundamental technical requirements for repair and inspection	
04.0265	检修工艺规程	technological regulations for repair and inspection	
04.0266	检修作业程序	repair procedure, shop program	
04.0267	磨耗限度	limit of wear	
04.0268	检修限度	locomotive repair limit	
04.0269	第一限度	1st limit	
04.0270	第二限度	2nd limit	
04.0271	使用限度	operation limit	
04.0272	中间工艺检验	intermediate inspection at the technological process	
04.0273	转向设备	turning facilities	
04.0274	转盘	turntable	
04.0275	机务设备通过能力	carrying capacity of locomotive facilities	
04.0276	机车专用设备	special equipment for locomotive operation	
04.0277	部件故障检测	inspection of component failure, failure diagnostic	又称"故障诊断"。
04.0278	机车验收	acceptance of locomotive	
04.0279	机车试运转	locomotive trial run	
04.0280	型式试验	type test	
04.0281	性能试验	performance test	
04.0282	鉴定试验	homologation test	
04.0283	出厂试验	predelivery test	
04.0284	制动试验	brake test	
04.0285	牵引试验	traction test	

序　码	汉　文　名	英　文　名	注　释
04.0286	负载试验	loaded test, load test	
04.0287	牵引热工试验	traction and thermodynamic test	
04.0288	定置试验	stationary test, test at standstill	
04.0289	动力学试验	dynamics test	
04.0290	强度试验	strength test	
04.0291	运用试验	service test, operation test	
04.0292	耐久性试验	durability test	
04.0293	可靠性试验	reliability test	
04.0294	例行试验	routine test	
04.0295	特殊试验	special test	
04.0296	研究性试验	investigation test	
04.0297	抽样试验	sampling test	
04.0298	模拟试验	analogue test	
04.0299	机车履历簿	locomotive logbook	
04.0300	直流电力机车	DC electric locomotive	
04.0301	单相交流电力机车	single-phase AC electric locomotive	
04.0302	单相工频交流电力机车	single-phase industrial frequency AC electric locomotive	
04.0303	双电压制电力机车	dual voltage electric locomotive	
04.0304	双频率制电力机车	dual frequency electric locomotive	
04.0305	多电流制电力机车	multiple system electric locomotive	
04.0306	硅整流器电力机车	silicon rectifier electric locomotive	
04.0307	晶闸管整流器电力机车	thyristor rectifier electric locomotive	
04.0308	晶闸管变流器电力机车	thyristor converter electric locomotive	
04.0309	电动车组	electric multiple unit, motor coach set, electric motor train unit	
04.0310	单相交流电动车组	single-phase industrial frequency AC motor train unit	
04.0311	地下铁道电动车组	subway motor train unit	

序　码	汉　文　名	英　文　名	注　释
04.0312	电流制	current system	
04.0313	直流制	DC system	
04.0314	单相工频交流制	single-phase industrial frequency AC system	
04.0315	单相低频交流制	single-phase low frequency AC system	
04.0316	受电弓标称电压	nominal voltage at pantograph	
04.0317	调压方式	voltage regulation mode	
04.0318	高压侧调压	high voltage regulation	
04.0319	低压侧调压	low voltage regulation	
04.0320	分级调压	stepped voltage regulation	
04.0321	无级调压　　一	stepless voltage regulation	
04.0322	相控调压	phase control	
04.0323	斩波调压	chopper control	
04.0324	变阻调压	rheostatic control	
04.0325	整流方式	mode of rectification	
04.0326	牵引变流器	traction convertor	
04.0327	牵引逆变器	traction invertor	
04.0328	牵引变频器	traction frequency convertor	
04.0329	可调牵引变频器	variable frequency convertor	
04.0330	直流斩波器	DC chopper	
04.0331	单相桥式整流器	single-phase bridge rectifier	
04.0332	三相桥式整流器	three-phase bridge rectifier	
04.0333	可控桥式整流器	controlled bridge rectifier	
04.0334	对称半控桥式整流器	symmetric half-controlled bridge rectifier	
04.0335	非对称半控桥式整流器	asymmetric half-controlled bridge rectifier	
04.0336	多段桥[联结]	bridges in cascade, multi rectifier bridge	
04.0337	硅整流装置	silicon rectifier device	
04.0338	晶闸管整流装置	thyristor rectifier device	
04.0339	励磁整流装置	excitation rectifier device	
04.0340	电力传动方式	mode of electric drive	
04.0341	直流传动	DC drive	
04.0342	交－直流传动	AC-DC drive	
04.0343	交－直－交流传	AC-DC-AC drive	

序 码	汉 文 名	英 文 名	注 释
	动		
04.0344	调速方式	mode of speed control	
04.0345	变压调速	variable voltage speed control	
04.0346	变频调速	variable frequency speed control	
04.0347	变极调速	pole changing speed control	
04.0348	车轴驱动方式	mode of axle drive	
04.0349	弹性齿轮驱动	resilient gear drive	
04.0350	刚性齿轮驱动	solid gear drive	
04.0351	单侧减速齿轮驱动	single reduction gear drive	
04.0352	双侧减速齿轮驱动	double reduction gear drive	
04.0353	单电动机驱动	monomotor drive	
04.0354	车轴空心轴驱动	quill drive, hollow axle drive	
04.0355	电机空心轴驱动	hollow shaft motor drive	
04.0356	万向轴驱动	cardan shaft drive	
04.0357	直接驱动	gearless drive, direct drive	
04.0358	独立驱动	individual drive	
04.0359	组合驱动	coupled axle drive	
04.0360	连杆驱动	rod drive	
04.0361	组合传动机车	coupled axle drive locomotive	
04.0362	独立传动机车	individual drive locomotive	
04.0363	齿轨[传动]机车	rack locomotive	又称"齿条[传动]机车"。
04.0364	牵引电动机供电制式	traction motor power supply system	
04.0365	机车集中供电	locomotive centralized power supply	
04.0366	转向架独立供电	bogie individual power supply	
04.0367	电动机独立供电	motor individual power supply	
04.0368	主电路	power circuit, main circuit	
04.0369	高压电路	high voltage circuit, high tension circuit	
04.0370	低压电路	low voltage circuit, low tension circuit	
04.0371	网侧电路	circuit on side of overhead contact line	

序　码	汉　文　名	英　文　名	注　释
04.0372	牵引电路	traction circuit	
04.0373	制动电路	braking circuit	
04.0374	滤波电路	filter circuit	
04.0375	辅助电路	auxiliary circuit	
04.0376	列车供电电路	power supply circuit for train	
04.0377	控制电路	control circuit	
04.0378	控制电源	control source	
04.0379	蓄电池电路	battery circuit	
04.0380	指令电路	command circuit	
04.0381	照明电路	lighting circuit	
04.0382	仪表电路	instrument circuit	
04.0383	联锁电路	interlocking circuit	
04.0384	信号电路	signal circuit	
04.0385	起动电路	starting circuit	
04.0386	保护电路	protective circuit	
04.0387	电子控制电路	electronic control circuit	
04.0388	电空制动电路	electropneumatic brake circuit, E-P brake circuit	
04.0389	防空转防滑行保护电路	anti-slip/slide protection circuit	
04.0390	音频通讯电路	audio communication circuit	
04.0391	电连接器	electric coupler	
04.0392	自动电连接器	electric automatic coupler	
04.0393	机车重联电连接器	multi-locomotive electric coupler	
04.0394	牵引电动机	traction motor	
04.0395	直流牵引电动机	DC traction motor	
04.0396	脉流牵引电动机	pulsating current traction motor	
04.0397	交流牵引电动机	AC traction motor	
04.0398	单相交流牵引电动机	single-phase AC traction motor	
04.0399	三相交流牵引电动机	three phase AC traction motor	
04.0400	抱轴式牵引电动机	axle hung traction motor, nose suspension traction motor	
04.0401	架承式牵引电动机	frame mounted traction motor	

序 码	汉 文 名	英 文 名	注 释
04.0402	转向架架承式牵引电动机	bogie mounted traction motor	
04.0403	底架架承式牵引电动机	underframe mounted traction motor	
04.0404	串励电动机	series excited motor	
04.0405	并励电动机	shunt excited motor	
04.0406	复励电动机	compound excited motor	
04.0407	他励电动机	separately excited motor	
04.0408	同步电动机	synchronous motor	
04.0409	异步电动机	asynchronous motor	
04.0410	交流换向器电动机	alternating current commutator motor	
04.0411	全封闭式电动机	totally-enclosed motor	
04.0412	通风式电动机	ventilated motor	
04.0413	自通风式电动机	self-ventilated motor	
04.0414	强迫通风式电动机	force ventilated motor	
04.0415	直线电动机	linear motor	又称"线性电动机"。
04.0416	直线同步电动机	linear synchronous motor	
04.0417	直线异步电动机	linear asynchronous motor, linear induction motor	又称"直线感应电动机"。
04.0418	电动机转速	motor speed	
04.0419	电动机超速	runaway speed	
04.0420	电动机特性	motor characteristic	
04.0421	电动机转矩	motor torque	又称"电动机扭矩"。
04.0422	起动转矩	starting torque	
04.0423	峰值转矩	peak torque	
04.0424	制动转矩	braking torque	
04.0425	最大输出功率	maximum output power	
04.0426	起动电流	starting current	
04.0427	峰值电流	peak current	
04.0428	标称电压	nominal voltage	
04.0429	变压器电势	transformer EMF	
04.0430	片间平均电压	mean voltage between segments	
04.0431	片间最高电压	maximum voltage between segments	
04.0432	满磁场	full field	又称"全磁场"。

序 码	汉 文 名	英 文 名	注 释
04.0433	削弱磁场	weakened field	
04.0434	最大磁场	maximum field	
04.0435	最小磁场	minimum field	
04.0436	磁场削弱率	field weakening	
04.0437	磁场削弱系数	coefficient of field weakening	
04.0438	恒功调速比	speed ratio on constant power	
04.0439	换向器	commutator	
04.0440	换向片	commutator segment	
04.0441	电枢线圈	armature coil	
04.0442	均压线	equalizer, cable bond	
04.0443	转轴	shaft	
04.0444	定子	stator	
04.0445	转子	rotor	
04.0446	主极铁心	mainpole core	
04.0447	主极线圈	mainpole coil	
04.0448	换向极铁心	interpole core	
04.0449	换向极线圈	interpole coil	
04.0450	补偿线圈	compensating coil	
04.0451	机座	frame	
04.0452	铸造机座	cast frame	
04.0453	焊接机座	welding frame	
04.0454	半叠片机座	semi-laminated frame	
04.0455	全叠片机座	full-laminated frame	
04.0456	电刷装置	brush gear	
04.0457	刷握	brush-holder	
04.0458	刷盒	brush box	
04.0459	电刷	brush	
04.0460	端盖	end shield	
04.0461	抱轴悬挂装置	suspension bearing	
04.0462	主发电机	main generator	
04.0463	直流主发电机	DC main generator	
04.0464	交流主发电机	main alternator	
04.0465	辅助发电机	auxiliary generator	
04.0466	直流辅助发电机	DC auxiliary generator	
04.0467	直流起动发电机	DC starting generator, dynastarter	
04.0468	直流控制发电机	DC control generator	
04.0469	直流励磁机	DC exciter	

序 码	汉 文 名	英 文 名	注 释
04.0470	劈相机	Arno converter, phase splitter	
04.0471	辅助电动机	auxiliary motor	
04.0472	直流辅助电动机	DC auxiliary motor	
04.0473	异步辅助电动机	asynchronous auxiliary motor	
04.0474	空压机电动机	air compressor motor	
04.0475	通风机电动机	blower motor	
04.0476	泵电动机	pump motor	
04.0477	控制电机	control electric machine	
04.0478	伺服电机	servomotor	又称"伺服马达"。
04.0479	自整角机	synchro	
04.0480	测速发电机	tachogenerator	
04.0481	机车牵引变压器	traction transformer of locomotive	
04.0482	心式牵引变压器	core-type traction transformer	
04.0483	壳式牵引变压器	shell-type traction transformer	
04.0484	自耦牵引变压器	traction autotransformer	
04.0485	调压牵引变压器	regulating traction transformer	
04.0486	分接牵引变压器	tapped traction transformer	
04.0487	油浸式牵引变压器	oil-immersed type traction transformer	
04.0488	整流变压器	rectifier transformer	
04.0489	电源变压器	supply transformer	
04.0490	隔离变压器	isolating transformer	又称"绝缘变压器"。
04.0491	同步变压器	synchronous transformer	
04.0492	脉冲变压器	pulse transformer	
04.0493	控制变压器	control transformer	
04.0494	信号变压器	signal transformer	
04.0495	励磁变压器	excitation transformer	
04.0496	高压绕组	high voltage winding, high tension winding	
04.0497	低压绕组	low voltage winding, low tension winding	
04.0498	调压绕组	regulating winding	
04.0499	励磁绕组	excitation winding	
04.0500	辅助绕组	auxiliary winding	
04.0501	列车供电绕组	train coach supply winding	
04.0502	自然循环	natural circulation	
04.0503	强迫循环	forced circulation	

序　码	汉　文　名	英　文　名	注　释
04.0504	强迫导向循环	forced guided circulation	
04.0505	牵引电抗器	traction reactor	
04.0506	平波电抗器	smoothing reactor	
04.0507	过渡电抗器	transition reactor	
04.0508	接地电抗器	earthing reactor, grounding reactor	
04.0509	制动电抗器	braking reactor	
04.0510	分流电抗器	divert shunt reactor, inductive shunt	又称"感应分路"。
04.0511	滤波电抗器	filter reactor	
04.0512	均流电抗器	sharing reactor	
04.0513	限流电抗器	inductive reactor, current limiting reactor	
04.0514	换相电抗器	commutation reactor	
04.0515	起动电阻器	starting resistor	
04.0516	制动电阻器	braking resistor	
04.0517	过渡电阻器	transition resistor	
04.0518	分流电阻器	divert shunt resistor, shunting resistor	
04.0519	稳定电阻器	stabilizing resistor	
04.0520	调节电阻器	regulating resistor	
04.0521	制动电阻柜	braking resistor cubicle	
04.0522	制动电阻元件	braking resistor grid	
04.0523	非线性电阻器	non-linear resistor	
04.0524	滤波电容器	filter capacitor	
04.0525	换相电容器	commutating capacitor	
04.0526	保护电容器	protective capacitor	
04.0527	起动电容器	starting capacitor	
04.0528	补偿电容器	compensation capacitor	
04.0529	主断路器	line circuit-breaker	
04.0530	真空主断路器	line vacuum circuit-breaker	
04.0531	空气主断路器	line air-blast circuit-breaker	
04.0532	直流高速断路器	DC high speed circuit-breaker	
04.0533	受电器	current collector	
04.0534	受电弓	pantograph	
04.0535	单臂受电弓	single arm pantograph	
04.0536	双臂受电弓	double arm pantograph	

序　码	汉　文　名	英　文　名	注　释
04.0537	弓头	pantograph bow	
04.0538	受电弓滑板	pantograph pan	
04.0539	弓角	pantograph horn	
04.0540	受电弓气缸	pantograph cylinder	
04.0541	支持绝缘子	supporting insulator	
04.0542	第三轨受电器	conductor rail collector	
04.0543	受电靴装置	shoegear	
04.0544	集电靴	collector shoe	
04.0545	电磁接触器	electromagnetic contactor	
04.0546	直流接触器	DC contactor	
04.0547	交流接触器	AC contactor	
04.0548	电空接触器	electropneumatic contactor	
04.0549	组合接触器	grouping contactor	
04.0550	线路接触器	line contactor	
04.0551	磁场削弱接触器	field weakening contactor	
04.0552	制动接触器	braking contactor	
04.0553	励磁接触器	excitation contactor	
04.0554	辅助接触器	auxiliary contactor	
04.0555	调压开关	tap changer	
04.0556	低压调压开关	low voltage tap changer, low tension tap changer	
04.0557	高压调压开关	high voltage tap changer, high tension tap changer	
04.0558	位置转换开关	position changeover switch	
04.0559	鼓形位置转换开关	drum position changeover switch	
04.0560	凸轮位置转换开关	cam position changeover switch	
04.0561	牵引-制动位转换开关	power/brake changeover switch	
04.0562	反向器	reverser	
04.0563	电流制转换开关	current system changeover switch	
04.0564	主电路库用转换开关	main circuit transfer switch for shed supply	
04.0565	辅助电路库用转换开关	auxiliary circuit transfer switch for shed supply	
04.0566	短路器	short-circuiting device	

序　码	汉　文　名	英　文　名	注　释
04.0567	接地开关	earthing switch	
04.0568	牵引电动机隔离开关	traction motor isolating switch	
04.0569	受电弓隔离开关	pantograph isolating switch	
04.0570	主整流柜隔离开关	isolating switch for main silicon rectifier cubicle	
04.0571	劈相机故障隔离开关	fault isolating switch for phase splitter	
04.0572	控制电源隔离开关	isolating switch for control supply	
04.0573	司机控制器	driver controller	
04.0574	调车控制器	shunting controller	
04.0575	电空制动控制器	E-P brake controller	
04.0576	按键开关	button switch, key switch	
04.0577	自复式按键开关	self-reset push-key switch	
04.0578	非自复式按键开关	non-self-reset push-key switch	
04.0579	主按键开关组	main button switch group, main key switch set	
04.0580	副按键开关组	secondary push-key switch group, secondary key switch set	
04.0581	电磁阀	electromagnetic valve	
04.0582	电空阀	electropneumatic valve	
04.0583	电液阀	electro-hydraulic valve	
04.0584	保护阀	protective valve	
04.0585	受电弓电空阀	pantograph valve	
04.0586	防空转撒砂电空阀	anti-slip sanding valve	
04.0587	防空转防滑行保护装置	anti-slip/slide protection device	
04.0588	轴重转移补偿装置	axle load transfer compensation device	
04.0589	功率因数补偿装置	power factor compensation device	
04.0590	电压继电器	voltage relay	
04.0591	电流继电器	current relay	
04.0592	时间继电器	time relay	

序 码	汉 文 名	英 文 名	注 释
04.0593	中间继电器	intermediate relay	
04.0594	速度继电器	speed relay	
04.0595	温度继电器	temperature relay	
04.0596	压力继电器	pressure relay	
04.0597	欠电压继电器	under-voltage relay	
04.0598	过电流继电器	over-current relay	
04.0599	转速继电器	tachometric relay	
04.0600	流速继电器	flow relay	
04.0601	接地继电器	earth fault relay	
04.0602	风压继电器	air pressure relay	
04.0603	油压继电器	oil-pressure relay	
04.0604	起动继电器	starting relay	
04.0605	主电路库用插座	main circuit socket for shed supply	
04.0606	辅助电路库用插座	auxiliary circuit socket for shed supply	
04.0607	控制电路库用插座	control circuit socket for shed supply	
04.0608	重联插座	multiple unit socket	
04.0609	印制电路板插座	printed circuit board socket	
04.0610	电压调整器	voltage regulator	
04.0611	温度调整器	temperature regulator	
04.0612	位置指示器	notch indicator	
04.0613	数字显示器	digital display	
04.0614	司机室取暖电炉	driver's cab electric heater	
04.0615	司机室热风装置	driver's cab air heater	
04.0616	司机室空调装置	driver's cab air conditioner	
04.0617	电热玻璃	electric heating glass	
04.0618	电测仪表	electrical measuring instrument	
04.0619	牵引电机电压表	traction motor voltmeter	
04.0620	牵引电机电流表	traction motor ammeter	
04.0621	励磁电流表	excitation ammeter	
04.0622	网侧电压表	voltmeter on side of overhead contact line, overhead side voltmeter	
04.0623	辅助电路电压表	auxiliary circuit voltmeter	
04.0624	控制电源电压表	control supply voltmeter	

序 码	汉 文 名	英 文 名	注 释
04.0625	控制电源电流表	control supply ammeter	
04.0626	机车速度表	locomotive speedmeter	
04.0627	单相电度表	single-phase wat-hour meter	
04.0628	速度记录仪	tachograph	
04.0629	传感器	sensor, transducer	
04.0630	电压传感器	voltage sensor	
04.0631	电流传感器	current sensor	
04.0632	压力传感器	pressure sensor	
04.0633	速度传感器	speed sensor	
04.0634	温度传感器	temperature sensor	
04.0635	直流电流互感器	DC current transformer	
04.0636	阀型避雷器	valve type arrester	
04.0637	放电器	discharger	
04.0638	轮缘喷油器	flange lubricator	
04.0639	接地安全棒	earthing pole	
04.0640	接地回流电刷	earth return brush	
04.0641	前照灯	head lamp, head light	
04.0642	副前照灯	subhead lamp, dim head light	
04.0643	标志灯	marker lamp	
04.0644	走行部灯	bogie lamp	
04.0645	车号灯	side number plate lamp	
04.0646	路签灯	train staff lamp	
04.0647	记事灯	writing lamp	
04.0648	照明灯	illuminating lamp	
04.0649	指示灯	indicator lamp	
04.0650	风喇叭	air horn	
04.0651	高音风喇叭	high tone air horn	
04.0652	低音风喇叭	low tone air horn	
04.0653	警笛	siren	
04.0654	警惕装置	vigilance device	
04.0655	机车组装后的检查与试验	inspection and test of locomotive after completion of construction	
04.0656	一般性检查	general inspection	
04.0657	称重试验	weighing test	
04.0658	压缩空气设备全面的气密性试验	test for over-all air-tightness of compressed air equipments	

序 码	汉 文 名	英 文 名	注 释
04.0659	车体及外部装备密封试验	test for sealing of body and external equipment	
04.0660	受电弓试验	pantograph test	
04.0661	介电强度试验	dielectric test	
04.0662	事故预防措施的检查	checks for prevention of accidents	
04.0663	安全设备试验	test on safety equipments	
04.0664	蓄电池充电系统试验	checks of battery charging-arrangement	
04.0665	通风冷却试验	test on ventilation and cooling	
04.0666	空气制动试验	test on air brake	
04.0667	曲线通过试验	curve negotiation test	
04.0668	机车振动参数测试	measurements of vibration parameters	
04.0669	辅助机组试验	test on auxiliary machines	
04.0670	机车调速试验	test on speed regulation	
04.0671	主电路短路保护系统试验	test on short-circuit protection system of main circuit	
04.0672	主电路过载保护系统试验	test on overload protection system of main circuit	
04.0673	内部过电压试验	test on internal overvoltage	
04.0674	外部过电压试验	test on external overvoltage	
04.0675	机车功率试验	locomotive [traction] power test	
04.0676	机车功率因数测定	measurement of power factor	
04.0677	机车效率测定	measurement of efficiency of locomotive	
04.0678	谐波电流百分比测定	measurement of percentage of harmonic current	
04.0679	起动加速试验	starting and acceleration test	
04.0680	运行阻力试验	test for running resistance	
04.0681	电气制动试验	electric braking test	
04.0682	滤尘效果试验	test on filter efficiency	
04.0683	撒砂装置试验	test on sanding gear	
04.0684	重联运行试验	test on coupled operation, test on multi unit operation	
04.0685	司机室工作条件	check on working conditions in the	

序　码	汉　文　名	英　文　名	注　释
	检查	driver's cab	
04.0686	机械传动内燃机车	diesel-mechanical locomotive	
04.0687	液力传动内燃机车	diesel-hydraulic locomotive	
04.0688	电力传动内燃机车	diesel-electric locomotive	
04.0689	内燃动车组	diesel coach set	
04.0690	柴油机	diesel engine	
04.0691	二冲程柴油机	two stroke diesel engine	
04.0692	四冲程柴油机	four stroke diesel engine	
04.0693	低速柴油机	low speed diesel engine	
04.0694	中速柴油机	medium speed diesel engine	
04.0695	高速柴油机	high speed diesel engine	
04.0696	增压柴油机	supercharged diesel engine	
04.0697	废气涡轮增压柴油机	turbocharged diesel engine	
04.0698	机械增压柴油机	engine-driven supercharged diesel engine	
04.0699	两级增压柴油机	two-stage supercharged diesel engine	
04.0700	复合式柴油机	compound-supercharged diesel engine	
04.0701	直列式柴油机	straight type diesel engine, in line type diesel engine	
04.0702	对活塞式柴油机	opposed-piston type diesel engine	
04.0703	V 型柴油机	V-type diesel engine	
04.0704	H 型柴油机	H-type diesel engine	
04.0705	水冷柴油机	water-cooled diesel engine	
04.0706	气缸直径	cylinder bore diameter	
04.0707	曲柄半径	crank radius	
04.0708	活塞行程	piston stroke	
04.0709	行程缸径比	stroke/bore ratio	
04.0710	连杆长度	length of connecting rod	
04.0711	连杆比	connecting rod length/crank radius ratio	
04.0712	上止点	top dead point, upper dead center	

序 码	汉 文 名	英 文 名	注 释
04.0713	下止点	bottom dead point, lower dead center	
04.0714	内止点	inner dead point, inner dead center	
04.0715	外止点	outer dead point, outer dead center	
04.0716	气缸最大容积	maximum stroke volume	
04.0717	气缸最小容积	minimum stroke volume	
04.0718	气缸工作容积	stroke volume	
04.0719	总排量	total displacement	
04.0720	充量	charge	
04.0721	工质	working substance, medium	
04.0722	工作循环	working cycle	
04.0723	进气行程	intake stroke	
04.0724	压缩行程	compression stroke	
04.0725	膨胀行程	expansion stroke	
04.0726	排气行程	exhaust stroke	
04.0727	配气相位	valve timing	又称"配气定时"。
04.0728	换气过程	scavenging period	
04.0729	扫气泵扫气	blower scavenging	
04.0730	扫气压力	scavenging pressure	
04.0731	进气温度	intake temperature	
04.0732	进气压力	intake pressure	
04.0733	缸盖出口废气温度	exhaust temperature at cylinder head outlet	
04.0734	涡轮入口废气温度	exhaust temperature at turbine inlet	
04.0735	涡轮入口废气压力	exhaust pressure at turbine inlet	
04.0736	涡轮膨胀比	expansion ratio of turbine	
04.0737	排气温度	exhaust temperature	
04.0738	排气背压	exhaust back pressure	
04.0739	残余废气	residual gas	
04.0740	残余废气系数	coefficient of residual gas	
04.0741	充量系数	coefficient of charge	
04.0742	扫气系数	coefficient of scavenging	
04.0743	过量空气系数	excess air factor	

序　码	汉　文　名	英　文　名	注　释
04.0744	空燃比	air/fuel ratio	
04.0745	扫气效率	scavenging efficiency	
04.0746	压缩始点压力	compression beginning pressure	
04.0747	压缩终点压力	compression terminal pressure	
04.0748	压缩始点温度	compression beginning temperature	
04.0749	压缩终点温度	compression terminal temperature	
04.0750	压缩多变指数	polytropic index of compression	
04.0751	压缩比	compression ratio	
04.0752	有效压缩比	effective compression ratio	
04.0753	最高燃烧温度	maximum combustion temperature	
04.0754	最高爆发压力	maximum explosive pressure, peak pressure, maximum firing pressure	
04.0755	压力升高比	pressure step-up ratio, rate of pressure rise	
04.0756	膨胀终点温度	expansion terminal temperature	
04.0757	膨胀终点压力	expansion terminal pressure	
04.0758	膨胀多变指数	polytropic index of expansion	
04.0759	放热规律	law of heat release	
04.0760	热平衡	heat balance, thermo balance	
04.0761	示功图	indicator diagram	
04.0762	示功图丰满系数	fullness coefficient of indicator diagram	
04.0763	工作过程	working process	
04.0764	指示功	indicated work	
04.0765	指示燃油消耗率	indicated specific fuel consumption	
04.0766	有效燃油消耗率	effective specific fuel consumption	
04.0767	燃油消耗量	fuel consumption	
04.0768	机油消耗率	specific oil consumption	
04.0769	机油消耗量	oil consumption	
04.0770	空气消耗量	air consumption	
04.0771	空气消耗率	specific air consumption	
04.0772	发火次序	firing order	
04.0773	定时	timing	
04.0774	燃烧过程	combustion process	
04.0775	点火	ignition	

序 码	汉 文 名	英 文 名	注 释
04.0776	雾化	atomization	
04.0777	喷油速率	injection rate	
04.0778	喷油规律	law of injection	
04.0779	喷油持续角	continuous injection angle	
04.0780	供油提前角	advance angle of fuel supply	
04.0781	喷油提前角	injection advance angle	
04.0782	滞燃期	combustion lagging period, combusting delay period	
04.0783	显燃期	sensible combustion period	
04.0784	后燃期	after burning period	
04.0785	增压	supercharging	
04.0786	增压压力	supercharging pressure	
04.0787	增压比	supercharging ratio	
04.0788	废气涡轮增压	exhaust turbocharging	
04.0789	机械增压	engine-driven supercharging	
04.0790	定压增压	constant pressure charging	
04.0791	脉冲增压	pulse charging	
04.0792	增压中冷	charge inter-cooling	
04.0793	增压器出口温度	discharge temperature of supercharger	
04.0794	增压度	degree of turbocharging, degree of supercharging	
04.0795	喘振	surge	
04.0796	时间－面积值	time-area value	
04.0797	直接喷射燃烧室	direct injection combustion chamber	
04.0798	涡流室式燃烧室	swirl combustion chamber	
04.0799	预燃室式燃烧室	precombustion chamber, pre-chamber	
04.0800	球型燃烧室	spherical combustion chamber	
04.0801	ω型燃烧室	toroidal combustion chamber	
04.0802	盆形燃烧室	bowl [shaped] combustion chamber	
04.0803	挤压涡流	extruding swirl	
04.0804	进气涡流	intake swirl	
04.0805	积炭	carbon deposit	
04.0806	结胶	caking	

序　码	汉　文　名	英　文　名	注　释
04.0807	柴油机爆燃	engine detonation	又称"柴油机工作粗暴"。
04.0808	敲缸	knock	
04.0809	废气净化	exhaust purification	
04.0810	排气烟度	exhaust smoke density	
04.0811	持续功率	continuous power	
04.0812	修正功率	corrected power	
04.0813	指示功率	indicated power	
04.0814	有效功率	effective power	
04.0815	机械损失功率	mechanical loss power	
04.0816	最大功率	maximum power	
04.0817	单缸功率	power per cylinder	
04.0818	升功率	volume power	
04.0819	活塞面积	piston area	
04.0820	单位活塞面积功率	piston unit area power	
04.0821	储备功率	reserve power	
04.0822	功率储备系数	coefficient of reserve power	
04.0823	平均指示压力	mean indicated pressure	
04.0824	平均有效压力	mean effective pressure	
04.0825	平均机械损失压力	mean mechanical loss pressure	
04.0826	机械效率	mechanical efficiency	
04.0827	指示热效率	indicated thermal efficiency	
04.0828	有效热效率	useful thermal efficiency	
04.0829	标称转速	nominal speed	
04.0830	最低空载稳定转速	minimum idling stabilized speed	
04.0831	转速波动率	rate of speed fluctuation	
04.0832	瞬时调速率	instantaneous speed change rate	
04.0833	活塞平均速度	mean piston speed	
04.0834	热负荷	thermal load	
04.0835	机械负荷	mechanical load	
04.0836	强化系数	coefficient of intensification	
04.0837	烟度	limit of smoke	
04.0838	柴油机净重	net weight of diesel engine	
04.0839	穴蚀	cavitation	

序　码	汉　文　名	英　文　名	注　释
04.0840	窜机油	lubricating oil carry-over	
04.0841	窜气	blow-by	
04.0842	拉缸	piston scraping	
04.0843	抱缸	piston seizure	
04.0844	飞车	run away	
04.0845	标准大气状况	standard atmospheric condition	
04.0846	环境温度	ambient temperature	
04.0847	相对湿度	relative humidity	
04.0848	柴油机特性	diesel engine characteristic	
04.0849	负荷特性	load characteristic	又称"负载特性"。
04.0850	速度特性	speed characteristic	
04.0851	万有特性	universal characteristic	
04.0852	调速特性	speed regulation characteristics	又称"外特性",即全负荷速度特性。
04.0853	空载特性	no-load characteristic	
04.0854	台架试验	stand test, bench test	
04.0855	磨合	running-in	
04.0856	柴油机起动试验	diesel engine starting test	
04.0857	各缸均匀性试验	cylinder power equalizing test	
04.0858	最低工作稳定转速试验	minimum steady speed test	
04.0859	最低空载转速试验	minimum no-load speed test	
04.0860	热平衡试验	heat balance test	
04.0861	超负荷试验	over-load test	
04.0862	增压器配机试验	turbocharger matching test	
04.0863	机体	engine block, main frame	
04.0864	气缸体	cylinder block	
04.0865	曲轴箱	crankcase	
04.0866	曲轴箱呼吸器	crankcase breather	
04.0867	柴油机支座	engine support	
04.0868	油底壳	oil pan, oil sump	
04.0869	曲轴箱防爆门	explosion-proof door of crankcase	
04.0870	油气分离器	oil separator	
04.0871	活塞冷却喷嘴	piston cooling nozzle	
04.0872	气缸	cylinder	
04.0873	气缸套	cylinder liner	

序　码	汉　文　名	英　文　名	注　释
04.0874	水套	water jacket	
04.0875	扫气箱	scavenging box	
04.0876	连接箱	connecting box	
04.0877	泵支承箱	pump supporting box	
04.0878	主轴承	main bearing	
04.0879	主轴承座	main bearing seat, main bearing housing	
04.0880	主轴承盖	main bearing cap	
04.0881	止推轴承盖	thrust bearing cap	
04.0882	主轴瓦	main bearing shell	
04.0883	止推轴瓦	thrust bearing shell	
04.0884	止推环	thrust ring	
04.0885	主轴承螺栓	main bearing stud	
04.0886	主轴承螺母	main bearing nut	
04.0887	曲轴	crankshaft	
04.0888	主轴颈	crank journal, main journal	
04.0889	曲柄销	crank pin	
04.0890	曲柄臂	crank arm	
04.0891	曲柄	crank	
04.0892	曲轴平衡块	crankshaft counter balance	
04.0893	飞轮	flywheel	
04.0894	曲轴减振器	crankshaft vibration damper	
04.0895	硅油减振器	silicon oil damper	
04.0896	硅油弹簧减振器	silicon oil spring damper	
04.0897	摆式减振器	pendulum damper	
04.0898	弹簧式减振器	spring damper	
04.0899	活塞	piston	俗称"勾贝"。
04.0900	整体活塞	one-piece piston	
04.0901	组合活塞	composite piston	
04.0902	活塞体	piston body	
04.0903	活塞头	piston head	
04.0904	活塞裙	piston skirt	
04.0905	活塞衬套	piston bush	
04.0906	活塞环	piston ring	
04.0907	气环	gas ring	
04.0908	油环	oil ring	
04.0909	活塞销	piston pin	

序　码	汉　文　名	英　文　名	注　释
04.0910	连杆	connecting rod	
04.0911	连杆体	connecting rod body	
04.0912	连杆衬套	connecting rod bush	
04.0913	连杆轴瓦	connecting rod bearing shell	
04.0914	连杆盖	connecting rod cap	
04.0915	连杆螺栓	connecting rod bolt	
04.0916	连杆螺母	connecting rod nut	
04.0917	主连杆	main connecting rod	
04.0918	副连杆	auxiliary connecting rod	
04.0919	气缸盖	cylinder head	
04.0920	气缸盖螺栓	cylinder head stud	
04.0921	气缸盖螺母	cylinder head nut	
04.0922	预燃室	precombustion chamber	
04.0923	预燃室喷嘴	precombustion chamber nozzle	
04.0924	示功阀	indicator valve	
04.0925	气门	valve	
04.0926	进气门	inlet valve	
04.0927	排气门	exhaust valve	
04.0928	气门弹簧	valve spring	
04.0929	气门座	valve seat	
04.0930	气门导管	valve guide	
04.0931	气门摇臂	valve rocker	
04.0932	气门横臂	valve cross arm	
04.0933	气门推杆	valve push rod	
04.0934	气门挺柱	valve tappet	
04.0935	滚轮摇臂	roller rocker	
04.0936	气门旋转机构	valve rotating mechanism	
04.0937	凸轮轴	camshaft	
04.0938	进气凸轮	inlet cam	
04.0939	排气凸轮	exhaust cam	
04.0940	喷油泵凸轮	injection pump cam	
04.0941	曲轴正时齿轮	crankshaft timing gear	
04.0942	凸轮轴正时齿轮	camshaft timing gear	
04.0943	调速器驱动齿轮	governor drive gear	又称"调速器传动齿轮"。
04.0944	喷油泵驱动齿轮	injection pump drive gear	又称"喷油泵传动齿轮"。

序 码	汉 文 名	英 文 名	注 释
04.0945	喷油泵传动装置	injection pump transmission mechanism	
04.0946	自动喷调装置	injection timer	
04.0947	喷油泵	injection pump	
04.0948	柱塞偶件	plunger matching parts	
04.0949	柱塞	plunger	
04.0950	柱塞套	plunger sleeve	
04.0951	出油阀偶件	delivery valve matching parts	
04.0952	出油阀	delivery valve	
04.0953	出油阀座	delivery valve seat	
04.0954	喷油器	fuel injector	
04.0955	喷油嘴偶件	injector nozzle matching parts	
04.0956	喷油嘴针阀	nozzle needle valve	
04.0957	喷油嘴针阀体	nozzle needle valve body	
04.0958	喷油器滤芯	injector filter core	
04.0959	高压油管	high pressure fuel pipe	
04.0960	调速器	governor	
04.0961	调速系统	speed-governing system	
04.0962	调速伺服机构	speed-governing servomechanism	
04.0963	功率调节系统	power regulating system	
04.0964	转速表传动装置	transmission gear of tachometer	
04.0965	柴油机转速表	diesel engine tachometer	
04.0966	起动加速器	starting accelerator	
04.0967	控制机构	control mechanism	
04.0968	超速停车装置	overspeed trip	
04.0969	紧急停车装置	emergency stop mechanism	
04.0970	盘车机构	barring mechanism, barring gear	
04.0971	滑动式盘车机构	sliding type barring mechanism	
04.0972	回转式盘车机构	rotary type barring mechanism	
04.0973	涡轮增压器	turbocharger	
04.0974	[增压器]压气机	blower	
04.0975	导风轮	inducer	
04.0976	压气机叶轮	blower impeller	
04.0977	废气涡轮	exhaust turbine	
04.0978	喷嘴环	nozzle ring	
04.0979	扩压器	diffuser	
04.0980	中冷器	intercooler	

序 码	汉 文 名	英 文 名	注 释
04.0981	进气管	air inlet pipe	
04.0982	进气稳压箱	air inlet pressure stabilizing chamber	
04.0983	排气总管	exhaust manifold	
04.0984	排气支管	exhaust branch pipe	
04.0985	排气箱	exhaust box	
04.0986	波纹管	bellows	
04.0987	消声器	noise silencer, muffler	
04.0988	扫气泵	scavenging pump	
04.0989	液压传动	hydrostatic drive	
04.0990	液力传动	hydraulic transmission	
04.0991	单循环液力传动	single-circuit hydraulic transmission	
04.0992	多循环液力传动	multi-circuit hydraulic transmission	
04.0993	液力机械传动	hydromechanical drive	
04.0994	单流液力机械传动	hydromechanical drive with inner ramification	
04.0995	双流液力机械传动	hydromechanical drive with outer ramification	
04.0996	混合液力机械传动	hydromechanical drive with direct step	
04.0997	液力传动系统	hydraulic transmission system	
04.0998	液力传动箱	hydraulic transmission gear box	
04.0999	液力换向传动箱	hydrodynamic reverser	
04.1000	主弹性联轴节	main elastic coupling	
04.1001	中间齿轮箱	intermediate gear box	
04.1002	车轴齿轮箱	axle gear box	
04.1003	起动变扭器	starting torque converter	又称"起动变矩器"。
04.1004	运转变扭器	running torque converter	又称"运转变矩器"。
04.1005	液力循环元件	hydraulic unit	
04.1006	液力偶合器	hydraulic coupler	
04.1007	液力制动器	hydraulic brake	
04.1008	液力变扭器	hydraulic torque converter	又称"液力变矩器"。
04.1009	变扭比	torque ratio	又称"变矩比"。
04.1010	泵轮	pump impeller	
04.1011	涡轮	turbine	

序　码	汉　文　名	英　文　名	注　释
04.1012	导轮	guide wheel	
04.1013	工作轮	blade wheel	
04.1014	输入轴	input shaft	
04.1015	变扭器轴	torque converter shaft	又称"变矩器轴"。
04.1016	输出轴	output shaft	
04.1017	换向轴	reversing shaft	
04.1018	供油泵	oil feed pump	
04.1019	换挡	changeover governor	
04.1020	液力制动阀	hydrodynamic brake valve	
04.1021	液力制动操纵阀	hydrodynamic brake operating valve	
04.1022	充量限制阀	filling limiting valve	
04.1023	闸板阀	gate valve	
04.1024	闸板转换阀	gate change-over valve	
04.1025	换向限止阀	standstill detector valve	
04.1026	温度调节阀	temperature regulating valve	
04.1027	控制泵	control pump	
04.1028	惰行泵	idle running pump	
04.1029	冷却水泵	cooling water pump	
04.1030	膨胀水箱	expansion tank, expansion drum	
04.1031	散热器	radiator	
04.1032	冷却风扇	cooling fan	
04.1033	预热锅炉	preheating boiler	
04.1034	燃油输送泵	fuel feed pump	
04.1035	燃油粗滤器	fuel prefilter	
04.1036	燃油精滤器	fuel precision filter	
04.1037	机油泵	lubricating oil pump	
04.1038	起动机油泵	starting lubricating oil pump	
04.1039	辅助机油泵	auxiliary lubricating oil pump	
04.1040	高压机油泵	high pressure lubricating oil pump	
04.1041	机油滤清器	lubricating oil filter	
04.1042	机油粗滤器	lubricating prefilter	
04.1043	离心式机油滤清器	centrifugal oil filter	
04.1044	增压器机油滤清器	oil precision filter for turbocharger	
04.1045	机油热交换器	lubricating oil heat exchanger	

序　码	汉　文　名	英　文　名	注　释
04.1046	油浴式空气滤清器	oil bath air filter	
04.1047	钢板网式空气滤清器	steel sheet mesh type air filter	
04.1048	旋风筒式空气滤清器	cyclone type air filter	
04.1049	摩擦离合器	friction clutch	
04.1050	离合器操纵装置	clutch operating device	
04.1051	静液压泵	hydrostatic pump	
04.1052	静液压马达	hydrostatic motor	
04.1053	安全阀	safety valve	
04.1054	百叶窗油缸	oil cylinder of shutter	
04.1055	车架	frame	
04.1056	[机车]牵引梁	draw beam	
04.1057	前窗	front window	
04.1058	侧窗	side window	
04.1059	司机座椅	driver's chair	
04.1060	动力室	power room	
04.1061	冷却室	cooling room	
04.1062	电器室	electric apparatus room	
04.1063	内走廊	inside corridor, gangway	又称"通道"。
04.1064	外走廊	outside corridor, running board	
04.1065	排障器	pilot, life guard	
04.1066	路签授受器	staff exchanger, tablet exchanger	
04.1067	刮雨器	windshield wiper, windscreen wiper	
04.1068	遮阳板	sun-shield	
04.1069	燃油箱	fuel tank	
04.1070	油位表	fuel level gage	
04.1071	油尺	fuel dipstick	
04.1072	轴箱轴承	axle box bearing	
04.1073	轴端轴承	axle end bearing	
04.1074	抱轴轴承	axle hung bearing, suspension bearing	
04.1075	抱轴瓦	axle suspension bush	
04.1076	传动齿轮	transmission gear	
04.1077	主动齿轮	driving gear	

序　码	汉　文　名	英　文　名	注　释
04.1078	从动齿轮	driven gear	
04.1079	牵引装置	draw gear	
04.1080	牵引杆	draw bar	
04.1081	燃煤机车	coal fired locomotive	
04.1082	煤粉机车	pulverized coal locomotive, fine coal locomotive	
04.1083	煤气机车	gas fired locomotive	
04.1084	单胀式蒸汽机车	single expansion steam locomotive	
04.1085	复胀式蒸汽机车	compound expansion steam loco-motive	
04.1086	模数牵引力	modulus of tractive effort	
04.1087	指示牵引力	indicated tractive effort	
04.1088	锅炉牵引力	boiler tractive effort	
04.1089	汽缸牵引力	cylinder tractive effort	
04.1090	临界牵引力	critical tractive effort	
04.1091	指示压力系数	coefficient of indicated pressure	
04.1092	遮断比	cut-off	俗称"断汽"。
04.1093	计算遮断比	calculated cut-off	
04.1094	蒸汽机车热工特性	thermo-characteristic of steam locomotive	
04.1095	蒸发量	evaporation capacity	
04.1096	供汽量	evaporation capacity for engine	
04.1097	蒸发率	rate of evaporation	
04.1098	供汽率	rate of evaporation for engine	
04.1099	计算供汽率	calculated rate of evaporation for engine	
04.1100	燃烧率	rate of combustion, rate of firing	
04.1101	放热率	rate of heat release	
04.1102	混煤	mixed coal	
04.1103	型煤	moulded coal	
04.1104	水煤浆	coal water slurry, coal water mix-ture	
04.1105	煤的技术当量	equivalence of coal	
04.1106	锅炉热平衡	boiler heat balance	
04.1107	不完全燃烧热损失	heat loss due to incomplete com-bustion	
04.1108	机械不完全燃烧	heat loss due to combustibles in	

序　码	汉　文　名	英　文　名	注　　释
	热损失	refuse	
04.1109	排烟热损失	heat loss due to exhaust gas	
04.1110	散热损失	heat dissipation	
04.1111	锅炉总效率	total boiler efficiency	
04.1112	锅炉净效率	net boiler efficiency	
04.1113	燃烧效率	combustion efficiency	
04.1114	通风效率	drafting efficiency	
04.1115	机车锅炉	locomotive boiler	
04.1116	火箱	firebox	
04.1117	外火箱	outside firebox, outer firebox	
04.1118	内火箱	inside firebox, inner firebox	
04.1119	外火箱顶板	roof sheet	
04.1120	内火箱顶板	crown sheet	
04.1121	后板	back sheet	
04.1122	喉板	throat sheet	
04.1123	侧板	side sheet	
04.1124	底圈	mud ring	
04.1125	炉床	grate	
04.1126	炉口	fire hole	
04.1127	炉门	fire door	
04.1128	拱砖	arch brick	
04.1129	拱砖管	arch tube	
04.1130	炉撑	stay	
04.1131	拉撑	brace	又称"斜撑"。
04.1132	火箱管板	firebox tube sheet	指后管板。
04.1133	灰箱	ashpan	
04.1134	锅胴	boiler barrel course	
04.1135	大烟管	flue [tube]	
04.1136	小烟管	[smoke] tube	
04.1137	烟箱	smokebox	
04.1138	烟箱管板	smokebox tube sheet	指前管板。
04.1139	通风装置	drafting apparatus	
04.1140	矩形通风装置	oblong ejector, Giesl ejector	又称"扁烟筒"。
04.1141	乏汽喷口	exhaust nozzle	
04.1142	乏汽喷口座	exhaust nozzle seat	
04.1143	外烟筒	smokestack, chimney	
04.1144	内烟筒	stack extension	

序　码	汉　文　名	英　文　名	注　　释
04.1145	裙筒	apron shell	
04.1146	送风器	blower	
04.1147	垂直板	diaphragm plate, deflecting plate	又称"垂直反射板"。
04.1148	水平板	horizontal plate, table plate	又称"水平反射板"。
04.1149	前垂板	draft plate	
04.1150	火星网	spark arrester netting	
04.1151	烟箱大门	smokebox front	
04.1152	烟箱小门	smokebox door	
04.1153	挡烟板	smoke deflector	
04.1154	过热管	superheater tube	
04.1155	过热箱	superheater header	
04.1156	饱和蒸汽室	saturated chamber	
04.1157	过热蒸汽室	superheater chamber	
04.1158	主蒸汽管	steam pipe	
04.1159	调整阀	regulator valve, throttle valve	又称"汽门"。
04.1160	先开阀	pilot valve	
04.1161	主阀	main valve	
04.1162	扇形齿板	quadrant	
04.1163	干燥管	dry pipe	
04.1164	空气预热装置	air preheater	
04.1165	给水装置	feed water rigging	
04.1166	吸上式注水器	attraction injector	
04.1167	非吸上式注水器	non-attraction injector	
04.1168	给水预热装置	feed water heater	
04.1169	混合式给水预热装置	mixed feed water heater	
04.1170	递热器	combining chamber	又称"混合室"。
04.1171	冷水泵	cold water pump	
04.1172	热水泵	hot water pump	
04.1173	止回阀	boiler check valve	
04.1174	放水阀	blow off valve	又称"排水阀"。
04.1175	洗炉堵	washout plug	
04.1176	锅炉安全装置	boiler safety device	
04.1177	水表	water gage	
04.1178	水柱	water column	
04.1179	易熔塞	fusible plug	俗称"铅堵"。
04.1180	蒸汽塔	turret	

序　码	汉 文 名	英 文 名	注　释
04.1181	汽笛	whistle	
04.1182	汽缸套	cylinder bushing	
04.1183	汽缸前盖	front cylinder head	
04.1184	汽缸后盖	back cylinder head	
04.1185	汽缸鞍	cylinder saddle	
04.1186	汽室	steam chest, valve chest	
04.1187	汽室套	steam chest bushing	
04.1188	填料	packing	
04.1189	汽缸排水阀	cylinder drain valve	
04.1190	活塞杆	piston rod	俗称"勾贝杆"。
04.1191	十字头	crosshead	
04.1192	十字头扁销	crosshead key	
04.1193	十字头圆销	crosshead pin	
04.1194	滑板	guide, slide bar, slip sheet	
04.1195	多滑面滑板	multiple bearing type guide	
04.1196	摇杆	main rod	
04.1197	蒸汽机车连杆	steam locomotive side rod	
04.1198	肘销	knuckle pin	
04.1199	阀装置	valve gear	又称"阀动装置"。
04.1200	华氏阀装置	Walschaerts valve gear	
04.1201	汽阀	piston valve	
04.1202	固定式汽阀	rigid piston valve	
04.1203	分动式汽阀	adjustable piston valve, Trofiemov piston valve	又称"特氏阀"。
04.1204	偏心杆	eccentric rod	
04.1205	月牙板	reverse link, quadrant link	
04.1206	半径杆	radius bar, radius rod	
04.1207	合并杆	combination lever	
04.1208	结合杆	union link, connecting link	
04.1209	月牙板滑块	reverse link block, die block	
04.1210	动力回动机	power reverse gear	
04.1211	回动手把	reverse lever	
04.1212	回动拉杆	reverse pull rod	
04.1213	板式车架	plate frame	
04.1214	棒式车架	bar frame	
04.1215	主车架片	main frame	
04.1216	后车架片	rear frame	

序 码	汉 文 名	英 文 名	注 释
04.1217	轴箱托板	pedestal brace	
04.1218	前缓冲铁	front bumper	
04.1219	后缓冲铁	rear bumper	
04.1220	锅腰托板	waist sheet	
04.1221	滑台	expansion shoe	
04.1222	后膨胀板	rear expansion sheet	
04.1223	动轴箱	driving box	又称"动轮轴箱"。
04.1224	动轴箱平铁	driving box shoe	又称"动轴箱槽铁"。
04.1225	动轴箱楔铁	driving box wedge	又称"动轴箱斜铁"。
04.1226	自动调整楔铁装置	automatic compensator	又称"自动楔铁调整器"。
04.1227	主动轮对	main driving wheel set	
04.1228	主曲拐销	main crank pin	
04.1229	偏心曲拐	eccentric crank	
04.1230	从动轮对	driven wheel set	又称"他动轮对"。
04.1231	导轮转向架	leading truck	
04.1232	导轮对	leading truck wheel set	
04.1233	[复原]摇枕	swing bolster	
04.1234	摇鞍	swing rocker	
04.1235	从轮转向架	trailing truck	
04.1236	从轮对	trailing wheel set	
04.1237	复原装置	centering device	
04.1238	粘着重量增加器	booster	
04.1239	煤水车	tender	
04.1240	水柜	water tank	
04.1241	煤槽	fuel space, coal bin	
04.1242	煤水车转向架	tender truck	
04.1243	水柜阀	tank valve	
04.1244	中间牵引装置	intermediate draw gear	
04.1245	中间缓冲装置	intermediate buffer	
04.1246	加煤机	stoker	
04.1247	推煤机	coal pusher	
04.1248	压油机	mechanical lubricator	
04.1249	涡轮发电机	turbo-generator	
04.1250	风动摇炉装置	pneumatic grate shaking rigging	
04.1251	锅炉蒸发面积	boiler evaporative heating surface	
04.1252	锅炉过热面积	boiler super heating surface	

序 码	汉 文 名	英 文 名	注 释
04.1253	锅炉散热面积	boiler heat dissipating surface	
04.1254	炉床面积	grate area	
04.1255	锅炉水位	boiler water level	
04.1256	阀行程	valve travel	
04.1257	导程	lead	
04.1258	进汽余面	steam lap	
04.1259	排汽余面	exhaust lap	
04.1260	阀调整	valve setting	
04.1261	阀动图	valve diagram, Zeuner valve diagram	
04.1262	阀动椭圆图	valve ellipse	
04.1263	锅炉借水	running at dropping water level	
04.1264	白水表行车	running without water in gage	
04.1265	汽水共腾	priming	
04.1266	水锤	water hammer	
04.1267	逆向牵引	backward haulage	
04.1268	单相工频交流电力牵引制	single-phase industrial frequency AC electric traction system	
04.1269	直流电力牵引制	DC electric traction system	
04.1270	电力牵引供电系统	power supply system of electric traction	
04.1271	电力牵引远动系统	electric traction telemechanical system, electric traction remote control system	
04.1272	牵引网	traction electric network	
04.1273	视在单位能耗	specific apparent energy consumption	
04.1274	列车带电运行时分	train running time on load	
04.1275	直接供电方式	direct feeding system	
04.1276	带回流线的直接供电方式	direct feeding system with return wire	
04.1277	吸流变压器供电方式	booster transformer feeding system	又称"BT 供电方式"。
04.1278	自耦变压器供电方式	autotransformer feeding system	又称"AT 供电方式"。
04.1279	单回路供电	single circuit power supply	

序　码	汉　文　名	英　文　名	注　释
04.1280	双回路供电	double circuit power supply	
04.1281	分支接线	branch connection	
04.1282	桥形接线	bridge connection	
04.1283	换相联接	exchange phase connection, phase alternating connection	
04.1284	牵引变压器	traction transformer	
04.1285	三相 YN,d11 接线牵引变压器	traction transformer of threephase YN, d11 connection	
04.1286	三相三绕组接线牵引变压器	traction transformer of threephase three winding connection	
04.1287	单相 V/V 接线牵引变压器	traction transformer of singlephase V/V connection	
04.1288	单相接线牵引变压器	traction transformer of singlephase connection	
04.1289	斯柯特接线牵引变压器	traction transformer of Scott connection	
04.1290	伍德布里奇接线牵引变压器	traction transformer of Wood Bridge connection	
04.1291	三相 YN,d11 d1 十字交叉接线牵引变压器	traction transformer of cross connection with threephase YN, d11 d1	
04.1292	供电臂	feeding section	
04.1293	单边供电	one way feeding	
04.1294	双边供电	two way feeding	
04.1295	越区供电	over-zone feeding	
04.1296	集中供电方式	centralized power supply system	
04.1297	分散供电方式	diversified power supply system	
04.1298	移动备用方式	movable reservation system	
04.1299	固定备用方式	fixed reservation system	
04.1300	串联电容补偿装置	compensator with series capacitance	
04.1301	并联电容补偿装置	compensator with parallel capacitance	
04.1302	有载自动调节	autoregulation on load	
04.1303	补偿装置的电抗比	reactance ratio of compensator	
04.1304	滤波装置	filter	

序 码	汉 文 名	英 文 名	注 释
04.1305	电压自动补偿装置	autoregulation voltage compensator	
04.1306	滞后相供电臂	lagging phase feeding section	
04.1307	引前相供电臂	leading phase feeding section	
04.1308	分相装置	neutral section	
04.1309	电分段装置	section point	
04.1310	供电线	feeder	
04.1311	自耦变压器供电线	AT-feeder	
04.1312	载流承力索	current-carrying catenary	
04.1313	牵引回流电路	traction return current circuits	
04.1314	回流线	return wire	
04.1315	分相回流线	split return wire	又称"裂相回流线"。
04.1316	保护线	protective wire	
04.1317	保护线用连接线	crossbond of protective wire	
04.1318	吸流变压器	booster transformer, BT	
04.1319	自耦变压器	autotransformer, AT	
04.1320	吸上线	boosting cable	
04.1321	列车平均电流	average current of train	
04.1322	列车有效电流	effective current of train, r.m.s current of train	
04.1323	列车带电平均电流	average current of charging train	
04.1324	供电臂平均电流	average current of feeding section	
04.1325	供电臂有效电流	r.m.s current of feeding section, effective current of feeding section	
04.1326	供电臂干扰计算电流	disturbing calculation current of feeding section	
04.1327	供电臂短路电流	short-circuit current of feeding section	
04.1328	供电臂瞬时最大电流	instantaneous maximum current of feeding section	
04.1329	供电臂最大负荷电流	maximum load current of feeding section	
04.1330	牵引变电所标称电压	nominal voltage of traction substation	

序 码	汉 文 名	英 文 名	注 释
04.1331	接触网最高电压	maximum voltage of overhead contact line	
04.1332	接触网标称电压	nominal voltage of overhead contact line	
04.1333	接触网最低电压	minimum voltage of overhead contact line	
04.1334	事故状态的接触网最低电压	minimum voltage of overhead contact line at accident condition	
04.1335	牵引网阻抗	impedance of traction electric network	
04.1336	电分段	sectioning	
04.1337	牵引回流轨	traction return current rail	
04.1338	钢轨对地电位	rail potential to ground	
04.1339	钢轨连接器	rail joint bond	
04.1340	火花间隙	spark gap	
04.1341	接地保护放电装置	earth protection discharger	
04.1342	牵引变电所	traction substation , SS	
04.1343	直流牵引变电所	DC traction substation	
04.1344	交流牵引变电所	AC traction substation	
04.1345	移动牵引变压器	movable traction transformer	
04.1346	移动牵引变电所	movable traction substation	
04.1347	开闭所	sub-section post , SSP	
04.1348	自耦变压器所	auto-transformer post, ATP	
04.1349	分区所	section post, SP	
04.1350	接触网	overhead contact line equipment	
04.1351	接触网悬挂	overhead contact line, catenary	
04.1352	链形悬挂	overhead contact line with catenary, longitudinal suspension	
04.1353	简单悬挂	tramway type suspension equipment	
04.1354	弹性简单悬挂	stitched tramway type suspension equipment	
04.1355	斜链形悬挂	inclined catenary, srew catenary	
04.1356	直链形悬挂	polygonal catenary	
04.1357	弹性链形悬挂	stitched catenary equipment	
04.1358	全补偿链形悬挂	auto-tensioned catenary equipment	

序　码	汉　文　名	英　文　名	注　释
04.1359	半补偿链形悬挂	semi-auto-tensioned catenary equipment	
04.1360	双链形悬挂	compound catenary equipment	
04.1361	承力索	catenary	
04.1362	辅助承力索	auxiliary catenary	
04.1363	横承力索	headspan wire	
04.1364	定位索	registration wire	
04.1365	接触线	contact wire	
04.1366	附加导线	additive wire	
04.1367	加强线	line feeder	
04.1368	承力索终端锚固线夹	termination fitting for catenary	
04.1369	接触线终端锚固线夹	termination fitting for contact wire	
04.1370	承力索接头线夹	catenary splice	
04.1371	接触线接头线夹	contact wire splice	
04.1372	横承力索线夹	headspan wire clamp	
04.1373	吊环	suspension ring	
04.1374	电连接	electrical connector	
04.1375	电连接线夹	electrically connecting clamp	
04.1376	接触线电连接线夹	electrically connecting clamp for contact wire	
04.1377	接触网架空地线	overhead earth wire	
04.1378	接地线夹	earth clamp	
04.1379	棒式绝缘子	strut insulator, rod insulator	
04.1380	双重绝缘棒式绝缘子	double strut insulator, double rod insulator	
04.1381	腕臂	cantilever	
04.1382	旋转腕臂	hinged cantilever	
04.1383	腕臂底座	cantilever base	
04.1384	拉杆[压管]	upper cantilever	
04.1385	拉杆底座	bracket base	
04.1386	调节板	regulating plate	
04.1387	套管铰环	sleeve with clevis and ring	
04.1388	钩头鞍子	hook-type saddle	
04.1389	杆座鞍子	socket-type saddle	
04.1390	定位装置	registration device	

序码	汉文名	英文名	注释
04.1391	定位管	registration arm	
04.1392	定位器	steady arm	
04.1393	软定位器	pull-off arm	
04.1394	定位环	steady ring	
04.1395	定位线夹	steady ear	
04.1396	正定位	pull-off mode	
04.1397	反定位	push-off mode	
04.1398	支持器	supporter	
04.1399	杵环杆	bar with ball and eye, ball-socket bar	
04.1400	拉出值	stagger	
04.1401	吊弦	hanger	
04.1402	吊弦线夹	hanger ear	
04.1403	吊索	dropper	
04.1404	线岔	overhead crossing	
04.1405	硬横跨	portal structure	
04.1406	软横跨	headspan suspension	
04.1407	中间柱	single suspension mast	
04.1408	转换柱	transition mast	
04.1409	绝缘转换柱	insulated transition mast	
04.1410	非绝缘转换柱	uninsulated transition mast	
04.1411	中心柱	center mast	
04.1412	锚柱	anchor mast	
04.1413	道岔柱	turnout mast	
04.1414	定位柱	registration mast	
04.1415	分段绝缘器	section insulator	
04.1416	分相绝缘器	neutral section insulator	
04.1417	接触网故障探测装置	fault locator for overhead contact line equipment	
04.1418	悬吊滑轮	suspension pulley	
04.1419	补偿滑轮	tension pulley	
04.1420	定滑轮装置	fixed pulley	
04.1421	坠铊	balance weight	
04.1422	坠铊补偿器	balance weight tensioner	
04.1423	弹簧补偿器	spring tensioner	
04.1424	锚段	tension length	
04.1425	中心锚结	mid-point anchor	

序 码	汉 文 名	英 文 名	注 释
04.1426	中心锚结线夹	mid-point anchor clamp	
04.1427	防串装置	anticreeping device	
04.1428	锚固装置	anchor fitting	
04.1429	锚段关节	overlap	
04.1430	绝缘锚段关节	insulated overlap	
04.1431	非绝缘锚段关节	uninsulated overlap	
04.1432	同侧下锚	same-side anchor	
04.1433	异侧下锚	different-side anchor	
04.1434	限界门	warning portal	
04.1435	导线安装曲线	wire installation curve	
04.1436	当量跨距	equivalent span length	
04.1437	张力增量	tension increment	
04.1438	绝缘间隙	insulation gap	
04.1439	支柱侧面限界	mast gauge	
04.1440	冷滑	cold-running	
04.1441	热滑	hot-running	
04.1442	硬点	hard spot	
04.1443	铅垂弹性	vertical elasticity, vertical resilience	
04.1444	横向弹性	lateral elasticity, lateral resilience	
04.1445	受电弓上抬力	pantograph upthrust, pantograph static contact force	
04.1446	受电弓空气动力效应	aerodynamic effects of pantograph	
04.1447	离线	contact loss	
04.1448	离线率	contact loss rate	
04.1449	承力索弛度	catenary sag	
04.1450	接触线弛度	contact wire sag	
04.1451	接触线预留弛度	contact wire pre-sag	
04.1452	接触线最大水平偏移值	maximum horizontal displacement of contact wire	
04.1453	结构高度	system height, encumbrance	
04.1454	供电段	section for power supply	
04.1455	供电领工区	fore work district for power supply	
04.1456	接触网工区	maintenance gang for catenary	
04.1457	远动工区	work district for telemechanical system	

序　码	汉 文 名	英 文 名	注 释
04.1458	公铁两用检修车	road-railway repairing vehicle	
04.1459	接触网测试车	measuring car for overhead contact line equipment	
04.1460	接触网检修车	repairing car for overhead contact line equipment	
04.1461	变电所测试车	substation testing car	
04.1462	绝缘子清洗车	insulator cleaning car	
04.1463	绝缘梯车	insulated ladder trolley	
04.1464	危险影响	dangerous influence	
04.1465	磁感应	magnetic induction	
04.1466	电感应	electric induction	
04.1467	地电流影响	influence of ground current	属于阻性耦合。
04.1468	影响电流	influencing current	
04.1469	静电[感应]电流	electrostatic induced current	
04.1470	漏泄电流	leakage current	又称"迷流"。
04.1471	纵电动势	longitudinal electro-motive force	
04.1472	对地电压	voltage to ground	
04.1473	互感系数	mutual inductance coefficient	又称"磁耦合系数"。
04.1474	互感阻抗	mutual inductive impedance	
04.1475	视在大地导电率	apparent earth conductivity	
04.1476	大地电阻率	earth resistivity	
04.1477	屏蔽系数	shielding factor	
04.1478	综合屏蔽系数	combination shielding factor	
04.1479	钢轨屏蔽系数	shielding factor of track	
04.1480	桥隧屏蔽系数	shielding factor of bridge and tunnel	
04.1481	电缆屏蔽系数	shielding factor of cable	
04.1482	架空地线屏蔽系数	shielding factor of aerial earth wire	
04.1483	终端效应	terminal effect	
04.1484	接近长度	approach length	
04.1485	接近距离	separation distance	
04.1486	平行接近	parallelism approach	
04.1487	斜接近	oblique exposure	
04.1488	斜接近段长度	oblique exposure length	
04.1489	斜接近段的等效距离	equivalent distance of the oblique exposure	

序　码	汉　文　名	英　文　名	注　释
04.1490	交叉跨越	crossing	
04.1491	电气化干扰	electrification interference	
04.1492	干扰影响	disturbing influence	
04.1493	等效干扰电流	equivalent disturbing current	
04.1494	敏感系数	sensitivity coefficient	
04.1495	中和变压器	neutralizing transformer	
04.1496	幻通谐振变压器	phantom resonant transformer	
04.1497	屏蔽变压器	reduction transformer	
04.1498	杂音抑制线圈	noise suppression coil	
04.1499	杂音抑制器	noise suppressor	
04.1500	防护滤波器	protection filter	
04.1501	有源降压装置	active degrade voltage apparatus	
04.1502	阳[电]极	anode	
04.1503	防护	protection	
04.1504	半段效应	half section effect	
04.1505	无线电噪声	radio noise	
04.1506	宽带干扰	broadband interference	
04.1507	窄带干扰	narrowband interference	
04.1508	干扰源	interference source	
04.1509	抗扰度	immunity to interference	
04.1510	防护率	protection ratio	
04.1511	[无线电]干扰测量仪	[radio] interference meter	
04.1512	屏蔽体	shield	
04.1513	屏蔽	shielding	
04.1514	横电磁波小室	cross electromagnetic wave small room	
04.1515	电磁屏蔽暗室	electromagnetic shielding darkroom	
04.1516	电磁环境	electromagnetic environment	
04.1517	频谱特性	spectrum character	
04.1518	横向传播特性	cross propagation characteristic	
04.1519	倍程衰减	double attenuation	
04.1520	倍频衰减	frequency doubling attenuation	
04.1521	背景噪声场强	background noise field strength	
04.1522	防护间距	protection distance	
04.1523	电耦合系数	electric coupling coefficient	

序 码	汉 文 名	英 文 名	注 释
04.1524	开式简单循环燃气轮机	open simple cycle gas turbine	
04.1525	开式回热循环燃气轮机	open regenerative cycle gas turbine	
04.1526	单轴燃气轮机	single-shaft gas turbine	
04.1527	分轴燃气轮机	split-shaft gas turbine	
04.1528	变几何燃气轮机	variable-geometry gas turbine	
04.1529	轴向推力	axial thrust	
04.1530	盘车	barring	
04.1531	热挂	thermal blockage	
04.1532	燃气发生器	gas generator	
04.1533	回热器	regenerator	
04.1534	报警保护系统	alarm and protection system	
04.1535	轴向位移保护装置	axial displacement limiting device	
04.1536	进气道	intake duct	
04.1537	排气道	exhaust duct	
04.1538	轴流式涡轮	axial flow turbine	
04.1539	动力涡轮	power turbine	
04.1540	压气涡轮	compressor turbine	
04.1541	轴流式压气机	axial flow compressor	
04.1542	排气速度	exhaust velocity [for a turbine]	
04.1543	反动度	degree of reaction	
04.1544	膨胀比	expansion ratio	
04.1545	通流部分	flow passage	
04.1546	叶型	blade profile	
04.1547	安装角	stagger angle	
04.1548	进口角	blade inlet angle	
04.1549	出口角	blade outlet angle	
04.1550	叶型折转角	camber	
04.1551	速度三角形	velocity triangle	
04.1552	二次流损失	secondary flow loss	
04.1553	静叶损失	stationary blade loss	
04.1554	动叶损失	moving blade loss	
04.1555	余速损失	leaving velocity loss	
04.1556	火焰管	flame tube	
04.1557	热疲劳	thermal fatigue	

序　码	汉　文　名	英　文　名	注　释
04.1558	热腐蚀	hot corrosion	
04.1559	可转静叶	variable stator blade	
04.1560	涡轮功率	turbine power	
04.1561	压气机耗功	power input to compressor	
04.1562	失速	stall	
04.1563	管形燃烧室	can-type combustor	
04.1564	直流燃烧	straight-flow combustion	
04.1565	逆流燃烧	counter-flow combustion	
04.1566	旋流燃烧	swirl-flow combustion	
04.1567	旋风燃烧	cyclone-combustion	
04.1568	一次空气	primary air	
04.1569	二次空气	secondary air	
04.1570	燃烧区	combustion zone	
04.1571	回流区	recirculating zone	
04.1572	掺混区	dilution zone	
04.1573	理论空气量	theoretical air	
04.1574	回热度	regenerator effectiveness	
04.1575	旋流器	swirler	
04.1576	燃气侧全压损失	total pressure loss for gas side	
04.1577	燃烧稳定性	combustion stability	
04.1578	磁[悬]浮	maglev, magnetic levitation	
04.1579	电磁悬浮系统	electromagnetic suspension system, EMS	
04.1580	电动排斥式系统	electrodynamic repulsion system, EDS	
04.1581	超导悬浮系统	superconducting suspension system	
04.1582	永磁悬浮系统	permanent magnet suspension system, PMM	
04.1583	电磁吸引式系统	electromagnetic attraction system	
04.1584	磁[悬]浮车辆	maglev vehicle	
04.1585	长定子	long stator	
04.1586	短定子	short stator	
04.1587	推进系统	propulsion system	
04.1588	导向系统	guidance system	
04.1589	悬浮系统	suspension system	
04.1590	可控电磁铁	controlled electromagnet	
04.1591	悬浮组件	suspension module	

序　码	汉　文　名	英　文　名	注　释
04.1592	反应板	reaction plate	
04.1593	电磁铁驱动器	magnet driver, chopper	又称"斩波器"。
04.1594	自适应控制	adaptive control	
04.1595	吸浮力	attraction [lift] force	
04.1596	排斥力	repulsion force	
04.1597	悬浮力	lift force	
04.1598	侧力	[maglev] lateral force	
04.1599	推力	thrusting force, propulsion force	
04.1600	多磁铁系统	multimagnet system	
04.1601	非接触式传感器	noncontacting proximity sensor	
04.1602	超导体	superconductor	
04.1603	反应轨	reaction rail	
04.1604	单边型直线感应电动机	single sided linear induction motor	
04.1605	双边型直线感应电动机	double sided linear induction motor	
04.1606	加速度反馈	acceleration feedback	
04.1607	绝对加速度反馈	absolute acceleration feedback	
04.1608	电感式加速度计	inductive accelerometer	
04.1609	电容式加速度计	capacitive accelerometer	
04.1610	伺服型加速度计	servo accelerometer	
04.1611	压电式加速度计	piezoelectric accelerometer	
04.1612	气垫	air-cushion	
04.1613	气隙磁通密度	air gap flux density	
04.1614	悬球	ball suspension	
04.1615	悬距	clearance, air-gap	
04.1616	闭环控制	closed-loop control	
04.1617	线圈高度	coil height	
04.1618	悬浮导向兼用	combined lift and guidance	
04.1619	去滞曲线	demagnetisation curve	
04.1620	动态耦合	dynamic coupling	
04.1621	悬浮系统动力学	dynamics of suspension system	
04.1622	霍尔片	Hall plate	
04.1623	铁磁吸力	ferromagnetic attraction force	
04.1624	超导体斥力	superconducting repulsion force	
04.1625	导向力	guidance force	
04.1626	电流 – 力特性	current-force characteristic	

序 码	汉 文 名	英 文 名	注 释
04.1627	气隙－力特性	distance-force characteristic	
04.1628	导轨与悬浮系统相互作用	guideway suspension interaction	
04.1629	横向磁通直线感应电动机	transverse flux linear induction motor	
04.1630	磁轮	magnet wheel	
04.1631	平均工作气隙	mean operating air gap	
04.1632	悬浮导向分别控制	separate control lift/guidance	
04.1633	交错排列的磁铁布置	staggered magnet configuration	
04.1634	双线圈结构	two coil configuration	

05. 铁 道 车 辆

序 码	汉 文 名	英 文 名	注 释
05.0001	铁道车辆	railway vehicle, railway car	
05.0002	二轴车	two-axle car	
05.0003	四轴车	four-axle car	
05.0004	多轴车	multi-axle car	
05.0005	转向架式车	bogie car	
05.0006	车辆纵向	longitudinal direction of car	
05.0007	车辆横向	lateral direction of car	
05.0008	一位端	"B" end of car	
05.0009	二位端	"A" end of car	
05.0010	一位侧	left side of car	
05.0011	二位侧	right side of car	
05.0012	空车	empty car	
05.0013	重车	loaded car	
05.0014	车辆标记	lettering and marking of car	
05.0015	车种	type of car	
05.0016	车型	model of car	
05.0017	车号	number of car	
05.0018	定员	seating capacity	
05.0019	车辆全长	length over pulling faces of couplers	又称"车钩连结线间长度"。

序 码	汉 文 名	英 文 名	注 释
05.0020	车体长度	length over ends of body, length of car body	
05.0021	底架长度	length over end sills, length of underframe	
05.0022	罐体长度	length of tank	
05.0023	车体内长	length inside car body	
05.0024	车辆最大宽度	maximum width of car	
05.0025	车体宽度	width over sides of car body	
05.0026	底架宽度	width over side sills	
05.0027	车体内宽	width inside car body	
05.0028	车钩中心线高度	height of coupler center from top of rail, coupler height	
05.0029	车辆最大高度	maximum height of car	
05.0030	车辆高度	height of car	
05.0031	车体内高	height inside car body	
05.0032	车体内中心处高度	height inside car body from floor to roof center	
05.0033	地板面高度	height of floor from rail top, floor height	
05.0034	车辆定距	length between truck centers	
05.0035	车辆长距比	ratio of car body length to length between truck centers	即车体长度与定距比。
05.0036	旁承间隙	side bearing clearance	
05.0037	门孔宽度	width of door opening	
05.0038	地板面积	floor area	
05.0039	总容积	total volume	
05.0040	有效容积	effective volume	
05.0041	罐车容积计表	tank volume table	
05.0042	心盘面自由高	free height of center plate wearing surface from rail top	
05.0043	转向架对角线	truck frame diagonal	
05.0044	轴距	wheelbase	
05.0045	固定轴距	rigid wheelbase	
05.0046	车辆全轴距	wheelbase of car	
05.0047	转向架组全轴距	wheelbase of combination truck	
05.0048	轮对内侧距	distance between backs of wheel flanges	

序　码	汉　文　名	英　文　名	注　释
05.0049	轴颈中心距	distance between journal centers	
05.0050	匀质列车	even mass train	
05.0051	稳定运行工况	steady running condition	
05.0052	运行过渡工况	transitional running condition	
05.0053	比容[积]	specific volume	
05.0054	比面积	specific floor area	
05.0055	旁承承载	side bearing loading	
05.0056	走行装置	running gear	又称"走行部"。
05.0057	转向架	truck, bogie	
05.0058	单轴转向架	single-axle truck	
05.0059	二轴转向架	two-axle truck	
05.0060	三轴转向架	three-axle truck	
05.0061	多轴转向架	multi-axle truck	
05.0062	三大件转向架	three-piece truck	
05.0063	一体构架转向架	rigid frame truck	
05.0064	无摇动台式转向架	truck with no swing bolster	
05.0065	导框式转向架	pedestal truck	
05.0066	无摇枕转向架	bolsterless truck	
05.0067	无导框式转向架	non-pedestal truck	
05.0068	曲梁式转向架	curved-beam truck	
05.0069	自导向径向转向架	self-steering radial truck	
05.0070	迫导向径向转向架	forced-steering radial truck	
05.0071	组合式转向架	combination truck	由两个以上的转向架组成为一单元的转向架。
05.0072	旁承支重转向架	side bearing truck	
05.0073	内构架转向架	inside-frame truck	
05.0074	转向架构架	truck frame	
05.0075	构架侧梁	truck side sill	
05.0076	构架端梁	truck end sill	
05.0077	构架横梁	truck transom	
05.0078	构架纵梁	truck longitudinal sill	
05.0079	构架辅助梁	truck auxiliary transom	
05.0080	转向架侧架	truck side frame	

序　码	汉　文　名	英　文　名	注　释
05.0081	侧架上弦杆	side frame top chord	
05.0082	侧架下弦杆	side frame bottom chord	
05.0083	侧架上斜弦杆	side frame top oblique chord	
05.0084	侧架下斜弦杆	side frame bottom oblique chord	
05.0085	侧架弹簧承台	side frame spring seat	
05.0086	侧架立柱	side frame column	
05.0087	轴箱导框	side frame pedestal	
05.0088	轴箱挡	axle box guide	
05.0089	轴箱承台	[side frame] pedestal bearing boss	
05.0090	转向架摇枕	truck bolster	
05.0091	摇枕挡	bolster guide	
05.0092	[货车]摇枕挡	column guide	
05.0093	弹性止挡	elastic bolster guide	
05.0094	轮对	wheelset, wheel pair	
05.0095	车轮	wheel	
05.0096	整体车轮	solid wheel, mono-bloc wheel	
05.0097	有箍车轮	tired wheel, tyred wheel	
05.0098	轮心	wheel center	
05.0099	轮箍	tire, tyre	
05.0100	车轮扣环	retaining ring [of tire]	
05.0101	铸钢车轮	cast steel wheel	
05.0102	辗钢车轮	wrought steel wheel, rolled steel wheel	
05.0103	一次磨耗车轮	one-wear wheel	
05.0104	多次磨耗车轮	multiple wear wheel	
05.0105	冷铸铁轮	chilled cast iron wheel	
05.0106	弹性车轮	elastic wheel	
05.0107	车轴	axle	
05.0108	空心车轴	tubular axle, hollow axle	
05.0109	轴领	end collar, axle collar	
05.0110	轴颈	journal	
05.0111	轴颈后肩	journal back fillet	
05.0112	防尘板座	dust guard seat	
05.0113	轮座	wheel seat	
05.0114	轴身	axle body	
05.0115	轴中央部	axle center	
05.0116	轮缘	wheel flange	

序　码	汉　文　名	英　文　名	注　释
05.0117	轮辋	wheel rim	
05.0118	轮毂	wheel hub	
05.0119	轮毂孔	wheel hub bore	
05.0120	踏面	wheel tread	
05.0121	辐板	plate, web	
05.0122	s形辐板	s-type wheel plate, s-plate	
05.0123	深盆辐板	deep dish wheel plate	
05.0124	波形辐板	corragated wheel plate	
05.0125	直辐板	straight wheel plate	
05.0126	辐板孔	plate hole, web hole	
05.0127	轴箱	journal box, axle box	
05.0128	滑动轴承	plain bearing	
05.0129	轴箱盖	journal box lid	用于滑动轴箱。
05.0130	轴箱前盖	journal box front cover	用于滚动轴箱。
05.0131	轴箱后盖	journal box rear cover	用于滚动轴箱。
05.0132	轴箱体	journal box body	
05.0133	轴瓦	plain journal bearing	
05.0134	轴瓦垫	journal bearing wedge	
05.0135	塑料轴瓦头	plastic wearing end for plain bearing	
05.0136	防尘板	dust guard	
05.0137	后挡板	back shield	
05.0138	油卷	lubricating roll	
05.0139	木前枕	wooden strut	
05.0140	圆锥滚子轴承	tapered roller bearing	
05.0141	圆柱滚子轴承	cylindrical roller bearing	
05.0142	球形滚子轴承	spherical roller bearing	
05.0143	轴承端盖	roller bearing end cap	
05.0144	迷宫式密封	labyrinth seal	
05.0145	涨圈式密封	piston ring type seal	
05.0146	密封圈	seal ring	
05.0147	承载鞍	adapter	
05.0148	油导筒式轴箱定位装置	oil guide cylinder type journal box positioning device	
05.0149	干摩擦式轴箱定位装置	dry friction type journal box positioning device	
05.0150	橡胶弹性导柱式	rubber elastic guide post type jou-	

序 码	汉 文 名	英 文 名	注 释
	轴箱定位装置	rnal box positioning device	
05.0151	拉杆式轴箱定位装置	tie-rod type journal box positioning device	
05.0152	转臂式轴箱定位装置	rocker type journal box positioning device	
05.0153	拉板式轴箱定位装置	tie-plate type journal box positioning device	
05.0154	球形轴箱定位装置	ball type journal box positioning device	
05.0155	轴箱弹簧支柱	journal box spring guide post	
05.0156	减振器	shock absorber, damper	
05.0157	摩擦式减振器	snubber	
05.0158	常摩擦式减震装置	constant friction type snubbing device	
05.0159	变摩擦式减震装置	variable friction type snubbing device	
05.0160	斜楔	wedge	
05.0161	摩擦板	friction plate	
05.0162	磨耗板	wearing plate	
05.0163	液压减振器	hydraulic damper, oil damper	又称"油压减振器"。
05.0164	弹簧悬挂装置	spring suspension	
05.0165	轴箱弹簧	journal box spring	
05.0166	摇枕弹簧	bolster spring	
05.0167	螺旋弹簧	coil spring, helical spring	
05.0168	叠板弹簧	laminated spring	
05.0169	椭圆弹簧	elliptic spring	
05.0170	扭杆弹簧	torsion bar spring	
05.0171	膜式空气弹簧	diaphragm type air spring	
05.0172	囊式空气弹簧	bellows type air spring	
05.0173	橡胶弹簧	rubber spring	
05.0174	橡胶堆	rubber-metal pad	
05.0175	橡胶垫	rubber pad	
05.0176	一系悬挂	single stage suspension	
05.0177	两系悬挂	two stage suspension	
05.0178	第一系悬挂	primary suspension	
05.0179	第二系悬挂	secondary suspension	
05.0180	自动高度调整装	automatic leveling device	

序 码	汉 文 名	英 文 名	注 释
	置		
05.0181	高度调整阀	leveling valve	
05.0182	旁承	side bearing	
05.0183	弹性旁承	elastic side bearing	
05.0184	转向架基础制动	truck brake rigging	
05.0185	固定杠杆	truck dead lever	
05.0186	移动杠杆	truck live lever	
05.0187	制动拉杆	lever connection	
05.0188	杠杆上拉杆	lever connecting rod	
05.0189	制动梁	brake beam	
05.0190	制动梁拉杆	brake beam pull rod	
05.0191	制动梁下拉杆	brake beam bottom rod	
05.0192	闸瓦托	brake head	
05.0193	闸瓦	brake shoe	
05.0194	铸铁闸瓦	cast iron brake shoe	
05.0195	中磷闸瓦	medium phosphor cast iron brake shoe	
05.0196	高磷闸瓦	high phosphor cast iron brake shoe	
05.0197	高摩合成闸瓦	high friction composite brake shoe	
05.0198	闸瓦背	brake shoe back	
05.0199	低摩合成闸瓦	low friction composite brake shoe	
05.0200	闸瓦插销	brake shoe key	
05.0201	制动盘	brake disc	
05.0202	闸片	brake lining, brake pad	
05.0203	制动梁槽钢	brake beam compression channel	
05.0204	制动梁弓形杆	brake beam tension rod	
05.0205	制动梁支柱	brake beam strut	
05.0206	摇枕吊	bolster hanger, swing hanger	
05.0207	摇枕吊轴	swing hanger cross beam	
05.0208	弹簧托梁	spring plank carrier	
05.0209	弹簧托板	spring plank	
05.0210	下心盘	truck center plate	
05.0211	中心销	center pin, king pin, pivot pin	
05.0212	锁紧中心销	locking center pin	
05.0213	横向拉杆	lateral connecting rod	
05.0214	均衡梁	equalizer	
05.0215	踏面外形	tread contour, tread profile	

序 码	汉 文 名	英 文 名	注 释
05.0216	磨耗型踏面	worn profile tread	
05.0217	踏面基点	tread taping point	
05.0218	踏面锥度	tread conicity	过去曾用踏面斜度表示。
05.0219	滚动圆	tread rolling circle	
05.0220	车轮直径	wheel diameter	
05.0221	轮箍厚度	tire thickness	
05.0222	轮辋宽度	rim width	
05.0223	轮缘高度	flange height	
05.0224	轮缘厚度	flange thickness	
05.0225	轮毂厚度	hub thickness	
05.0226	轮毂直径	hub diameter	
05.0227	轮毂长度	hub length	
05.0228	轮毂孔直径	hub bore diameter	
05.0229	轴箱导框间隙	axle box play	
05.0230	摇枕挡间隙	bogie bolster play	
05.0231	轴载荷	load on axle journals	静态时,沿垂直方向加在车轴轴颈上的力。
05.0232	允许轴载荷	allowable load on journals of same axle	车轴所允许负担的最大重量。
05.0233	心盘载荷	load on center plate	
05.0234	旁承载荷	load on side bearing	
05.0235	斜对称载荷	diagonally symmetrical loading force	
05.0236	弹簧柔度	spring flexibility	
05.0237	弹簧刚度	spring stiffness	
05.0238	弹簧静挠度	static spring deflection	
05.0239	弹簧动挠度	dynamic spring deflection	
05.0240	第一第二系弹簧挠度比	ratio of spring deflections of primary and secondary suspension	
05.0241	挠度裕量系数	coefficient of spring deflection reservation	
05.0242	转向架扭曲刚度	truck rigidity against distorsion	
05.0243	相对摩擦系数	relative friction coefficient	
05.0244	阻尼系数	damping coefficient	
05.0245	临界阻尼	critical damping	
05.0246	减振指数	damping index	

序 码	汉 文 名	英 文 名	注 释
05.0247	临界阻尼值	critical damping value	
05.0248	相对阻尼系数	damping factor	阻尼系数与临界阻尼之比。
05.0249	踏面磨耗	tread wear	
05.0250	轮缘垂直磨耗	vertical flange	
05.0251	踏面擦伤	flat sliding, tread slid flat	
05.0252	踏面剥离	shelled tread	
05.0253	轮辋烧伤	burnt rim	制造过程中过热造成轮辋轮缘局部脱落，断面粗大。
05.0254	轮辋辗出	spread rim	指局部辗出，并非全部外周之翻卷。
05.0255	车轮不圆	wheel out of round	
05.0256	热裂纹	thermal crack	
05.0257	轮毂破裂	burst hub	
05.0258	轮毂松动	wheel loose on axle	
05.0259	辐板径向裂纹	radial crack in plate	
05.0260	辐板圆周裂纹	circumferential crack in plate	
05.0261	车轴弯曲	bent axle	
05.0262	制动系统	brake system	为控制机车、车辆、列车速度或使其停止的所有构成制动机的零部件所组成的一整套机构。
05.0263	制动装置	brake equipment, brake gear	制动系统中除基础制动和手制动机以外所有对制动缸进行供气和排气所必需的机构。
05.0264	制动机	brake	安排在一个系统中由人手或风动控制的，用以对机车、车辆、列车进行减速，控制速度或使其停止的一组器具。
05.0265	手制动机	hand brake	用人力作为原动力的制动机。

序 码	汉 文 名	英 文 名	注 释
05.0266	空气制动装置	air brake equipment	以压缩空气为原动力的制动装置。
05.0267	真空制动装置	vacuum brake equipment	利用负压空气为控制介质,大气压力为动力的制动装置。
05.0268	空气真空两用制动装置	air and vacuum dual brake equipment	
05.0269	电空制动装置	electropneumatic brake equipment	用电操纵,用压缩空气为原动力的制动装置。
05.0270	自动空气制动装置	automatic air brake equipment	
05.0271	直通空气制动装置	straight air brake equipment	
05.0272	基础制动装置	foundation brake rigging	制动系统中用以传递和放大制动力的一系列杠杆连接装置。
05.0273	制动管路	brake piping	机车车辆为输送制动装置所需压缩(或负压)空气的管路。
05.0274	制动主管	brake pipe	贯穿于机车车辆两端的制动管,包括端接管。
05.0275	制动支管	brake branch pipe	制动主管与分配阀或三通阀之间的连接管。
05.0276	端接管	pipe nipple	
05.0277	列车管	train pipe	贯通列车前后的制动管路。
05.0278	压力主管	pressure pipe, second brake pipe	
05.0279	直通制动管	direct air brake pipe	
05.0280	制动软管连接器	[brake] hose coupling	
05.0281	制动软管	hose	
05.0282	折角塞门	angle cock	
05.0283	端部塞门	end cock	在装转动车钩的敞车上具有端部塞门。
05.0284	球芯折角塞门	ball type angle cock	

序　码	汉　文　名	英　文　名	注　释
05.0285	制动支管三通	branch pipe tee	
05.0286	联合集尘截断塞门	combined dirt collector and cut-out cock	
05.0287	集尘器	dirt collector	
05.0288	截断塞门	cut-out cock	
05.0289	空重车转换塞门	empty and load changeover cock	
05.0290	紧急制动阀	emergency brake valve	
05.0291	分配阀	distributing valve	
05.0292	作用部	service portion	
05.0293	紧急部	emergency portion	
05.0294	缓解部	release portion	
05.0295	加速制动阀部	accelerated application valve portion	
05.0296	中间体	pipe bracket	
05.0297	滑阀	slide valve	
05.0298	[制动]柱塞	spool	
05.0299	膜板	diaphragm	
05.0300	提升阀	poppet valve	
05.0301	空重车制动装置	empty/load brake equipment	
05.0302	载荷传感阀	load sensor valve	
05.0303	载荷比例阀	load proportional valve	
05.0304	均衡风缸	equalizing reservoir	
05.0305	限压阀	limiting valve	
05.0306	三通阀	triple valve	
05.0307	制动缸	brake cylinder	
05.0308	副风缸	auxiliary reservoir	
05.0309	附加风缸	supplementary reservoir	
05.0310	降压气室	pressure reducing reservoir	
05.0311	双室风缸	two-compartment reservoir	
05.0312	紧急风缸	emergency reservoir	
05.0313	定压风缸	constant pressure reservoir	
05.0314	压力风缸	pressure reservoir	
05.0315	缓解阀	release valve	
05.0316	放风阀	vent valve	
05.0317	车长阀	caboose valve, conductor's valve, guard's valve	
05.0318	减压中继阀	reduction relay valve	

序 码	汉 文 名	英 文 名	注 释
05.0319	[制动缸压力]保持阀	retaining valve	
05.0320	防滑器	anti-skid device	
05.0321	踏面清扫器	tread cleaner	
05.0322	摩擦制动	friction braking	
05.0323	粘着制动	adhesion braking	
05.0324	非粘着制动	non-adhesion braking	
05.0325	转向架制动组件	truck-mounted brake assembly	货车转向架上制动梁与制动缸结合一体的组件。
05.0326	非踏面制动	off tread braking	
05.0327	单元制动	brake unit	
05.0328	轮装盘形制动	wheel-mounted disc brake	
05.0329	轴装盘形制动	axle-mounted disc brake	
05.0330	单侧[踏面]制动	single shoe brake	
05.0331	双侧[踏面]制动	clasp brake	
05.0332	二压力机构	two-pressure equalizing system	
05.0333	三压力机构	three-pressure equalizing system	
05.0334	二三压力混合机构	composite two and three-pressure equalizing system	
05.0335	闸瓦压力	brake shoe pressure	闸瓦压在车轮踏面上的力。
05.0336	制动力	braking force	
05.0337	制动率	braking ratio	车辆的闸瓦总压力与该车的重力之比。
05.0338	制动倍率	leverage ratio	理论闸瓦总压力与压缩空气给予制动缸活塞的推力之比值。
05.0339	制动效率	brake efficiency	实际闸瓦压力与理论闸瓦压力之比值。
05.0340	制动杠杆传动效率	brake rigging efficiency	
05.0341	净制动率	net braking ratio	
05.0342	制动缸活塞行程	piston travel	俗称"勾贝行程"。
05.0343	制动时间	braking time	
05.0344	空走时间	idling braking time	
05.0345	实制动时间	actual braking time, instantaneous	

序　码	汉　文　名	英　文　名	注　释
		application time, IAT	
05.0346	制动距离	stopping distance	
05.0347	实制动距离	actual stopping distance	
05.0348	空走距离	idling stopping distance	
05.0349	制动初速	initial speed at brake application	
05.0350	平均减速度	average retardation rate	
05.0351	制动灵敏度	sensitivity	
05.0352	制动稳定性	insensitivity	
05.0353	制动安定性	service stability	
05.0354	制动衰竭性	exhaustibility	
05.0355	制动不衰竭性	inexhaustibility	
05.0356	制动缸前杠杆	cylinder lever	
05.0357	制动缸后杠杆	auxiliary lever	
05.0358	连接拉杆	cylinder lever connecting rod	
05.0359	上拉杆	top rod	
05.0360	闸瓦间隙自动调节器	automatic slack adjuster	
05.0361	均衡拉杆	equalizing pull rod	
05.0362	均衡杠杆	equalizing lever	
05.0363	手制动曲拐	hand brake bell crank	
05.0364	手制动链轮	hand brake sheave wheel	
05.0365	停驻制动	parking braking	
05.0366	直接进入阀	direct admission valve	
05.0367	滚圈	rolling ring	
05.0368	真空缸	vacuum chamber	
05.0369	高鹅头弯管	high level bend, swan neck	
05.0370	分配阀试验台	distributing valve test rack	
05.0371	单车试验器	single car testing device	
05.0372	列车制动试验器	train brake tester	
05.0373	列车尾部风压反馈	train rear end air pressure feedback	
05.0374	车体	car body	
05.0375	流线型车体	streamlined car body	
05.0376	摆式车体	pendulum type car body	
05.0377	侧倾车体	tilting type car body	
05.0378	车体侧倾装置	car body tilting device	
05.0379	钢结构	steel structure	

序 码	汉 文 名	英 文 名	注 释
05.0380	车体骨架	body framing	
05.0381	板梁式钢结构	plate girder type steel structure	
05.0382	桁架式钢结构	truss type steel structure	
05.0383	薄壁筒体结构	thin-shelled tubular structure	
05.0384	整体承载结构	monocoque structure, integral loadcarrying structure	
05.0385	构件	structural member	
05.0386	杆件	member	
05.0387	底架	underframe	
05.0388	有中梁底架	underframe with center sill	
05.0389	无中梁底架	underframe without center sill	
05.0390	活动中梁底架	underframe with sliding center sill, cushioning underframe	
05.0391	裙板	apron	
05.0392	中梁	center sill	
05.0393	中梁悬臂部	overhang, cantilever portion of center sill	
05.0394	侧梁	side sill	
05.0395	端梁	end sill	
05.0396	枕梁	body bolster	
05.0397	大横梁	cross bearer	
05.0398	小横梁	crosstie	
05.0399	辅助梁	floor stringer	
05.0400	缓冲梁	buffer beam	
05.0401	对角撑	diagonal brace	又称"斜撑"。
05.0402	平板地板	flat floor	
05.0403	波纹地板	corrugated floor	
05.0404	可钉地板	nailable floor	
05.0405	[客车]脚蹬	entrance door step	
05.0406	[货车]脚蹬	sill step	
05.0407	上心盘	body center plate	
05.0408	球形心盘	spherical center plate	
05.0409	心盘座	center filler	
05.0410	上旁承	body side bearing	
05.0411	前从板座	front draft lug, front draft stop	
05.0412	后从板座	rear draft lug, rear draft stop	
05.0413	一体从板座	rear draft check casting	

序 码	汉 文 名	英 文 名	注 释
05.0414	冲击座	striker, striking casting	
05.0415	柱插	stake pocket	
05.0416	绳栓	rope lug	
05.0417	走板	running board	
05.0418	扶手	hand hold, grab iron	
05.0419	梯	ladder	
05.0420	牵引钩	hauling hook	
05.0421	牵引梁	draft sill	
05.0422	鱼腹梁	fish-belly sill	
05.0423	刀把梁	lowered draft sill	
05.0424	钳形梁	schnabel	
05.0425	顶车座	jacking pad	
05.0426	侧墙	side wall	
05.0427	侧柱连铁	side post connecting rail	
05.0428	侧柱内补强	inside reinforcement of side post	
05.0429	端柱	end post	
05.0430	侧柱	side post	
05.0431	角柱	corner post	
05.0432	门柱	door post	
05.0433	外端墙	outside end wall	
05.0434	内端墙	inside end wall	
05.0435	大腰带	waist rail	
05.0436	小腰带	window lintel	
05.0437	横带	rail	
05.0438	[客车]上侧梁	cant rail	
05.0439	[货车]上侧梁	top chord	
05.0440	侧墙包板	side sheathing	又称"包板"。
05.0441	上挠度	camber	
05.0442	下垂	drooping	
05.0443	旁弯	sidewise bending	
05.0444	盖板	cover plate	
05.0445	端墙包板	end sheathing	
05.0446	上墙板	cornice sheathing	
05.0447	下墙板	wainscot sheathing	
05.0448	窗间板	pier sheathing	
05.0449	墙板压筋	flute or rib on sheathing	
05.0450	窗台	window sill, window rail	

序　码	汉　文　名	英　文　名	注　释
05.0451	[客车]内墙板	panel	
05.0452	[货车]内墙板	lining	
05.0453	渡板	gangway foot plate	
05.0454	车顶	roof	
05.0455	外顶板	roof sheet	
05.0456	内顶板	ceiling	又称"天花板"。
05.0457	车顶弯梁	carline	
05.0458	车顶纵梁	purline	
05.0459	车顶侧梁	roof cant rail	
05.0460	车顶端横梁	roof end rail	
05.0461	车顶横梁	roof cross beam	
05.0462	装货口	hatch	又称"吊装口"。
05.0463	活顶盖	removable hatch cover	
05.0464	顶盖座	hatch cover seat	
05.0465	雨檐	eaves	
05.0466	翻板	platform trap door	
05.0467	漏斗	hopper	
05.0468	罐体	tank	
05.0469	罐体鞍座	tank saddle	
05.0470	罐端板	tank head	
05.0471	空气包	tank dome	
05.0472	人孔	manhole, manway	
05.0473	人孔盖	manhole cover	
05.0474	液面高度指示牌	liquid level indicating plate, tell-tale	
05.0475	内梯	manhole ladder	
05.0476	上鞍	tank anchor	
05.0477	聚液窝	liquid trap	
05.0478	进气阀	air inlet valve	
05.0479	抽液管座	unloading pipe connection	
05.0480	通气口	vent	
05.0481	卡带	tank band	又称"罐带"。
05.0482	加温套	steam jacket	
05.0483	中心排油阀	central oil outlet valve	
05.0484	侧排油阀	side oil outlet valve	
05.0485	快速接头	quick coupling	
05.0486	端门	end door	

序　码	汉　文　名	英　文　名	注　释
05.0487	外端门	gangway door	又称"折棚门"。
05.0488	[客车]脚蹬门	vestibule entrance door	
05.0489	[客车]隔门	partition door	
05.0490	[客车]摇门	swing door	
05.0491	[客车]摆门	spring butt rocking door	
05.0492	拉门	sliding door	又称"滑门"。
05.0493	[货车]上开门	upward swing door	
05.0494	[货车]下开门	downward swing door	
05.0495	底门	bottom door	
05.0496	侧门	side door	
05.0497	矮侧板	low side	
05.0498	矮端板	low end	
05.0499	塞入门	plug door	
05.0500	[塞入门]导轮	door guide wheel	
05.0501	门滑轮	door guide roller	
05.0502	[客车]门锁	door lock	
05.0503	[货车]门锁	door latch	
05.0504	上开窗	uplifting window	
05.0505	下开窗	dropping window	
05.0506	天窗	skylight	
05.0507	了望窗	observation window	
05.0508	百叶窗	louver	
05.0509	窗卷帘	window blind, window shade	
05.0510	窗帘	window curtain	
05.0511	窗框	window sash	
05.0512	窗锁	window sash lock	
05.0513	漏斗门风手动传动装置	pneumatic/manual hopper door operating device	
05.0514	空气控制装置	pneumatic control device	
05.0515	变位阀	changeover valve	
05.0516	作用阀	application valve	
05.0517	给气调整阀	pressure regulating valve	
05.0518	边走边卸阀	ballast flow control valve	
05.0519	双向风缸	two way cylinder	
05.0520	传动轴	transmission shaft	
05.0521	压紧装置	pinch device	
05.0522	双联杠杆	toggle linkage	

序　码	汉　文　名	英　文　名	注　释
05.0523	拉杆	toggle arm	
05.0524	拨叉	poking fork	
05.0525	离水格子	floor rack	
05.0526	木结构	wood structure	
05.0527	车钩缓冲装置	coupler and draft gear	
05.0528	密接式车钩	tight-lock coupler	
05.0529	转动式车钩	rotary dump coupler	
05.0530	上作用车钩	top operation coupler	
05.0531	下作用车钩	bottom operation coupler	
05.0532	车钩轮廓	coupler contour	
05.0533	钩体	coupler body	
05.0534	钩身	coupler shank	
05.0535	钩尾	coupler tail	
05.0536	钩颈	coupler neck	
05.0537	钩肩	coupler horn	
05.0538	钩腕外臂	coupler guard arm	
05.0539	钩头正面	coupler front face	
05.0540	钩耳	coupler pivot lug	
05.0541	上防跳台	top operation anticreep ledge	
05.0542	下防跳台	rotary operation anticreep ledge	
05.0543	钩舌	coupler knuckle	
05.0544	钩锁	coupler lock	
05.0545	钩锁销	coupler lock lift	
05.0546	钩舌销	knuckle pivot pin	
05.0547	钩尾销	draft key	
05.0548	尾框	coupler yoke	
05.0549	橡胶缓冲器	rubber draft gear	
05.0550	橡胶摩擦式缓冲器	rubber friction draft gear	
05.0551	弹簧摩擦式缓冲器	spring friction draft gear	
05.0552	液压式缓冲器	hydraulic draft gear	
05.0553	液力气动式缓冲器	hydropneumatic draft gear	
05.0554	盘形缓冲器	side buffer	
05.0555	中央缓冲器	central draft gear	用于活动中梁车。
05.0556	链子钩	screw coupling	

序　码	汉　文　名	英　文　名	注　释
05.0557	无间隙牵引杆	slackless drawbar	
05.0558	车端缓冲器	end-of-car cushioning device	用于活动中梁车。
05.0559	缓冲器箱体	draft gear housing	
05.0560	压块	pressing block	
05.0561	中央楔块	center wedge	
05.0562	动板	movable plate	
05.0563	缩短装置	shortening device	
05.0564	解钩装置	uncoupling device	
05.0565	车钩三态作用	three states of coupler operation	
05.0566	闭锁位置	locked position of coupler	
05.0567	开锁位置	lockset position of coupler	
05.0568	全开位置	full open position of coupler	
05.0569	车钩复原装置	coupler centering device	
05.0570	摆块	centering block	
05.0571	摆块吊	centering block hanger	
05.0572	车钩托梁	coupler carrier	
05.0573	从板	follower	
05.0574	车钩连接线	coupling line	
05.0575	风挡	vestibule diaphram	
05.0576	风挡面板	vestibule diaphram face plate	
05.0577	风挡缓冲板	vestibule diaphram buffer plate	
05.0578	缓冲杆	buffer rod	
05.0579	缓冲器容量	draft gear capacity	
05.0580	缓冲器行程	draft gear travel	
05.0581	缓冲器预紧力	draft gear initial compression	
05.0582	缓冲器阻抗力	draft gear reaction force at rating travel	
05.0583	缓冲器能量吸收率	rate of energy absorbed by draft gear	
05.0584	缓冲器反弹	draft gear recoil	
05.0585	车钩间隙	coupler slack	
05.0586	列车纵向动力	longitudinal dynamic force of train	
05.0587	[车钩缓冲装置] 压缩与拉伸	[coupler and draft gear] running-in and running-out	
05.0588	车辆冲击	car impact	
05.0589	车辆互撞	car collision	
05.0590	落锤试验	drop test	

序 码	汉 文 名	英 文 名	注 释
05.0591	采暖装置	heating system	
05.0592	蒸汽采暖装置	steam heating equipment	
05.0593	直压式采暖装置	direct pressure steam heating equipment	
05.0594	丁形离水阀	tee trap	又称"重力除水阀"。
05.0595	直压式暖汽调整阀	car pressure regulater	
05.0596	大气压式采暖装置	atmospheric pressure steam heating equipment	
05.0597	转换塞门	cut-out cock	
05.0598	大气压式暖汽调整阀	vapor regulater	
05.0599	温水采暖装置	hot water heating equipment	
05.0600	燃煤独立温水采暖装置	coal-burning [heater type] hot water heating equipment	
05.0601	燃油独立温水采暖装置	oil-burning [heater type] hot water heating equipment	
05.0602	电动水泵	electric water pump	
05.0603	手动水泵	hand water pump	
05.0604	燃煤温水锅炉	coal-burning heater	
05.0605	燃油温水锅炉	oil burning heater	
05.0606	水位表	water level gage	
05.0607	散热管	radiator	
05.0608	控制箱	control box	
05.0609	降压电阻	step down resistance	
05.0610	点火器	spark lighter	
05.0611	雾化轮	atomizing wheel	
05.0612	雾化杯	atomizing cup	
05.0613	风轮	fan	
05.0614	水温控制器	water temperature regulater	
05.0615	电热采暖装置	electric heating equipment	
05.0616	热风采暖装置	hot air heating equipment	
05.0617	自然通风装置	natural ventilation equipment	
05.0618	机械通风装置	mechanical ventilation equipment	
05.0619	通风器	ventilator	
05.0620	排[气]风扇	exhaust fan	
05.0621	暖汽端阀	train pipe end valve	

序　码	汉　文　名	英　文　名	注　释
05.0622	暖汽主管	heater train pipe	
05.0623	暖汽支管	heater branch pipe	
05.0624	暖汽软管	heater hose	
05.0625	暖汽软管连接器	heater hose coupler	
05.0626	空气调节装置	air conditioning equipment	
05.0627	空气自动控制装置	automatic air control device	
05.0628	往复式制冷压缩机	reciprocating type refrigeration compressor	
05.0629	螺杆式制冷压缩机	screw type refrigeration compressor	
05.0630	热泵	heat pump	
05.0631	冷凝器	condenser	
05.0632	蒸发器	evaporator	
05.0633	贮液桶	liquid receiver	
05.0634	恒温膨胀阀	thermostatic expansion valve	
05.0635	干燥器	dryer	
05.0636	过滤器	filter	
05.0637	油分离器	oil extractor	
05.0638	进风道	air inlet duct	
05.0639	送风道	air delivery duct	
05.0640	回风道	air return duct	
05.0641	排风道	air exhausting duct	
05.0642	风口	air port	
05.0643	空气预冷装置	air precooler	
05.0644	循环挡板	circulating baffle plate	
05.0645	室温控制器	room thermostat	又称"恒温控制器"。
05.0646	控制室	control cabin	
05.0647	制冷加温装置	refrigeration and heating equipment	
05.0648	直通截止阀	through shut-off valve	
05.0649	吸入压力调节阀	suction pressure regulating valve	
05.0650	冷凝器风道	condenser air concentrator	
05.0651	直角截止阀	angle shut-off valve	
05.0652	导热系数	coefficient of thermal conductivity	
05.0653	气密性	air-tightness	
05.0654	给水系统	water supply system	

序　码	汉　文　名	英　文　名	注　释
05.0655	车上给水装置	water supply equipment with roof tank	
05.0656	车上水箱	roof water tank	
05.0657	水盘	drip pan	
05.0658	车下给水装置	water supply equipment with lower tank	
05.0659	车下水箱	lower water tank	
05.0660	给水风缸	water supply air reservoir	
05.0661	温水箱	hot water tank	
05.0662	延长杆	extension rod	
05.0663	[给水]减压阀	[pressure] reducing valve	
05.0664	给水调整阀	water supply governer valve	
05.0665	注水口	filling pipe end	
05.0666	水箱验水阀	water tank test cock	
05.0667	五通塞门	five-way cock	
05.0668	弯形止阀	bend stop valve	
05.0669	地板排水装置	floor draining device	
05.0670	茶炉	drinking water boiler	
05.0671	电气系统	electric system	
05.0672	照明装置	illumination equipment	
05.0673	播音连接器	public address coupling	
05.0674	控制电缆连接器	control cable coupling	
05.0675	电力电缆及连接器	power supply cable and coupling	
05.0676	配电盘	switch board	
05.0677	配电柜	electric power distribution cabinet	
05.0678	顶灯	ceiling lamp	
05.0679	侧灯	side lamp	
05.0680	尾灯	tail lamp	
05.0681	席别灯	car class indicating lamp	
05.0682	角灯	corner lamp	
05.0683	床头灯	berth lamp	
05.0684	阅读灯	reading lamp	
05.0685	终夜灯	whole night lamp	
05.0686	半夜灯	evening lamp	
05.0687	单管荧光灯	single tube fluorescent lamp	
05.0688	双管荧光灯	double tube fluorescent lamp	

序　码	汉　文　名	英　文　名	注　释
05.0689	枕形壁灯	pillow shaped wall lamp	
05.0690	单筒壁灯	single cylindrical shade wall lamp	
05.0691	双筒壁灯	double cylindrical shade wall lamp	
05.0692	播音装置	public address system	
05.0693	四用广播机	four-function public address equip-ment	
05.0694	呼叫装置	call button	
05.0695	电加热器	electric heater	
05.0696	电热管	tubular electric heating element	
05.0697	电加湿器	electric moistening device	
05.0698	交流感应子发电机	inductor type alternator	
05.0699	单管逆变器	individual inverter	
05.0700	集中式逆变器	centralized inverter	
05.0701	照明稳压器	illumination voltage stabilizer	
05.0702	车底电线管	electric wire conduit underneath the car	
05.0703	接线盒	connection box	
05.0704	车电分线盒	junction box	
05.0705	接线端子	connection terminal	
05.0706	电扇	electric fan	
05.0707	蓄电池	storage battery, accumulator	
05.0708	蓄电池箱	storage battery box, accumulator box	
05.0709	车下电气插座	car power receptacle	
05.0710	充电插头	charging plug	
05.0711	车轴发电机	axle generator	
05.0712	车轴发电机控制箱	axle generator control box	
05.0713	单线制	single wire system	
05.0714	双线制	two wire system	
05.0715	车辆集中供电	centralized power supply [system] for car	
05.0716	车辆分散供电	separate power supply [system] for car	
05.0717	车辆交直流供电	AC-DC power supply for car	
05.0718	客车	passenger car, carriage, coach	

序 码	汉 文 名	英 文 名	注 释
05.0719	全金属客车	all metal passenger car, all metal coach	
05.0720	空调客车	air conditioned passenger car, air conditioned coach	
05.0721	双层客车	double deck passenger car, double deck coach, bi-level coach	
05.0722	合造客车	composite passenger car, composite coach	
05.0723	关节客车	articulated passenger car, articulated coach	
05.0724	市郊客车	suburban passenger car, suburban coach	
05.0725	硬座车	semi-cushioned seat coach	
05.0726	软座车	cushioned-seat coach, upholstered-seat coach	
05.0727	硬卧车	semi-cushioned berth sleeping car, semi-cushioned couchette	
05.0728	软卧车	cushioned berth sleeping car, upholstered couchette	
05.0729	开敞式卧车	open type sleeping car	
05.0730	包间式卧车	[corridor] compartment [type] sleeping car	
05.0731	高级包房卧车	superclass [corridor] compartment [type] sleeping car	
05.0732	餐车	dining car	
05.0733	酒吧车	buffet car	
05.0734	俱乐部车	club car	
05.0735	厨房车	kitchen car	
05.0736	行李车	baggage car, luggage van	
05.0737	邮政车	postal car, mail van	
05.0738	客厅车	saloon car	
05.0739	了望车	observation car	
05.0740	宿营车	dormitory car	
05.0741	发电车	generator car	
05.0742	公务车	officer's car, service car	
05.0743	教育车	education car	
05.0744	试验车	test car	

序　码	汉　文　名	英　文　名	注　释
05.0745	动力试验车	dynamometer car	
05.0746	维修车	maintenance car	
05.0747	检衡车	weigh bridge test car, track scale test car	
05.0748	医疗车	hospital car	
05.0749	隧道摄影车	tunnel photographing car	
05.0750	母车	car with axle generator	
05.0751	子车	car without axle generator	
05.0752	客室	passenger room, passenger compartment	
05.0753	卧室	bedroom, sleeping compartment	
05.0754	特等卧室	superclass bedroom, superclass sleeping compartment	
05.0755	餐室	dining room	
05.0756	炉灶	range	
05.0757	厨房	kitchen	
05.0758	炊事室	cooking room	
05.0759	小卖部	snack counter, buffet	
05.0760	邮政间	post office compartment	
05.0761	分检室	mail sorting room	
05.0762	行李室	baggage room, luggage compartment	
05.0763	储藏室	storage room	
05.0764	工具室	tool room	
05.0765	发电室	power plant compartment	
05.0766	列车员室	attendant's room	
05.0767	播音室	public address room	
05.0768	行李架	baggage rack, luggage rack	
05.0769	毛巾杆	towel hanging rod, towel rail	
05.0770	茶桌	tea table	
05.0771	固定座椅	fixed seat	
05.0772	转动座椅	rotating seat	
05.0773	可躺座椅	reclining seat	
05.0774	折叠座椅	folding seat	
05.0775	活动座椅	self-folding seat	
05.0776	餐车洗池	sink	
05.0777	厕所	lavatory, toilet	

序　码	汉　文　名	英　文　名	注　　释
05.0778	洗面间	washing room, washing compartment	
05.0779	洗手器	wash bowl	
05.0780	下作用水阀	under lever faucet	
05.0781	压式水阀	compression faucet	
05.0782	蹲式便器	eastern type toilet, squat-across type water closet	
05.0783	坐式便器	western type toilet, seat-type water closet	
05.0784	冲便阀	flush valve	
05.0785	洗面器	wash basin	
05.0786	梳妆架	comb rack	
05.0787	走廊	corridor	
05.0788	[客室]通道	aisle, gangway	
05.0789	通过台	vestibule	
05.0790	间壁	partition	
05.0791	货车	freight car, wagon	
05.0792	通用货车	general-purpose freight car	
05.0793	专用货车	special-purpose freight car	
05.0794	敞车	gondola car, open goods wagon	
05.0795	棚车	box car, covered goods wagon	
05.0796	活顶棚车	sliding roof box car, sliding roof goods van	
05.0797	活墙棚车	sliding side box car	
05.0798	平车	flat car	
05.0799	罐车	tank car	为轻油罐车、粘油罐车、浓硫酸罐车、水罐车等的统称。
05.0800	高压罐车	high pressure tank car	
05.0801	粉末货物车	powdered goods car	
05.0802	守车	caboose, brake van, guard's van	
05.0803	煤车	coal car	
05.0804	转动车钩煤车	rotary dumping coal car	
05.0805	自卸车	dumping car	
05.0806	砂石车	gravel car	
05.0807	矿石车	ore car	
05.0808	漏斗车	hopper car	为煤漏斗车、矿石漏

序 码	汉 文 名	英 文 名	注 释
			斗车等的统称。
05.0809	有盖漏斗车	covered hopper car	粮食漏斗车、糖漏斗车等的统称。
05.0810	散装水泥车	bulk cement car	
05.0811	隔热棚车	insulated box car	
05.0812	加温车	heater car	
05.0813	冷藏加温车	refrigerator and heater car	
05.0814	车端冰箱冷藏车	ice-bunker refrigerator car	
05.0815	车顶冰箱冷藏车	overhead brine tank refrigerator car	
05.0816	冷冻板冷藏车	freezing-plate refrigerator car	
05.0817	机械冷藏车	mechanical refrigerator car	又称"单节机械冷藏车"。
05.0818	机械冷藏车组	mechanical refrigerator car group	
05.0819	通风车	ventilated box car	
05.0820	家畜车	stock car	
05.0821	家禽车	poultry car	
05.0822	活鱼车	live fish car	
05.0823	甘蔗车	sugar cane car	
05.0824	纸浆车	pulp car	
05.0825	刨花车	wood chip car	
05.0826	集装箱车	container car	
05.0827	背负运输车	piggyback car	
05.0828	半拖车	semi-trailer	
05.0829	双层轿车平车	double deck sedan car	
05.0830	风动石碴[漏斗]车	pneumatic ballast hopper car	
05.0831	毒品车	poison car, poisonous goods wagon	
05.0832	零担办公车	office car for peddler train	
05.0833	长大货物车	oversize commodity car	
05.0834	凹底平车	depressed center flat car	
05.0835	落下孔车	well-hole car	
05.0836	钳夹车	schnabel car	
05.0837	双联平车	twinned flat car	
05.0838	关节货车	articulated freight car	
05.0839	电站列车[车组]	power plant car train-set	
05.0840	除雪车	snow plow, snow plough	

序　码	汉　文　名	英　文　名	注　释
05.0841	整体列车	integral train	
05.0842	车辆构造速度	design speed of car, construction speed of car	
05.0843	车辆最大容许速度	maximum permissible speed of car	
05.0844	通过最小曲线半径	minimum radius of curvature negotiable	
05.0845	曲线通过	curve negotiating	
05.0846	蠕滑	creep	
05.0847	最大外移位置	maximum outward position	
05.0848	最大倾斜位置	maximum inclining position	
05.0849	乘座舒适度	riding comfortableness, ride comfort	
05.0850	横向稳定性	lateral stability	
05.0851	平稳性指标	riding index	
05.0852	货物积累损伤指数	rate of accumulated freight damage	
05.0853	旅客疲劳时间	passenger fatigue time	
05.0854	抗倾覆稳定性	stability against overturning	
05.0855	抗脱轨稳定性	stability against derailment	
05.0856	抗挤出稳定性	stability against forcing out during train buckling	
05.0857	套车	telescoping	
05.0858	轮重减载率	rate of wheel load reduction	
05.0859	自重系数	ratio of light weight to loading capacity	
05.0860	每延米重量	load per meter of track	
05.0861	自重	light weight, tare weight	
05.0862	载重	loading capacity	
05.0863	总重	gross weight	
05.0864	整备品重量	servicing weight	
05.0865	簧上重量	suspended weight, sprung weight	
05.0866	簧下重量	non-suspended weight, unsprung weight	
05.0867	线路试验	road test, running test	
05.0868	运用考验	service test	
05.0869	车辆强度试验	car strength test	

序 码	汉 文 名	英 文 名	注 释
05.0870	车辆冲击试验	car impact test	
05.0871	车辆动力学试验	car dynamics test	
05.0872	车体弯曲振动试验	test of vibration caused by carbody bending	
05.0873	冲角	angle of attack	
05.0874	爬轨	climb on rail	
05.0875	挤轨	gage widening	
05.0876	脱轨	derailment	
05.0877	跳轨	jump on rail	
05.0878	车轮贴靠	flanging	
05.0879	[列车制动]溜放试验	coasting braking test	
05.0880	弹性定位轮对	elastically positioned wheelset	
05.0881	刚性定位轮对	rigidly positioned wheelset	
05.0882	自由车轮	independent wheel	
05.0883	动力系数	coefficient of dynamic force	又称"动荷系数"。
05.0884	垂直载荷	vertical load	
05.0885	垂直动载荷	vertical dynamic load	
05.0886	横向力	lateral force	
05.0887	纵向力	longitudinal force	
05.0888	运行品质	riding quality	
05.0889	稳定性安全系数	safety factor of stability	
05.0890	车辆检修	car inspection and maintenance	
05.0891	车辆运营	car operation	
05.0892	车辆段	car depot	
05.0893	车辆检修设备	car repair facilities	
05.0894	滚动轴承间	roller bearing shop	
05.0895	油线室	journal box packing room	
05.0896	挂瓦室	journal bearing babbit metal lining room	
05.0897	制动室	brake repair room	
05.0898	机械保温车辆段	mechanical refrigerator car depot	
05.0899	修车库	freight car temporary repairing shed	又称"修车棚"。
05.0900	客车整备库	coach servicing shed	又称"客车整备棚"。
05.0901	调梁设备	beam straightening equipment	
05.0902	洗罐棚	tank washing shed	又称"洗罐库"。

序　码	汉　文　名	英　文　名	注　释
05.0903	修车台位长度	length of repair position	
05.0904	洗罐设备	tank washing equipment	
05.0905	货车洗刷所	freight car washing point	
05.0906	货车站修所	freight car repairing point	简称"站修所"。
05.0907	货物列车检修所	freight train inspection and service point	又称"列检所"。
05.0908	货物列车主要检修所	freight train main inspection and service point	
05.0909	货物列车区段检修所	transit freight train inspection and service point	
05.0910	货物列车一般检修所	freight train ordinary inspection and service point	
05.0911	装卸检修所	inspection and service point for car before loading or after unloading	
05.0912	制动检修所	brake inspection point	
05.0913	旅客列车检修所	passenger train inspection and service point	
05.0914	技术整备	technical servicing	
05.0915	车辆运用维修	car operation and maintenance	
05.0916	客车技术整备所	passenger train technical servicing point	
05.0917	货车技术交接所	freight car technical condition handing-over post	
05.0918	定期修	periodic repair	
05.0919	状态修	repair according to condition	
05.0920	车辆厂修	car repair in works	
05.0921	车辆段修	car repair in depot	
05.0922	车辆辅修	car auxiliary repair	
05.0923	车辆轴检	car journal and box examination, car journal and box inspection	
05.0924	车辆制检	car brake examination, car brake inspection	
05.0925	车辆大修	car heavy repair	
05.0926	车辆中修	car medium repair	
05.0927	车辆年修	car yearly repair	
05.0928	车辆装卸修	car repair before loading or after unloading	

序　码	汉　文　名	英　文　名	注　释
05.0929	站修	repair track maintenance	
05.0930	临修	temporary repair	
05.0931	摘车修	car detached repair	
05.0932	不摘车修	in-train repair	
05.0933	库检	examination in depot	
05.0934	列检	train examination	
05.0935	架车	jack up car body	
05.0936	顶车	lift one end of car	
05.0937	脱轨器	derailer	
05.0938	站修线	repair track in station, repair siding	
05.0939	洗罐线	tank washing siding	
05.0940	转向线	turn-around wye, Y-track	
05.0941	修车线	repair siding	
05.0942	临修线	temporary repair siding	
05.0943	车辆修理厂	car repair works	
05.0944	车辆制造厂	car manufacturing works	
05.0945	洗罐站	tank washing point	
05.0946	车轴超声探伤	ultrasonic inspection for axle	
05.0947	车轴电磁探伤	magnetic particle inspection for axle, magnaflux inspection for axle	
05.0948	车辆检修限度	car repair limit	
05.0949	车辆运用限度	car road service limit	
05.0950	车辆报废限度	car condemning limit	
05.0951	列车制动全部试验	train brake overall test	
05.0952	列车制动简易试验	train brake simplified test	
05.0953	单车试验	single car test	
05.0954	爱车点	car caring point	
05.0955	燃轴	severe hot box	
05.0956	热轴	hot box	
05.0957	轴温	journal temperature	
05.0958	红外线轴温检测所	infrared journal temperature detection point	
05.0959	红外线轴温探测	infrared journal temperature detec-	

序　码	汉　文　名	英　文　名	注　释
	系统	tion system	
05.0960	客车轴温报警	passenger car journal temperature warning	
05.0961	滚动轴承故障自动检测	automatic roller bearing defect detection	
05.0962	轮对动平衡检验	wheelset dynamic balance test	
05.0963	车轮静平衡检验	car wheel static balance test	
05.0964	扣车条件	specified conditions for detaining cars	
05.0965	台位利用系数	utility factor of the position	
05.0966	整备线配置系数	allocation factor of service track	
05.0967	车辆计算长度	calculated length of car	
05.0968	段修循环系数	circulating factor of repair in depot	
05.0969	拉钩检查距离	car spacing for uncoupled inspection	
05.0970	车辆平均长度	average length of car	
05.0971	车辆检修率	rate of cars under repair	
05.0972	车辆检修停留时间	down time for holding cars for repairing	
05.0973	车辆检修在修时间	down time for car under repair	
05.0974	车辆段检修台位利用率	rate of utilization of repair positions in car depot	
05.0975	残车率	rate of bad order cars	
05.0976	非运用车系数	coefficient of cars not in service	
05.0977	客车保有量	number of passenger cars on hand	
05.0978	客车配属辆数	number of allocated passenger cars	
05.0979	车底数	number of allocated passenger trains	
05.0980	货车保有量	number of freight cars on hand	
05.0981	车辆技术履历簿	technical record book of car	
05.0982	车轮厂	[car] wheelset repair factory	
05.0983	轮对存放场	wheelset storing yard	
05.0984	转向架换装所	truck changing point	
05.0985	客货车厂段修规程	regulations for passenger/freight car repair in factory/depot	
05.0986	车辆设计规范	specifications for design of cars	

序　码	汉　文　名	英　文　名	注　释
05.0987	中间试验	intermediate test	
05.0988	K 值试验	test for K value of complete car	
05.0989	静强度试验台	static strength test rack	
05.0990	滚动试验台	rolling rig	
05.0991	热工试验室	heat engineering laboratory	
05.0992	伺服疲劳试验机	servo fatigue testing machine	
05.0993	全车振动试验台	full car vibration test rig	
05.0994	滚柱轴承试验台	roller bearing test stand	
05.0995	闸瓦试验台	[brake shoe] inertia dynamometer	
05.0996	车轴模拟试验台	axle analogy test machine	
05.0997	减振器试验台	damper test stand	

06. 铁道运输管理与运输经济

序　码	汉　文　名	英　文　名	注　释
06.0001	铁路运输	railway transportation, railway traffic	
06.0002	铁路运输管理	railway transport administration	
06.0003	铁路运营	railway operation	
06.0004	铁路运输组织	railway traffic organization	
06.0005	铁路运输质量管理	quality control of railway transportation	
06.0006	铁路旅客运输规程	regulations for railway passenger traffic	
06.0007	铁路货物运输规程	regulations for railway freight traffic	
06.0008	铁路重载运输	railway heavy haul traffic	
06.0009	铁路高速运输	railway high speed traffic	
06.0010	铁路保险运输	insured rail traffic	
06.0011	铁路保价运输	value insured rail traffic	
06.0012	铁路军事运输	railway military service	
06.0013	铁路旅客运输	railway passenger traffic	
06.0014	铁路客运组织	railway passenger traffic organization	
06.0015	行李	luggage, baggage	
06.0016	包裹	parcel	

序 码	汉 文 名	英 文 名	注 释
06.0017	广厅	public hall, concourse	
06.0018	行李房	luggage office, baggage office	
06.0019	售票处	booking office, ticket office	
06.0020	候车室	waiting room, waiting hall	
06.0021	高架候车厅	overhead waiting hall	
06.0022	问讯处	information office, inquiry office	
06.0023	客流	passenger flow	
06.0024	直通客流	through passenger flow·	
06.0025	管内客流	local passenger flow	
06.0026	市郊客流	suburban passenger flow	
06.0027	客流量	passenger flow volume	
06.0028	客流调查	passenger flow investigation	
06.0029	客流图	passenger flow diagram	
06.0030	旅客发送人数	number of passengers despatched, number of passengers originated	
06.0031	旅客到达人数	number of passengers arrived	
06.0032	旅客运送人数	number of passengers transported	
06.0033	旅客最高聚集人数	maximum number of passengers in peak hours	
06.0034	车票	ticket	
06.0035	客票	passenger ticket	
06.0036	加快票	fast extra ticket	
06.0037	特快加快票	express extra ticket	
06.0038	卧铺票	berth ticket	
06.0039	站台票	platform ticket	
06.0040	减价票	reduced-fare ticket	
06.0041	学生票	student ticket	
06.0042	小孩票	child ticket	
06.0043	残废军人票	disabled armyman ticket	
06.0044	国际联运旅客车票	passenger ticket for international through traffic	
06.0045	册页[客]票	coupon ticket	又称"联票"。
06.0046	代用票	substituting ticket	
06.0047	定期票	periodical ticket	
06.0048	公用乘车证	service pass	
06.0049	行李票	luggage ticket, baggage ticket	
06.0050	车票有效期	ticket availability	

序 码	汉 文 名	英 文 名	注 释
06.0051	行李包裹托运	consigning of luggages and parcels	
06.0052	行李包裹承运	acceptance of luggages and parcels	
06.0053	行李包裹交付	dilivery of luggages and parcels	
06.0054	旅客换乘	passenger transference	
06.0055	变更径路	route diversion	
06.0056	错乘	taking wrong train	
06.0057	漏乘	missing a train	
06.0058	越站乘车	overtaking the station	
06.0059	旅客列车乘务组	passenger train crew	
06.0060	旅客列车乘务制度	crew working system of passenger train	
06.0061	旅客列车轮乘制	crew pooling system of passenger train	
06.0062	旅客列车包乘制	assigning crew system of passenger train	
06.0063	旅客列车包车制	responsibility crew system of passenger train	
06.0064	列车员	train attendant	
06.0065	列车长	train conductor	
06.0066	乘警	train police	
06.0067	客运密度	passenger traffic density	
06.0068	旅客列车直达速度	through speed of passenger train	
06.0069	旅客列车车底周转时间	turnround time of passenger train set	
06.0070	列车车底需要数	number of passenger train set required	
06.0071	客车平均日车公里	average car-kilometers per car-day	
06.0072	列车平均载客人数	average number of passengers carried per train	
06.0073	列车客座利用率	percentage of passenger seats utilization per train	
06.0074	客车客座利用率	percentage of passenger seats utilization per car	
06.0075	铁路货物运输	railway freight traffic	
06.0076	铁路货运组织	railway freight traffic organization	

序 码	汉 文 名	英 文 名	注 释
06.0077	综合性货运站	general freight station, general goods station	
06.0078	专业性货运站	specialized freight station	
06.0079	零担货物中转站	less-than-carload freight transhipment station, part-load transhipment station	
06.0080	营业站	operating station	
06.0081	非营业站	non-operating station	
06.0082	货场	freight yard, goods yard	
06.0083	尽头式货场	stub-end type freight yard	
06.0084	通过式货场	through-type freight yard	
06.0085	混合式货场	mixed-type freight yard	
06.0086	装卸线	loading and unloading track	
06.0087	轨道衡线	weight bridge track	
06.0088	货区	freight area, goods area	
06.0089	场库	storage yard and warehouse	
06.0090	堆货场	storage yard	
06.0091	货物站台	freight platform, goods platform	
06.0092	货棚	freight shed, goods shed	
06.0093	仓库	warehouse	
06.0094	货位	freight section, goods section	
06.0095	企业自备车	private car	
06.0096	月度货物运输计划	monthly freight traffic plan	
06.0097	旬间装车计划	ten day car loading plan	
06.0098	要车计划表	car planned requisition list	
06.0099	日要车计划	daily car requisition plan	
06.0100	货物品类	goods category	
06.0101	计划内运输	planned freight traffic	
06.0102	计划外运输	out-of-plan freight traffic, unplanned freight traffic	
06.0103	直达运输	through traffic	
06.0104	成组装车	car loading by groups	
06.0105	合理运输	rational traffic	
06.0106	对流运输	cross-haul traffic	
06.0107	过远运输	excessively long-distance traffic	
06.0108	重复运输	repeated traffic	

序　码	汉　文　名	英　文　名	注　释
06.0109	迂回运输	round about traffic, circuitous traffic	
06.0110	无效运输	ineffective traffic	
06.0111	整车货物	car load freight	
06.0112	零担货物	less-than-carload freight	
06.0113	大宗货物	mass freight	
06.0114	散装货物	bulk freight	
06.0115	堆装货物	stack-loading freight	
06.0116	成件包装货物	packed freight	
06.0117	鲜活货物	fresh and live freight	
06.0118	罐装货物	tank car freight	
06.0119	易燃货物	inflammable freight	
06.0120	易冻货物	freezable freight	
06.0121	轻浮货物	light and bulk freight	
06.0122	重质货物	heavy freight	
06.0123	整车分卸	car load freight unloaded at two or more stations	
06.0124	一批货物	consignment	
06.0125	货物运到期限	freight transit period	
06.0126	货物运单	consignment note	
06.0127	货票	way bill, freight invoice	
06.0128	货车装载清单	car loading list	
06.0129	货物托运	consigning of freight	
06.0130	货物承运	acceptance of freight	
06.0131	货物交付	dilivery of freight	
06.0132	货主	owner of freight, consignor, consignee	
06.0133	货物发送作业	freight operation at originated station	
06.0134	货物到达作业	freight operation at destination station	
06.0135	货物途中作业	freight operation en route	
06.0136	货物标记	freight label	
06.0137	运输条件	traffic condition	
06.0138	运输限制	traffic limitation, traffic restriction	
06.0139	货车施封	car seal	
06.0140	货物换装整理	transhipment and rearrangement	

序 码	汉 文 名	英 文 名	注 释
		of goods	
06.0141	货物运输变更	traffic diversion	
06.0142	货源	freight traffic source	
06.0143	货流	freight flow	
06.0144	货流量	freight flow volume	
06.0145	货流图	freight flow diagram	
06.0146	货物发送吨数	tonnage of freight despatched	
06.0147	货物到达吨数	tonnage of freight arrived	
06.0148	货物运送吨数	tonnage of freight transported	
06.0149	计费吨公里	tonne-kilometers charged	
06.0150	运营吨公里	tonne-kilometers operated	
06.0151	货运密度	density of freight traffic	
06.0152	货车标记载重量	marked loading capacity of car	
06.0153	货车静载重	static load of car	
06.0154	货车动载重	dynamic load of car	
06.0155	货车载重量利用率	coefficient of utilization for car loading capacity	
06.0156	货车日产量	serviceable work-done per car day	
06.0157	超限货物	out-of-gauge freight	
06.0158	超限货物等级	classification of out-of-gauge freight	
06.0159	超限货物检查架	examining rack for out-of-gauge freight	
06.0160	阔大货物	exceptional dimension freight	
06.0161	超长货物	exceptional length freight	
06.0162	货物转向架	freight turning rack	
06.0163	货物转向架支距	distance between centers of freight turning rack	
06.0164	跨装	straddle	
06.0165	车钩缓冲停止器	device for stopping buffer action	
06.0166	游车	idle car	
06.0167	货物重心的横向位移	lateral shift for center of gravity of goods	
06.0168	货物重心的纵向位移	longitudinal shift for center of gravity of goods	
06.0169	集重货物	concentrated weight goods	
06.0170	重车重心	center of gravity for car loaded	

序 码	汉 文 名	英 文 名	注 释
06.0171	重车重心高	center height of gravity for car loaded	
06.0172	危险货物	dangerous freight, dangerous goods	
06.0173	易腐货物	perishable freight	
06.0174	冻结货物	frozen freight	
06.0175	冷却货物	cooled freight	
06.0176	加冰所	re-icing point	
06.0177	控温运输	transport under controlled temperature	
06.0178	保温运输	insulated transport	
06.0179	冷藏运输	refrigerated transport	
06.0180	加温运输	heating transport	
06.0181	通风运输	ventilated transport	
06.0182	容许运输期限	permissive period of transport	
06.0183	国际货物联运	international through freight traffic	
06.0184	铁路的连带责任	joint responsibility of railway	
06.0185	发送路	originating railway	
06.0186	到达路	destination railway	
06.0187	过境路	transit railway	
06.0188	国际铁路协定	agreement of frontier railway	
06.0189	国际铁路货物联运协定	agreement of international railway through freight traffic	
06.0190	国际联运货物票据	international through freight shipping documents	
06.0191	国际联运货物交接单	acceptance and delivery list of freight for international through traffic	
06.0192	国际联运车辆交接单	acceptance and delivery list of car for international through traffic	
06.0193	国际联运货物换装	transhipment of international through goods	
06.0194	国际联运车辆过轨	transferring of car from one railway to another for international through traffic	
06.0195	货物交接所	freight transfer point	

序 码	汉 文 名	英 文 名	注 释
06.0196	铁路行车组织	organization of train operation	
06.0197	铁路行车组织规则	rules for organization of train operation	
06.0198	车站行车工作细则	instructions for train operation at station	
06.0199	列车	train	
06.0200	车列	train set	
06.0201	旅客列车编组	passenger train formation	
06.0202	旅客列车	passenger train	
06.0203	旅客快车	fast passenger train	
06.0204	旅客特别快车	express train	
06.0205	旅客直达特别快车	through express train	
06.0206	国际联运旅客特别快车	international express train	
06.0207	直通旅客列车	through passenger train	
06.0208	管内旅客列车	local passenger train	
06.0209	市郊旅客列车	suburban passenger train	
06.0210	混合列车	mixed train	
06.0211	旅游列车	tourist train	
06.0212	临时旅客列车	extra passenger train, additional passenger train	
06.0213	军用列车	military train, troop train	
06.0214	货物列车	freight train, goods train	
06.0215	始发直达列车	through train originated from one loading point	
06.0216	阶梯直达列车	through train originated from several adjoining loading points	
06.0217	空车直达列车	through train with empty cars	
06.0218	循环直达列车	shuttled block train	
06.0219	单元列车	unit train	
06.0220	组合列车	combined train	
06.0221	技术直达列车	technical through train	
06.0222	直通列车	transit train	
06.0223	区段列车	district train	
06.0224	摘挂列车	pick-up and drop train	
06.0225	区段小运转列车	district transfer train	

序　码	汉　文　名	英　文　名	注　释
06.0226	枢纽小运转列车	junction terminal transfer train	
06.0227	路用列车	railway service train	
06.0228	列车重量标准	railway train load norm	
06.0229	车辆换算长度	converted car length	
06.0230	车站工作组织	organization of station operation	
06.0231	站界	station limit	
06.0232	车站等级	class of station	
06.0233	无调中转车	transit car without resorting	
06.0234	有调中转车	transit car with resorting	
06.0235	本站作业车	local car	
06.0236	接发列车	train reception and departure	
06.0237	行车闭塞法	train block system	
06.0238	空间间隔法	space-interval method	
06.0239	时间间隔法	time-interval method	
06.0240	书面联络法	written liaison method	
06.0241	行车凭证	running token	
06.0242	办理闭塞	blocking	
06.0243	进路	route	
06.0244	准备进路	preparation of the route	
06.0245	列车进路	train route	
06.0246	调车进路	shunting route	
06.0247	通过进路	through route	
06.0248	接车进路	receiving route	
06.0249	发车进路	departure route	
06.0250	平行进路	parallel route	
06.0251	敌对进路	conflicting route	
06.0252	开放信号	clearing signal	
06.0253	关闭信号	closing signal	
06.0254	调车	shunting, resorting, car classi-fication	
06.0255	解体调车	break-up of trains	
06.0256	编组调车	make-up of trains	
06.0257	摘挂调车	detaching and attaching of cars	
06.0258	取送调车	taking-out and placing-in of cars	
06.0259	推送调车	push-pull shunting	
06.0260	溜放调车	fly-shunting, coasting, jerking	
06.0261	驼峰调车	humping	

序 码	汉 文 名	英 文 名	注 释
06.0262	有调中转车停留时间	detention time of car in transit with resorting	
06.0263	集结时间	car detention time under accumulation	
06.0264	无调中转车停留时间	detention time of car in transit without resorting	
06.0265	中转车平均停留时间	average detention time of car in transit	
06.0266	双重作业	double freight operations	
06.0267	一次货物作业平均停留时间	average detention time of local car for loading or unloading	
06.0268	车站办理车数	number of inbound and outbound car handled at station	
06.0269	车站技术作业表	station technical working diagram	
06.0270	现在车	cars on hand	
06.0271	运用车	serviceable car, car for traffic use, cars open to traffic	
06.0272	非运用车	non-serviceable car, car not for traffic use	
06.0273	列车编组顺序表	train consist list, train list	
06.0274	列车预报	train list information in advance	
06.0275	列车确报	train list information after departure	
06.0276	车流	car flow	
06.0277	车流组织	organization of car flow	
06.0278	货物列车编组计划	freight train formation plan	
06.0279	车流径路	car flow routing	
06.0280	列车去向	train destination	
06.0281	列车编成辆数	number of cars in a train	
06.0282	列车运行时刻表	timetable	
06.0283	列车运行线	train path	
06.0284	上行方向	up direction	
06.0285	下行方向	down direction	
06.0286	列车车次	train number	
06.0287	核心车次	scheduled train number	
06.0288	机车周转图	locomotive working diagram	

序 码	汉 文 名	英 文 名	注 释
06.0289	平行运行图	parallel train diagram	
06.0290	非平行运行图	non-parallel train diagram	
06.0291	单线运行图	train diagram for singletrack	
06.0292	双线运行图	train diagram for doubletrack	
06.0293	成对运行图	train diagram in pairs	
06.0294	不成对运行图	train diagram not in pairs	
06.0295	追踪运行图	train diagram for automatic block signals	
06.0296	基本运行图	primary train diagram	
06.0297	分号运行图	variant train diagram	
06.0298	车站间隔时间	time interval between two adjacent trains at station	
06.0299	不同时到达间隔时间	time interval between two opposing trains arriving at station not at the same time	
06.0300	会车间隔时间	time interval for two meeting trains at station	
06.0301	同方向列车连发间隔时间	time interval for two trains despatching in succession in the same direction	
06.0302	追踪列车间隔时间	time interval between trains spaced by automatic block signals	
06.0303	运输能力	transport capacity	
06.0304	通过能力	carrying capacity	
06.0305	输送能力	traffic capacity	
06.0306	货运波动系数	fluctuating coefficient of freight traffic	
06.0307	能力储备系数	coefficient of reserved capacity	
06.0308	区间通过能力	carrying capacity of the block section	
06.0309	运行图周期	period in the train diagram	
06.0310	通过能力限制区间	restriction section of carrying capacity	
06.0311	列车扣除系数	coefficient of train removal	
06.0312	运输工作技术计划	plan of technical indices for freight traffic	
06.0313	装车数	number of car loadings	

序 码	汉 文 名	英 文 名	注 释
06.0314	卸车数	number of car unloadings	
06.0315	接运重车数	number of loaded cars received	
06.0316	交出重车数	number of loaded cars delivered	
06.0317	接入空车数	number of empty cars received	
06.0318	交出空车数	number of empty cars delivered	
06.0319	运用车工作量	number of serviceable cars turn-round	
06.0320	管内工作车	local cars to be unloaded	
06.0321	移交车	loaded cars to be delivered at junction stations	
06.0322	空车走行率	percentage of empty to loaded car kilometers	
06.0323	货车周转距离	average car-kilometers in one turn-round	
06.0324	货车中转距离	average car-kilometers per transit operation	
06.0325	管内装卸率	local loading and unloading rate	
06.0326	货车周转时间	car turnround time	
06.0327	运用车保有量	number of serviceable cars held kept	
06.0328	货车日车公里	car kilometers per car per day	
06.0329	列车密度	train density	
06.0330	技术速度	technical speed	
06.0331	旅行速度	travelling speed, commercial speed	
06.0332	列车出发正点率	percentage of punctuality of trains despatched to total trains	
06.0333	列车运行正点率	percentage of punctuality of trains running to total trains	
06.0334	铁路运输调度	railway traffic control, railway traffic dispatching	
06.0335	调度所	traffic controller's office, dispatcher's office	
06.0336	调度区段	train dispatching section, train control section	
06.0337	调度命令	traffic [dispatching] order, train [dispatching] order	
06.0338	车流调整	adjustment of car flow	

序　码	汉　文　名	英　文　名	注　释
06.0339	装车调整	adjustment of car loading	
06.0340	空车调整	adjustment of empty cars	
06.0341	备用货车	reserved cars	
06.0342	运输工作日常计划	day-to-day traffic working plan	
06.0343	调度日班计划	daily and shift traffic plans	
06.0344	运行图'天窗'	'sky-light' in the train diagram, 'gap' in the train diagram	
06.0345	车站作业计划	station operating plan	
06.0346	车站班计划	station shift operating plan	
06.0347	车站阶段计划	station stage operating plan	
06.0348	调车作业计划	shunting operation plan	
06.0349	列车运行调整	train operation adjustment	
06.0350	运转车长	train guard	
06.0351	列车等级	train class	
06.0352	反向行车	train running in reverse direction	
06.0353	列车运缓	train running delay	
06.0354	列车等线	train waiting for a receiving track	
06.0355	列车保留	train stock reserved	
06.0356	列车停运	withdrawal of train	
06.0357	列车加开	running of extra train	
06.0358	运输方案	traffic program	
06.0359	分界点	train spacing point	
06.0360	线路所	block post	
06.0361	辅助所	auxiliary block post	
06.0362	车站	station	
06.0363	会让站	passing station	
06.0364	越行站	overtaking station	
06.0365	中间站	intermediate station	
06.0366	区段站	district station	
06.0367	横列式区段站	transversal type district station	
06.0368	纵列式区段站	longitudinal type district station	
06.0369	编组站	marshalling station, marshalling yard	
06.0370	路网性编组站	network marshalling station	
06.0371	区域性编组站	regional marshalling station	
06.0372	地方性编组站	local marshalling station	

序 码	汉 文 名	英 文 名	注 释
06.0373	单向横列式编组站	unidirectional transversal type marshalling station	
06.0374	单向纵列式编组站	unidirectional longitudinal type marshalling station	
06.0375	单向混合式编组站	unidirectional combined type marshalling station	
06.0376	双向横列式编组站	bidirectional transversal type marshalling station	
06.0377	双向纵列式编组站	bidirectional longitudinal type marshalling station	
06.0378	双向混合式编组站	bidirectional combined type marshalling station	
06.0379	主要编组站	main marshalling station	
06.0380	辅助编组站	auxiliary marshalling station	
06.0381	自动化编组站	automatic marshalling station	
06.0382	客运站	passenger station	
06.0383	通过式客运站	through-type passenger station	
06.0384	尽头式客运站	stub-end passenger station	
06.0385	客货运站	mixed passenger and freight station	
06.0386	货运站	freight station	
06.0387	尽头式货运站	stub-end freight station	
06.0388	直通式货运站	through-type freight station	
06.0389	换装站	transhipment station	
06.0390	工业站	industrial station	
06.0391	港湾站	harbour station	
06.0392	国境站	frontier station	
06.0393	国际联运站	international through traffic station	
06.0394	联轨站	junction station	
06.0395	技术站	technical station	
06.0396	铁路枢纽	railway junction terminal	
06.0397	三角形枢纽	triangle-type junction terminal	
06.0398	十字形枢纽	cross-type junction terminal	
06.0399	顺列式枢纽	longitudinal arrangement type junction terminal	
06.0400	并列式枢纽	parallel arrangement type junction	

序 码	汉 文 名	英 文 名	注 释
06.0401	环形枢纽	loop-type junction terminal	
06.0402	混合形枢纽	combined type junction terminal	
06.0403	尽端式枢纽	stub-end type junction terminal	
06.0404	站线	siding, station track, yard track	
06.0405	到发线	arrival and departure track	
06.0406	到达线	receiving track, arriving track	
06.0407	出发线	departure track	
06.0408	编发线	marshalling-departure track	
06.0409	调车线	shunting track, classification track	
06.0410	牵出线	switching lead, shunting neck, lead track	
06.0411	存车线	storage siding	
06.0412	机车走行线	locomotive running track	
06.0413	机待线	locomotive waiting track	
06.0414	安全线	catch siding	
06.0415	避难线	refuge siding	
06.0416	尽头线	stub-end siding	
06.0417	专用线	private siding	
06.0418	客车洗车线	washing siding for passenger vehicle	
06.0419	联络线	connecting line	
06.0420	迂回线	round about line	
06.0421	环线	loop	
06.0422	枢纽直径线	diametrical line of junction terminal	
06.0423	段管线	depot siding	
06.0424	整备线	servicing siding	
06.0425	线路中心线	central lines of track	
06.0426	驼峰推送线	pushing track of hump	
06.0427	驼峰溜放线	hump lead, rolling track of hump	
06.0428	驼峰迂回线	round about line of hump	
06.0429	难行线	hard running track	
06.0430	易行线	easy running track	
06.0431	线束	track group	
06.0432	线路全长	total track length	
06.0433	线路有效长	effective track length	
06.0434	坡度牵出线	draw-out track at grade	

序　码	汉　文　名	英　文　名	注　释
06.0435	道岔绝缘段	insulated switch section	
06.0436	道岔配列	switch layout	
06.0437	禁溜车停留线	no-humping car storage	
06.0438	车场	yard	
06.0439	到达场	receiving yard, arriving yard	
06.0440	出发场	departure yard	
06.0441	到发场	receiving-departure yard	
06.0442	直通场	through yard	
06.0443	调车场	marshalling yard, shunting yard, classification yard	
06.0444	辅助车场	auxiliary yard	
06.0445	箭翎线	herringbone track	
06.0446	调车设备	marshalling facilities, classification facilities	
06.0447	驼峰	hump	
06.0448	峰顶	hump crest	
06.0449	峰高计算点	calculate point of hump height	
06.0450	驼峰推送部分	pushing section of hump	
06.0451	驼峰溜放部分	rolling section of hump	
06.0452	峰顶平台	platform of hump crest	
06.0453	驼峰溜车方向	rolling direction of hump	
06.0454	简易驼峰	simplified hump	
06.0455	非机械化驼峰	non-mechanized hump	
06.0456	机械化驼峰	mechanized hump	
06.0457	半自动化驼峰	semi-automatic hump	
06.0458	自动化驼峰	automatic hump	
06.0459	调速设备	speed control device	
06.0460	减速器	retarder	
06.0461	减速顶	dowty retarder	
06.0462	加速顶	dowty accelerator	
06.0463	加减速顶	dowty accelerator-retarder	
06.0464	可控减速顶	dowty controllable retarder	
06.0465	制动铁鞋	brake shoe, skate	
06.0466	脱鞋器	skate throw-off device	
06.0467	停车器	stopping device	
06.0468	挡车器	stop buffer	
06.0469	制动位	retarder location	

序　码	汉　文　名	英　文　名	注　释
06.0470	间隔制动	spacing braking	
06.0471	目的制动	target braking	
06.0472	溜车有利条件	favorable condition for car rolling	
06.0473	溜车不利条件	unfavorable condition for car rolling	
06.0474	货车溜放基本阻力	basic rolling car resistance	
06.0475	货车溜放风阻力	rolling car resistance due to wind effects	
06.0476	道岔阻力	switch resistance	
06.0477	制动能高	velocity hump crest of retarder	
06.0478	双推单溜	single rolling on double pushing track	
06.0479	双推双溜	double rolling on double pushing track	
06.0480	压钩坡	coupler compression grade	
06.0481	推送速度	pushing speed	
06.0482	连挂速度	coupling speed	
06.0483	减速器入口速度	entrance speed at retarder	
06.0484	减速器出口速度	release speed at retarder	
06.0485	空档	stop short	
06.0486	点式调速系统	point type speed control system	
06.0487	连续式调速系统	continued type speed control system	
06.0488	点连式调速系统	point-continued type speed control system	
06.0489	难行车	hard rolling car	
06.0490	中行车	middle rolling car	
06.0491	易行车	easy rolling car	
06.0492	最易行车	easiest rolling car	
06.0493	驼峰高度	hump height	简称"峰高"。
06.0494	驼峰纵断面	hump profile	
06.0495	加速坡	accelerating grade	
06.0496	中间坡	intermediate grade	
06.0497	道岔区坡	gradient within the switching area	
06.0498	溜放速度	rolling speed	
06.0499	脱钩点	separation point	

序　码	汉　文　名	英　文　名	注　释
06.0500	钩车	cars per cut	
06.0501	驼峰解体能力	break-up capacity of hump	
06.0502	车站咽喉通过能力	carrying capacity of station throat	
06.0503	牵出线改编能力	resorting capacity of lead track	
06.0504	到发线通过能力	carrying capacity of receiving-departure track	
06.0505	车站通过能力	carrying capacity of station	
06.0506	车站咽喉	station throat	
06.0507	编组能力	make-up capacity	
06.0508	平面调车场	flat marshalling yard	
06.0509	驼峰调车场	hump yard	
06.0510	驼峰调车场头部	hump yard classification throat	
06.0511	驼峰调车场尾部	tail throat of a hump yard	
06.0512	咽喉道岔	throat point	
06.0513	咽喉区长度	throat length	
06.0514	警冲标	fouling post	
06.0515	车挡	bumper post	
06.0516	站坪	station site	
06.0517	进路交叉	crossing of routes	
06.0518	交叉疏解	crossing untwining	
06.0519	旅客站舍	passenger building	
06.0520	站台	platform	
06.0521	天桥	over-bridge, passenger foot-bridge	
06.0522	地道	underground path	
06.0523	行包邮政地道	tunnel for luggage and postbag	
06.0524	站前广场	station square	
06.0525	客车整备所	passenger car servicing depot	
06.0526	旅客乘降所	passenger stop point	
06.0527	站场排水	yard drainage	
06.0528	零担仓库	scattered freight storehouse	
06.0529	运转室	operation office for train receiving-departure	
06.0530	站调楼	yard controller's tower	
06.0531	峰顶调车员室	shunter's cabin at hump crest	
06.0532	驼峰连结员室	couper's cabin at hump crest	
06.0533	制动员室	brakeman's cabin	

序 码	汉 文 名	英 文 名	注 释
06.0534	扳道房	switchman's cabin	
06.0535	道岔清扫房	switch cleaner's cabin	
06.0536	信号楼	signal tower	
06.0537	机车库	locomotive shed	
06.0538	集装箱	freight container	
06.0539	国家标准集装箱	GB freight container	
06.0540	国际标准集装箱	ISO freight container	
06.0541	通用集装箱	general purpose container	
06.0542	专用集装箱	specific purpose container	
06.0543	封闭式通风集装箱	closed ventilated container	
06.0544	敞顶集装箱	open top container	
06.0545	台架式集装箱	platform based container	
06.0546	平台集装箱	platform container	
06.0547	保温集装箱	thermal container	
06.0548	绝热集装箱	insulated container	
06.0549	冷藏集装箱	refrigerated container	
06.0550	加热集装箱	heated container	
06.0551	罐式集装箱	tank container	
06.0552	干散货集装箱	dry bulk container	
06.0553	空陆水联运集装箱	air/surface container, air/inter-modal container	
06.0554	集装箱额定质量	rating of freight container, gross mass of freight container	又称"集装箱总重"。
06.0555	集装箱自重	tare mass of freight container	
06.0556	集装箱载重	payload of freight container	
06.0557	堆码能力	stacking capability	
06.0558	栓固能力	restraint capability	
06.0559	箱底承载能力	floor loading capability	
06.0560	风雨密性	weatherproofness	
06.0561	角件	corner fittings	
06.0562	叉槽	fork pockets	
06.0563	箱门封条	door seal gasket	
06.0564	鹅颈槽	goose neck tunnel	
06.0565	托盘	pallet	
06.0566	平托盘	flat pallet	
06.0567	单面托盘	single-deck pallet	

序　码	汉　文　名	英　文　名	注　释
06.0568	双面托盘	double-deck pallet	
06.0569	箱式托盘	box pallet	
06.0570	立柱式托盘	post pallet	
06.0571	运输包装	transport package	
06.0572	集装袋	flexible freight container	
06.0573	包装储运图示标志	pictorial markings for handling of packages	
06.0574	危险货物包装标志	labels for packages of dangerous goods	
06.0575	包装运输试验	transporting test for package	
06.0576	集装化运输	containerized traffic	
06.0577	托盘运输	pallet traffic	
06.0578	集装箱运输	freight container traffic	
06.0579	装卸搬运	handling	
06.0580	装卸作业机械化	handling mechanization	
06.0581	装卸作业自动化	handling automation	
06.0582	装卸能力	handling capacity	
06.0583	装卸作业组织	organization of handling	
06.0584	装卸作业量	handling volume	
06.0585	装卸定额	handling quota	
06.0586	装卸工作单	handling sheet	
06.0587	装卸自然吨	actual tons of handling	
06.0588	装卸换算吨	converted tons of handling	
06.0589	装卸机械作业量	handling volume by machine	
06.0590	装卸机械完好率	percentage of machine handling in good condition	
06.0591	装卸机械利用率	utilization ratio of machine handling	
06.0592	自理装卸	handling by shipper-self	
06.0593	装卸费率	rate of handling charge	
06.0594	堆码作业	stacking operation	
06.0595	货垛	stack of freight	
06.0596	牵引车及挂车	tractor and trailer	
06.0597	链斗装车机	loading machine with chain buckets	
06.0598	链斗卸车机	unloading machine with chain buckets	

序　码	汉　文　名	英　文　名	注　释
06.0599	螺旋卸车机	spiral unloading machine	
06.0600	叉车	fork-lift truck	
06.0601	翻车机	tipper, tipping plant, dumper	
06.0602	铁路运输安全	safety of railway traffic	
06.0603	运输安全系统工程	safety system engineering of traffic	
06.0604	运输安全管理	safety management of traffic	
06.0605	运输安全检查	safety inspection of traffic	
06.0606	运输安全评估	safety evaluation of traffic	
06.0607	运输安全监察	safety supervision of traffic	
06.0608	运输安全控制系统	safety control system of traffic	
06.0609	事故处理	settlement of accident, accident disposal	
06.0610	事故记录	accident record	
06.0611	事故调查	accident investigation	
06.0612	事故预测	accident forecast	
06.0613	事故赔偿	accident indemnity	
06.0614	事故信息管理	accident information management	
06.0615	事故分析	accident analysis	
06.0616	责任事故	responsible accident	
06.0617	非责任事故	nonresponsible accident	
06.0618	行车事故	train operation accident	
06.0619	货运事故	freight traffic accident	
06.0620	客运事故	passenger traffic accident	
06.0621	行包事故	luggage and parcel traffic accident	
06.0622	重大事故	grave accident	
06.0623	大事故	serious accident	
06.0624	险性事故	bad accident, dangerous accident	
06.0625	一般事故	ordinary accident	
06.0626	事故隐患	accident threat	
06.0627	列车事故	train accident	
06.0628	调车事故	accident in shunting operation	
06.0629	冲突	collision	
06.0630	挤岔	forcing open of the point	
06.0631	冒进信号	overrunning of signal	
06.0632	机车车辆破损	rolling stock damage	

序 码	汉 文 名	英 文 名	注 释
06.0633	设备故障	breakdown of equipment, equipment failure	
06.0634	机车车辆溜逸	runaway of locomotive or car	
06.0635	行车中断	traffic interruption	
06.0636	爆炸事故	explosion accident	
06.0637	火灾事故	fire accident	
06.0638	被盗事故	robbery accident, burglary accident	
06.0639	丢失事故	loss accident	
06.0640	损失事故	damage accident	
06.0641	腐坏事故	decay accident	
06.0642	污染事故	contamination accident	
06.0643	湿损事故	wet damage accident	
06.0644	票货分离	separation of waybill from shipment	
06.0645	误交付	delivery mistake	
06.0646	货损率	damage rate of goods	
06.0647	货差率	mistake rate of goods	
06.0648	装卸事故	loading and unloading accident	
06.0649	工伤事故	accident on duty	
06.0650	路内人员伤亡	casualty of railway man, on-duty casualty	
06.0651	路外人员伤亡	casualty of non-railway man, not on-duty casualty	
06.0652	旅客伤亡	passenger casualty	
06.0653	事故率	accident rate	
06.0654	伤亡率	casualty rate	
06.0655	事故预防	prevention of accident, accident averting	
06.0656	防护信号	protection signal	
06.0657	事故信号	accident signal	
06.0658	告警信号	warning signal	
06.0659	呼救信号	calling help signal	
06.0660	火警信号	fire alarm signal	
06.0661	响墩信号	torpedo	
06.0662	火炬信号	torch	
06.0663	事故救援	accident rescue	

序　码	汉　文　名	英　文　名	注　释
06.0664	救援列车	breakdown train, rescue train	
06.0665	救援队	breakdown gang	
06.0666	救援机车	breakdown locomotive	
06.0667	救援起重机	wrecking crane	又称"救援吊车"。
06.0668	止轮器	wheel skid	
06.0669	复轨器	re-railer, rerailing device	
06.0670	封锁区间	closing the section	
06.0671	恢复通车	restoring traffic	
06.0672	铁路运输经济	railway transport economy	
06.0673	铁路计划	railway plan	
06.0674	铁路长期计划	long-term railway plan	
06.0675	铁路年度计划	annual railway plan	
06.0676	旅客运输计划	passenger traffic plan	
06.0677	旅客运输量	volume of passenger traffic	
06.0678	旅客周转量	turnover of passenger traffic	
06.0679	旅客平均运程	average journey per passenger	
06.0680	旅客乘车系数	coefficient of passengers travelling by trains	
06.0681	货物运输计划	freight traffic plan	
06.0682	货物运输量	freight traffic volume	
06.0683	货物周转量	turnover of freight traffic	
06.0684	货物平均运程	average haul of freight traffic	
06.0685	货物运输系数	coefficient of freight traffic	
06.0686	货运经济调查	economic investigation of freight traffic	
06.0687	换算周转量	converted turnover	
06.0688	运量预测	forecast of traffic volume	
06.0689	机车车辆运用计划	rolling stock utilization plan	
06.0690	车辆公里	car kilometers	
06.0691	总重吨公里	gross ton-kilometers	
06.0692	机车公里	locomotive kilometers	
06.0693	列车公里	train kilometers	
06.0694	货车检修率	ratio of freight cars under repair	
06.0695	列车平均总重	average train gross weight	
06.0696	机车检修率	ratio of locomotives under repair	
06.0697	铁路固定资产	railway fixed assets	

序　码	汉　文　名	英　文　名	注　释
06.0698	低值易耗品	low value and easily wornout articles	
06.0699	有形损耗	tangible wear	
06.0700	无形损耗	intangible wear	
06.0701	固定资产大修	capital repair of fixed assets	
06.0702	大修计划	plan of capital repair	
06.0703	固定资产更新改造	renewal and reconstruction of fixed assets, renewal and upgrading of fixed assets	
06.0704	更新改造计划	plan of renewal and upgrading	
06.0705	基本折旧率	basic depreciation rate	
06.0706	综合折旧率	composite depreciation rate	
06.0707	分类折旧率	classified depreciation rate	
06.0708	基本建设计划	plan of capital construction	
06.0709	基本建设投资	capital construction investment	
06.0710	固定资产投资	fixed asset investment	
06.0711	投资效果	effect of investment	
06.0712	项目管理	project management	
06.0713	项目评估	project appraisal	
06.0714	劳动工资计划	plan of labor and wages	
06.0715	铁路职工数	number of railway staff and workers	
06.0716	铁路运输全员劳动生产率	labor productivity of railway transport	
06.0717	劳动定额	labor norm, labor ratings	
06.0718	材料消耗定额	material consumption norm, material consumption ratings	
06.0719	材料供应计划	material supply plan	
06.0720	材料申请计划	material requisition plan	
06.0721	运输成本	traffic cost	
06.0722	运输成本计划	traffic cost plan	
06.0723	运输支出	traffic expenditure	
06.0724	旅客人公里成本	cost per passenger kilometer	
06.0725	计费吨公里成本	cost of charged ton-kilometer	
06.0726	换算吨公里成本	cost of converted ton-kilometer	
06.0727	变动支出	variable expense	
06.0728	机车能耗	locomotive energy consumption	

序　码	汉　文　名	英　文　名	注　释
06.0729	营业外支出	non-operating outlay	
06.0730	铁路运价	railway tariff, railway rate	
06.0731	旅客票价	passenger fare	
06.0732	基本票价	basic fare	
06.0733	保险费	insurance charge	
06.0734	货物运价	freight rate	
06.0735	普通运价	general rate	
06.0736	特定运价	special rate	
06.0737	货物运价里程	tariff kilometerage	
06.0738	[货物]计费重量	charged weight	
06.0739	货物运价号	freight tariff No.	
06.0740	货物运价率	freight rate	
06.0741	货运杂费	miscellaneous fees of goods traffic	
06.0742	铁路财务	railway finance	
06.0743	固定资金	fixed capital	
06.0744	固定资产原价	original value of fixed assets	
06.0745	固定资产残值	scrap value of fixed assets	
06.0746	固定资产退废率	rate of fixed assets retirement	
06.0747	固定资产更新率	rate of fixed assets renewal	
06.0748	流动资金	current capital, liquid fund	
06.0749	流动资金周转	turnover of current capital	
06.0750	运输收入	traffic revenue	
06.0751	运输收入率	rate of traffic revenue	
06.0752	清算收入	clearing revenue	
06.0753	专用基金	special fund	
06.0754	铁路建设基金	railway construction fund	
06.0755	铁路运输周转基金	railway traffic turnover fund	
06.0756	经济核算	economic accounting	
06.0757	经济效果	economic effects	
06.0758	铁路财务成果	railway financial result	
06.0759	铁路财务状况	railway financial condition	
06.0760	铁路运输利润	railway traffic profit	
06.0761	财务决算审查	financial statements review	
06.0762	财务管理信息系统	financial management information system	

07. 铁 道 通 信

序　码	汉　文　名	英　文　名	注　释
07.0001	铁道通信	railway communication, railroad communication	又称"铁路通信"。
07.0002	通信	communication	
07.0003	信息	information	
07.0004	通路	path	
07.0005	频道	frequency channel	
07.0006	频带	frequency band	
07.0007	频率	frequency	
07.0008	话路	telephone channel	
07.0009	多路复用	multiplex	
07.0010	时分复用	time division multiplex	
07.0011	频分复用	frequency division multiplex	
07.0012	音频	audio frequency	
07.0013	话频	voice frequency	
07.0014	射频	radio frequency	
07.0015	奇偶校验码	odd-even check code	
07.0016	编码	coding	
07.0017	解码	decode	
07.0018	调制	modulation	
07.0019	解调	demodulation	
07.0020	检波	detection	
07.0021	载波	carrier [wave]	
07.0022	调幅	amplitude modulation	
07.0023	调频	frequency modulation	
07.0024	调相	phase modulation	
07.0025	脉冲	impulse, pulse	
07.0026	增量调制	delta modulation	
07.0027	脉冲编码调制	pulse code modulation	
07.0028	差分脉码调制	differential pulse-code modulation	
07.0029	网络	network	
07.0030	电平	level	
07.0031	增益	gain	
07.0032	衰减	attenuation	

序 码	汉 文 名	英 文 名	注 释
07.0033	损耗	loss	
07.0034	灵敏度	sensitivity	
07.0035	选择性	selectivity	
07.0036	自动增益控制	automatic gain control	
07.0037	噪声	noise	
07.0038	串音	crosstalk	
07.0039	近端串音	near-end crosstalk	
07.0040	远端串音	far-end crosstalk	
07.0041	串音测试器	crosstalk meter	
07.0042	干扰	disturbance, interference	
07.0043	信噪比	signal to noise ratio	
07.0044	分贝	decibel	
07.0045	比特	bit	
07.0046	比特率	bit rate	
07.0047	波特	baud	
07.0048	整流器	rectifier	
07.0049	逆变器	inverter	
07.0050	稳压器	voltage stabilizer	
07.0051	扼流圈	choke	
07.0052	放大器	amplifier	
07.0053	变频器	frequency converter	
07.0054	混频器	mixer	
07.0055	振荡器	oscillator	
07.0056	检波器	detecter	
07.0057	滤波器	filter	
07.0058	传输线	transmission line	
07.0059	通信网	communication network	
07.0060	数字电话网	digital telephone network	
07.0061	数据通信网	data communication network	
07.0062	专用通信网	private communication network	
07.0063	电话	telephone	
07.0064	可视电话	videophone	
07.0065	电视会议	video conference	
07.0066	静止图象可视电话	still picture videophone	
07.0067	响度	loudness	
07.0068	峰值话音功率	peak speech power	

序　码	汉 文 名	英 文 名	注　释
07.0069	音量	volume	
07.0070	送话器	transmitter	
07.0071	受话器	receiver	
07.0072	耳机	earphone	
07.0073	扬声器	loud-speaker	
07.0074	手机	handset	
07.0075	感应线圈	induction coil	
07.0076	消侧音器	anti-sidetone device	
07.0077	信令电路	signaling circuit	
07.0078	叉簧	hook switch	
07.0079	信号发生器	signal generator	
07.0080	磁石发电机	magneto	
07.0081	信号铃流发生器	tone and ringing generator	
07.0082	告警信号电路	alarm circuit	
07.0083	拨号盘	dial	
07.0084	旋转式拨号盘	rotary dial	
07.0085	按键式拨号盘	key pad	
07.0086	信令电流	signaling current	
07.0087	馈电电流	feeding current	
07.0088	中央电池	central battery	
07.0089	本地电池	local battery	
07.0090	电话机	telephone set	
07.0091	磁石电话机	magneto telephone set	
07.0092	共电电话机	common battery telephone set	
07.0093	自动电话机	automatic telephone set	
07.0094	旋转号盘电话机	rotary dial telephone set	
07.0095	按键电话机	key pad telephone set	
07.0096	可视电话机	videophone set	
07.0097	携带电话机	portable telephone set	
07.0098	录音电话机	recording phone set	
07.0099	扬声电话机	loud speaking telephone set	
07.0100	头戴送受话器	head set	
07.0101	插塞	plug	
07.0102	插孔	jack	又称"插座"。
07.0103	插孔排	jack strip	
07.0104	信号灯	signal lamp	
07.0105	通话计数器	message register	又称"电磁计数器"。

序　码	汉　文　名	英　文　名	注　释
07.0106	电话网	telephone network	
07.0107	公用电话网	public telephone network	
07.0108	专用电话网	private telephone network	
07.0109	公用电话交换网	public switched telephone network	
07.0110	电路	circuit	
07.0111	租用电路	leased circuit	
07.0112	专用电路	private circuit	
07.0113	[电话]用户	[telephone] subscriber	
07.0114	主叫用户	calling party	
07.0115	被叫用户	called party	
07.0116	连接	connection	
07.0117	挂断	hanging up	
07.0118	摘机状态	off-hook	
07.0119	挂机状态	on-hook	
07.0120	拨号	dialling	
07.0121	拨号脉冲	dial impulse	
07.0122	振铃	ringing	
07.0123	再振铃	re-ringing	
07.0124	铃流	ring current	
07.0125	分隔	sever	
07.0126	监听	monitoring	
07.0127	强拆	forced releasing	
07.0128	走线架	chute, chamfer	
07.0129	中间配线架	intermadiate distributing frame	
07.0130	总配线架	main distribution frame	
07.0131	测试台	test desk	
07.0132	用户引入线	subscriber's lead-in	
07.0133	用户主机	subscriber's main station	
07.0134	分机	extension	
07.0135	传输性能	transmission performance	
07.0136	主串通路	disturbing channel	
07.0137	被串通路	disturbed channel	
07.0138	可懂串音	intelligible crosstalk	
07.0139	不可懂串音	unintelligible crosstalk	
07.0140	信串比	signal to crosstalk ratio	又称"串音防卫度"。
07.0141	近端串音衰减	near-end crosstalk attenuation	
07.0142	远端串音衰减	far-end crosstalk attenuation	

序　码	汉　文　名	英　文　名	注　释
07.0143	回声	echo	
07.0144	侧音	side tone	
07.0145	话音侧音	speech side tone	
07.0146	环境噪声侧音	ambient noise side tone	
07.0147	音量表	volume meter	
07.0148	可懂度	intelligibility	
07.0149	清晰度	articulation	
07.0150	音控防鸣电路	voice-operated anti-singing circuit	
07.0151	音控门限电平	threshold level of voice-operated circuit	
07.0152	话务量	telephone traffic	
07.0153	呼损率	percent of call loss	
07.0154	线路利用率	line efficiency	
07.0155	接通率	percent of call completed	
07.0156	分品复接	grading	
07.0157	架空明线	open wire	
07.0158	铜线	copper wire	
07.0159	钢线	steel wire	曾用名"铁线"。
07.0160	铜包钢线	copper-clad steel wire	
07.0161	钢芯铝绞线	steel-cored aluminum stranded wire	
07.0162	绝缘子	insulator	
07.0163	横担	cross arm	
07.0164	电杆	pole	
07.0165	钢筋混凝土电杆	reinforced concrete pole	
07.0166	试验杆	test pole	
07.0167	终端杆	terminal pole	
07.0168	跨越杆	cross-over pole	
07.0169	角杆	angular pole	
07.0170	拉线	stay	
07.0171	电缆	cable	
07.0172	同轴电缆	coaxial cable	
07.0173	漏泄同轴电缆	leaky coaxial cable	
07.0174	对称电缆	symmetrical cable	
07.0175	局用电缆	central office cable	
07.0176	铠装电缆	armoured cable	
07.0177	综合电缆	composite cable	

序　码	汉　文　名	英　文　名	注　释
07.0178	架空电缆	aerial cable	
07.0179	地下电缆	ground cable	
07.0180	海底电缆	submarine cable	
07.0181	屏蔽电缆	shielded cable	
07.0182	应急电缆	emergency cable	
07.0183	管道电缆	duct cable	
07.0184	直埋电缆	buried cable	
07.0185	水底电缆	subaqueous cable	
07.0186	进局电缆	entrance cable	
07.0187	主干电缆	main cable	
07.0188	配线电缆	distribution cable	
07.0189	链接电缆	link cable	
07.0190	成端电缆	formed cable	
07.0191	尾巴电缆	stub cable	
07.0192	加感电缆	loaded cable	
07.0193	交接箱	cross-connecting box	
07.0194	加感箱	loading coil box	
07.0195	分线箱	distribution box with protectors	
07.0196	分线盒	distribution box without protectors	
07.0197	高频分线盒	high frequency terminal box	
07.0198	[阻抗]匹配变压器	impedance matching transformer	
07.0199	气闭头	gas-tight block	
07.0200	电缆管道	cable duct	
07.0201	电缆套管	cable sleeve	
07.0202	手孔	hand hole	
07.0203	电缆标石	cable marking stake	
07.0204	电缆充气维护设备	cable gas-feeding equipment	
07.0205	炭精避雷器	carbon arrester	
07.0206	充气避雷器	gas filled arrester	
07.0207	陶瓷避雷器	ceramic arrester	
07.0208	保安器	protector	
07.0209	纵向扼流线圈	longitudinal choke coil	
07.0210	排流线圈	drainage coil	
07.0211	避雷线	lightning conductor	
07.0212	架空地线	aerial earth wire	

序　码	汉　文　名	英　文　名	注　释
07.0213	杆上工作台	pole balcony	
07.0214	线障脉冲测试器	pulse echo fault locator	
07.0215	卤素查漏仪	halogen leak detector	
07.0216	超声波查漏仪	ultrasonic leak detector	
07.0217	电缆障碍探测器	cable fault detector	
07.0218	气敏查漏仪	gas-sensitive leak detector	
07.0219	兆欧表	megger	
07.0220	直流电桥	direct current bridge	
07.0221	阻抗电桥	impedance bridge	
07.0222	导纳电桥	admittance bridge	
07.0223	环路电阻	loop resistance	
07.0224	绝缘电阻	insulation resistance	
07.0225	不平衡电阻	unbalanced resistance	
07.0226	特性阻抗	characteristic impedance	
07.0227	垂度	sag	
07.0228	角深	pull	
07.0229	交叉制式	transposition system	
07.0230	基本交叉间隔	fundamental transposition interval	
07.0231	交叉偏差	deviation from transposition interval	
07.0232	交叉指数	transposition index	
07.0233	交叉区	transposition section	
07.0234	集总加感	lumped loading	
07.0235	加感节距	loading coil spacing	
07.0236	分级保护	cascade protection	
07.0237	参考当量	reference equivalent	
07.0238	长途通信网	toll communication network	
07.0239	通信枢纽	communication center	
07.0240	局通信枢纽	communication center of railway administration	
07.0241	局间通信枢纽	communication center between several railway administration	
07.0242	分通信枢纽	sectional communication center of railway branch administration	
07.0243	通信总枢纽	master communication center of railway whole administration	
07.0244	通信端站	terminal toll office	

序　码	汉　文　名	英　文　名	注　释
07.0245	干线长途通信网	trunk communication network	
07.0246	局线长途通信网	railway administration toll communication network	
07.0247	干线长途通信	trunk communication	
07.0248	局线长途通信	toll communication within railway administration	
07.0249	铁路长途电话网	railway long distance telephone network	
07.0250	铁路长途字冠	prefix number for railway toll call	
07.0251	长途电话交换机	toll telephone switching system	
07.0252	长途接续台	manual toll switching board	
07.0253	长途半自动接续台	toll switch board for semi-automatic operation	
07.0254	长途业务台	toll service desk	
07.0255	长途调度台	trunk dispatcher switchboard	
07.0256	记录台	recording desk	
07.0257	班长台	chief operator's desk	
07.0258	查号台	information desk	
07.0259	绳路	cord circuit	
07.0260	座席电路	operator's circuit	
07.0261	中继电路	trunk circuit	又称"中继器"。
07.0262	出中继电路	outgoing trunk circuit	又称"出中继器"。
07.0263	入中继电路	incoming trunk circuit	又称"入中继器"。
07.0264	溢呼	overflow call	
07.0265	迂回中继	alternative trunking	
07.0266	长途自动接续	toll automatic dialling	
07.0267	长途半自动接续	toll semi-automatic dialling	
07.0268	点对点长途自动接续	point to point toll automatic dialling	
07.0269	长途自动电话	toll automatic telephone	
07.0270	长途自动电话中继器	toll automatic switching repeater	
07.0271	长途电话所	toll telephone office	
07.0272	长途中继线	toll junction line	
07.0273	记录制	record [demand] working	
07.0274	立接制	no-delay [demand] working	
07.0275	呼叫信号	calling signal	

序　码	汉　文　名	英　文　名	注　　释
07.0276	拨号音	dialling tone	
07.0277	忙音	busy tone	
07.0278	回铃音	ring back tone	
07.0279	通知音	warning tone	
07.0280	喀音	click	
07.0281	高频转接段	high frequency section	
07.0282	音频转接段	audio frequency section	
07.0283	低频干扰防卫度	signal to low frequency interference rate	
07.0284	频[率]偏[差]	frequency offset	
07.0285	通路线性	channel linearity	
07.0286	载波遥接话路	carrier channel connected to telephone line	
07.0287	载波调度电话中继器	carrier adaptor for dispatching telephone	
07.0288	导频无人增音机	pilot unattended repeater	
07.0289	地温折返无人增音机	unattended repeater with ground temperature compensation and powerfeed loop back	
07.0290	地温通过无人增音机	unattended repeater with ground temperature compensation and powerpassing	
07.0291	中间线路滤波器	intermediate line filter	
07.0292	基群配线架	basic group distribution frame	
07.0293	超群配线架	supergroup distribution frame	
07.0294	进局设备	incoming equipment	
07.0295	干线调度电话	trunk dispatching telephone	
07.0296	局线调度电话	dispatching telephone within railway administration	
07.0297	干线会议电话	trunk conference telephone	
07.0298	局线会议电话	telephone conference within railway administration	
07.0299	全分配制会议电话	conference telephone of full-distribution system	
07.0300	半分配制会议电话	telephone conference of semi-distribution system	
07.0301	会议电话总机	conference telephone central board	

序　码	汉文名	英　文　名	注　释
07.0302	会议电话汇接机	conference telephone tandem board	
07.0303	会议电话分机	conference telephone subset	
07.0304	载频	carrier frequency	
07.0305	导频	pilot freqency	
07.0306	载漏	carrier leak	
07.0307	前群	pregroup	
07.0308	基群	basic group	
07.0309	超群	super group	
07.0310	主群	master group	
07.0311	平调	flat regulation	
07.0312	斜调	slope regulation	
07.0313	载频同步	carrier frequency synchronization	
07.0314	通路净衰耗	channel net loss	
07.0315	载波通信	carrier communication	
07.0316	明线通信	open-wire communication	
07.0317	对称电缆通信	symmetrical cable communication	
07.0318	同轴电缆通信	coaxial cable communication	
07.0319	增音站	repeater station	
07.0320	群转接站	group through-connection station	
07.0321	终端站	terminal station	
07.0322	载波电话终端机	carrier telephone terminal	
07.0323	载波电话增音机	carrier telephone repeater	
07.0324	有人增音机	attended repeater	
07.0325	无人增音机	unattended repeater	
07.0326	载供系统	carrier supply system	
07.0327	自动电平调节系统	automatic level regulating system	
07.0328	均衡系统	equalizing system	
07.0329	远供系统	remote power feeding system	
07.0330	业务通信系统	service communication system	
07.0331	告警系统	alarm system	
07.0332	音频终端装置	audio frequency terminating set	
07.0333	振铃器	signaling equipment	
07.0334	群调制器	group modulator	
07.0335	群解调器	group demodulator	
07.0336	群放大器	group amplifier	
07.0337	线路放大器	line amplifier	

序 码	汉 文 名	英 文 名	注 释
07.0338	载频放大器	carrier amplifier	
07.0339	导频放大器	pilot amplifier	
07.0340	调节放大器	regulating amplifier	
07.0341	发信放大器	transmitting amplifier	
07.0342	收信放大器	receiving amplifier	
07.0343	主振器	master oscillator	
07.0344	振铃信号振荡器	ringing signal oscillator	
07.0345	假线	building-out network	
07.0346	仿真线	artificial line	
07.0347	混合线圈	hybrid coil	
07.0348	方向滤波器	directional filter	
07.0349	线路滤波器	line filter	
07.0350	串音抑制滤波器	crosstalk suppression filter	
07.0351	引入架	lead-in rack	
07.0352	试验架	test rack	
07.0353	引入试验架	lead-in test rack	
07.0354	杂音测试器	psophometer	
07.0355	白噪声测试器	white noise test set	
07.0356	电平表	level meter	
07.0357	选频电平表	selective level meter	
07.0358	通路固有杂音	channel basic noise	
07.0359	忙时串杂音	busy hour crosstalk and noise	
07.0360	线路杂音	line noise	
07.0361	制际串音	inter-system crosstalk	
07.0362	路际串音	inter-channel crosstalk	
07.0363	通路频率特性	channel frequency characteristic	
07.0364	通路振幅特性	channel amplitude characteristic	
07.0365	通路串杂音防卫度	channel signal to crosstalk and noise ratio	
07.0366	通路振鸣边际	channel singing margin	
07.0367	脉码调制终端机	PCM terminal	
07.0368	数字复接设备	digital multiplex equipment	
07.0369	误码测试仪	code error tester	
07.0370	相位抖动测量仪	phase jitter tester	
07.0371	码型发生器	pattern generator	
07.0372	抽样	sample	
07.0373	量化	quantizing	

序　码	汉　文　名	英　文　名	注　释
07.0374	判决值	decision value	
07.0375	压扩律	companding law	
07.0376	一次群	primary group	
07.0377	二次群	secondary group	
07.0378	码速调整	code rate justification	
07.0379	再生中继	regeneration and repetition	
07.0380	定时抖动	timing jitter	
07.0381	地区电话网	local telephone network	
07.0382	铁路地区电话	railway local telephone	
07.0383	单局制	single-office system	
07.0384	多局制	multi-office system	
07.0385	分电话所	branch telephone office	
07.0386	中心电话所	central telephone office	
07.0387	汇接电话所	tandem telephone office	
07.0388	支电话所	minor telephone office	
07.0389	人工电话所	manual telephone office	
07.0390	地区电话交换机	local telephone switching system	
07.0391	磁石电话交换机	magneto telephone switch board	
07.0392	共电电话交换机	common battery telephone switch board	
07.0393	步进制电话交换机	step-by-step telephone switching system	
07.0394	纵横制电话交换机	crossbar telephone switching system	
07.0395	电子电话交换机	electronic telephone switching system	
07.0396	程控电话交换机	stored program controlled telephone switching system	
07.0397	记发器	register	
07.0398	发码器	code sender	
07.0399	译码器	decoder	
07.0400	标志器	marker	
07.0401	直插用户	direct plug-in subscriber	
07.0402	限制用户	limited subscriber	
07.0403	特种用户	special subscriber	
07.0404	远距用户	distant subscriber	
07.0405	台间联络线	interposition trunk	又称"座席间中继

序　码	汉　文　名	英　文　名	注　释
			线"。
07.0406	局间直通中继方式	inter office through trunk	
07.0407	主叫控制复原方式	calling subscriber release	
07.0408	被叫控制复原方式	called subscriber release	
07.0409	互不控制复原方式	called and calling subscriber release	
07.0410	区段通信	division communication	
07.0411	中间站公务电话	interstation telephone	又称"各站电话"。
07.0412	列车调度电话	train dispatching telephone	
07.0413	货运调度电话	freight dispatching telephone	
07.0414	电力调度电话	power dispatching telephone	
07.0415	站间行车电话	interstation train operation telephone, blocking telephone	又称"闭塞电话"。
07.0416	养路电话	track maintenance telephone	
07.0417	区间电话	track-side telephone	
07.0418	桥隧守护电话	bridge and tunnel guarder telephone	
07.0419	区间电话转接机	track-side telephone switching device	
07.0420	共线自动电话	party-line automatic telephone	
07.0421	音频选叫	VF selective calling	
07.0422	全呼	general calling	
07.0423	组呼	group calling	
07.0424	单呼	individual calling	
07.0425	音选调度电话总机	dispatching telephone control board with VF selective calling	
07.0426	音选调度电话分机	dispatching telephone subset with VF selective calling	
07.0427	音选调度电话汇接分配器	tandem distributor for dispatching telephone with VF selective calling, tandem distributor	简称"调度分配器"。
07.0428	调度所选叫通话箱	selective calling and talking box for dispatching office	
07.0429	音选同线电话总	party-line telephone control board	

序 码	汉 文 名	英 文 名	注 释
	机	with VF selective calling	
07.0430	音选同线电话分配器	party-line telephone distributor for VF selective calling	
07.0431	音选同线电话分机	party-line telephone subset with VF selective calling	
07.0432	音选双向增音机	two-way repeater for VF selective calling, two-way repeater	简称"双向增音机"。
07.0433	音选调度电话滤波器	bridging filter for dispatching telephone with VF selective calling	
07.0434	指挥电话总机	command telephone control board	
07.0435	列车确报电报	train out report telegraph	
07.0436	站场通信	station-yard communication	
07.0437	铁路站内电话	railway station telephone	
07.0438	扳道电话	switchman's telephone	
07.0439	电话集中器	concentrated telephone unit	
07.0440	扩音转接机	control set for sound amplifying in yard	
07.0441	扩音柱	speaking post in yard	
07.0442	光纤	optical fiber	
07.0443	单模光纤	single-mode optical fiber	
07.0444	多模渐变型光纤	graded index multimode optical fiber	
07.0445	芯径	core diameter	
07.0446	包层直径	cladding diameter	
07.0447	被覆层	coating	
07.0448	折射率分布	refractive-index profile	
07.0449	数值孔径	numerical aperture	
07.0450	模场直径	mode field diameter	
07.0451	芯[包层表面]同心度	core [cladding] concentricity	
07.0452	芯[包层表面]不圆度	non-circularity of core [cladding]	
07.0453	全反射	total internal reflection	
07.0454	传导模	guided modes	
07.0455	辐射模	radiation modes	
07.0456	漏泄模	leaky modes	
07.0457	衰减常数	attenuation constant	

序　码	汉　文　名	英　文　名	注　　释
07.0458	色散常数	chromatic dispersion constant	
07.0459	近场	near field	
07.0460	远场	far field	
07.0461	截止波长	cut-off wavelength	
07.0462	光纤带宽	bandwidth of an optical fiber	
07.0463	光线路终端设备	optical line terminal equipment	
07.0464	平均输出光功率	average optical output power	
07.0465	接收机灵敏度	receiver sensitivity	
07.0466	接收机动态范围	receiver dynamic range	
07.0467	工作波长	operating wavelength	
07.0468	光传输模式	optical transmission mode	
07.0469	光源	optical source	
07.0470	激光器	laser [diode], LD	
07.0471	发光二极管	light-emitting diode, LED	
07.0472	光探测器	optical detector	
07.0473	光连接器	optical connector	
07.0474	光适配器	optical adapter	
07.0475	光衰减器	optical attenuator	
07.0476	系统余度	system margin	
07.0477	光[再生]中继器	optical [regenerative] repeater	
07.0478	光接口	optical interface	
07.0479	光纤分配架	optical fiber distribution frame	
07.0480	光缆	optical fiber cable	
07.0481	综合光缆	combined optical fiber cable	
07.0482	加强构件	tension member	
07.0483	护套	sheath	
07.0484	缓冲层	buffer layer	
07.0485	填充线	interstitial wire	
07.0486	充油型光缆	jelly filled type optical fiber cable	
07.0487	充气维护型光缆	gas maintenance type optical fiber cable	
07.0488	光纤融接机	optical fiber fusion splicing machine	
07.0489	光纤切断器	optical fiber cutter	
07.0490	光纤剥除器	optical fiber stripper	
07.0491	光缆接头	optical fiber cable joint closure	
07.0492	保护套管	protection sleeve	

序　码	汉　文　名	英　文　名	注　释
07.0493	光功率计	optical power meter	
07.0494	稳定光源	stabilized light source	
07.0495	光时域反射仪	optical time domain reflectometer	
07.0496	瑞利散射	Rayleigh scattering	
07.0497	菲涅尔反射	Fresnel reflection	
07.0498	反向反射信号	backscattered signal	
07.0499	光纤接续损耗	optical fiber splice loss	
07.0500	光纤通信系统	optical fiber communication system	
07.0501	光纤数字线路系统	optical fiber digital line system	
07.0502	光再生[中继]段	optical regenerator section	
07.0503	光数字段	optical digital section	
07.0504	光线路保护切换设备	optical line protection switching equipment	
07.0505	监视系统	supervision system	
07.0506	波分复用	wavelength devision multiplex	
07.0507	铁路无线电通信	railway radio communication	
07.0508	铁路移动无线电通信	railway mobile communication	
07.0509	铁路无线电中继通信	relay radio communication for railway	
07.0510	铁路感应无线电通信	inductive radio communication for railway	
07.0511	单边带通信	single side band communication	
07.0512	单工无线电通信	simplex radio communication	
07.0513	同频单工无线电通信	same-frequency simplex radio communication	
07.0514	异频单工无线电通信	different-frequency simplex radio communication	
07.0515	双工无线电通信	duplex radio communication	
07.0516	半双工无线电通信	semi-duplex radio communication	
07.0517	单双工兼容无线电通信	compatible simplex-duplex radio communication	
07.0518	最小可用接收电平	minimum available receiving level	
07.0519	最小接收场强	minimum receiving field strength	

序 码	汉 文 名	英 文 名	注 释
07.0520	场强中值	median of field strength	
07.0521	干扰电压	interference voltage	
07.0522	干扰场强	interference field strength	
07.0523	载噪比	carrier-to-noise ratio	
07.0524	无线电干扰	radio interference, radio distur-bance	
07.0525	传导干扰	conducted interference.	
07.0526	辐射干扰	radiated interference	
07.0527	互调干扰	intermodulation interference	
07.0528	交调干扰	cross modulation interference	
07.0529	同频干扰	same frequency interference	
07.0530	邻道干扰	adjacent channel interference	
07.0531	阻塞干扰	block interference	
07.0532	越区干扰	overshooting interference	
07.0533	电力牵引干扰	electric traction interference	
07.0534	工业干扰	industrial interference	
07.0535	正向传播	forward propagation	
07.0536	反向传播	reverse propagation	
07.0537	多径传播	multipath propagation	
07.0538	衰落	fading	
07.0539	干涉场	interference field	
07.0540	场强覆盖区	field strength coverage	
07.0541	弱电场区	weak electric field area	
07.0542	隧道弱电场区	weak electric field area in tunnel	
07.0543	通信盲区	communication blind district	
07.0544	同频干扰区	same frequency interference area	
07.0545	近场区	near field area	
07.0546	远场区	far field area	
07.0547	空间[波]传播方式	space-wave propagation mode	
07.0548	感应传输方式	inductive transmission mode	
07.0549	波导线传输方式	transmission mode with waveguide line	
07.0550	漏泄同轴电缆传输方式	transmission mode with leaky coaxible cable	
07.0551	中继传输方式	relay transmission mode	
07.0552	无线电台	radio station	

序 码	汉 文 名	英 文 名	注 释
07.0553	固定电台	base station	又称"基地台"。
07.0554	移动电台	mobile station	
07.0555	便携电台	portable radio set	
07.0556	机车电台	locomotive station	
07.0557	车长电台	train conductor's station	
07.0558	车站电台	station radio set	
07.0559	中心电台	center station	
07.0560	无线话筒	wireless transmitter	
07.0561	无线寻呼机	radio paging set	
07.0562	差转电台	radio repeating set	
07.0563	隧道中继器	tunnel repeater	
07.0564	频道选择器	channel selector	
07.0565	频组选择器	group frequency selector	
07.0566	定向天线	directional antenna	
07.0567	全向天线	omnidirectional antenna	
07.0568	加顶圆盘机车天线	disc-loading locomotive antenna	
07.0569	加顶垂直折合机车天线	vertical loading folded locomotive antenna	
07.0570	峰值检波器	peak detector	
07.0571	准峰值检波器	quasi-peak detector	
07.0572	平均值检波器	average detector	
07.0573	均方根值检波器	root mean square detector	
07.0574	无线电干扰测量仪	radio interference meter	
07.0575	电磁兼容[性]	electromagnetic compatibility	
07.0576	无线电台守侯状态	radio set in stand-by state	
07.0577	分集接收	diversity reception	
07.0578	空间分集	space diversity	
07.0579	频率分集	frequency diversity	
07.0580	列车无线电通信	train radio communication	
07.0581	感应式列车无线电通信	inductive train radio communication	
07.0582	列车无线电通信系统	train radio communication system	
07.0583	列车无线电调度	radio dispatching communication	

序　码	汉　文　名	英　文　名	注　释
	通信	for train	
07.0584	接发车无线电通信	radio communication for train reception and starting	
07.0585	列车业务无线电通信	radio communication for train service	
07.0586	列车广播	broadcasting for train	
07.0587	多机牵引无线电通信	radio communication for multiple-operated locomotive units	
07.0588	列车旅客无线电话	passenger radiotelephone on train	
07.0589	列车无线电调度总机	office equipment for radio train-dispatching	
07.0590	列车无线电调度转接分机	transfer branch set for radio train-dispatching	
07.0591	隧道功率分配器	power divider in tunnel	
07.0592	控制信号检波器	control signal detector	
07.0593	波导线	waveguide line	
07.0594	通话请求	calling request	
07.0595	通话命令	calling order	
07.0596	通话锁闭	calling block	
07.0597	呼叫检查	calling check	
07.0598	自环检测	self-loop test	
07.0599	遥控转发	telecontrol repeat	
07.0600	频组方式	frequency group mode	
07.0601	四频组方式	four-frequency group mode	
07.0602	独立同步方式	individual synchronized mode	
07.0603	传送同步方式	transmission synchronized mode	
07.0604	跟踪接收方式	tracking-receiving mode	
07.0605	频率跟踪切换方式	frequency tracking switching mode	
07.0606	频道切换阈值	threshold for channel switching	
07.0607	频率定点切换	fixed-point frequency switching	
07.0608	控制信号	control signal	
07.0609	监测信号	monitor signal	
07.0610	频道空闲信号	path free signal	
07.0611	信道空闲信号	channel free signal	
07.0612	导音频信号	pilot audio fequency signal	

序 码	汉 文 名	英 文 名	注 释
07.0613	选控信号	selectivity signal	
07.0614	紧急信号	emergency signal	
07.0615	紧急呼叫	emergency call	
07.0616	选呼	selective call	
07.0617	音频组合选呼	selective call with audio frequency coding	
07.0618	数字编码选呼	selective call with digital pulse coding	
07.0619	站场无线电通信	station-yard radio communication	
07.0620	站场感应通信	inductive communication in yard and station	
07.0621	客运业务无线电通信	radio communication for passenger service	
07.0622	调车无线电通信	radio communication for shunting	
07.0623	列检无线电通信	radio communication for train inspection	
07.0624	车号员无线电通信	radio communication for number taker	
07.0625	站场无线电中继转发台	radio relay set in yard and station	
07.0626	调车信号音	shunting tone	
07.0627	紧急制动信号音	emergency braking tone	
07.0628	调车呼叫信号音	shunting calling tone	
07.0629	铁路告警无线电通信	radio communication for railway warning	
07.0630	铁路防护无线电通信	radio communication for railway protection	
07.0631	施工防护无线电通信	radio communication for protection of construction	
07.0632	道口防护无线电通信	radio communication for highway crossing protection	
07.0633	列车防护无线电通信	radio communication for train protection	
07.0634	列车接近传感器	train approaching sensor	
07.0635	列车接近告警无线电通信	radio communication for warning of train approaching	
07.0636	工务维修无线电	radio communication for track	

序　码	汉　文　名	英　文　名	注　释
	通信	maintenance	
07.0637	机务维修无线电通信	radio communication for maintenance of locomotive	
07.0638	电务维修无线电通信	radio communication for maintenance of signal and communication equipment	
07.0639	工程施工无线电通信	radio communication for engineering construction	
07.0640	勘测设计无线电通信	radio communication for survey and design	
07.0641	巡道工无线电通信	radio communication for track walker	
07.0642	客站货场无线电通信	station and freight yard radio communication	
07.0643	铁路公安无线电通信	radio communication for railway public security	
07.0644	应急救灾无线电通信	radio communication for emergency purpose	
07.0645	救援列车无线电通信	radio communication for train relieving	
07.0646	专运列车无线电通信	radio communication for special train	
07.0647	铁路短波通信	short wave communication for railway	
07.0648	铁路应急短波通信	short wave communication for railway emergency	
07.0649	短波单边带无线电台	single side-band short wave station	
07.0650	短波通信车	short wave radio communication vehicle	
07.0651	铁路电视	railway TV	
07.0652	客运监视电视	monitor TV for passenger service	
07.0653	旅客问讯电视	TV for passenger information service	
07.0654	检票监视电视	monitor TV for ticket check	
07.0655	站场监视电视	monitor TV for yard and station	
07.0656	货场监视电视	monitor TV for freight yard	

序　码	汉文名	英　文　名	注　释
07.0657	商务检查电视	TV for railway commerce inspection	
07.0658	检车电视	TV for inspection	
07.0659	车号抄录电视	TV for record vehicle number	
07.0660	道口监视电视	monitor TV for highway crossing	
07.0661	列车闭路电视	cable TV on train	
07.0662	水下监视电视	monitor TV under water	
07.0663	铁路微波中继通信	microwave relay communication for railway	
07.0664	模拟微波中继通信	analog microwave relay communication	
07.0665	数字微波中继通信	digital microwave relay communication	
07.0666	微波通信	microwave communication	
07.0667	移动微波通信	mobile microwave communication	
07.0668	微波通信车	microwave communication vehicle	
07.0669	卫星通信	satellite communication	
07.0670	应急卫星通信	emergency satellite communication	
07.0671	通信卫星	communication satellite	
07.0672	地球站	earth station	
07.0673	车载地球站	vehicle earth station	
07.0674	扩频通信	spread spectrum communication	
07.0675	跳频通信	frequency hopping communication	
07.0676	移动通信	mobile communication	
07.0677	个人通信	personal communication	
07.0678	铁路无线电遥控	radio telecontrol for railway	
07.0679	机车无线电遥控	radio telecontrol for locomotive, locomotive radio-control	又称"机车无线电操纵"。
07.0680	驼峰调车机车无线电遥控	radio telecontrol of locomotive for shunting at hump	
07.0681	无线电测距	radio distance-measurement	
07.0682	无线电机车信号	radio locomotive signal	
07.0683	无线电调车信号	radio shunting signal	
07.0684	驼峰无线电调车信号	radio shunting signal at hump	
07.0685	平面无线电调车信号	radio operated signal for level shunting	

序　码	汉　文　名	英　文　名	注　释
07.0686	电报[通信]网	telegraph network	
07.0687	电报通信	telegraph communication	
07.0688	电码	code	
07.0689	电传机	teletype	
07.0690	人工电报机	manual telegraph set	
07.0691	电键	key	
07.0692	纸带发报机	tape transmitter	
07.0693	中文译码机	Chinese character code translation equipment	
07.0694	键盘	keyboard	
07.0695	载波电报终端机	carrier telegraph terminal	
07.0696	时分多路电报设备	time division multiplex telegraph equipment	
07.0697	自动检错重发设备	automatic retransmission on request equipment, ARQ equipment	
07.0698	电报交换机	telegraph switching equipment	
07.0699	起止信号发生器	start-stop signal generator	
07.0700	起止信号畸变测试器	start-stop signal distortion tester	
07.0701	用户电报	telex subscriber's telegraph	
07.0702	莫尔斯电码	Morse code	
07.0703	五单位数字保护电码	protected 5-unit numerical code	
07.0704	单工	simplex operation	
07.0705	双工	duplex operation	
07.0706	半双工	half-duplex operation	
07.0707	单流	single current	
07.0708	双流	double current	
07.0709	通报速率	telegraph rate	
07.0710	偏畸变	bias distortion	
07.0711	特性畸变	characteristic distortion	
07.0712	不规则畸变	fortuitous distortion	
07.0713	起止式	start-stop type	
07.0714	回车	carriage return	
07.0715	换行	line feed	
07.0716	退格	back space	

序　码	汉　文　名	英　文　名	注　释
07.0717	间隔	space	
07.0718	空白	blank	
07.0719	输纸孔	feed holes	
07.0720	电码孔	code holes	
07.0721	电路交换	circuit switching	
07.0722	电文交换	message switching	
07.0723	传真机	facsimile apparatus, Fax	
07.0724	传真发送机	facsimile transmitter	
07.0725	传真接收机	facsimile receiver	
07.0726	传真收发机	facsimile transceiver	
07.0727	相片传真机	photographic facsimile apparatus	
07.0728	文件传真机	document facsimile apparatus	
07.0729	数据通信	data communication	
07.0730	数据信号速率	data signaling rate	
07.0731	数据传送率	data transfer rate	
07.0732	频移键控	frequency shift keying, FSK	
07.0733	相移键控	phase shift keying, PSK	
07.0734	曼彻斯特编码	Manchester encoding	
07.0735	二进制编码	binary encoding	
07.0736	双二进制编码	doubinary encoding	
07.0737	多电平编码	multilevel encoding	
07.0738	报文	message	
07.0739	标题	heading	
07.0740	报头	header	
07.0741	正文	text	
07.0742	包	packet, package	又称"分组"。
07.0743	码组	block	
07.0744	互换	interchange	
07.0745	交换	exchange, switching	
07.0746	透明传送	transparent transfer	
07.0747	吞吐量	throughput	
07.0748	状态转移图	state transition diagram	
07.0749	访问	access	又称"接入"。
07.0750	选件	option	又称"选项"。
07.0751	缺省值	default value	
07.0752	数据网[络]	data network	
07.0753	公用数据网	public data network	

序 码	汉 文 名	英 文 名	注 释
07.0754	专用数据网	private data network	
07.0755	广域网	wide area network, WAN	
07.0756	局域网	local area network, LAN	
07.0757	分级网[络]	hierarchical network	又称"分层网[络]"。
07.0758	树形网[络]	tree network	
07.0759	网形网[络]	mesh network	
07.0760	星形网[络]	star network	
07.0761	环形网[络]	ring network	
07.0762	集中[式]网[络]	centralized network	
07.0763	分布[式]网[络]	distributed network	
07.0764	计算机网[络]	computer network	
07.0765	通信子网	communication subnet	
07.0766	包交换网	packet switching network	又称"分组交换网"。
07.0767	结点	node	
07.0768	网络互连	inter operation	
07.0769	网络管理	network management	
07.0770	开放系统互连	open system interconnection, OSI	
07.0771	网关	gateway	
07.0772	总线网[络]	bus network	
07.0773	以太网	ethernet	
07.0774	通信协议	communication protocol	
07.0775	双向同时通信	two-way simultaneous communication	
07.0776	双向交替通信	two-way alternate communication	
07.0777	单向通信	one-way communication	
07.0778	同步通信	synchronous communication	
07.0779	异步通信	asynchronous communication	
07.0780	虚电路	virtual circuit	
07.0781	双工传输	duplex transmission	
07.0782	半双工传输	half-duplex transmission	
07.0783	单工传输	simplex transmission	
07.0784	并行传输	parallel transmission	
07.0785	串行传输	serial transmission	
07.0786	同步传输	synchronous transmission	
07.0787	异步传输	asynchronous transmission	
07.0788	双向传输	bidirectional transmission	
07.0789	单向传输	unidirectional transmission	

序 码	汉 文 名	英 文 名	注 释
07.0790	突发传输	burst transmission	
07.0791	基带	baseband	
07.0792	基带传输	baseband transmission	
07.0793	差错控制	error control	
07.0794	差错恢复	error recovery	
07.0795	检错码	error detecting code	
07.0796	纠错码	error correcting code	
07.0797	循环冗余检验	cyclic redundancy check, CRC	
07.0798	奇偶检验	parity check	
07.0799	纵横奇偶检验	vertical horizontal parity	
07.0800	多点连接	multipoint connection	
07.0801	点对点连接	point to point connection	
07.0802	正向信道	forward channel	
07.0803	反向信道	backward channel	
07.0804	信道容量	channel capacity	
07.0805	数据链路	data link	
07.0806	链路管理	link management	
07.0807	多链路	multilink	
07.0808	确认	acknowledge, ACK	
07.0809	否认	negative acknowledge, NAK	
07.0810	码元	element	
07.0811	时隙	time slot	
07.0812	自动应答	automatic answering	
07.0813	自动呼叫	automatic calling	
07.0814	畸变	distortion	
07.0815	码组校验	block check	
07.0816	广播呼叫	broadcast call	
07.0817	主叫	calling	又称"主呼"。
07.0818	被叫	called	又称"被呼"。
07.0819	字符	character	
07.0820	字符检验	character check	
07.0821	相干调制	coherrent modulation	
07.0822	组合站	combined station	
07.0823	控制字符	control character	
07.0824	控制站	control station	
07.0825	循环码	cyclic code	
07.0826	数据	data	

序　码	汉　文　名	英　文　名	注　　释
07.0827	数据信道	data channel	
07.0828	数据电路端接设备	data circuit terminating equipment, DCE	
07.0829	数据终端设备	data terminal equipment	
07.0830	集中器	concentrator	
07.0831	数据站	data station	
07.0832	差分调制	differential modulation	
07.0833	差分移相键控	differential phase shift keying, DPSK	
07.0834	回波效应	echo effect	
07.0835	位同步	bit synchronizing	
07.0836	共线信令系统	common channel signaling system	
07.0837	码组结束信号	end-of-block signal	
07.0838	正文结束信号	end-of-text signal	
07.0839	传输结束信号	end-of-transmission signal	
07.0840	突发差错	burst error	
07.0841	信息反馈重发纠错	error correction by information feed-back repetition	
07.0842	反馈重发纠错	error correction by feed-back repetition	
07.0843	差错检测	error detection	
07.0844	固有畸变	inherrent distortion	
07.0845	链路协议	link protocol	
07.0846	规程	procedure	
07.0847	主站	master station	
07.0848	从站	slave station	
07.0849	调制解调器	modem	
07.0850	包交换	packet switching	又称"分组交换"。
07.0851	永久虚[拟]电路	permanent virtual circuit	
07.0852	轮询	polling	又称"探询"。
07.0853	选择	selecting	
07.0854	主控站	primary station	
07.0855	恢复规程	recovery procedure	
07.0856	请求数据传送	request data transfer	
07.0857	差错漏检率	residual error-rate	
07.0858	受控站	secondary station	
07.0859	信号质量检测	signal quality detection	

序 码	汉 文 名	英 文 名	注 释
07.0860	码组起始信号	start-of-block signal	
07.0861	报头开始信号	start-of-heading signal	
07.0862	正文开始信号	start-of-text signal	
07.0863	起止信号	start-stop signal	
07.0864	存储转发	store and forward	
07.0865	发送信道	transmit channel	
07.0866	接收信道	receive channel	
07.0867	流量控制	flow control	
07.0868	窗口	window	
07.0869	抖动	jitter	
07.0870	阻塞	congestion	
07.0871	未检出差错	undetected error	
07.0872	宽带信道	wide-band channel	
07.0873	虚呼叫	virtual call	
07.0874	缩位地址	abbreviated address	
07.0875	误比特率	bit-error rate	又称"误码率"。
07.0876	误字率	character-error rate	
07.0877	误组率	block-error rate	
07.0878	中间设备	intermediate equipment	
07.0879	逻辑信道	logical channel	
07.0880	重定向	redirection	
07.0881	协议	protocol	
07.0882	交互式协议	interactive protocol	
07.0883	面向字符协议	character-oriented protocol	
07.0884	面向比特协议	bit-oriented protocol	
07.0885	协议规范	protocol specification	
07.0886	链路控制规程	link control procedure	
07.0887	基本型链路控制规程	basic link control procedure	
07.0888	通信控制字符	communication control character	
07.0889	重传	retransmission	
07.0890	一致性测试	conformance testing	
07.0891	前置机	front-end processor	
07.0892	通信处理机	communication processor	
07.0893	通信控制器	communication controller	
07.0894	通信接口	communication interface	
07.0895	包装/拆器	packet assembler/disassembler,	又称"分组装拆器"。

序　码	汉　文　名	英　文　名	注　　释
		PAD	
07.0896	包式终端	packet mode terminal	又称"分组式终端"。
07.0897	帧格式	frame format	
07.0898	定界符	delimiter	
07.0899	帧首定界符	frame start delimiter	
07.0900	地址	address	
07.0901	源点地址	source address	
07.0902	终点地址	destination address	
07.0903	连接方式	connection mode	
07.0904	无连接方式	connectionless mode	
07.0905	始发者	originator	
07.0906	接受者	receptor	
07.0907	本地终端	local terminal	
07.0908	远程终端	remote terminal	
07.0909	拨号终端	dial-up terminal	
07.0910	虚拟终端	virtual terminal	

08. 铁 道 信 号

序　码	汉　文　名	英　文　名	注　　释
08.0001	铁道信号	railway signaling, railroad signaling	又称"铁路信号"。
08.0002	区间信号	wayside signaling	
08.0003	车站信号	signaling at stations	
08.0004	信号	signal	
08.0005	视觉信号	visual signal	
08.0006	听觉信号	audible signal	
08.0007	昼间信号	day signal	
08.0008	夜间信号	night signal	
08.0009	昼夜通用信号	signal for day and night	
08.0010	固定信号	fixed signal	
08.0011	移动信号	movable signal	
08.0012	机车信号	cab signal	又称"机车自动信号"。
08.0013	地面信号	ground signal	
08.0014	手信号	hand signal	

序　码	汉　文　名	英　文　名	注　释
08.0015	闪光信号	flashing signal	
08.0016	速差制信号	speed signaling	
08.0017	选路制信号	route signaling	
08.0018	行车信号	train signal	
08.0019	调车信号	shunting signal	
08.0020	进行信号	proceed signal	又称"允许信号"。
08.0021	注意信号	caution signal	
08.0022	减速信号	restriction signal	
08.0023	停车信号	stop signal	
08.0024	绝对信号	absolute signal	
08.0025	容许信号	permissive signal	
08.0026	接车信号	receiving signal	
08.0027	发车信号	departure signal	
08.0028	通过信号	through signal	
08.0029	引导信号	calling-on signal	
08.0030	预告信号	approaching signal	
08.0031	复示信号	repeating signal	
08.0032	遮断信号	obstruction signal	
08.0033	驼峰信号	humping signal	
08.0034	推送信号	start humping signal	
08.0035	允许预推信号	permissive prehumping signal	
08.0036	加速推送信号	humping fast signal	
08.0037	减速推送信号	humping slow signal	
08.0038	下峰信号	down hump trimming signal	
08.0039	后退信号	backing signal	
08.0040	去禁溜线信号	shunting signal to prohibitive humping line	
08.0041	最大限制信号	most restrictive signal	
08.0042	较大限制信号	more restrictive signal	
08.0043	最大允许信号	most favorable signal	
08.0044	较大允许信号	more favorable signal	
08.0045	敌对信号	conflicting signal	
08.0046	信号显示	signal aspect and indication	
08.0047	错误显示	wrong indication	
08.0048	乱显示	false indication	
08.0049	显示距离	range of a signal	
08.0050	信号开放	signal at clear	

序　码	汉　文　名	英　文　名	注　释
08.0051	信号关闭	signal at stop	
08.0052	信号机	signal	
08.0053	色灯信号机	color-light signal	
08.0054	透镜式色灯信号机	multi-lenses signal	又称"多灯信号机"。
08.0055	探照式色灯信号机	searchlight signal	
08.0056	臂板信号机	semaphore signal	
08.0057	机械臂板信号机	mechanically operated semaphore signal	
08.0058	电动臂板信号机	electric semaphore signal	
08.0059	单线臂板信号机	single wire semaphore signal	
08.0060	双线臂板信号机	double wire semaphore signal	
08.0061	高柱信号机	high signal	
08.0062	矮型信号机	dwarf signal	
08.0063	信号托架	signal bracket	
08.0064	信号桥	signal bridge	
08.0065	进站信号机	home signal	
08.0066	出站信号机	starting signal	又称"出发信号机"。
08.0067	总出站信号机	advance starting signal	又称"总出发信号机"。
08.0068	线群出站信号机	group starting signal	又称"线群出发信号机"。
08.0069	进路信号机	route signal	
08.0070	接车进路信号机	route signal for receiving	
08.0071	发车进路信号机	route signal for departure	
08.0072	接发车进路信号	route signal for receiving-departure	
08.0073	通过信号机	block signal	
08.0074	主体信号机	main signal	
08.0075	从属信号机	dependent signal	
08.0076	预告信号机	distant signal	
08.0077	行车信号机	train signal	
08.0078	调车信号机	shunting signal	
08.0079	驼峰信号机	hump signal	
08.0080	驼峰复示信号机	humping signal repeater	
08.0081	遮断信号机	obstruction signal	
08.0082	遮断预告信号机	approach obstruction signal	

序 码	汉 文 名	英 文 名	注 释
08.0083	复示信号机	repeating signal	
08.0084	双面调车信号机	signal for shunting forward and backward	
08.0085	咽喉信号机	signal in throat section	
08.0086	尽头信号机	signal for stub-end track	
08.0087	组合式信号机构	modular type signal mechanism	
08.0088	表示器	indicator	
08.0089	进路表示器	route indicator	
08.0090	发车表示器	departure indicator	
08.0091	调车表示器	shunting indicator	
08.0092	发车线路表示器	departure track indicator	
08.0093	标志	sign, marker	
08.0094	道岔表示器	switch indicator	
08.0095	脱轨表示器	derail indicator	
08.0096	线路遮断表示器	track obstruction indicator	
08.0097	水鹤表示器	water crane indicator	
08.0098	车档表示器	buffer stop indicator	
08.0099	带柄道岔表示器	switch indicator with level	
08.0100	站界标	station limit sign	
08.0101	预告标	warning signs for approaching a station	
08.0102	信号无效标	signal out of order sign	
08.0103	联锁	interlocking	
08.0104	联锁设备	interlocking equipment	
08.0105	集中联锁	centralized interlocking	
08.0106	机械集中联锁	mechanical interlocking	
08.0107	电机集中联锁	electro-mechanical interlocking	
08.0108	电气集中[联锁]	electric interlocking	
08.0109	继电式电气集中联锁	all-relay interlocking	
08.0110	单独操纵继电式电气集中联锁	individual level type all-relay interlocking	
08.0111	进路继电式电气集中联锁	route type all-relay interlocking	
08.0112	微机－继电式电气集中联锁	microcomputer-relay interlocking	
08.0113	微机联锁	microcomputer interlocking	

序　码	汉　文　名	英　文　名	注　释
08.0114	大站电气集中联锁	relay interlocking for large station	
08.0115	小站电气集中联锁	relay interlocking for small station	
08.0116	调车区电气集中联锁	interlocking for shunting area	
08.0117	组合式电气集中联锁	unit-block type relay interlocking	
08.0118	组匣式电气集中联锁	modular type relay interlocking	
08.0119	非集中联锁	non-centralized interlocking	
08.0120	联锁箱联锁	interlocking by point detector	
08.0121	电锁器联锁	interlocking by electric locks	
08.0122	色灯电锁器联锁	interlocking by electric locks with color light-signals	
08.0123	臂板电锁器联锁	interlocking by electric locks with semaphore	
08.0124	电动臂板电锁器联锁	interlocking by electric locks with electric semaphore	
08.0125	自动点灯	automatic lighting	
08.0126	锁闭	locking	
08.0127	进路锁闭	route locking	
08.0128	接近锁闭	approach locking	
08.0129	照查锁闭	check locking	
08.0130	区段锁闭	section locking	
08.0131	道岔锁闭	switch [point] locking	
08.0132	道岔密贴	switch [point] closure	
08.0133	4 毫米锁闭	check 4mm opening of a switch point	
08.0134	防止重复	prevention for repetitive clear of a signal	
08.0135	电气锁闭	electric locking	
08.0136	机械锁闭	mechanical locking	
08.0137	定位锁闭	normal locking	
08.0138	反位锁闭	reverse locking	
08.0139	定反位锁闭	normal and reverse locking	
08.0140	解锁	release	

序 码	汉 文 名	英 文 名	注 释
08.0141	进路解锁	route release	
08.0142	进路一次解锁	route release at once	
08.0143	进路分段解锁	sectional release of a locked route	
08.0144	人工解锁	manual release	
08.0145	进路人工解锁	manual route release	
08.0146	道岔人工解锁	manual release of a locked switch	
08.0147	限时人工解锁	manual time release	
08.0148	不限时人工解锁	manual non-time release	
08.0149	自动限时解锁	automatic time release	
08.0150	两点检查	released by checking two sections	
08.0151	三点检查	released by checking three sections	
08.0152	四点检查	released by checking four sections	
08.0153	基本进路	basic route	
08.0154	迂回进路	detour route, alternative route	
08.0155	对向重叠进路	route with overlapped section in the opposite direction	
08.0156	顺向重叠进路	route with overlapped section in the same direction	
08.0157	延续进路	succesisve route	
08.0158	表示	indication	
08.0159	接近表示	approach indication	
08.0160	离去表示	departure indication	
08.0161	道岔表示	switch indication	
08.0162	信号表示	signal indication	
08.0163	信号开放表示	cleared signal indication	
08.0164	信号关闭表示	stop signal indication	
08.0165	进路锁闭表示	route locking indication	
08.0166	人工解锁表示	manual release indication	
08.0167	按钮表示	button indication	
08.0168	道岔定位表示	switch normal indication	
08.0169	道岔反位表示	switch reverse indication	
08.0170	线路占用表示	track occupancy indication	
08.0171	区段占用表示	section occupancy indication	
08.0172	道岔锁闭表示	switch locked indication	
08.0173	光点式表示	spotted indication light	
08.0174	光带式表示	strip indication light	
08.0175	报警	alarm	

序　码	汉　文　名	英　文　名	注　　释
08.0176	主灯丝断丝报警	alarm for burnout of a main filament	
08.0177	灯丝断丝报警	alarm for burnout of filaments	
08.0178	熔断器断丝报警	fuse break alarm	
08.0179	接地报警	grounding alarm	
08.0180	挤岔报警	alarm for a trailed switch	
08.0181	道岔启动	switch starting	
08.0182	道岔转换	switch in transition	
08.0183	道岔顺序启动	sequential starting of switches	
08.0184	道岔顺序转换	sequential transiting of switches	
08.0185	局部控制	local control	
08.0186	双重控制	dual control	
08.0187	排列进路	route setting	
08.0188	预排进路	presetting of a route	
08.0189	取消进路	to cancel a route	
08.0190	解锁进路	released route	
08.0191	选路	route selection	
08.0192	非进路调车	to hold route for shunting	
08.0193	中途返回	midway return operation	又称"中途折返"。
08.0194	防护道岔	protective turnout	
08.0195	带动道岔	switch with follow up movement	
08.0196	联锁道岔	interlocked switch	
08.0197	非联锁道岔	non-interlocked switch	
08.0198	集中道岔	centrally operated switch	
08.0199	非集中道岔	locally operated switch	
08.0200	联锁区	interlocking area	
08.0201	非联锁区	non-interlocking area	
08.0202	道岔区段	section with a switch or switches	
08.0203	无岔区段	section without a switch	
08.0204	线路区段	track section	
08.0205	接近区段	approach section	
08.0206	第一接近区段	first approach section	
08.0207	第二接近区段	second approach section	
08.0208	离去区段	departure section	
08.0209	第一离去区段	first departure section	
08.0210	第二离去区段	second departure section	
08.0211	联锁图表	interlocking chart and table	

序　码	汉　文　名	英　文　名	注　释
08.0212	联锁表	interlocking table	
08.0213	进路表	route sheet	
08.0214	安全电路	vital circuit, safety circuit	
08.0215	非安全电路	non-vital circuit	
08.0216	励磁电路	energizing circuit	
08.0217	自闭电路	self-stick circuit	又称"自保电路"。
08.0218	站场型网路	geographical circuitry	
08.0219	继电并联网路	parallel relay network	
08.0220	继电串联网路	series relay network	
08.0221	继电并联传递网路	successively worked parallel relay network	
08.0222	单断	single break	
08.0223	双断	double break	
08.0224	重复检查	repeated checking	
08.0225	极性检查电路	polarity checking circuit	
08.0226	送受分开电路	sending and receiving separated circuit	
08.0227	灭火花电路	spark extinguishing circuit	
08.0228	基本联锁电路	fundamental interlocking circuit	
08.0229	进路电路	route selecting circuit	
08.0230	锁闭电路	locking circuit	
08.0231	解锁电路	release circuit	
08.0232	信号控制电路	signal control circuit	
08.0233	信号机点灯电路	signal lighting circuit	
08.0234	道岔控制电路	switch control circuit	
08.0235	表示电路	indication circuit	
08.0236	联系电路	liaison circuit	
08.0237	区间联系电路	liaison circuit with block signaling	
08.0238	自动闭塞联系电路	liaison circuit with automatic blocks	
08.0239	半自动闭塞联系电路	liaison circuit with semi-automatic blocks	
08.0240	站间联系电路	liaison circuit between stations	
08.0241	场间联系电路	liaison circuit between yards	
08.0242	机务段联系电路	liaison circuit with a locodepot	
08.0243	站内道口联系电路	liaison circuit with highway cros-sings within the station	

序　码	汉　文　名	英　文　名	注　释
08.0244	非进路调车电路	circuit to hold a route for shunting	
08.0245	局部控制电路	local control circuit	
08.0246	下坡道防护电路	protection circuit for approaching heavy down grade	
08.0247	到发线出岔电路	protection circuit with switch lying in receiving-departure track	
08.0248	进路表示器电路	route indicator circuit	
08.0249	发车表示器电路	departure indicator circuit	
08.0250	调车表示器电路	shunting indicator circuit	
08.0251	通过按钮电路	through button circuit	
08.0252	自动通过按钮电路	automatically through button circuit	
08.0253	控制台	control desk	
08.0254	操纵台	operating console	
08.0255	表示盘	indicating panel	
08.0256	控制盘	control panel	
08.0257	解锁按钮盘	manual release button panel	
08.0258	同意按钮盘	agreement button panel	
08.0259	组合	unit block	
08.0260	组匣	modular block	
08.0261	组合架	unit block assembly rack	
08.0262	综合架	composite rack	
08.0263	组合柜	modular block rack	
08.0264	组匣柜	modular block rack	
08.0265	分线盘	distributing terminal board	
08.0266	控制台单元	control desk element	
08.0267	手柄	handle, knob	
08.0268	拉钮	pull-out button	
08.0269	按钮	push-in button	
08.0270	自复式按钮	nonstick button	
08.0271	非自复式按钮	stick button	
08.0272	表示灯	indication lamp	
08.0273	信号复示器	signal repeater	
08.0274	光带	light strip	
08.0275	汇流条	bus-bar	
08.0276	分线盘端子	terminals on distributing board	
08.0277	电源端子	terminals for power supplies	

序 码	汉 文 名	英 文 名	注 释
08.0278	零层端子	terminals of layer 0 of a relay rack	
08.0279	组合端子	terminals of a unit block	
08.0280	组匣端子	terminals of a modular block	
08.0281	继电器箱	relay case	
08.0282	继电器防震架	shock absorber base for relays	
08.0283	变压器箱	transformer box	
08.0284	局部控制盘	local control panel	
08.0285	地中电缆盒	underground cable terminal box	
08.0286	杆上电缆盒	cable terminal box on a post	
08.0287	分向电缆盒	cable branching terminal box	
08.0288	中间电缆盒	intermediate cable terminal box	
08.0289	终端电缆盒	cable terminal box	
08.0290	联锁箱	point detector	
08.0291	道岔握柄	switch lever	
08.0292	信号握柄	signal lever	
08.0293	导管装置	pipe installation	
08.0294	导管调整器	pipe compensator	
08.0295	接杆	pipe jaw	
08.0296	连接杆	pipe link	
08.0297	转向杆	deflecting bar	
08.0298	直角拐肘	right angle crank	
08.0299	可调拐肘	adjustable crank	
08.0300	牵纵拐肘	escapement	
08.0301	转辙锁闭器	plunger lock	
08.0302	转换锁闭器	switch-and-lock mechanism	
08.0303	尖端杆	front rod of a point	
08.0304	密贴调整杆	adjustable switch operating rod	
08.0305	导线装置	wire installation	
08.0306	导线反正扣	wire-adjusting screw	
08.0307	导线调整器	wire compensator	
08.0308	导线平轮	horizontal wheel	
08.0309	导线立轮	vertical wheel	
08.0310	导线导轮	wire carrier	
08.0311	导线平轮组	horizontal wheel assembly	
08.0312	信号选别器	signal slot	
08.0313	轨道接触器	track treadle	
08.0314	电锁器	electric lock	

序 码	汉 文 名	英 文 名	注 释
08.0315	臂板接触器	contacts operated by semaphore	
08.0316	臂板转极器	pole changer operated by sema-phore	
08.0317	闭塞	block system	
08.0318	区间闭塞	section blocked	
08.0319	区间空闲	section cleared	
08.0320	区间占用	section occupied	
08.0321	电话闭塞	telephone block system	
08.0322	电气路签闭塞	electric staff system	
08.0323	电气路牌闭塞	electric tablet block system	
08.0324	半自动闭塞	semi-automatic block system	
08.0325	继电半自动闭塞	all-relay semi-automatic block system	
08.0326	单线继电半自动闭塞	single track all-relay semi-automatic block system	
08.0327	双线继电半自动闭塞	double track all-relay semi-automatic block system	
08.0328	自动闭塞	automatic block system	
08.0329	单向自动闭塞	single-directional running automatic block	
08.0330	双向自动闭塞	double-direction running automatic block	
08.0331	电码自动闭塞	automatic block with coded track circuit	
08.0332	脉冲自动闭塞	automatic block with impulse track circuit	
08.0333	移频自动闭塞	automatic block with audio frequency shift modulated track circuit	
08.0334	极频自动闭塞	automatic block with polar frequency impulse track circuit	
08.0335	交流计数电码自动闭塞	automatic block with AC counting code track circuit	
08.0336	计轴自动闭塞	automatic block with axle counter	
08.0337	二显示自动闭塞	two-aspect automatic block	
08.0338	三显示自动闭塞	three-aspect automatic block	
08.0339	四显示自动闭塞	four-aspect automatic block	

序　码	汉　文　名	英　文　名	注　释
08.0340	闭塞分区	block section	
08.0341	重叠区段	overlap section	
08.0342	保护区段	overlap protection block section	
08.0343	防护区段	protected section	
08.0344	接近发码	coding during train approaching	
08.0345	信号点	signal location	
08.0346	单置信号点	single signal location	
08.0347	并置信号点	double signal location	
08.0348	预办闭塞	preworking a block	
08.0349	解除闭塞	block cleared	
08.0350	取消闭塞	to cancel a block	
08.0351	灯光转移	to transfer of lighting indication	
08.0352	闭塞机	block instrument	
08.0353	电气路签机	electric staff instrument	
08.0354	电气路牌机	electric tablet instrument	
08.0355	半自动闭塞机	semi-automatic block machine	
08.0356	路签	train staff	
08.0357	路牌	tablet	
08.0358	路签携带器	staff pouch	
08.0359	路签自动授收机	automatic staff exchanger	
08.0360	路牌携带器	tablet pouch	
08.0361	路牌自动授收机	automatic tablet exchanger	
08.0362	钥匙路签	staff with a key	
08.0363	驼峰溜放控制系统	humping control system	
08.0364	自动化驼峰系统	automatic hump yard system	
08.0365	半自动化驼峰系统	semi-automatic hump yard system	
08.0366	机械化驼峰设备	mechanized hump yard equipment	
08.0367	非机械化驼峰设备	unmechanized hump yard equipment	
08.0368	错溜	miseroute	
08.0369	静态长度	car space	
08.0370	追钩	catch up	
08.0371	侧撞	conering	
08.0372	动态长度	distance-to-go	
08.0373	分路道岔	branching turnout	又称"分歧道岔"。

序 码	汉 文 名	英 文 名	注 释
08.0374	驼峰电气集中	electric interlocking for hump yard	
08.0375	单独操纵作业	manual operation	又称"手动作业"。
08.0376	进路操纵作业	semi-automatic operation by route	又称"半自动作业"。
08.0377	自动操纵作业	automatic operation	
08.0378	调车线始端减速器	tangent retarder	
08.0379	峰下减速器	master retarder	
08.0380	线束减速器	group retarder	
08.0381	减速器制动状态	retarder in closed state	
08.0382	减速器工作状态	retarder in working state	
08.0383	减速器缓解状态	retarder released	
08.0384	减速器接近限界	clearance of a retarder	
08.0385	溜放速度自动控制	automatic rolling down speed control	
08.0386	测速	speed measurement	
08.0387	测长	distance-to-coupling measurement	
08.0388	测重	weight sensing	
08.0389	测阻	rollability measurement	
08.0390	自动抄车号	automatic car identification	又称"车号自动识别"。
08.0391	驼峰机车遥控	remote control of hump engines	
08.0392	溜放进路自动控制	automatic switching control of humping yard by routes	又称"溜放进路程序控制"。
08.0393	推峰速度自动控制	automatic control for humping speed	
08.0394	目标打靶控制	target shooting	
08.0395	票据传送设备	classification list conveyer system	
08.0396	雷达测速器	radar speedometer	
08.0397	车辆减速器	car retarder	
08.0398	车辆加速器	car accelerator	
08.0399	推送小车辆	propelling trolley	
08.0400	储风罐	air reservoir	
08.0401	限界检查器	clearance treadle	
08.0402	水分离器	water separator	
08.0403	风压调整器	manometer regulator	
08.0404	风管路调压设备	air pipeline pressure governor	
08.0405	进路储存器	route storaging devices	

序 码	汉 文 名	英 文 名	注 释
08.0406	电空传送设备	electropneumatic conveyer	
08.0407	电动传送设备	electric motor operated conveyer	
08.0408	车辆存在监测器	presence monitor	
08.0409	机车信号设备	cab signaling equipment	
08.0410	点式机车信号	intermittent type cab signaling	
08.0411	连续式机车信号	continuous type cab signaling	
08.0412	接近连续式机车信号	approach continuous cab signaling	
08.0413	自动停车	automatic train stop	
08.0414	自动停车装置	automatic train stop equipment	
08.0415	感应式机车信号	inductive cab signaling	
08.0416	机车信号测试区段	cab signaling testing section	
08.0417	机车信号作用点	cab signaling inductor location	
08.0418	地面设备	wayside equipment	
08.0419	机车设备	locomotive equipment	
08.0420	接收线圈	receiving coil	
08.0421	机车感应器	locomotive inductor	
08.0422	地面感应器	wayside inductor	
08.0423	车轮检测器	wheel detector	
08.0424	共用箱	cab signal box	
08.0425	警惕手柄	acknowledgment lever	
08.0426	警惕按钮	acknowledgment button	
08.0427	方向转接器	directional switch	
08.0428	单频感应器	single frequency inductor	
08.0429	双频感应器	double frequency inductor	
08.0430	测试环线	test loop	
08.0431	列车运行控制系统	train operation control system	
08.0432	列车自动限速	automatic train speed restriction	
08.0433	列车自动运行	automatic train operation	
08.0434	定点停车	stopping a train at a target point	
08.0435	车门自动控制	automatic train door control	
08.0436	列车自动调速	automatic train speed regulation	
08.0437	机车接近通知	approaching announcing in cab	
08.0438	道口通知设备	highway level crossing announcing device	

序 码	汉 文 名	英 文 名	注 释
08.0439	道口自动信号	automatic level crossing signal	
08.0440	自动栏木	automatic operated barrier	
08.0441	道口室外音响器	highway level crossing out door audible device	
08.0442	道口接近区段	approach section of a highway level crossing	
08.0443	道口遮断信号	highway level crossing obstruction signal	
08.0444	道口信号机	highway level crossing signal	
08.0445	道口信号控制盘	highway level crossing signal control panel	
08.0446	切断音响按钮	button for cut-off an audible signal	
08.0447	道口闪光信号	highway level crossing flashing signal	
08.0448	道口遥信遥测设备	remote surveillance and telemetering for highway level crossing	
08.0449	桥梁遮断信号	bridge obstruction signal	
08.0450	桥梁通知设备	bridge announciating device	
08.0451	隧道遮断信号	tunnel obstruction signal	
08.0452	隧道通知设备	tunnel announciating device	
08.0453	遮断信号按钮	obstruction signal button	
08.0454	遥控	remote control	
08.0455	遥信	remote surveillance	
08.0456	遥测	telemetry	
08.0457	遥调	remote regulation	
08.0458	区段遥控	remote control for a section	
08.0459	枢纽遥控	remote control of a junction terminal	
08.0460	调度集中	centralized traffic control, CTC	
08.0461	调度集中总机	control office equipment of CTC	
08.0462	调度集中分机	field equipment of CTC	
08.0463	中继站	repeater station	
08.0464	线路点	field location	
08.0465	遥控区段	remotely controlled section	
08.0466	控制对象	controlled object	
08.0467	表示对象	indicated object	
08.0468	调度控制	dispatcher's control	

序　码	汉　文　名	英　文　名	注　　释
08.0469	中心控制	centralized control	
08.0470	车站控制	station master control	
08.0471	调度监督	dispatcher's supervision system	指调度表示系统。
08.0472	遥信区段	remotely surveillanced section	
08.0473	监督对象	surveillanced object	
08.0474	循环检查制	cyclic scanning system	
08.0475	近程监督分区	directly surveillanced subsection	又称"近程遥信分区"。
08.0476	远程监督分区	relayed surveillanced subsection	又称"远程遥信分区"。
08.0477	近程网路	directly surveillanced network	
08.0478	远程网路	relayed surveillanced network	
08.0479	控制点	controlling point	
08.0480	被控点	controlled point	
08.0481	控制周期	control cycle	
08.0482	表示周期	indication cycle	
08.0483	列车位置表示	train position indication	
08.0484	车次表示	train number indication	
08.0485	占线表示	occupancy indication	
08.0486	晚点表示	delaying time indication	
08.0487	行车记录设备	train movement recording equipment	
08.0488	运行图描绘仪	train diagram plotter	
08.0489	轨道电路	track circuit	
08.0490	直流轨道电路	DC track circuit	
08.0491	交流轨道电路	AC track circuit	
08.0492	阀式轨道电路	valve type track circuit	
08.0493	连续式轨道电路	continuous track circuit	
08.0494	脉冲式轨道电路	pulse track circuit	
08.0495	电码轨道电路	coded track circuit	
08.0496	叠加轨道电路	overlap track circuit	
08.0497	极频轨道电路	polar-frequency pulse track circuit	
08.0498	无绝缘轨道电路	jointless track circuit	
08.0499	调频轨道电路	frequency modulated track circuit	
08.0500	移频轨道电路	frequency-shift modulated track circuit	
08.0501	不对称脉冲轨道	asymmetrical impulse track circuit	

序 码	汉 文 名	英 文 名	注 释
	电路		
08.0502	交流计数电码轨道电路	AC counting coded track circuit	
08.0503	相敏轨道电路	phase detecting track circuit	
08.0504	开路式轨道电路	open type track circuit	
08.0505	闭路式轨道电路	close type track circuit	
08.0506	单轨条式轨道电路	single rail track circuit	
08.0507	双轨条式轨道电路	double rail track circuit	
08.0508	轨道电路电码化	coding of continuous track circuit	
08.0509	股道空闲	track clear	
08.0510	股道占用	track occupied	
08.0511	分路效应	shunting effect	
08.0512	轨道电路分路状态	shunted state of a track circuit	
08.0513	轨道电路调整状态	regulated state of a track circuit	
08.0514	送电端	feed end	
08.0515	受电端	receiving end	
08.0516	一送多受	single feeding and multiple receiving track circuit	
08.0517	道碴电阻	ballast resistance	又称"道碴漏泄电阻"。
08.0518	钢轨阻抗	rail impedance	
08.0519	分路灵敏度	shunting sensitivity	又称"分流感度"。
08.0520	标准分路灵敏度	standard shunting sensitivity	又称"标准分流感度"。
08.0521	断轨保障	broken rail protection	
08.0522	串联式轨道电路	serially connected track circuit	
08.0523	并联式轨道电路	multiply connected track circuit	
08.0524	轨道电路蓄电现象	track storage effect	
08.0525	轨道生电现象	track galvanic effect	
08.0526	分路	shunt	
08.0527	人工分路	manual shunt	
08.0528	瞬时分路	instantaneous shunt	

序　码	汉　文　名	英　文　名	注　释
08.0529	死区段	dead section	
08.0530	轨道电路分割	cut-section of a track circuit	
08.0531	分割区段	cut section	
08.0532	极性交叉	polar transposition	
08.0533	一次参数	primary parameter	
08.0534	二次参数	secondary parameter	
08.0535	钢轨绝缘	rail insulation	
08.0536	岔中绝缘	insulated joint within a turnout	
08.0537	侵入限界绝缘	insulated joint located within the clearance limit	
08.0538	钢轨引接线	track lead	
08.0539	跳线	jumper	
08.0540	钢轨接续线	rail bond	
08.0541	塞钉式钢轨接续线	plug bond	
08.0542	焊接式钢轨接续线	welded bond	
08.0543	轨道送电变压器	track transformer feed end	
08.0544	轨道受电变压器	track relay transformer	
08.0545	轨道电抗器	track reactor	
08.0546	轨道变阻器	track rheostat	
08.0547	防护变压器	protective transformer	
08.0548	扼流变压器	impedance transformer	
08.0549	轨道变压器箱	track transformer box	
08.0550	继电器	relay	
08.0551	动接点	contact heel, movable contact	
08.0552	前接点	front contact	
08.0553	后接点	back contact	
08.0554	定位接点	normal contact	
08.0555	反位接点	reverse contact	
08.0556	接点闭合	contact closed	
08.0557	接点断开	contact open	
08.0558	继电器吸起	relay energized	又称"继电器励磁"。
08.0559	继电器释放	relay released	又称"继电器失磁"。
08.0560	吸起时间	pick-up time	
08.0561	释放时间	drop away time	又称"落下时间"。
08.0562	转换时间	transfer time	

序 码	汉 文 名	英 文 名	注 释
08.0563	转极时间	pole-changing time	
08.0564	缓吸时间	slow pick-up time	
08.0565	缓放时间	slow release time	
08.0566	吸起值	pick-up value	
08.0567	工作值	working value	又称"完全吸起值"。
08.0568	额定值	rated value	
08.0569	释放值	release value	又称"落下值"。
08.0570	转极值	pole-changing value	
08.0571	返还系数	release factor	
08.0572	继电器灵敏度	relay sensitivity	
08.0573	接点压力	contact pressure	
08.0574	磁路［系统］	magnetic circuit	
08.0575	接点系统	contact system	
08.0576	前圈	front coil	
08.0577	后圈	back coil	
08.0578	电磁继电器	electromagnetic relay	
08.0579	电码继电器	code relay	
08.0580	直流继电器	DC relay	
08.0581	交流继电器	AC relay	
08.0582	交直流继电器	AC-DC relay	
08.0583	脉冲继电器	impulse relay	
08.0584	传输继电器	transmitting relay, transmission relay	
08.0585	座式继电器	shelf-type relay	
08.0586	插入式继电器	plug-in type relay	
08.0587	重力继电器	gravitation type relay	
08.0588	弹力继电器	spring-type relay	
08.0589	热［力］继电器	thermal relay	
08.0590	整流继电器	rectifier relay	
08.0591	无极继电器	neutral relay	
08.0592	有极继电器	polarized relay	
08.0593	偏极继电器	polar biased relay	
08.0594	组合继电器	combination relay	
08.0595	交流二元二位继电器	AC two element two position relay	
08.0596	缓动继电器	slow-acting relay	
08.0597	缓吸继电器	slow pick-up relay	

序　码	汉　文　名	英　文　名	注　释
08.0598	缓放继电器	slow release relay	
08.0599	快动继电器	quick-acting relay	
08.0600	快吸继电器	quick pick-up relay	
08.0601	正常动作继电器	normal acting relay	
08.0602	转辙机	switch machine	
08.0603	电动转辙机	electric switch machine	
08.0604	电空转辙机	electropneumatic switch machine	
08.0605	电液转辙机	electrohydraulic switch machine	
08.0606	大功率转辙机	heavy duty switch machine	
08.0607	快速转辙机	quick-acting switch machine	
08.0608	传动系统	driving system	
08.0609	锁闭系统	locking system	
08.0610	[自动]开闭器	switch circuit controller	
08.0611	摩擦联结器	frictional clutch	
08.0612	摩擦电流	frictional working current	又称"故障电流"。
08.0613	动作杆	throw rod	
08.0614	表示杆	indication rod	
08.0615	锁闭杆	locking rod	
08.0616	正装	right-handed machine	
08.0617	反装	left-handed machine	
08.0618	动程	stroke	
08.0619	挤脱	trailable	
08.0620	挤切	dissectible	
08.0621	挤切销	dissectible pin	
08.0622	牵引力	tractive force	
08.0623	工作电流	working current	
08.0624	锁闭力	locking force	
08.0625	解锁力	releasing force	
08.0626	动作连接杆	operating rod for driving a switch	
08.0627	表示连接杆	connecting rod for indication	
08.0628	安全接点	safety contact, power off contact	
08.0629	转辙机安装装置	switch machine installation	
08.0630	直流供电制	DC power supply system	
08.0631	交流供电制	AC power supply system	
08.0632	混合供电制	AC-battery power supply system	
08.0633	浮充供电	floating charge power supply	
08.0634	主电源	main power source	

序 码	汉 文 名	英 文 名	注 释
08.0635	备电源	stand-by power source	
08.0636	集中供电	centrally connected power supply	
08.0637	干线供电	main linely connected power supply	
08.0638	环状供电	looply connected power supply	
08.0639	一次电池供电	primary cell power supply	
08.0640	蓄电池供电	storage battery power supply	
08.0641	整流器供电	rectifier power supply	
08.0642	混合电源	AC-battery power source	
08.0643	条件电源	conditional power source	
08.0644	方向电源	directional traffic power source	
08.0645	集中电源	centrally connected power source	
08.0646	局部电源	locally supplied power source	
08.0647	信号机点灯电源	signal lighting power source	
08.0648	事故照明	emergency lighting	
08.0649	自动转接	automatic switching over	又称"自动切换"。
08.0650	电源屏	power supply panel	
08.0651	交流电源屏	AC power supply panel	
08.0652	直流电源屏	DC power supply panel	
08.0653	条件电源屏	conditional power supply panel	
08.0654	电源转换屏	power switching over panel	
08.0655	电压自动调整器	automatic voltage regulator	
08.0656	自动调压	automatic voltage regulation	
08.0657	手动调压	manual voltage regulation	
08.0658	道岔控制电源	power source for switch control	
08.0659	道岔动作电源	power source for switch operation	
08.0660	道岔表示电源	power source for switch indication	
08.0661	继电器控制电源	power source for relay control	
08.0662	表示灯电源	power source for indication lamp	
08.0663	闪光电源	flashing power source	
08.0664	熔断器	fuse	
08.0665	避雷器	lightning arrester	
08.0666	故障－安全	fail-safe	
08.0667	可靠性	reliability	
08.0668	安全性	safety	
08.0669	错误办理	wrong handling	
08.0670	导向安全	failure to the safe side	

序 码	汉 文 名	英 文 名	注 释
08.0671	瞬时分路不良	instantaneous loss of shunting	
08.0672	邻线干扰	interference from neighboring line	
08.0673	强电干扰	high voltage interference	
08.0674	谐波干扰	harmonic interference	
08.0675	内燃牵引干扰	diesel traction interference	
08.0676	雷电干扰	lightning interference	
08.0677	错误锁闭	false locking	
08.0678	错误解锁	false release	
08.0679	错误开放信号	wrong clearing of a signal	
08.0680	错误关闭信号	false stopping of a signal	
08.0681	道岔失去表示	loss of indication of a switch	
08.0682	道岔错误表示	false indication of a switch	
08.0683	自动缓解	false release by itself	
08.0684	漏锁闭	missing locking	
08.0685	漏解锁	missing release	
08.0686	失去联锁	loss of interlocking	
08.0687	停电	power failure	
08.0688	电压过低	voltage below level	
08.0689	灯丝断丝	filament burn-out	
08.0690	熔断器断丝	fuse burn-out	
08.0691	插接不良	plug-in trouble	
08.0692	线头脱落	wire lead drop out	
08.0693	维修不良	not well maintained	
08.0694	检修不良	not well inspected and repaired	
08.0695	施工妨碍	construction interference	
08.0696	材质不良	bad material	
08.0697	绝缘不良	bad insulation	
08.0698	钢轨绝缘不良	bad rail insulation	
08.0699	接触不良	bad contact	
08.0700	道岔中途转换	switch thrown under moving cars	
08.0701	人为故障	human failure	
08.0702	故障办理	emergency treatment after failure	
08.0703	故障复原	restoration after a failure	
08.0704	故障积累	failure accumulation	
08.0705	故障升级	progression of failure	
08.0706	封锁	close up	
08.0707	区间封锁	section closed up	

序　码	汉 文 名	英 文 名	注　释
08.0708	道岔封锁	switch closed up	
08.0709	设备停用	equipment out-of use	
08.0710	开通	put into operation	
08.0711	加封	sealing	
08.0712	破封	break a seal	
08.0713	信号机前方	in advance of a signal	又称"信号机外方"。
08.0714	信号机后方	in rear of a signal	又称"信号机内方"。
08.0715	继电器室	relay room	
08.0716	电源室	power supply room	
08.0717	蓄电池室	battery room	
08.0718	储酸室	acid store room	
08.0719	控制台室	control room	
08.0720	空气压缩机室	air compressor room	
08.0721	驼峰机械修理室	hump mechanics repair room	
08.0722	油压动力室	hydraulic pressure engine room	
08.0723	信号维修	signal maintenance	
08.0724	定期维修	periodical maintenance	
08.0725	日常维修	current maintenance	
08.0726	预防维修	preventive maintenance	
08.0727	故障修	corrective maintenance	
08.0728	计划修	planned maintenance	
08.0729	在线修	on line maintenance	
08.0730	脱线修	off line maintenance	
08.0731	轮修	alternative maintenance	
08.0732	信号集中修	signal centralized maintenance	
08.0733	信号检修	signal inspection, signal check-out maintenance	
08.0734	信号整治	signal renovation	
08.0735	检测	check-out, inspection and measurement	
08.0736	信号中修	signal intermediate repair	
08.0737	信号大修	signal overhaul repair, signal major repair	
08.0738	信号故障	signal fault	
08.0739	误用故障	misuse fault	
08.0740	妨害故障	hindrance fault	

英 汉 索 引

A

abandoned blasting　扬弃爆破　02.0275

abandoned construction work　废弃工程　02.0841

abandoned project　废弃工程　02.0841

abandonment blasting　扬弃爆破　02.0275

abbreviated address　缩位地址　07.0874

absolute acceleration feedback　绝对加速度反馈
　04.1607

absolute signal　绝对信号　08.0024

abutment　桥台　02.0547

abutment body　台身　02.0553

abutment coping　台帽　02.0555

abutment with cantilevered retaining wall　耳墙式桥
　台　02.0552

AC-battery power source　混合电源　08.0642

AC-battery power supply system　混合供电制
　08.0632

accelerated application valve portion　加速制动阀部
　05.0295

accelerating grade　加速缓坡　02.0199, 加速坡
　06.0495

acceleration　加速　04.0106

acceleration feedback　加速度反馈　04.1606

acceleration force　加速力　04.0109

acceptance and delivery list of car for international
　through traffic　国际联运车辆交接单　06.0192

acceptance and delivery list of freight for international
　through traffic　国际联运货物交接单　06.0191

acceptance following construction　随工验收
　02.1052

acceptance of engineering quality　工程质量验收
　02.1051

acceptance of freight　货物承运　06.0130

acceptance of locomotive　机车验收　04.0278

acceptance of luggages and parcels　行李包裹承运
　06.0052

access　访问，＊接入　07.0749

accident analysis　事故分析　06.0615

accident averting　事故预防　06.0655

accident disposal　事故处理　06.0609

accident forecast　事故预测　06.0612

accident indemnity　事故赔偿　06.0613

accident information management　事故信息管理
　06.0614

accident in shunting operation　调车事故　06.0628

accident investigation　事故调查　06.0611

accident on duty　工伤事故　06.0649

accident rate　事故率　06.0653

accident record　事故记录　06.0610

accident report　事故报告　02.0989

accident rescue　事故救援　06.0663

accident signal　事故信号　06.0657

accident threat　事故隐患　06.0626

AC contactor　交流接触器　04.0547

AC counting coded track circuit　交流计数电码轨道
　电路　08.0502

accumulator　蓄电池　05.0707

accumulator box　蓄电池箱　05.0708

accurate traverse survey　精密导线测量　02.0140

AC-DC-AC drive　交－直－交流传动　04.0343

AC-DC drive　交－直流传动　04.0342

AC-DC power supply for car　车辆交直流供电
　05.0717

AC-DC relay　交直流继电器　08.0582

acid store room　储酸室　08.0718

ACK　确认　07.0808

acknowledge　确认　07.0808

acknowledgment button　警惕按钮　08.0426

acknowledgment lever　警惕手柄　08.0425

AC power supply panel　交流电源屏　08.0651

AC power supply system　交流供电制　08.0631

AC relay　交流继电器　08.0581

active degrade voltage apparatus　有源降压装置

04.1501

AC track circuit 交流轨道电路 08.0491

AC traction motor 交流牵引电动机 04.0397

AC traction substation 交流牵引变电所 04.1344

actual braking time 实制动时间 05.0345

actual length of turnout 道岔实际长度 03.0327

actual point of frog 辙叉心轨尖端 03.0321

actual point of switch rail 尖轨尖端 03.0317

actual stopping distance 实制动距离 05.0347

actual tons of handling 装卸自然吨 06.0587

AC two element two position relay 交流二元二位继
电器 08.0595

acute frog 锐角辙叉 03.0352

adapter 承载鞍 05.0147

adaptive control 自适应控制 04.1594

additional passenger train 临时旅客列车 06.0212

additional resistance 附加阻力 04.0099

additional stake 加桩 02.0114

additive wire 附加导线 04.1366

address 地址 07.0900

adhesion braking 粘着制动 05.0323

adhesion coefficient 粘着系数 01.0067

adhesive tractive effort 粘着牵引力 04.0093

adhesive weight utility factor 粘着重量利用系数
04.0048

adjacent channel interference 邻道干扰 07.0530

adjacent curves in one direction 同向曲线 02.0177

adjustable crank 可调拐肘 08.0299

adjustable piston valve 分动式汽阀，＊特氏阀
04.1203

adjustable switch operating rod 密贴调整杆
08.0304

adjusted sum of approximate estimate 调整总概算
02.0024

adjusting joint gaps up to standard 整正轨缝
03.0209

adjusting of cross level 整正水平 03.0204

adjusting of rail gaps 调整轨缝，＊均匀轨缝
03.0208

adjustment of car flow 车流调整 06.0338

adjustment of car loading 装车调整 06.0339

adjustment of empty cars 空车调整 06.0340

admittance bridge 导纳电桥 07.0222

adobe blasting 裸露药包爆破 02.0279

advance anchor bolt 超前锚杆 02.0773

advance angle of fuel supply 供油提前角
04.0780

advance heading 超前导坑 02.0743

advance starting signal 总出站信号机，＊总出发信
号机 08.0067

advance support 超前支护 02.0772

"A" end of car 二位端 05.0009

aerial cable 架空电缆 07.0178

aerial earth wire 架空地线 07.0212

aerial ropeway 空中索道 02.0900

aerial surveying alignment 航测选线 02.0135

aerodynamic effects of pantograph 受电弓空气动力
效应 04.1446

after burning period 后燃期 04.0784

agreement button panel 同意按钮盘 08.0258

agreement of frontier railway 国际铁路协定
06.0188

agreement of international railway through freight traffic 国际铁路货物联运协定 06.0189

agreement of international through traffic 国际联运
协定 01.0123

air and vacuum dual brake equipment 空气真空两用
制动装置 05.0268

air brake 空气制动 04.0113

air brake equipment 空气制动装置 05.0266

air compressor 空气压缩机，＊风泵 04.0161

air compressor motor 空压机电动机 04.0474

air compressor room 空气压缩机室 08.0720

air conditioned coach 空调客车 05.0720

air conditioned passenger car 空调客车 05.0720

air conditioning equipment 空气调节装置
05.0626

air consumption 空气消耗量 04.0770

air curtain method for sinking caisson 空气幕沉井法
02.0590

air-cushion 气垫 04.1612

air delivery duct 送风道 05.0639

air dryer 空气干燥器 04.0176

air exhausting duct 排风道 05.0641

air/fuel ratio 空燃比 04.0744

air-gap 悬距 04.1615

air gap flux density 气隙磁通密度 04.1613

air hammer drill 凿岩机 02.0890

air horn 风喇叭 04.0650

air inlet duct 进风道 05.0638

air inlet pipe 进气管 04.0981

air inlet pressure stabilizing chamber 进气稳压箱 04.0982

air inlet valve 进气阀 05.0478

air/intermodal container 空陆水联运集装箱 06.0553

air pick 风镐 02.0961

air pipeline pressure governor 风管路调压设备 08.0404

air port 风口 05.0642

air precooler 空气预冷装置 05.0643

air preheater 空气预热装置 04.1164

air pressure relay 风压继电器 04.0602

air reservoir 储风罐 08.0400

air resistance 空气阻力 04.0102

air return duct 回风道 05.0640

air/surface container 空陆水联运集装箱 06.0553

air-tightness 气密性 05.0653

aisle [客室]通道 05.0788

alarm 报警 08.0175

alarm and protection system 报警保护系统 04.1534

alarm circuit 告警信号电路 07.0082

alarm for a trailed switch 挤岔报警 08.0180

alarm for burnout of a main filament 主灯丝断丝报警 08.0176

alarm for burnout of filaments 灯丝断丝报警 08.0177

alarm system 告警系统 07.0331

algebraic difference between adjacent gradients 坡度[代数]差 02.0194

alignment 定测 02.0101

alignment guiding line 导向线 02.0160

all metal coach 全金属客车 05.0719

all metal passenger car 全金属客车 05.0719

allocated locomotive 配属机车 04.0019

allocation factor of service track 整备线配置系数 05.0966

allowable load on journals of same axle 允许轴载荷 05.0232

allowable scour 容许冲刷 02.0095

allowable stress design method 容许应力设计法 02.0205

alloy steel rail 合金轨 03.0052

all-relay interlocking 继电式电气集中联锁 08.0109

all-relay semi-automatic block system 继电半自动闭塞 08.0325

all welded steel bridge 全焊钢桥 02.0365

altered design 变更设计 02.0018

alternate joint 相错式接头，＊相互式接头 03.0067

alternating current commutator motor 交流换向器电动机 04.0410

alternative maintenance 轮修 08.0731

alternative route 迂回进路 08.0154

alternative trunking 迂回中继 07.0265

[alumino-]thermit welding 铝热焊 03.0240

ambient noise side tone 环境噪声侧音 07.0146

ambient temperature 环境温度 04.0846

amended sum of approximate estimate 修正总概算 02.0023

amplifier 放大器 07.0052

amplitude modulation 调幅 07.0022

analog microwave relay communication 模拟微波中继通信 07.0664

analogue test 模拟试验 04.0298

anchorage 锚碇 02.0498

anchor bolt 钩螺栓 03.0084

anchor bolt support 锚杆支护 02.0769

anchored bolt retaining wall 锚杆挡墙 02.0289

anchored bulkhead retaining wall 锚定板挡土墙 02.0287

anchored plate retaining wall 锚定板挡土墙 02.0287

anchored retaining wall by tie rods 锚定板挡土墙 02.0287

anchored retaining wall by tie rods 锚杆挡墙 02.0289

anchor fitting 锚固装置 04.1428

anchor mast 锚柱 04.1412

anchor-section lining 下锚段衬砌，＊接触网锚段
　衬砌 02.0691

anchor slab abutment 锚定板式桥台 02.0550

anchor span 锚跨，＊锚孔 02.0441

angle cock 折角塞门 05.0282

angle of attack 冲角 05.0873

angle shut-off valve 直角截止阀 05.0651

angular pole 角杆 07.0169

annual railway plan 铁路年度计划 06.0675

anode 阳[电]极 04.1502

anti-creeper 防爬器 03.0100

anticreeping device 防串装置 04.1427

anti-creep strut 防爬支撑 03.0103

anti-sidetone device 消侧音器 07.0076

anti-skid device 防滑器 05.0320

antiskid shoe 防滑鞋 02.1026

anti-skid-type gallery 抗滑明洞 02.0699

anti-skid-type open cut tunnel 抗滑明洞
　02.0699

anti-skid-type tunnel without cover 抗滑明洞
　02.0699

anti-slide pile 抗滑桩 02.0292

anti-slip sanding valve 防空转撒砂电空阀
　04.0586

anti-slip/slide protection circuit 防空转防滑行保护
　电路 04.0389

anti-slip/slide protection device 防空转防滑行保护
　装置 04.0587

anti splitting device 木枕防裂装置 03.0222

aperture of culvert 涵洞孔径 02.0626

apparent earth conductivity 视在大地导电率
　04.1475

application valve 作用阀 05.0516

appraisal of design 设计鉴定 02.0034

approach continuous cab signaling 接近连续式机车
　信号 08.0412

approach indication 接近表示 08.0159

approaching announcing in cab 机车接近通知
　08.0437

approaching signal 预告信号 08.0030

approach length 接近长度 04.1484

approach locking 接近锁闭 08.0128

approach obstruction signal 遮断预告信号机
　08.0082

approach section 接近区段 08.0205

approach section of a highway level crossing 道口接
　近区段 08.0442

approach spans 引桥 02.0424

approximate estimate of design 设计概算
　02.0019

approximate railway location 铁路选线 02.0146

apron 裙板 05.0391

aproned fish plate 带裙鱼尾板 03.0081

apron shell 裙筒 04.1145

arch [隧道]拱圈 02.0677

arch brick 拱砖 04.1128

arch bridge 拱桥 02.0396

arch crown 拱顶 02.0484

arch culvert 拱涵 02.0629

archfirst lining method 先拱后墙法 02.0751

arch gallery 拱形明洞 02.0695

arch open cut tunnel 拱形明洞 02.0695

arch rib 拱肋 02.0483

arch ring [桥涵]拱圈 02.0482

arch tube 拱砖管 04.1129

arch tunnel without cover 拱形明洞 02.0695

armature coil 电枢线圈 04.0441

armoured cable 铠装电缆 07.0176

Arno converter 劈相机 04.0470

ARQ equipment 自动检错重发设备 07.0697

arrival and departure track 到发线 06.0405

arriving track 到达线 06.0406

arriving yard 到达场 06.0439

articulated coach 关节客车 05.0723

articulated freight car 关节货车 05.0838

articulated passenger car 关节客车 05.0723

articulation 清晰度 07.0149

artificial line 仿真线 07.0346

ascent of elevation 拔起高度，＊克服高度
　02.0161

ashpan 灰箱 04.1133

asphalt cemented ballast bed 沥青道床 03.0122

assembled frog 钢轨组合辙叉 03.0348

assembly pier 拼装式桥墩 02.0564

assembly type girder-erecting machine 组拼式架桥
　机 02.0950

assigning crew system of passenger train 旅客列车包乘制 06.0062

assistant driver 副司机 04.0202

assisting grade 加力牵引坡度 02.0183

assisting locomotive 辅机 04.0227

asymmetrical impulse track circuit 不对称脉冲轨道电路 08.0501

asymmetrical rail head profile grinding 轨头非对称断面打磨 03.0214

asymmetric half-controlled bridge rectifier 非对称半控桥式整流器 04.0335

asynchronous auxiliary motor 异步辅助电动机 04.0473

asynchronous communication 异步通信 07.0779

asynchronous motor 异步电动机 04.0409

asynchronous transmission 异步传输 07.0787

AT 自耦变压器 04.1319

AT-feeder 自耦变压器供电线 04.1311

atmospheric pressure steam heating equipment 大气压式采暖装置 05.0596

atomization 雾化 04.0776

atomizing cup 雾化杯 05.0612

atomizing wheel 雾化轮 05.0611

ATP 自耦变压器所 04.1348

attached type vibrator 附着式混凝土振捣器 02.0941

attached type vibrator for concrete 附着式混凝土振捣器 02.0941

attendant's room 列车员室 05.0766

attended repeater 有人增音机 07.0324

attenuation 衰减 07.0032

attenuation constant 衰减常数 07.0457

attraction injector 吸上式注水器 04.1166

attraction [lift] force 吸浮力 04.1595

audible signal 听觉信号 08.0006

audio communication circuit 音频通讯电路 04.0390

audio frequency 音频 07.0012

audio frequency section 音频转接段 07.0282

audio frequency terminating set 音频终端装置 07.0332

auger injected pile 螺旋喷射桩 02.0596

autocrane 汽车式起重机 02.0894

auto-leveling-lifting-lining-tamping machine 自动液压大型捣固车，＊自动整平－起道－拨道－捣固车 03.0418

automatic air brake equipment 自动空气制动装置 05.0270

automatic air control device 空气自动控制装置 05.0627

automatically through button circuit 自动通过按钮电路 08.0252

automatic answering 自动应答 07.0812

automatic block system 自动闭塞 08.0328

automatic block with AC counting code track circuit 交流计数电码自动闭塞 08.0335

automatic block with audio frequency shift modulated track circuit 移频自动闭塞 08.0333

automatic block with axle counter 计轴自动闭塞 08.0336

automatic block with coded track circuit 电码自动闭塞 08.0331

automatic block with impulse track circuit 脉冲自动闭塞 08.0332

automatic block with polar frequency impulse track circuit 极频自动闭塞 08.0334

automatic brake valve 自动制动阀，＊大闸 04.0167

automatic calling 自动呼叫 07.0813

automatic car identification 自动抄车号，＊车号自动识别 08.0390

automatic compensator 自动调整楔铁装置，＊自动楔铁调整器 04.1226

automatic control for humping speed 推峰速度自动控制 08.0393

automatic gain control 自动增益控制 07.0036

automatic hump 自动化驼峰 06.0458

automatic hump yard system 自动化驼峰系统 08.0364

automatic level crossing signal 道口自动信号 08.0439

automatic leveling device 自动高度调整装置 05.0180

automatic level regulating system 自动电平调节系统 07.0327

automatic lighting 自动点灯 08.0125

automatic locomotive operation 机车自动操纵 04.0237

automatic marshalling station 自动化编组站 06.0381

automatic operated barrier 自动栏木 08.0440

automatic operation 自动操纵作业 08.0377

automatic retransmission on request equipment 自动检错重发设备 07.0697

automatic roller bearing defect detection 滚动轴承故障自动检测 05.0961

automatic rolling down speed control 溜放速度自动控制 08.0385

automatic sanding 自动撒砂 04.0182

automatic slack adjuster 闸瓦间隙自动调节器 05.0360

automatic staff exchanger 路签自动授收机 08.0359

automatic switching control of humping yard by routes 溜放进路自动控制，*溜放进路程序控制 08.0392

automatic switching over 自动转接，*自动切换 08.0649

automatic tablet exchanger 路牌自动授收机 08.0361

automatic telephone set 自动电话机 07.0093

automatic time release 自动限时解锁 08.0149

automatic train door control 车门自动控制 08.0435

automatic train operation 列车自动运行 08.0433

automatic train speed regulation 列车自动调速 08.0436

automatic train speed restriction 列车自动限速 08.0432

automatic train stop 自动停车 08.0413

automatic train stop equipment 自动停车装置 08.0414

automatic voltage regulation 自动调压 08.0656

automatic voltage regulator 电压自动调整器 08.0655

automation of synthetic operations at freight station 货运站综合作业自动化 01.0110

automation of synthetic operations in marshalling yard 编组场综合作业自动化 01.0112

automation of traffic control 行车指挥自动化 01.0111

automobile crane 汽车式起重机 02.0894

autoregulation on load 有载自动调节 04.1302

autoregulation voltage compensator 电压自动补偿装置 04.1305

autostress rating 自动力矩检定 03.0520

auto-tensioned catenary equipment 全补偿链形悬挂 04.1358

autotransformer 自耦变压器 04.1319

autotransformer feeding system 自耦变压器供电方式，*AT 供电方式 04.1278

auto-transformer post 自耦变压器所 04.1348

auxiliary block post 辅助所 06.0361

auxiliary catenary 辅助承力索 04.1362

auxiliary circuit 辅助电路 04.0375

auxiliary circuit socket for shed supply 辅助电路库用插座 04.0606

auxiliary circuit transfer switch for shed supply 辅助电路库用转换开关 04.0565

auxiliary circuit voltmeter 辅助电路电压表 04.0623

auxiliary connecting rod 副连杆 04.0918

auxiliary contactor 辅助接触器 04.0554

auxiliary deflection angle 分转向角 02.0123

auxiliary facilities 辅助设施 02.0862

auxiliary generator 辅助发电机 04.0465

auxiliary girder for controlling angle change 梁端缓冲梁 02.0467

auxiliary intersection point 副交点 02.0121

auxiliary lever 制动缸后杠杆 05.0357

auxiliary lubricating oil pump 辅助机油泵 04.1039

auxiliary marshalling station 辅助编组站 06.0380

auxiliary motor 辅助电动机 04.0471

auxiliary reservoir 副风缸 05.0308

auxiliary running kilometers 辅助走行公里 04.0214

auxiliary winding 辅助绕组 04.0500

auxiliary yard 辅助车场 06.0444

average car-kilometers in one turnround 货车周转距离 06.0323

average car-kilometers per car-day 客车平均日车公

里 06.0071

average car-kilometers per transit operation 货车中
转距离 06.0324

average current of charging train 列车带电平均电
流 04.1323

average current of feeding section 供电臂平均电流
04.1324

average current of train 列车平均电流 04.1321

average daily locomotive running kilometers 机车日
车公里 04.0218

average daily output of locomotive 机车日产量
04.0220

average detector 平均值检波器 07.0572

average detention time of car in transit 中转车平均
停留时间 06.0265

average detention time of local car for loading or unloading 一次货物作业平均停留时间 06.0267

average gross weight hauled by locomotive 机车平均
牵引总重 04.0219

average haul of freight traffic 货物平均运程
06.0684

average journey per passenger 旅客平均运程
06.0679

average length of car 车辆平均长度 05.0970

average number of passengers carried per train 列车
平均载客人数 06.0072

average optical output power 平均输出光功率
07.0464

average retardation rate 平均减速度 05.0350

average train gross weight 列车平均总重
06.0695

axial displacement limiting device 轴向位移保护装

置 04.1535

axial flow compressor 轴流式压气机 04.1541

axial flow pump 轴流水泵 02.0919

axial flow turbine 轴流式涡轮 04.1538

axial thrust 轴向推力 04.1529

axle 车轴 05.0107

axle analogy test machine 车轴模拟试验台
05.0996

axle arrangement 轴列式，＊轴式 04.0034

axle body 轴身 05.0114

axle box 轴箱 05.0127

axle box bearing 轴箱轴承 04.1072

axle box guide 轴箱挡 05.0088

axle box play 轴箱导框间隙 05.0229

axle center 轴中央部 05.0115

axle collar 轴领 05.0109

axle end bearing 轴端轴承 04.1073

axle gear box 车轴齿轮箱 04.1002

axle generator 车轴发电机 05.0711

axle generator control box 车轴发电机控制箱
05.0712

axle hung bearing 抱轴轴承 04.1074

axle hung traction motor 抱轴式牵引电动机
04.0400

axle load 轴重 01.0050

axle load limited 限制轴重 01.0052

axle load transfer 轴重转移，＊轴载荷转移
04.0047

axle load transfer compensation device 轴重转移补偿
装置 04.0588

axle-mounted disc brake 轴装盘形制动 05.0329

axle suspension bush 抱轴瓦 04.1075

B

back coil 后圈 08.0577

back contact 后接点 08.0553

back cylinder head 汽缸后盖 04.1184

back gage 护背距离 03.0339

background noise field strength 背景噪声场强
04.1521

backhoe 反铲挖土机 02.0880

backing signal 后退信号 08.0039

backscattered signal 反向反射信号 07.0498

back sheet 后板 04.1121

back shield 后挡板 05.0137

back space 退格 07.0716

backward channel 反向信道 07.0803

backward haulage 逆向牵引 04.1267

backwater height in front of bridge 桥前壅水高度
02.0083

bad accident 险性事故 06.0624

bad contact 接触不良 08.0699

bad insulation 绝缘不良 08.0697

bad material 材质不良 08.0696

bad rail insulation 钢轨绝缘不良 08.0698

baggage 行李 06.0015

baggage car 行李车 05.0736

baggage office 行李房 06.0018

baggage rack 行李架 05.0768

baggage room 行李室 05.0762

baggage ticket 行李票 06.0049

balanced grade 均衡坡度 02.0188

balanced type retaining wall 衡重式挡土墙 02.0286

balance weight 坠铊 04.1421

balance weight retaining wall 衡重式挡土墙 02.0286

balance weight tensioner 坠铊补偿器 04.1422

balancing speed 均衡速度 01.0077

ballast 道碴 03.0124

ballast bed 道床 03.0120

ballast box 道碴箱 03.0487

ballast cleaning 清筛道床 03.0225

ballast cleaning machine 道碴清筛机械 03.0375

ballast cleaning machine with removed track panels 大揭盖清筛机 03.0431

ballast coefficient 道床系数 03.0038

ballast consolidating 夯实道床 03.0224

ballast consolidating machine 夯实机械 03.0374

ballast distributing and regulating machine 配碴整形机 03.0419

ballasted deck 道碴桥面 02.0501

ballasted floor 道碴桥面 02.0501

ballasted track 有碴轨道 03.0013

ballast flow control valve 边走边卸阀 05.0518

ballast grading 道碴级配 03.0125

ballast layer 道碴层, * 道碴床 03.0123

ballastless deck 明桥面 02.0499

ballastless track 无碴轨道 03.0014

ballast machine 道碴机械 03.0372

ballast modulus 道床系数 03.0038

ballast nest 道碴巢 03.0489

ballast plow 道碴犁 03.0420

ballast pocket 道碴袋 03.0488

ballast ramming 夯实道床 03.0224

ballast recollecting machine 收碴机 03.0422

ballast resistance 道床阻力 03.0044, 道碴电阻, * 道碴漏泄电阻 08.0517

ballast shoulder cleaning machine 边坡清筛机 03.0428

ballast shoulder consolidating machine 道床边坡夯实机 03.0424

ballast tamping 捣固道床 03.0223

ballast trimming 整理道床 03.0226

ballast trough 桥梁道碴槽 02.0500

ballast tub 道碴槽 03.0486

ball-socket bar 杵环杆 04.1399

ball suspension 悬球 04.1614

ball type angle cock 球芯折角塞门 05.0284

ball type journal box positioning device 球形轴箱定位装置 05.0154

bandwidth of an optical fiber 光纤带宽 07.0462

bankette 弃土堆 02.0283

banking locomotive 补机 04.0228

bar bender 钢筋弯曲机 02.0938

bar cold-drawing machine 钢筋冷拉机 02.0939

bar cutter 钢筋切断机 02.0937

bar frame 棒式车架 04.1214

barring 盘车 04.1530

barring gear 盘车机构 04.0970

barring mechanism 盘车机构 04.0970

bar straightener 钢筋调直机 02.0936

bar with ball and eye 杵环杆 04.1399

bascule bridge 竖旋桥 02.0409

base 基底 02.0247

baseband 基带 07.0791

baseband transmission 基带传输 07.0792

base station 固定电台, * 基地台 07.0553

basic depreciation rate 基本折旧率 06.0705

basic fare 基本票价 06.0732

basic group 基群 07.0308

basic group distribution frame 基群配线架 07.0292

basic intensity of earthquake 地震基本烈度 02.0066

basic link control procedure 基本型链路控制规程 07.0887

basic resistance 基本阻力 04.0095

basic rolling car resistance 货车溜放基本阻力 06.0474

basic route 基本进路 08.0153

bateau bridge 浮桥 02.0395, 舟桥 02.0953

battered joint of rail 钢轨低接头 03.0156

batter pile 斜桩 02.0612

battery circuit 蓄电池电路 04.0379

battery room 蓄电池室 08.0717

baud 波特 07.0047

beam straightening equipment 调梁设备 05.0901

bearing 支座 02.0514

bearing capacity of foundation 地基承载力 02.0047

bearing capacity of ground 地基承载力 02.0047

bearing capacity of subgrade 地基承载力 02.0047

bearing pile 支承桩 02.0598

bearing plate test on subgrade 路基承载板测定 02.0263

bearing platform 承台 02.0575

bearing slab method for subgrade testing 路基承载 板测定 02.0263

beater 捣镐, ＊道镐 03.0467

bedroom 卧室 05.0753

beginning of turnout 道岔始端 03.0311

bellows 波纹管 04.0986

bellows type air spring 囊式空气弹簧 05.0172

belt conveyer 带式输送机 02.0885

benching tunnelling method 台阶法 02.0730

benchmark leveling 水准点高程测量, ＊基平 02.0116

bench test 台架试验 04.0854

bench tunnel portal 台阶式洞门 02.0672

"B" end of car 一位端 05.0008

bend stop valve 弯形止阀 05.0668

bent 排架 02.0622

bent axle 车轴弯曲 05.0261

bent-up bar 弯起钢筋 02.0476

berm 护道 02.0250

berm for back pressure 反压护道 02.0265

berm in cutting 路堑平台 02.0282

berm with superloading 反压护道 02.0265

berth lamp 床头灯 05.0683

berth ticket 卧铺票 06.0038

best density 最佳密度 02.0261

best moisture content 最佳含水量 02.0260

betterment and improvement of railway 铁路技术改 造 02.0002

bias distortion 偏畸变 07.0710

bidding for project 工程投标 02.0846

bidirectional combined type marshalling station 双向 混合式编组站 06.0378

bidirectional longitudinal type marshalling station 双 向纵列式编组站 06.0377

hidirectional transmission 双向传输 07.0788

bidirectional transversal type marshalling station 双向 横列式编组站 06.0376

bi-level coach 双层客车 05.0721

binary encoding 二进制编码 07.0735

bit 比特 07.0045

bit-error rate 误比特率, ＊误码率 07.0875

bit-oriented protocol 面向比特协议 07.0884

bit rate 比特率 07.0046

bit synchronizing 位同步 07.0835

blade 尖轨 03.0313

blade inlet angle 进口角 04.1548

blade outlet angle 出口角 04.1549

blade profile 叶型 04.1546

blade wheel 工作轮 04.1013

blank 空白 07.0718

blasting discharging sedimentation 爆破排淤 02.0268

blasting for loosening rock 松动爆破 02.0277

blasting protection facilities 防爆设施 02.1036

blind ditch 盲沟 02.0800

blind drain 渗水暗沟 02.0303

block 码组 07.0743

block check 码组校验 07.0815

block cleared 解除闭塞 08.0349

block-error rate 误组率 07.0877

blocking 办理闭塞 06.0242

blocking telephone 站间行车电话, ＊闭塞电话

07.0415

block instrument　闭塞机　08.0352

block interference　阻塞干扰　07.0531

block post　线路所　06.0360

block section　闭塞分区　08.0340

block signal　通过信号机　08.0073

block system　闭塞　08.0317

block tie　短枕　03.0118

blow-by　窜气　04.0841

blower　[增压器]压气机　04.0974, 送风器
04.1146

blower impeller　压气机叶轮　04.0976

blower motor　通风机电动机　04.0475

blower scavenging　扫气泵扫气　04.0729

blow off valve　放水阀, ＊排水阀　04.1174

body bolster　枕梁　05.0396

body center plate　上心盘　05.0407

body framing　车体骨架　05.0380

body side bearing　上旁承　05.0410

bogie　转向架　05.0057

bogie bolster play　摇枕挡间隙　05.0230

bogie car　转向架式车　05.0005

bogie individual power supply　转向架独立供电
04.0366

bogie lamp　走行部灯　04.0644

bogie mounted traction motor　转向架架承式牵引
电动机　04.0402

bogie pivot center　转向架中心　04.0035

bogie pivot pitch　转向架中心距离　04.0036

boiler barrel course　锅胴　04.1134

boiler check valve　止回阀　04.1173

boiler evaporative heating surface　锅炉蒸发面积
04.1251

boiler heat balance　锅炉热平衡　04.1106

boiler heat dissipating surface　锅炉散热面积
04.1253

boiler safety device　锅炉安全装置　04.1176

boiler super heating surface　锅炉过热面积
04.1252

boiler tractive effort　锅炉牵引力　04.1088

boiler water level　锅炉水位　04.1255

bolster guide　摇枕挡　05.0091

bolster hanger　摇枕吊　05.0206

bolsterless truck　无摇枕转向架　05.0066

bolster spring　摇枕弹簧　05.0166

bolted and welded steel bridge　栓焊钢桥
02.0364

bolted rigid frog　钢轨组合辙叉　03.0348

bolt hole chamfering　螺栓孔削角, ＊螺栓孔倒棱
03.0184

bolt hole cold-working strenthening　螺栓孔加强
03.0185

bolt hole crack　螺孔裂纹　03.0193

bolt-screw spike　螺栓螺纹钉　03.0093

bolt wrench with indicator　螺栓示功扳手　02.0368

booking office　售票处　06.0019

booster　粘着重量增加器　04.1238

booster transformer　吸流变压器　04.1318

booster transformer feeding system　吸流变压器供电
方式, ＊BT供电方式　04.1277

boosting cable　吸上线　04.1320

bored pile　钻孔桩　02.0599

boring [prospecting]　钻探　02.0052

borrow pit　取土坑　02.0251

bottom dead point　下止点　04.0713

bottom door　底门　05.0495

bottom layer of subgrade　基床底层　02.0243

bottom layer of subgrade bed　基床底层
02.0243

bottom operation coupler　下作用车钩　05.0531

bouncing [vibration]　浮沉振动　01.0100

bowl [shaped] combustion chamber　盆形燃烧室
04.0802

box car　棚车　05.0795

box culvert　箱涵　02.0628

box girder　箱形梁　02.0386

box pallet　箱式托盘　06.0569

brace　拉撑, ＊斜撑　04.1131

bracket base　拉杆底座　04.1385

brake　制动机　05.0264

brake beam　制动梁　05.0189

brake beam bottom rod　制动梁下拉杆　05.0191

brake beam compression channel　制动梁槽钢
05.0203

brake beam pull rod　制动梁拉杆　05.0190

brake beam strut　制动梁支柱　05.0205

brake beam tension rod　制动梁弓形杆　05.0204

brake branch pipe　制动支管　05.0275

brake cylinder　制动缸　05.0307

brake disc　制动盘　05.0201

brake efficiency　制动效率　05.0339

brake equipment　制动装置　05.0263

brake gear　制动装置　05.0263

brake head　闸瓦托　05.0192

[brake] hose coupling　制动软管连接器
　　05.0280

brake inspection point　制动检修所　05.0912

brake lining　闸片　05.0202

brakeman's cabin　制动员室　06.0533

brake mode　制动方式　04.0112

brake pad　闸片　05.0202

brake pipe　制动主管　05.0274

brake pipe pressure reduction　制动管减压量
　　04.0147

brake piping　制动管路　05.0273

brake repair room　制动室　05.0897

brake rigging efficiency　制动杠杆传动效率
　　05.0340

brake shoe　闸瓦　05.0193，制动铁鞋　06.0465

brake shoe back　闸瓦背　05.0198

brake shoe force per axle of locomotive　机车每轴闸
　　瓦作用力　04.0127

[brake shoe] inertia dynamometer　闸瓦试验台
　　05.0995

brake shoe key　闸瓦插销　05.0200

brake shoe pressure　闸瓦压力　05.0335

brake system　制动系统　05.0262

brake test　制动试验　04.0284

brake unit　单元制动　05.0327

brake van　守车　05.0802

braking　制动　04.0128

braking bracing　制动撑架　02.0460

braking circuit　制动电路　04.0373

braking contactor　制动接触器　04.0552

braking force　制动力　05.0336

braking force of train　列车制动力　02.0341

braking pier　制动墩　02.0565

braking propagation rate　制动波速　04.0159

braking ratio　制动率　05.0337

braking reactor　制动电抗器　04.0509

braking resistor　制动电阻器　04.0516

braking resistor cubicle　制动电阻柜　04.0521

braking resistor grid　制动电阻元件　04.0522

braking time　制动时间　05.0343

braking torque　制动转矩　04.0424

branch connection　分支接线　04.1281

branching turnout　分路道岔，＊分歧道岔
　　08.0373

branch line　支线　01.0038

branch line of turnout　道岔侧线　03.0302

branch pipe tee　制动支管三通　05.0285

branch telephone office　分电话所　07.0385

branch track of turnout　道岔侧线　03.0302

branding and stamping of rails　钢轨打标记
　　03.0182

break a seal　破封　08.0712

breakdown gang　救援队　06.0665

breakdown locomotive　救援机车　06.0666

breakdown of equipment　设备故障　06.0633

breakdown train　救援列车　06.0664

break in grade　变坡点　02.0191

breaking up ice floe prevent collision　防撞破凌
　　03.0588

breaking up ice run　防撞破凌　03.0588

break-up capacity of hump　驼峰解体能力
　　06.0501

break-up of trains　解体调车　06.0255

breathing zone　伸缩区，＊呼吸区　03.0234

bridge and other equipments　工务设备台帐
　　03.0262

bridge and tunnel guarder telephone　桥隧守护电话
　　07.0418

bridge and tunnel patrolling　桥隧巡守　03.0571

bridge and tunnel patrolling man　桥隧巡守工
　　03.0009

bridge and tunnel watchman　桥隧巡守工　03.0009

bridge announciating device　桥梁通知设备
　　08.0450

bridge connection　桥形接线　04.1282

bridge construction clearance　桥梁建筑限界
　　01.0057

bridge erection by floating　浮运架桥法　02.0530

bridge fabrication machine　造桥机　02.0951

bridge foundation　桥梁基础　02.0580

bridge load spectrum　桥梁荷载谱　02.0327

bridge obstruction signal　桥梁遮断信号　08.0449

bridge post　桥梁标　03.0252

bridge rating test　桥梁检定试验　03.0517

bridge shallow foundation protection　桥梁浅基防护　03.0569

bridges in cascade　多段桥[联结]　04.0336

bridge sleeper　桥枕　02.0504

bridge structure gauge　桥梁建筑限界　01.0057

bridge test car　桥梁实验车　03.0448

bridge tie　桥枕　02.0504

bridge tower　桥塔　02.0490

bridging filter for dispatching telephone with VF selective calling　音选调度电话滤波器　07.0433

brittle fractures of rail　钢轨折断　03.0191

broadband interference　宽带干扰　04.1506

broadcast call　广播呼叫　07.0816

broadcasting for train　列车广播　07.0586

broad concrete tie　宽混凝土轨枕, * 轨枕板　03.0109

broad-gage railway　宽轨铁路　01.0014

broken chain　断链　02.0127

broken height　断高　02.0129

broken joint　相错式接头, * 相互式接头　03.0067

broken rail base　轨底崩裂　03.0190

broken rail protection　断轨保障　08.0521

brush　电刷　04.0459

brush box　刷盒　04.0458

brush cleaning　钢丝刷除锈　03.0550

brush cutter　灌木切割机　03.0462

brush cutting machine　灌木切割机　03.0462

brush gear　电刷装置　04.0456

brush-holder　刷握　04.0457

BT　吸流变压器　04.1318

bucket elevator　斗式提升机　02.0986

bucket loader　斗车　02.0968

budgetary estimate of design　设计概算　02.0019

buffer beam　缓冲梁　05.0400

buffer layer　缓冲层　07.0484

buffer rail　调节轨　03.0237

buffer rod　缓冲杆　05.0578

buffer stop indicator　车档表示器　08.0098

buffer zone　缓冲区　03.0235

buffet　小卖部　05.0759

buffet car　酒吧车　05.0733

building-out network　假线　07.0345

bulk cement car　散装水泥车　05.0810

bulk freight　散装货物　06.0114

bulldozer　推土机　02.0882

bumper post　车挡　06.0515

bundled steel wires　钢丝束　02.0471

bunker train　槽式列车　02.0966

buoyancy　浮力　02.0334

burglary accident　被盗事故　06.0638

buried abutment　埋置式桥台　02.0549

buried cable　直埋电缆　07.0184

buried depth of tunnel　隧道埋置深度, * 埋深　02.0714

burnt rim　轮辋烧伤　05.0253

burst error　突发差错　07.0840

burst hub　轮毂破裂　05.0257

burst of rail base　轨底崩裂　03.0190

burst of rail bottom　轨底崩裂　03.0190

burst transmission　突发传输　07.0790

bus-bar　汇流条　08.0275

bus network　总线网[络]　07.0772

busy hour crosstalk and noise　忙时串杂音　07.0359

busy tone　忙音　07.0277

button for cut-off an audible signal　切断音响按钮　08.0446

button indication　按钮表示　08.0167

button switch　按键开关　04.0576

buttress wall for intercepting falling rocks　拦石墙　02.0318

C

cable 缆索 02.0492，电缆 07.0171

cable band 索夹 02.0496

cable bond 均压线 04.0442

cable branching terminal box 分向电缆盒 08.0287

cable clamp 索夹 02.0496

cable crane 缆索起重机 02.0899

cable duct 电缆管道 07.0200

cable fault detector 电缆障碍探测器 07.0217

cable gas-feeding equipment 电缆充气维护设备 07.0204

cable marking stake 电缆标石 07.0203

cable plane 索平面 02.0491

cable saddle 索鞍 02.0495

cable sleeve 电缆套管 07.0201

cable-stayed bridge 斜拉桥，*斜张桥 02.0394

cable terminal box 终端电缆盒 08.0289

cable terminal box on a post 杆上电缆盒 08.0286

cable TV on train 列车闭路电视 07.0661

caboose 守车 05.0802

caboose valve 车长阀 05.0317

cab signal 机车信号，*机车自动信号 08.0012

cab signal box 共用箱 08.0424

cab signaling equipment 机车信号设备 08.0409

cab signaling inductor location 机车信号作用点 08.0417

cab signaling testing section 机车信号测试区段 08.0416

caisson retaining wall 沉井挡墙 02.0291

caking 结胶 04.0806

calculated cut-off 计算遮断比 04.1093

calculated length of car 车辆计算长度 05.0967

calculated rate of evaporation for engine 计算供汽率 04.1099

calculated weight of locomotive 机车计算重量 04.0043

calculate point of hump height 峰高计算点 06.0449

call button 呼叫装置 05.0694

called 被叫，*被呼 07.0818

called and calling subscriber release 互不控制复原方式 07.0409

called party 被叫用户 07.0115

called subscriber release 被叫控制复原方式 07.0408

calling 主叫，*主呼 07.0817

calling block 通话锁闭 07.0596

calling check 呼叫检查 07.0597

calling for tenders of project 工程招标 02.0845

calling for tending of project 工程招标 02.0845

calling help signal 呼救信号 06.0659

calling-on signal 引导信号 08.0029

calling order 通话命令 07.0595

calling party 主叫用户 07.0114

calling request 通话请求 07.0594

calling signal 呼叫信号 07.0275

calling subscriber release 主叫控制复原方式 07.0407

camber 拱度 02.0438，叶型折转角 04.1550，上挠度 05.0441

camber of bridge span 桥梁上拱度 03.0531

cam position changeover switch 凸轮位置转换开关 04.0560

camshaft 凸轮轴 04.0937

camshaft timing gear 凸轮轴正时齿轮 04.0942

cant 曲线超高 03.0026

cantilever 腕臂 04.1381

cantilever base 腕臂底座 04.1383

cantilever beam bridge 悬臂梁桥 02.0381

cantilevered assembling construction 悬臂拼装法 02.0537

cantilever erection 悬臂架设法 02.0535

cantilever portion of center sill 中梁悬臂部

05.0393

cantilever shed gallery 悬臂式棚洞 02.0702

cantilever shed tunnel 悬臂式棚洞 02.0702

cant rail ［客车］上侧梁 05.0438

can-type combustor 管形燃烧室 04.1563

capacitive accelerometer 电容式加速度计 04.1609

capital construction investment 基本建设投资 06.0709

capital repair of bridge and tunnel 桥隧大修 03.0545

capital repair of fixed assets 固定资产大修 06.0701

car accelerator 车辆加速器 08.0398

car auxiliary repair 车辆辅修 05.0922

car body 车体 05.0374

car body tilting device 车体侧倾装置 05.0378

carbon arrester 炭精避雷器 07.0205

carbon deposit 积炭 04.0805

carbon equivalent 碳当量 02.0469

car brake examination 车辆制检 05.0924

car brake inspection 车辆制检 05.0924

car caring point 爱车点 05.0954

car classification 调车 06.0254

car class indicating lamp 席别灯 05.0681

car collision 车辆互撞 05.0589

car condemning limit 车辆报废限度 05.0950

cardan shaft drive 万向轴驱动 04.0356

car depot 车辆段 05.0892

car detached repair 摘车修 05.0931

car detention time under accumulation 集结时间 06.0263

car dynamics test 车辆动力学试验 05.0871

car flow 车流 06.0276

car flow routing 车流径路 06.0279

car for traffic use 运用车 06.0271

car heavy repair 车辆大修 05.0925

car impact 车辆冲击 05.0588

car impact test 车辆冲击试验 05.0870

car inspection and maintenance 车辆检修 05.0890

car journal and box examination 车辆轴检 05.0923

car journal and box inspection 车辆轴检 05.0923

car kilometers 车辆公里 06.0690

car kilometers per car per day 货车日车公里 06.0328

carline 车顶弯梁 05.0457

car load freight 整车货物 06.0111

car load freight unloaded at two or more stations 整车分卸 06.0123

car loading by groups 成组装车 06.0104

car loading list 货车装载清单 06.0128

car manufacturing works 车辆制造厂 05.0944

car medium repair 车辆中修 05.0926

car not for traffic use 非运用车 06.0272

car operation 车辆运营 05.0891

car operation and maintenance 车辆运用维修 05.0915

car planned requisition list 要车计划表 06.0098

car power receptacle 车下电气插座 05.0709

car pressure regulater 直压式暖汽调整阀 05.0595

car repair before loading or after unloading 车辆装卸修 05.0928

car repair facilities 车辆检修设备 05.0893

car repair in depot 车辆段修 05.0921

car repair in works 车辆厂修 05.0920

car repair limit 车辆检修限度 05.0948

car repair works 车辆修理厂 05.0943

car retarder 车辆减速器 08.0397

carriage 客车 05.0718

carriage return 回车 07.0714

carrier adaptor for dispatching telephone 载波调度电话中继器 07.0287

carrier amplifier 载频放大器 07.0338

carrier channel connected to telephone line 载波遥接话路 07.0286

carrier communication 载波通信 07.0315

carrier frequency 载频 07.0304

carrier frequency synchronization 载频同步 07.0313

carrier leak 载漏 07.0306

carrier supply system 载供系统 07.0326

carrier telegraph terminal 载波电报终端机 07.0695

carrier telephone repeater 载波电话增音机 07.0323

carrier telephone terminal 载波电话终端机 07.0322

carrier-to-noise ratio 载噪比 07.0523

carrier [wave] 载波 07.0021

car road service limit 车辆运用限度 05.0949

carrying capacity 通过能力 06.0304

carrying capacity of locomotive facilities 机务设备通过能力 04.0275

carrying capacity of receiving-departure track 到发线通过能力 06.0504

carrying capacity of station 车站通过能力 06.0505

carrying capacity of station throat 车站咽喉通过能力 06.0502

carrying capacity of the block section 区间通过能力 06.0308

carrying scraper 铲运机 02.0883

car seal 货车施封 06.0139

cars on hand 现在车 06.0270

cars open to traffic 运用车 06.0271

car space 静态长度 08.0369

car spacing for uncoupled inspection 拉钩检查距离 05.0969

cars per cut 钩车 06.0500

car strength test 车辆强度试验 05.0869

car turnround time 货车周转时间 06.0326

[car] wheelset repair factory 车轮厂 05.0982

car wheel static balance test 车轮静平衡检验 05.0963

car with axle generator 母车 05.0750

car without axle generator 子车 05.0751

car yearly repair 车辆年修 05.0927

cascade protection 分级保护 07.0236

cast frame 铸造机座 04.0452

cast-in-place cantilever construction 悬臂灌注法 02.0536

cast-in-place concrete pile 就地灌注桩 02.0595

cast-in-place method 就地灌筑法 02.0541

cast-in-situ concrete pile 就地灌注桩 02.0595

cast-in-situ method 就地灌筑法 02.0541

cast iron brake shoe 铸铁闸瓦 05.0194

cast manganese steel frog 高锰钢整铸辙叉 03.0347

cast steel wheel 铸钢车轮 05.0101

casualty of non-railway man 路外人员伤亡 06.0651

casualty of railway man 路内人员伤亡 06.0650

casualty rate 伤亡率 06.0654

catch-drain 截水沟 02.0300

catchment area 汇水面积 02.0089

catchment basin of debris flow 泥石流流域 02.0087

catch siding 安全线 06.0414

catch up 追钩 08.0370

category of traction 牵引种类 01.0070

catenary 接触网悬挂 04.1351, 承力索 04.1361

catenary sag 承力索弛度 04.1449

catenary splice 承力索接头线夹 04.1370

caterpiller crane 履带式起重机 02.0896

caution signal 注意信号 08.0021

cavitation 穴蚀 04.0839

ceiling 内顶板, *天花板 05.0456

ceiling lamp 顶灯 05.0678

cement pump 水泥输送泵 02.0924

center filler 心盘座 05.0409

center height of gravity for car loaded 重车重心高 06.0171

centering block 摆块 05.0570

centering block hanger 摆块吊 05.0571

centering device 复原装置 04.1237

center line stake 中线桩 02.0113

center line survey 中线测量 02.0112

center mast 中心柱 04.1411

center of gravity for car loaded 重车重心 06.0170

center of turnout 道岔中心 03.0306

center pin 中心销 05.0211

center sill 中梁 05.0392

center stake leveling 中桩高程测量, *中平 02.0117

center station 中心电台 07.0559

center to center distance between main girder 主梁中心距 02.0435

center wedge 中央楔块 05.0561

central battery 中央电池 07.0088

central draft gear 中央缓冲器 05.0555

centralization of repair　集中化修理，＊集中修
　04.0263

centralized interlocking　集中联锁　08.0105

centralized inverter　集中式逆变器　05.0700

centralized network　集中[式]网[络]　07.0762

centralized power supply system　集中供电方式
　04.1296

centralized power supply [system] for car　车辆集中
　供电　05.0715

central lines of track　线路中心线　06.0425

centralized control　中心控制　08.0469

centralized traffic control　调度集中　08.0460

centrally connected power source　集中电源
　08.0645

centrally connected power supply　集中供电
　08.0636

centrally operated switch　集中道岔　08.0198

central office cable　局用电缆　07.0175

central oil outlet valve　中心排油阀　05.0483

central telephone office　中心电话所　07.0386

centrifugal force of train　列车离心力　02.0336

centrifugal oil filter　离心式机油滤清器　04.1043

centrifugal pump　离心水泵　02.0916

ceramic arrester　陶瓷避雷器　07.0207

certification of design　设计鉴定　02.0034

chamber explosive package blasting　洞室药包爆破
　02.0274

chamfer　走线架　07.0128

chamher blasting　洞室药包爆破　02.0274

changeover governor　换挡　04.1019

changeover valve　变位阀　05.0515

change side of double line　换侧，＊换边　02.0204

change soil　换土　02.0267

channel amplitude characteristic　通路振幅特性
　07.0364

channel basic noise　通路固有杂音　07.0358

channel capacity　信道容量　07.0804

channel erosion　河床冲刷　03.0585

channel free signal　信道空闲信号　07.0611

channel frequency characteristic　通路频率特性
　07.0363

channel linearity　通路线性　07.0285

channel net loss　通路净衰耗　07.0314

channel selector　频道选择器　07.0564

channel signal to crosstalk and noise ratio　通路串杂
　音防卫度　07.0365

channel singing margin　通路振鸣边际　07.0366

character　字符　07.0819

character check　字符检验　07.0820

character-error rate　误字率　07.0876

characteristic distortion　特性畸变　07.0711

characteristic impedance　特性阻抗　07.0226

character-oriented protocol　面向字符协议
　07.0883

charge　充量　04.0720

charged weight　[货物]计费重量　06.0738

charge inter-cooling　增压中冷　04.0792

charge position　充风位，＊充气位　04.0141

charging　充风，＊充气　04.0140

charging plug　充电插头　05.0710

check 4mm opening of a switch point　4毫米锁闭
　08.0133

checking of clearance　限界检查　03.0514

checking of investment　投资检算　02.0025

check locking　照查锁闭　08.0129

check on working conditions in the driver's cab　司机
　室工作条件检查　04.0685

check-out　检测　08.0735

check rail　护轨　03.0060

checks for prevention of accidents　事故预防措施的
　检查　04.0662

checks of battery charging-arrangement　蓄电池充电
　系统试验　04.0664

chief operator's desk　班长台　07.0257

child ticket　小孩票　06.0042

chilled cast iron wheel　冷铸铁轮　05.0105

chimney　外烟筒　04.1143

China railway standard live loading　中-活载
　02.0325

Chinese character code translation equipment　中文译
　码机　07.0693

choke　扼流圈　07.0051

chopper　电磁铁驱动器，＊斩波器　04.1593

chopper control　斩波调压　04.0323

chord member　弦杆　02.0447

chromatic dispersion constant　色散常数　07.0458

chute 急流槽 02.0301, 走线架 07.0128

circuit 电路 07.0110

circuit on side of overhead contact line 网侧电路 04.0371

circuitous traffic 迂回运输 06.0109

circuit switching 电路交换 07.0721

circuit to hold a route for shunting 非进路调车电路 08.0244

circular curve 圆曲线 02.0172

circular pier 圆形桥墩 02.0569

circular routing 环形交路 04.0192

circulating baffle plate 循环挡板 05.0644

circulating factor of repair in depot 段修循环系数 05.0968

circumferential crack in plate 辐板圆周裂纹 05.0260

cladding diameter 包层直径 07.0446

clamshell bucket 蛤壳式抓斗 02.0881

clasp brake 双侧[踏面]制动 05.0331

classification facilities 调车设备 06.0446

classification list conveyer system 票据传送设备 08.0395

classification of filling material 填料分类 02.0253

classification of locomotive repair 机车检修修程 04.0240

classification of out-of-gauge freight 超限货物等级 06.0158

classification of soil and rock 土石分类 02.0046

classification of track 轨道类型 03.0011

classification of tunnel surrounding rock 隧道围岩分级 02.0048

classification of waterproof 防水等级 02.0794

classification track 调车线 06.0409

classification yard 调车场 06.0443

classified depreciation rate 分类折旧率 06.0707

class of station 车站等级 06.0232

claw bar 撬棍 03.0469

claw bolt 钩螺栓 03.0084

clearance 限界 01.0053, 悬距 04.1615

clearance between wheel flange and gage line 轮轨游间 03.0135

clearance car 限界检测车 03.0455

clearance check measurement 限界检查 03.0514

clearance diagram 限界图 01.0054

clearance improvement 限界改善 03.0572

clearance inspection car 限界检测车 03.0455

clearance limit for freight with exceptional dimension 阔大货物限界 01.0063

clearance limit for lower part of rolling stock 机车车辆下部限界 01.0061

clearance limit for overhead catenary system 接触网限界 01.0064

clearance limit for overhead contact wire 接触网限界 01.0064

clearance limit for oversize commodities 阔大货物限界 01.0063

clearance limit for upper part of rolling stock 机车车辆上部限界 01.0060

clearance limit frame 限界架 03.0280

clearance of a retarder 减速器接近限界 08.0384

clearance treadle 限界检查器 08.0401

cleared signal indication 信号开放表示 08.0163

clearing revenue 清算收入 06.0752

clearing signal 开放信号 06.0252

clear span 净跨 02.0432

click 喀音 07.0280

climbing shuttering 爬升模板 02.0959

climb on rail 爬轨 05.0874

close contact between switch point and stock rail 密贴尖轨 03.0355

closed joint 接头瞎缝 03.0157

closed-loop control 闭环控制 04.1616

closed switch rail 密贴尖轨 03.0355

closed ventilated container 封闭式通风集装箱 06.0543

close type track circuit 闭路式轨道电路 08.0505

close up 封锁 08.0706

closing signal 关闭信号 06.0253

closing the section 封锁区间 06.0670

closing the top of lining 拱圈封顶 02.0759

closure 桥梁合龙 02.0540

closure of track 线路封锁 02.0999

clothoid curve 螺旋曲线 03.0024

club car 俱乐部车 05.0734

clutch operating device 离合器操纵装置 04.1050

coach 客车 05.0718

coach servicing shed 客车整备库，＊客车整备棚 05.0900

coal bin 煤槽 04.1241

coal-burning heater 燃煤温水锅炉 05.0604

coal-burning [heater type] hot water heating equipment 燃煤独立温水采暖装置 05.0600

coal car 煤车 05.0803

coal fired locomotive 燃煤机车 04.1081

coal for locomotive 机车用煤 04.0054

coal pusher 推煤机 04.1247

coal water mixture 水煤浆 04.1104

coal water slurry 水煤浆 04.1104

coarse-grained soil fill 粗粒土填料 02.0255

coarse-grained soil filler 粗粒土填料 02.0255

coasting 溜放调车 06.0260

coasting braking test [列车制动]溜放试验 05.0879

coasting resistance 惰行阻力 04.0097

coating 被覆层 07.0447

coaxial cable 同轴电缆 07.0172

coaxial cable communication 同轴电缆通信 07.0318

code 电码 07.0688

coded track circuit 电码轨道电路 08.0495

code error tester 误码测试仪 07.0369

code holes 电码孔 07.0720

code rate justification 码速调整 07.0378

code relay 电码继电器 08.0579

code sender 发码器 07.0398

coding 编码 07.0016

coding during train approaching 接近发码 08.0344

coding of continuous track circuit 轨道电路电码化 08.0508

coefficient of cars not in service 非运用车系数 05.0976

coefficient of charge 充量系数 04.0741

coefficient of developed line 展线系数 02.0164

coefficient of dynamic force 动力系数，＊动荷系数 05.0883

coefficient of extension line 展线系数 02.0164

coefficient of field weakening 磁场削弱系数 04.0437

coefficient of freight traffic 货物运输系数 06.0685

coefficient of impact 冲击系数 02.0338

coefficient of indicated pressure 指示压力系数 04.1091

coefficient of intensification 强化系数 04.0836

coefficient of locomotive requirment 机车需要系数 04.0222

coefficient of passengers travelling by trains 旅客乘车系数 06.0680

coefficient of reserved capacity 能力储备系数 06.0307

coefficient of reserve power 功率储备系数 04.0822

coefficient of residual gas 残余废气系数 04.0740

coefficient of scavenging 扫气系数 04.0742

coefficient of spring deflection reservation 挠度裕量系数 05.0241

coefficient of thermal conductivity 导热系数 05.0652

coefficient of train removal 列车扣除系数 06.0311

coefficient of utilization for car loading capacity 货车载重量利用率 06.0155

cofferdam 围堰 02.0586

coherent modulation 相干调制 07.0821

coil height 线圈高度 04.1617

coil spring 螺旋弹簧 05.0167

cold-running 冷滑 04.1440

cold water pump 冷水泵 04.1171

cold weather construction 冬季施工 02.0843

collapse 崩塌 03.0480

collector shoe 集电靴 04.0544

collision 冲突 06.0629

collision force of ship or raft 船只或排筏的撞击力 02.0349

collision prevention around pier 墩台防撞 03.0573

color-light signal 色灯信号机 08.0053

column crane　塔式起重机　02.0893

column diagram of stratum　地层柱状图　02.0068

column guide　[货车]摇枕挡　05.0092

column pier　柱式桥墩　02.0566

combination lever　合并杆　04.1207

combination relay　组合继电器　08.0594

combination shielding factor　综合屏蔽系数　04.1478

combination truck　组合式转向架　05.0071

combined bridge　公铁两用桥　02.0356

combined dirt collector and cut-out cock　联合集尘截断塞门　05.0286

combined highway and railway bridge　公铁两用桥　02.0356

combined lift and guidance　悬浮导向兼用　04.1618

combined optical fiber cable　综合光缆　07.0481

combined rail-cum-road bridge　公铁两用桥　02.0356

combined station　组合站　07.0822

combined steel formwork　组合钢模板　02.0957

combined train　组合列车　06.0220

combined type junction terminal　混合形枢纽　06.0402

combining chamber　递热器，＊混合室　04.1170

comb rack　梳妆架　05.0786

combusting delay period　滞燃期　04.0782

combustion efficiency　燃烧效率　04.1113

combustion lagging period　滞燃期　04.0782

combustion process　燃烧过程　04.0774

combustion stability　燃烧稳定性　04.1577

combustion zone　燃烧区　04.1570

command circuit　指令电路　04.0380

command telephone control board　指挥电话总机　07.0434

commencement report of construction work　开工报告　02.0866

commercial speed　旅行速度　06.0331

common battery telephone set　共电电话机　07.0092

common battery telephone switch board　共电电话交换机　07.0392

common channel signaling system　共线信令系统　07.0836

communication　通信　07.0002

communication blind district　通信盲区　07.0543

communication center　通信枢纽　07.0239

communication center between several railway administration　局间通信枢纽　07.0241

communication center of railway administration　局通信枢纽　07.0240

communication control character　通信控制字符　07.0888

communication controller　通信控制器　07.0893

communication interface　通信接口　07.0894

communication network　通信网　07.0059

communication processor　通信处理机　07.0892

communication protocol　通信协议　07.0774

communication satellite　通信卫星　07.0671

communication subnet　通信子网　07.0765

commutating capacitor　换相电容器　04.0525

commutation reactor　换相电抗器　04.0514

commutator　换向器　04.0439

commutator segment　换向片　04.0440

compacting coefficient　压实系数　02.0259

compacting criteria　压实标准　02.0257

compacting factor　压实系数　02.0259

companding law　压扩律　07.0375

comparable horizon of river　河流比降　02.0077

compatible simplex-duplex radio communication　单双工兼容无线电通信　07.0517

compensating coil　补偿线圈　04.0450

compensation capacitor　补偿电容器　04.0528

compensation grade in tunnel　隧道坡度折减　02.0202

compensation of curve　曲线折减　02.0201

compensation of gradient　坡度折减　02.0200

compensation of gradient in tunnel　隧道坡度折减　02.0202

compensator with parallel capacitance　并联电容补偿装置　04.1301

compensator with series capacitance　串联电容补偿装置　04.1300

complete turnround of locomotive　机车全周转　04.0207

completion report of construction work　竣工报告　02.0867

component exchange repair　换件大修　04.0253

composite beam bridge　结合梁桥　02.0375

composite cable　综合电缆　07.0177

composite coach　合造客车　05.0722

composite depreciation rate　综合折旧率　06.0706

composite lining　组合衬砌，＊复合衬砌　02.0686

composite passenger car　合造客车　05.0722

composite piston　组合活塞　04.0901

composite rack　综合架　08.0262

composite two and three-pressure equalizing system
　二三压力混合机构　05.0334

compound catenary equipment　双链形悬挂
　04.1360

compound curve　复曲线　02.0176

compound excited motor　复励电动机　04.0406

compound expansion steam locomotive　复胀式蒸汽
　机车　04.1085

compound-supercharged diesel engine　复合式柴油机
　04.0700

comprehensive approximate estimate　综合概算
　02.0021

comprehensive maintenance of bridge and tunnel struc-
　ture　桥隧综合维修　03.0566

comprehensive transport　综合运输　01.0120

compression beginning pressure　压缩始点压力
　04.0746

compression beginning temperature　压缩始点温度
　04.0748

compression faucet　压式水阀　05.0781

compression ratio　压缩比　04.0751

compression stroke　压缩行程　04.0724

compression terminal pressure　压缩终点压力
　04.0747

compression terminal temperature　压缩终点温度
　04.0749

compressor turbine　压气涡轮　04.1540

compromise joint　异型接头　03.0074

compromise joint bar　异型接头夹板　03.0082

compromise rail　异形轨　03.0059

compulsory inscribing　强制内接　03.0138

computer aided track maintenance and management
　system　养路管理电脑系统　03.0263

computer network　计算机网[络]　07.0764

concentrated telephone unit　电话集中器　07.0439

concentrated weight goods　集重货物　06.0169

concentrator　集中器　07.0830

concourse　广厅　06.0017

concrete bridge　混凝土桥　02.0369

concrete cracks　混凝土裂纹　03.0557

concrete lifting bucket　混凝土吊斗　02.0935

concrete mixer　混凝土搅拌机　02.0925

concrete mixing and transporting car　混凝土搅拌
　运输车　02.0932

concrete mixing plant　混凝土搅拌楼　02.0930

concrete pump　混凝土输送泵　02.0933

concrete pump truck　混凝土泵车　02.0934

concrete tie　混凝土枕　03.0116

concrete tie dowel drilling and pulling machine　混凝
　土枕螺栓钻取机　03.0410

concrete vibrating stand　混凝土振动台　02.0943

concreting with tremie method　灌筑水下混凝土
　02.0616

condenser　冷凝器　05.0631

condenser air concentrator　冷凝器风道　05.0650

conditional power source　条件电源　08.0643

conditional power supply panel　条件电源屏
　08.0653

conducted interference　传导干扰　07.0525

conductor rail collector　第三轨受电器　04.0542

conductor's valve　车长阀　05.0317

cone penetration test　静力触探　02.0056

conering　侧撞　08.0371

conference telephone central board　会议电话总机
　07.0301

conference telephone of full-distribution system　全分
　配制会议电话　07.0299

conference telephone subset　会议电话分机
　07.0303

conference telephone tandem board　会议电话汇接
　机　07.0302

conflicting route　敌对进路　06.0251

conflicting signal　敌对信号　08.0045

conformance testing　一致性测试　07.0890

congestion　阻塞　07.0870

connecting bar　道岔连接杆　03.0364

connecting box　连接箱　04.0876

connecting line 联络线 06.0419

connecting link 结合杆 04.1208

connecting rod 连杆 04.0910

connecting rod bearing shell 连杆轴瓦 04.0913

connecting rod body 连杆体 04.0911

connecting rod bolt 连杆螺栓 04.0915

connecting rod bush 连杆衬套 04.0912

connecting rod cap 连杆盖 04.0914

connecting rod for indication 表示连接杆
08.0627

connecting rod length/crank radius ratio 连杆比
04.0711

connecting rod nut 连杆螺母 04.0916

connection 连接 07.0116

connection box 接线盒 05.0703

connectionless mode 无连接方式 07.0904

connection mode 连接方式 07.0903

connection terminal 接线端子 05.0705

consignee 货主 06.0132

consigning of freight 货物托运 06.0129

consigning of luggages and parcels 行李包裹托运
06.0051

consignment 一批货物 06.0124

consignment note 货物运单 06.0126

consignor 货主 06.0132

constant friction type snubbing device 常摩擦式减震
装置 05.0158

constant pressure charging 定压增压 04.0790

constant pressure reservoir 定压风缸 05.0313

constant speed 恒速 04.0108

constructional loading 施工荷载 02.0353

constructional reinforcement 构造钢筋 02.0478

construction detail design 施工图设计 02.0017

construction figure progress 施工形象进度
02.0876

construction interference 施工妨碍 08.0695

construction investigation 施工调查 02.0830

construction length of railway 铁路建筑长度
01.0045

construction level 施工水位 02.0093

construction method system 工法制度 02.0871

construction output value 施工产值 02.0872

construction preparation 施工准备 02.0831

construction process 施工工艺流程 02.0870

construction profit 施工利润 02.0873

construction protection 施工防护 02.0998

construction repetition 施工复测 02.0836

construction repetition survey 施工复测 02.0836

construction scheme 施工组织方案 02.0865

construction speed 构造速度 01.0078

construction speed of car 车辆构造速度 05.0842

construction starting report 开工报告 02.0866

construction technology process 施工工艺流程
02.0870

construction ventilation of tunnel 隧道施工通风
02.0810

construction water level 施工水位 02.0093

contact blasting 裸露药包爆破 02.0279

contact closed 接点闭合 08.0556

contact heel 动接点 08.0551

contact loss 离线 04.1447

contact loss rate 离线率 04.1448

contact open 接点断开 08.0557

contact pressure 接点压力 08.0573

contacts operated by semaphore 臂板接触器
08.0315

contact system 接点系统 08.0575

contact wire 接触线 04.1365

contact wire pre-sag 接触线预留驰度 04.1451

contact wire sag 接触线弛度 04.1450

contact wire splice 接触线接头线夹 04.1371

container car 集装箱车 05.0826

containerized traffic 集装化运输 06.0576

contamination accident 污染事故 06.0642

continued type speed control system 连续式调速
系统 06.0487

continuous beam bridge 连续梁桥 02.0380

continuous injection angle 喷油持续角 04.0779

continuously welded rail track 无缝线路
03.0017

continuous power 持续功率 04.0811

continuous speed 持续速度 01.0075

continuous track circuit 连续式轨道电路
08.0493

continuous tractive effort 持续牵引力 04.0094

continuous type cab signaling 连续式机车信号

08.0411

contracting of project 工程承包 02.0849

contracting out of project 工程发包 02.0848

control box 控制箱 05.0608

control cabin 控制室 05.0646

control cable coupling 控制电缆连接器 05.0674

control character 控制字符 07.0823

control circuit 控制电路 04.0377

control circuit socket for shed supply 控制电路库用插座 04.0607

control cycle 控制周期 08.0481

control desk 控制台 08.0253

control desk element 控制台单元 08.0266

control electric machine 控制电机 04.0477

controlled blasting 控制爆破 02.0747

controlled bridge rectifier 可控桥式整流器 04.0333

controlled electromagnet 可控电磁铁 04.1590

controlled object 控制对象 08.0466

controlled point 被控点 08.0480

controlling point 控制点 08.0479

controlling section 控制区间 02.0170

control mechanism 控制机构 04.0967

control office equipment of CTC 调度集中总机 08.0461

control panel 控制盘 08.0256

control pump 控制泵 04.1027

control room 控制台室 08.0719

control section 控制区间 02.0170

control set for sound amplifying in yard 扩音转接机 07.0440

control signal 控制信号 07.0608

control signal detector 控制信号检波器 07.0592

control source 控制电源 04.0378

control station 控制站 07.0824

control supply ammeter 控制电源电流表 04.0625

control supply voltmeter 控制电源电压表 04.0624

control transformer 控制变压器 04.0493

convention of international railway through traffic 国际铁路联运公约 01.0125

converted car length 车辆换算长度 06.0229

converted coal for locomotive 机车用换算煤 04.0057

converted resistance 换算阻力，*加算阻力 04.0104

converted running kilometers 换算走行公里 04.0212

converted tons of handling 装卸换算吨 06.0588

converted turnover 换算周转量 06.0687

cooking room 炊事室 05.0758

cooled freight 冷却货物 06.0175

cooling fan 冷却风扇 04.1032

cooling room 冷却室 04.1061

cooling water pump 冷却水泵 04.1029

copper-clad steel wire 铜包钢线 07.0160

copper wire 铜线 07.0158

cord circuit 绳路 07.0259

core [cladding] concentricity 芯[包层表面]同心度 07.0451

core diameter 芯径 07.0445

core-type traction transformer 心式牵引变压器 04.0482

corner fittings 角件 06.0561

corner lamp 角灯 05.0682

corner post 角柱 05.0431

cornice sheathing 上墙板 05.0446

corragated wheel plate 波形辐板 05.0124

corrected power 修正功率 04.0812

corrective maintenance 故障修 08.0727

corridor 走廊 05.0787

[corridor] compartment [type] sleeping car 包间式卧车 05.0730

corrosion cracking of steel bridge 钢梁腐蚀裂纹 03.0558

corrosion resistant rail 耐腐蚀轨 03.0055

corrugated floor 波纹地板 05.0403

corrugation of rail head 轨头波纹磨损 03.0165

cost of charged ton-kilometer 计费吨公里成本 06.0725

cost of converted ton-kilometer 换算吨公里成本 06.0726

cost per passenger kilometer 旅客人公里成本 06.0724

counter-flow combustion 逆流燃烧 04.1565

counter-sliding pile 抗滑桩 02.0292

counter superelevation 反超高 03.0029

counter swelling berm 反压护道 02.0265

couper's cabin at hump crest 驼峰连结员室 06.0532

coupled axle drive 组合驱动 04.0359

coupled axle drive locomotive 组合传动机车 04.0361

coupler and draft gear 车钩缓冲装置 05.0527

[coupler and draft gear] running-in and running-out [车钩缓冲装置]压缩与拉伸 05.0587

coupler body 钩体 05.0533

coupler carrier 车钩托梁 05.0572

coupler centering device 车钩复原装置 05.0569

coupler compression grade 压钩坡 06.0480

coupler contour 车钩轮廓 05.0532

coupler front face 钩头正面 05.0539

coupler guard arm 钩腕外臂 05.0538

coupler height 车钩中心线高度 05.0028

coupler horn 钩肩 05.0537

coupler knuckle 钩舌 05.0543

coupler lock 钩锁 05.0544

coupler lock lift 钩锁销 05.0545

coupler neck 钩颈 05.0536

coupler pivot lug 钩耳 05.0540

coupler shank 钩身 05.0534

coupler slack 车钩间隙 05.0585

coupler tail 钩尾 05.0535

coupler yoke 尾框 05.0548

coupling line 车钩连接线 05.0574

coupling speed 连挂速度 06.0482

coupon ticket 册页[客]票，＊联票 06.0045

cover arch of tunnel 隧道套拱 02.0829

covered goods wagon 棚车 05.0795

covered hopper car 有盖漏斗车 05.0809

covered protection 覆盖防护 02.1001

cover plate 盖板 05.0444

crab rock loader 蟹爪式装岩机 02.0964

cracking of concrete and masonry beam 圬工梁裂损 03.0525

crane on-track 轨行式起重机 03.0440

crank 曲柄 04.0891

crank arm 曲柄臂 04.0890

crankcase 曲轴箱 04.0865

crankcase breather 曲轴箱呼吸器 04.0866

crank journal 主轴颈 04.0888

crank pin 曲柄销 04.0889

crank radius 曲柄半径 04.0707

crankshaft 曲轴 04.0887

crankshaft counter balance 曲轴平衡块 04.0892

crankshaft timing gear 曲轴正时齿轮 04.0941

crankshaft vibration damper 曲轴减振器 04.0894

crawl 蠕变 03.0479

crawler crane 履带式起重机 02.0896

crawler pile frame 履带式桩架 02.0911

CRC 循环冗余检验 07.0797

creep 蠕变 03.0479，蠕滑 05.0846

crew pooling system of passenger train 旅客列车轮乘制 06.0061

crew working system of passenger train 旅客列车乘务制度 06.0060

crib 枕盒，＊枕木盒 03.0119

crib ballast removers 扒碴机 03.0421

crib consolidating machine 枕间夯实机 03.0423

critical damping 临界阻尼 05.0245

critical damping value 临界阻尼值 05.0247

critical grade 临界坡度 02.0185

critical height 临界高度 02.0246

critical speed 临界速度 01.0080

critical temperature of track buckling 轨道鼓出临界温度 03.0233

critical tractive effort 临界牵引力 04.1090

CR-live loading 中－活载 02.0325

cross arm 横担 07.0163

cross barrier at grade crossing 道口栏木 03.0277

crossbar telephone switching system 纵横制电话交换机 07.0394

cross bearer 大横梁 05.0397

crossbond of protective wire 保护线用连接线 04.1317

cross-connecting box 交接箱 07.0193

cross-curved arch 双曲拱 02.0402

cross electromagnetic wave small room 横电磁波小室 04.1514

cross-haul traffic 对流运输 06.0106

crosshead　十字头　04.1191

crosshead key　十字头扁销　04.1192

crosshead pin　十字头圆销　04.1193

crossing　交叉　03.0290，辙叉　03.0307，交叉跨越　04.1490

crossing of routes　进路交叉　06.0517

crossing untwining　交叉疏解　06.0518

cross leveling　横断面测量　02.0105

cross modulation interference　交调干扰　07.0528

crossover　渡线　03.0296

cross-over pole　跨越杆　07.0168

cross propagation characteristic　横向传播特性　04.1518

cross-section leveling　横断面测量　02.0105

cross-section method for location of line　横断面选线　02.0162

cross-section method of railway location　横断面选线　02.0162

cross-section survey　横断面测量　02.0105

cross stay derrick girder erecting machine　斜撑桅杆式架梁起重机　02.0945

crosstalk　串音　07.0038

crosstalk meter　串音测试器　07.0041

crosstalk suppression filter　串音抑制滤波器　07.0350

crosstie　小横梁　05.0398

cross tie　轨枕　03.0108

cross-type junction terminal　十字形枢纽　06.0398

crown sheet　内火箱顶板　04.1120

crusher　碎石机　02.0891

crushing of rail head　轨头压溃　03.0178

crustal stress　地应力　02.0716

CTC　调度集中　08.0460

cubic parabola curve　三次抛物线曲线　03.0025

current capital　流动资金　06.0748

current-carrying catenary　载流承力索　04.1312

current collector　受电器　04.0533

current-force characteristic　电流－力特性　04.1626

current limiting reactor　限流电抗器　04.0513

current maintenance　日常维修　08.0725

current relay　电流继电器　04.0591

current sensor　电流传感器　04.0631

current system　电流制　04.0312

current system changeover switch　电流制转换开关　04.0563

curve adjusting　整正曲线　03.0205

curve compensation　曲线折减　02.0201

curve control point　曲线控制点　02.0118

curved beam　曲梁　02.0415

curved-beam truck　曲梁式转向架　05.0068

curved bridge　曲线桥　02.0414

curved frog　曲线辙叉　03.0353

curved switch　曲线尖轨　03.0315

curve lining　整正曲线　03.0205

curve negotiating　曲线通过　05.0845

curve negotiation test　曲线通过试验　04.0667

curve of opposite sense　反向曲线　02.0178

curve post　曲线标　03.0250

curve resistance　曲线阻力　04.0101

curves of same sense　同向曲线　02.0177

curve versine　曲线正矢　03.0031

cushioned berth sleeping car　软卧车　05.0728

cushioned-seat coach　软座车　05.0726

cushioning underframe　活动中梁底架　05.0390

cut　路堑　02.0239

cut and fill section　半堤半堑　02.0240

cut-fill transition　土[石]方调配　02.0838

cut-off　遮断比，＊断汽　04.1092

cut-off wavelength　截止波长　07.0461

cut-out cock　截断塞门　05.0288，转换塞门　05.0597

cut section　分割区段　08.0531

cut-section of a track circuit　轨道电路分割　08.0530

cut the vertical earthwork bottom　掏底开挖，＊挖神仙土　02.0997

cutting edge of open caisson　沉井刃脚　02.0585

cutting slope　路堑边坡　02.0280

cut-type gallery　路堑式明洞　02.0697

cut-type open cut tunnel　路堑式明洞　02.0697

cut-type tunnel without cover　路堑式明洞　02.0697

cyclic code　循环码　07.0825

cyclic redundancy check　循环冗余检验　07.0797

cyclic scanning system　循环检查制　08.0474

cyclone-combustion　旋风燃烧　04.1567

cyclone type air filter　旋风筒式空气滤清器　04.1048

cylinder　气缸　04.0872

cylinder block　气缸体　04.0864

cylinder bore diameter　气缸直径　04.0706

cylinder bushing　汽缸套　04.1182

cylinder drain valve　汽缸排水阀　04.1189

cylinder head　气缸盖　04.0919

cylinder head nut　气缸盖螺母　04.0921

cylinder head stud　气缸盖螺栓　04.0920

cylinder lever　制动缸前杠杆　05.0356

cylinder lever connecting rod　连接拉杆　05.0358

cylinder liner　气缸套　04.0873

cylinder power equalizing test　各缸均匀性试验　04.0857

cylinder saddle　汽缸鞍　04.1185

cylinder tractive effort　汽缸牵引力　04.1089

cylindrical roller bearing　圆柱滚子轴承　05.0141

cylindrical shaft retaining wall　管柱挡墙　02.0290

D

daily and shift traffic plans　调度日班计划　06.0343

daily car requisition plan　日要车计划　06.0099

daily precipitation　日降雨量　03.0582

damage accident　损失事故　06.0640

damage rate of goods　货损率　06.0646

damage repair for bridge and tunnel　桥隧病害整治　03.0567

damper　减振器　05.0156

damper test stand　减振器试验台　05.0997

damping coefficient　阻尼系数　05.0244

damping factor　阻尼因数　03.0537,相对阻尼系数　05.0248

damping index　减振指数　05.0246

dangerous accident　险性事故　06.0624

dangerous freight　危险货物　06.0172

dangerous goods　危险货物　06.0172

dangerous influence　危险影响　04.1464

data　数据　07.0826

data channel　数据信道　07.0827

data circuit terminating equipment　数据电路端接设备　07.0828

data communication　数据通信　07.0729

data communication network　数据通信网　07.0061

data link　数据链路　07.0805

data network　数据网[络]　07.0752

data signaling rate　数据信号速率　07.0730

data station　数据站　07.0831

data terminal equipment　数据终端设备　07.0829

data transfer rate　数据传送率　07.0731

day signal　昼间信号　08.0007

day-to-day traffic working plan　运输工作日常计划　06.0342

DC auxiliary generator　直流辅助发电机　04.0466

DC auxiliary motor　直流辅助电动机　04.0472

DC chopper　直流斩波器　04.0330

DC contactor　直流接触器　04.0546

DC control generator　直流控制发电机　04.0468

DC current transformer　直流电流互感器　04.0635

DC drive　直流传动　04.0341

DCE　数据电路端接设备　07.0828

DC electric locomotive　直流电力机车　04.0300

DC electric traction system　直流电力牵引制　04.1269

DC exciter　直流励磁机　04.0469

DC high speed circuit-breaker　直流高速断路器　04.0532

DC main generator　直流主发电机　04.0463

DC power supply panel　直流电源屏　08.0652

DC power supply system　直流供电制　08.0630

DC relay　直流继电器　08.0580

DC starting generator　直流起动发电机　04.0467

DC system　直流制　04.0313

DC track circuit　直流轨道电路　08.0490

DC traction motor　直流牵引电动机　04.0395

DC traction substation　直流牵引变电所　04.1343

dead load　恒载　02.0331

dead section　死区段　08.0529

debris flow　泥石流　03.0483

decay accident　腐坏事故　06.0641

deceleration 减速 04.0107

deceleration force 减速力 04.0110

decibel 分贝 07.0044

decision value 判决值 07.0374

deck bridge 上承式桥 02.0425

decode 解码 07.0017

decoder 译码器 07.0399

deep burying tunnel 深埋隧道 02.0650

deep-depth tunnel 深埋隧道 02.0650

deep dish wheel plate 深盆辐板 05.0123

deep hole blasting 深孔爆破 02.0273

deep tunnel 深埋隧道 02.0650

deep well pump 深井水泵 02.0918

default value 缺省值 07.0751

defect project 缺陷工程 02.1054

deficient superelevation 欠超高 03.0027

deflecting bar 转向杆 08.0297

deflecting plate 垂直板，＊垂直反射板 04.1147

deflection 挠度 02.0439

deflection angle 转向角 02.0122

deflection of bridge span 桥梁挠度 03.0532

deformation allowance 预留变形量 02.0717

deformation-free zone 固定区 03.0236

deformation pressure 形变压力 02.0708

degree of reaction 反动度 04.1543

degree of supercharging 增压度 04.0794

degree of turbocharging 增压度 04.0794

delaying time indication 晚点表示 08.0486

delimiter 定界符 07.0898

delivery mistake 误交付 06.0645

delivery-receiving acceptance 施工验收 02.1058

delivery-receiving stake 交接桩 02.0835

delivery valve 出油阀 04.0952

delivery valve matching parts 出油阀偶件 04.0951

delivery valve seat 出油阀座 04.0953

delta modulation 增量调制 07.0026

demagnetisation curve 去滞曲线 04.1619

demodulation 解调 07.0019

demountable truss 拆装式桁架 02.0389

density of freight traffic 货运密度 06.0151

departure indication 离去表示 08.0160

departure indicator 发车表示器 08.0090

departure indicator circuit 发车表示器电路 08.0249

departure route 发车进路 06.0249

departure section 离去区段 08.0208

departure signal 发车信号 08.0027

departure track 出发线 06.0407

departure track indicator 发车线路表示器 08.0092

departure yard 出发场 06.0440

dependent signal 从属信号机 08.0075

depot siding 段管线 06.0423

depressed center flat car 凹底平车 05.0834

depressed joint 钢轨低接头 03.0156

depth of ballast 道床厚度 03.0132

depth of girder 梁高 02.0437

derailer 脱轨器 05.0937

derail indicator 脱轨表示器 08.0095

derailment 脱轨 05.0876

design current velocity 设计流速 02.0075

design discharge 设计流量 02.0091

designed flood hydrograph 设计洪水过程线 02.0094

design elevation 设计高程，＊设计标高 02.0076

design for guiding construction organization 指导性施工组织设计 02.0854

design for practical construction organization 实施性施工组织设计 02.0855

design load 设计荷载 02.0329

design of flight strip 航带设计 02.0132

design phase 设计阶段 02.0010

design speed 构造速度 01.0078

design speed of car 车辆构造速度 05.0842

design stage 设计阶段 02.0010

design water level 设计水位 02.0092

destination address 终点地址 07.0902

destination railway 到达路 06.0186

destressing 放散温度力 03.0230

detaching and attaching of cars 摘挂调车 06.0257

detail fracture of rail head 轨头微细裂纹 03.0180

detecter 检波器 07.0056

detection 检波 07.0020

detention time of car in transit without resorting 无调中转车停留时间 06.0264

detention time of car in transit with resorting 有调中

转车停留时间 06.0262

detention time of locomotive at depot 机车在段停留时间 04.0208

determination of bearing slab of subgrade 路基承载板测定 02.0263

determination of nuclear density-moisture 核子密度湿度测定 02.0262

detouring section 绕行地段 02.0203

detour route 迂回进路 08.0154

development of line 展线 02.0163

deviation from transposition interval 交叉偏差 07.0231

device for stopping buffer action 车钩缓冲停止器 06.0165

diagonal brace 对角撑, *斜撑 05.0401

diagonally symmetrical loading force 斜对称载荷 05.0235

diagonal member 斜杆 02.0449

dial 拨号盘 07.0083

dial impulse 拨号脉冲 07.0121

dialling 拨号 07.0120

dialling tone 拨号音 07.0276

dial-up terminal 拨号终端 07.0909

diametrical line of junction terminal 枢纽直径线 06.0422

diamond crossing 菱形交叉 03.0291

diaphragm 膜板 05.0299

diaphragm plate 垂直板, *垂直反射板 04.1147

diaphragm type air spring 膜式空气弹簧 05.0171

diaphragm wall method 地下连续墙法 02.0727

die block 月牙板滑块 04.1209

dielectric test 介电强度试验 04.0661

diesel coach set 内燃动车组 04.0689

diesel-electric locomotive 电力传动内燃机车 04.0688

diesel engine 柴油机 04.0690

diesel engine characteristic 柴油机特性 04.0848

diesel engine starting test 柴油机起动试验 04.0856

diesel engine tachometer 柴油机转速表 04.0965

diesel-hydraulic locomotive 液力传动内燃机车 04.0687

diesel locomotive 内燃机车, *柴油机车 04.0006

diesel-mechanical locomotive 机械传动内燃机车 04.0686

diesel oil for locomotive 机车用柴油 04.0055

diesel pile driver 柴油打桩机 02.0906

diesel traction interference 内燃牵引干扰 08.0675

different-frequency simplex radio communication 异频单工无线电通信 07.0514

differential modulation 差分调制 07.0832

differential phase shift keying 差分移相键控 07.0833

differential pulse-code modulation 差分脉码调制 07.0028

different-side anchor 异侧下锚 04.1433

diffuser 扩压器 04.0979

digital display 数字显示器 04.0613

digital microwave relay communication 数字微波中继通信 07.0665

digital multiplex equipment 数字复接设备 07.0368

digital telephone network 数字电话网 07.0060

dilivery of freight 货物交付 06.0131

dilivery of luggages and parcels 行李包裹交付 06.0053

dilution zone 掺混区 04.1572

dim head light 副前照灯 04.0642

dining car 餐车 05.0732

dining room 餐室 05.0755

direct admission valve 直接进入阀 05.0366

direct air brake pipe 直通制动管 05.0279

direct cost of project 工程直接费 02.0031

direct current bridge 直流电桥 07.0220

direct drive 直接驱动 04.0357

direct expense of project 工程直接费 02.0031

direct feeding system 直接供电方式 04.1275

direct feeding system with return wire 带回流线的直接供电方式 04.1276

direct holding fastening 不分开式扣件 03.0087

direct injection combustion chamber 直接喷射燃烧室 04.0797

directional antenna 定向天线 07.0566

directional blasting 定向爆破 02.0271

directional filter 方向滤波器 07.0348

directional switch 方向转接器 08.0427

directional traffic power source 方向电源 08.0644

directly surveillanced network 近程网路 08.0477

directly surveillanced subsection 近程监督分区，
* 近程遥信分区 08.0475

direct plug-in subscriber 直插用户 07.0401

direct pressure steam heating equipment 直压式采暖
装置 05.0593

direct release 直接缓解，* 一次缓解 04.0136

dirt collector 集尘器 05.0287

disabled armyman ticket 残废军人票 06.0043

disaster geology 灾害性地质 02.0990

disc brake 盘型制动 04.0124

discharge of sedimentation by blasting 爆破排淤
02.0268

discharger 放电器 04.0637

discharge temperature of supercharger 增压器出口温
度 04.0793

disc-loading locomotive antenna 加顶圆盘机车天线
07.0568

dispatcher's control 调度控制 08.0468

dispatcher's office 调度所 06.0335

dispatcher's supervision system 调度监督 08.0471

dispatching telephone control board with VF selective
calling 音选调度电话总机 07.0425

dispatching telephone subset with VF selective calling
音选调度电话分机 07.0426

dispatching telephone within railway administration
局线调度电话 07.0296

dispersal of rail gaps 整正轨缝 03.0209

disposal locomotive 支配机车 04.0021

disrepair of flow regulating and shore protecting structure 河调－防护失修 03.0515

dissectible 挤切 08.0620

dissectible pin 挤切销 08.0621

distance between backs of wheel flanges 轮对内侧距
05.0048

distance between bogie pivot centers 转向架中心距
离 04.0036

distance between centers of freight turning rack 货物
转向架支距 06.0163

distance between centers of tracks 线间距[离]
02.0152

distance between journal centers 轴颈中心距

05.0049

distance-force characteristic 气隙－力特性
04.1627

distance of one complete turnround of locomotive 机
车全周转距离 04.0209

distance-to-coupling measurement 测长 08.0387

distance-to-go 动态长度 08.0372

distant signal 预告信号机 08.0076

distant subscriber 远距用户 07.0404

distortion 畸变 07.0814

distributed network 分布[式]网[络] 07.0763

distributing terminal board 分线盘 08.0265

distributing valve 分配阀 05.0291

distributing valve test rack 分配阀试验台 05.0370

distribution box without protectors 分线盒
07.0196

distribution box with protectors 分线箱 07.0195

distribution cable 配线电缆 07.0188

distribution of stations 车站分布 02.0153

district 区段 01.0047

district station 区段站 06.0366

district train 区段列车 06.0223

district transfer train 区段小运转列车 06.0225

disturbance 干扰 07.0042

disturbed channel 被串通路 07.0137

disturbing calculation current of feeding section 供电
臂干扰计算电流 04.1326

disturbing channel 主串通路 07.0136

disturbing influence 干扰影响 04.1492

ditcher 开沟机 03.0460

ditching and grading machine 开沟平路机
03.0461

diversified power supply system 分散供电方式
04.1297

diversion dike 导流堤 02.0317

diversity reception 分集接收 07.0577

divert shunt reactor 分流电抗器，* 感应分路
04.0510

divert shunt resistor 分流电阻器 04.0518

dividing ridge 分水岭 02.0088

diving drill 潜孔钻机 02.0889

diving drill machine 潜水钻机 02.0913

diving pump 潜水泵 02.0920

division communication　区段通信　07.0410

document facsimile apparatus　文件传真机　07.0728

dog spike　道钉，＊狗头钉，＊钩头钉　03.0090

dominant project　控制工程　02.0860

door guide roller　门滑轮　05.0501

door guide wheel　［塞入门］导轮　05.0500

door latch　［货车］门锁　05.0503

door lock　［客车］门锁　05.0502

door post　门柱　05.0432

door seal gasket　箱门封条　06.0563

dormitory car　宿营车　05.0740

doubinary encoding　双二进制编码　07.0736

double arm pantograph　双臂受电弓　04.0536

double-arm routing　双肩回交路　04.0189

double attenuation　倍程衰减　04.1519

double beam girder-erecting machine　双梁式架桥机　02.0949

double break　双断　08.0223

double cantilever girder-erecting machine　双悬臂式架桥机　02.0947

double circuit power supply　双回路供电　04.1280

double crossover　交叉渡线　03.0297

double current　双流　07.0708

double cylindrical shade wall lamp　双筒壁灯　05.0691

double-deck bridge　双层桥　02.0428

double deck coach　双层客车　05.0721

double-deck pallet　双面托盘　06.0568

double deck passenger car　双层客车　05.0721

double deck sedan car　双层轿车平车　05.0829

double-direction running automatic block　双向自动闭塞　08.0330

double freight operations　双重作业　06.0266

double frequency inductor　双频感应器　08.0429

double locomotive traction　双机牵引，＊双机重联牵引　04.0224

double rail track circuit　双轨条式轨道电路　08.0507

double reduction gear drive　双侧减速齿轮驱动　04.0352

double rod insulator　双重绝缘棒式绝缘子　04.1380

double rolling on double pushing track　双推双溜　06.0479

double sided linear induction motor　双边型直线感应电动机　04.1605

double signal location　并置信号点　08.0347

double slip switches　复式交分道岔　03.0295

double spur grade　人字坡　02.0181

double strut insulator　双重绝缘棒式绝缘子　04.1380

double track all-relay semi-automatic block system　双线继电半自动闭塞　08.0327

double track bridge　双线桥　02.0421

double track railway　双线铁路　01.0016

double track tunnel　双线隧道　02.0652

double tube fluorescent lamp　双管荧光灯　05.0688

double wall steel cofferdam bored foundation　双壁钢围堰钻孔基础　02.0587

double wire semaphore signal　双线臂板信号机　08.0060

down direction　下行方向　06.0285

down hump trimming signal　下峰信号　08.0038

down time for car under repair　车辆检修在修时间　05.0973

down time for holding cars for repairing　车辆检修停留时间　05.0972

downward swing door　［货车］下开门　05.0494

dowty accelerator　加速顶　06.0462

dowty accelerator-retarder　加减速顶　06.0463

dowty controllable retarder　可控减速顶　06.0464

dowty retarder　减速顶　06.0461

DPSK　差分移相键控　07.0833

draft gear capacity　缓冲器容量　05.0579

draft gear housing　缓冲器箱体　05.0559

draft gear initial compression　缓冲器预紧力　05.0581

draft gear reaction force at rating travel　缓冲器阻抗力　05.0582

draft gear recoil　缓冲器反弹　05.0584

draft gear travel　缓冲器行程　05.0580

drafting apparatus　通风装置　04.1139

drafting efficiency　通风效率　04.1114

draft key　钩尾销　05.0547

draft plate　前垂板　04.1149

draft sill　牵引梁　05.0421

dragline 索铲挖土机 02.0879

drainage area 汇水面积 02.0089

drainage channel 排水槽 02.0302

drainage coil 排流线圈 07.0210

drainage ditch 排水沟 02.0295

drainage opening 泄水孔 02.0511

drainage sand blanket 排水砂垫层 02.0314

drainage tunnel 渗水隧洞 02.0304

drain cavern 泄水洞 02.0801

drain ditch 排水沟 02.0295

drain tunnel 泄水洞 02.0801

drawback project 缺陷工程 02.1054

draw bar 牵引杆 04.1080

drawbar pull 车钩牵引力 04.0091

draw beam [机车]牵引梁 04.1056

draw gear 牵引装置 04.1079

draw-out track at grade 坡度牵出线 06.0434

drift sand 流砂 02.0609

drilling and blasting method 钻[眼]爆[破]法 02.0721

drilling foundation for checking purpose 墩台基础钻探 03.0524

drill log of stratum 地层柱状图 02.0068

drill machine with casing 套管钻机 02.0914

drinking water boiler 茶炉 05.0670

drip pan 水盘 05.0657

driven gear 从动齿轮 04.1078

driven wheel set 从动轮对，＊他动轮对 04.1230

driver 司机 04.0201

driver controller 司机控制器 04.0573

driver instructor 指导司机 04.0203

driver's cab 司机室 04.0233

driver's cab air conditioner 司机室空调装置 04.0616

driver's cab air heater 司机室热风装置 04.0615

driver's cab electric heater 司机室取暖电炉 04.0614

driver's chair 司机座椅 04.1059

driver's desk 司机操纵台 04.0234

driver's log 司机运转报单 04.0232

driver's service-report 司机运转报单 04.0232

driving box 动轴箱，＊动轮轴箱 04.1223

driving box shoe 动轴箱平铁，＊动轴箱槽铁 04.1224

driving box wedge 动轴箱楔铁，＊动轴箱斜铁 04.1225

driving gear 主动齿轮 04.1077

driving system 传动系统 08.0608

drooping 下垂 05.0442

drop away time 释放时间，＊落下时间 08.0561

dropper 吊索 04.1403

dropping window 下开窗 05.0505

drop test 落锤试验 05.0590

drop test of rail 钢轨落锤试验 03.0192

drum concrete mixer 鼓筒式混凝土搅拌机 02.0926

drum position changeover switch 鼓形位置转换开关 04.0559

dry bridge 旱桥 02.0359

dry bulk container 干散货集装箱 06.0552

dryer 干燥器 05.0635

dry friction type journal box positioning device 干摩擦式轴箱定位装置 05.0149

dry pipe 干燥管 04.1163

dry shotcreting 干喷混凝土 02.0779

dry shotcreting machine 干喷混凝土机 02.0971

dry tamped cement mortar 干捣水泥砂浆 03.0556

dual control 双重控制 08.0186

dual frequency electric locomotive 双频率制电力机车 04.0304

dual voltage electric locomotive 双电压制电力机车 04.0303

duct cable 管道电缆 07.0183

dug pile 挖孔桩 02.0600

dumper 翻车机 06.0601

dumping car 自卸车 05.0805

dumping truck 自卸汽车 02.0985

duplex operation 双工 07.0705

duplex radio communication 双工无线电通信 07.0515

duplex transmission 双工传输 07.0781

durability test 耐久性试验 04.0292

dust control 防尘 02.1025

dust guard 防尘板 05.0136

dust guard seat 防尘板座 05.0112

dust prevention 防尘 02.1025

dwarf signal　矮型信号机　08.0062

dynamic brake　动力制动　04.0115

dynamic coupling　动态耦合　04.1620

dynamic inscribing　动力内接　03.0141

dynamic load of car　货车动载重　06.0154

dynamic penetration test　动力触探试验　02.0057

dynamics of suspension system　悬浮系统动力学　04.1621

dynamic spring deflection　弹簧动挠度　05.0239

dynamics test　动力学试验　04.0289

dynamic test　动载试验　03.0534

dynamic track irregularity　动态不平顺　03.0144

dynamic track stabilizer　动力稳定机　03.0427

dynamometer car　动力试验车　05.0745

dynastarter　直流起动发电机　04.0467

E

earphone　耳机　07.0072

earth clamp　接地线夹　04.1378

earth fault relay　接地继电器　04.0601

earth flow　泥石流　03.0483

earthing pole　接地安全棒　04.0639

earthing reactor　接地电抗器　04.0508

earthing switch　接地开关　04.0567

earth pressure　土压力　02.0332

earth protection discharger　接地保护放电装置　04.1341

earthquake intensity　地震烈度　02.0351

earthquake magnitude　地震震级　02.0352

earth resistivity　大地电阻率　04.1476

earth return brush　接地回流电刷　04.0640

earth station　地球站　07.0672

ear wall tunnel portal　耳墙式洞门　02.0671

easement curve　缓和曲线　02.0174

easiest rolling car　最易行车　06.0492

eastern type toilet　蹲式便器　05.0782

easy grade　缓和坡度　02.0197

easy gradient for acceleration　加速缓坡　02.0199

easy rolling car　易行车　06.0491

easy running track　易行线　06.0430

eaves　雨檐　05.0465

eccentric crank　偏心曲拐　04.1229

eccentric rod　偏心杆　04.1204

echo　回声　07.0143

echo effect　回波效应　07.0834

economic accounting　经济核算　06.0756

economic effects　经济效果　06.0757

economic investigation　经济调查　02.0042

economic investigation of freight traffic　货运经济调查　06.0686

economic survey　经济调查　02.0042

eddy current brake　涡流制动　04.0121

EDS　电动排斥式系统　04.1580

education car　教育车　05.0743

effective compression ratio　有效压缩比　04.0752

effective current of feeding section　供电臂有效电流　04.1325

effective current of train　列车有效电流　04.1322

effective power　有效功率　04.0814

effective specific fuel consumption　有效燃油消耗率　04.0766

effective track length　线路有效长　06.0433

effective volume　有效容积　05.0040

effect of investment　投资效果　06.0711

efficiency of locomotive at wheel rim　机车轮周效率　04.0076

elastically positioned wheelset　弹性定位轮对　05.0880

elastic bolster guide　弹性止挡　05.0093

elastic rail fastening　弹性扣件　03.0089

elastic rail spike　弹簧道钉　03.0091

elastic resistance　弹性抗力　02.0711

elastic side bearing　弹性旁承　05.0183

elastic squeeze-out　弹性挤开　03.0148

elastic wheel　弹性车轮　05.0106

electrical connector　电连接　04.1374

electrically connecting clamp　电连接线夹　04.1375

electrically connecting clamp for contact wire　接触线电连接线夹　04.1376

electrical measuring instrument　电测仪表　04.0618

electric apparatus room　电器室　04.1062

electric automatic coupler 自动电连接器 04.0392

electric braking test 电气制动试验 04.0681

electric coupler 电连接器 04.0391

electric coupling coefficient 电耦合系数 04.1523

electric fan 电扇 05.0706

electric heater 电加热器 05.0695

electric heating equipment 电热采暖装置 05.0615

electric heating glass 电热玻璃 04.0617

electric hoist 电动吊车，＊电动葫芦 02.0902

electric induction 电感应 04.1466

electric interlocking 电气集中[联锁] 08.0108

electric interlocking for hump yard 驼峰电气集中 08.0374

electricity for locomotive 机车用电 04.0056

electric lock 电锁器 08.0314

electric locking 电气锁闭 08.0135

electric locomotive 电力机车 04.0007

electric moistening device 电加湿器 05.0697

electric motor operated conveyer 电动传送设备 08.0407

electric motor train unit 电动车组 04.0309

electric multiple unit 电动车组 04.0309

electric power distribution cabinet 配电柜 05.0677

electric railway 电气化铁路，＊电力铁道 01.0020

electric semaphore signal 电动臂板信号机 08.0058

electric shock protector 触电保安器 02.1035

electric staff instrument 电气路签机 08.0353

electric staff system 电气路签闭塞 08.0322

electric switch machine 电动转辙机 08.0603

electric system 电气系统 05.0671

electric tablet block system 电气路牌闭塞 08.0323

electric tablet instrument 电气路牌机 08.0354

electric traction interference 电力牵引干扰 07.0533

electric traction remote control system 电力牵引远动系统 04.1271

electric traction telemechanical system 电力牵引远动系统 04.1271

electric water pump 电动水泵 05.0602

electric winch 电动卷扬机 02.0901

electric wire conduit underneath the car 车底电线管 05.0702

electrification interference 电气化干扰 04.1491

electrified railway 电气化铁路，＊电力铁道 01.0020

electro-chemical treatment of soil 电化学加固土壤 03.0512

electrodynamic repulsion system 电动排斥式系统 04.1580

electrohydraulic switch machine 电液转辙机 08.0605

electro-hydraulic valve 电液阀 04.0583

electromagnetic attraction system 电磁吸引式系统 04.1583

electromagnetic compatibility 电磁兼容[性] 07.0575

electromagnetic contactor 电磁接触器 04.0545

electromagnetic environment 电磁环境 04.1516

electromagnetic rail brake 磁轨制动 04.0122

electromagnetic relay 电磁继电器 08.0578

electromagnetic shielding darkroom 电磁屏蔽暗室 04.1515

electromagnetic suspension system 电磁悬浮系统 04.1579

electromagnetic valve 电磁阀 04.0581

electro-mechanical interlocking 电机集中联锁 08.0107

electronic control circuit 电子控制电路 04.0387

electronic telephone switching system 电子电话交换机 07.0395

electropneumatic brake 电空制动 04.0119

electropneumatic brake circuit 电空制动电路 04.0388

electropneumatic brake equipment 电空制动装置 05.0269

electropneumatic contactor 电空接触器 04.0548

electropneumatic conveyer 电空传送设备 08.0406

electropneumatic switch machine 电空转辙机 08.0604

electropneumatic valve 电空阀 04.0582

electrostatic induced current 静电[感应]电流 04.1469

element 码元 07.0810

elevated railway 高架铁路 01.0027

elevation of curve 曲线超高 03.0026

elliptic spring 椭圆弹簧 05.0169

embankment 路堤 02.0238

embankment crossing reservoir 水库路基 02.0223

embankment filler 路堤填料 02.0252

embankment fill material 路堤填料 02.0252

embankment on plain river beach 河滩路堤 02.0221

embankment on river bank 滨河路堤 02.0222

emergency application 紧急制动 04.0133

emergency brake valve 紧急制动阀 05.0290

emergency braking 紧急制动 04.0133

emergency braking tone 紧急制动信号音 07.0627

emergency cable 应急电缆 07.0182

emergency call 紧急呼叫 07.0615

emergency lighting 事故照明 08.0648

emergency portion 紧急部 05.0293

emergency rail stored along the way 备用轨 03.0242

emergency repairs 临险抢护 03.0580

emergency reservoir 紧急风缸 05.0312

emergency rush engineering 临险抢护 03.0580

emergency sanding 紧急撒砂 04.0181

emergency satellite communication 应急卫星通信 07.0670

emergency signal 紧急信号 07.0614

emergency staircase 安全梯 02.1013

emergency stop mechanism 紧急停车装置 04.0969

emergency treatment after failure 故障办理 08.0702

empty and load changeover cock 空重车转换塞门 05.0289

empty car 空车 05.0012

empty/load brake equipment 空重车制动装置 05.0301

EMS 电磁悬浮系统 04.1579

encumbrance 结构高度 04.1453

end cock 端部塞门 05.0283

end collar 轴领 05.0109

end door 端门 05.0486

end floor beam 端横梁 02.0465

end frog 锐角辙叉 03.0352

end-of-block signal 码组结束信号 07.0837

end-of-car cushioning device 车端缓冲器 05.0558

end-of-text signal 正文结束信号 07.0838

end-of-transmission signal 传输结束信号 07.0839

end of turnout 道岔终端 03.0312

end post 端柱 05.0429

end sheathing 端墙包板 05.0445

end shield 端盖 04.0460

end sill 端梁 05.0395

end wall tunnel portal 端墙式洞门 02.0668

energizing circuit 励磁电路 08.0216

energy consumption per 10 000 t·km of locomotive 机车万吨公里能耗 04.0053

energy-storing brake 蓄能制动 04.0120

engine block 机体 04.0863

engine burn 钢轨擦伤 03.0172

engine detonation 柴油机爆燃，＊柴油机工作粗暴 04.0807

engine-driven supercharged diesel engine 机械增压柴油机 04.0698

engine-driven supercharging 机械增压 04.0789

engineering geological map 工程地质图 02.0067

engineering geologic condition 工程地质条件 02.0063

engineering geologic location of line 工程地质选线 02.0151

engineering geologic requirement 工程地质条件 02.0063

engineering geology 工程地质 02.0060

engineering ship 工程船舶 02.0952

engineering transportation 工程运输 02.0842

engineering vessel 工程船舶 02.0952

engine support 柴油机支座 04.0867

enlarged preliminary design 扩大初步设计 02.0016

enlargement 扩大 02.0744

entrance cable 进局电缆 07.0186

entrance door step ［客车］脚蹬 05.0405

entrance speed at retarder 减速器入口速度 06.0483

E-P brake circuit 电空制动电路 04.0388

E-P brake controller 电空制动控制器 04.0575

equalizer 均压线 04.0442,均衡梁 05.0214

equalizing lever 均衡杠杆 05.0362

equalizing pull rod 均衡拉杆 05.0361

equalizing reservoir 均衡风缸 05.0304

equalizing system 均衡系统 07.0328

equilateral turnout 单式对称道岔,＊双开道岔 03.0283

equipment failure 设备故障 06.0633

equipment out-of use 设备停用 08.0709

equivalence of coal 煤的技术当量 04.1105

equivalent distance of the oblique exposure 斜接近段的等效距离 04.1489

equivalent disturbing current 等效干扰电流 04.1493

equivalent span length 当量跨距 04.1436

equivalent track kilometerage 换算线路长度 03.0268

equivalent uniform live load 换算均布活载 02.0328

erection bar 架立钢筋 02.0477

erection by bridge girder erecting equipment 架桥机架设法 02.0531

erection by incremental launching method 顶推式架设法 02.0532

erection by protrusion 悬臂架设法 02.0535

erection by swing method 旋转法施工,＊转体施工 02.0544

erection with scaffolding 鹰架式架设法 02.0534

erosion 冲刷 03.0481

error control 差错控制 07.0793

error correcting code 纠错码 07.0796

error correction by feed-back repetition 反馈重发纠错 07.0842

error correction by information feed-back repetition 信息反馈重发纠错 07.0841

error detecting code 检错码 07.0795

error detection 差错检测 07.0843

error recovery 差错恢复 07.0794

escapement 牵纵拐肘 08.0300

ethernet 以太网 07.0773

evaluating fatigue residual life of bridge 桥梁疲劳剩余寿命评估 03.0541

evaluation by track inspection car 轨检车评分 03.0269

evaluation of engineering quality 工程质量评定 02.1050

evaporation capacity 蒸发量 04.1095

evaporation capacity for engine 供汽量 04.1096

evaporator 蒸发器 05.0632

evening lamp 半夜灯 05.0686

evenly distributing joint gaps 调整轨缝,＊均匀轨缝 03.0208

even mass train 匀质列车 05.0050

examination in depot 库检 05.0933

examining rack for out-of-gauge freight 超限货物检查架 06.0159

excavated surface 开挖工作面 02.0752

excavated work face 开挖工作面 02.0752

excavating foundation for checking purpose 墩台基础挖验 03.0523

excavating machine 挖掘机 02.0877

excavation of side wall at intervals 马口 02.0746

excavation prospecting 挖探 02.0051

excavator 挖掘机 02.0877

exceptional dimension freight 阔大货物 06.0160

exceptional length freight 超长货物 06.0161

exceptionally long and heavy train 超长超重列车 01.0083

excess air factor 过量空气系数 04.0743

excess elevation 过超高 03.0028

excessive joint gap 大轨缝 03.0158

excessively long-distance traffic 过远运输 06.0107

exchange 交换 07.0745

exchange phase connection 换相联接 04.1283

excitation ammeter 励磁电流表 04.0621

excitation contactor 励磁接触器 04.0553

excitation rectifier device 励磁整流装置 04.0339

excitation transformer 励磁变压器 04.0495

excitation winding 励磁绕组 04.0499

exhaust back pressure 排气背压 04.0738

exhaust box 排气箱 04.0985

exhaust branch pipe 排气支管 04.0984

exhaust cam 排气凸轮 04.0939

exhaust duct 排气道 04.1537

exhaust fan 排[气]风扇 05.0620

exhaustibility 制动衰竭性 05.0354

exhaust lap 排汽余面 04.1259

exhaust manifold 排气总管 04.0983

exhaust nozzle 乏汽喷口 04.1141

exhaust nozzle seat 乏汽喷口座 04.1142

exhaust pressure at turbine inlet 涡轮入口废气压力 04.0735

exhaust purification 废气净化 04.0809

exhaust smoke density 排气烟度 04.0810

exhaust stroke 排气行程 04.0726

exhaust temperature 排气温度 04.0737

exhaust temperature at cylinder head outlet 缸盖出口废气温度 04.0733

exhaust temperature at turbine inlet 涡轮入口废气温度 04.0734

exhaust turbine 废气涡轮 04.0977

exhaust turbocharging 废气涡轮增压 04.0788

exhaust valve 排气门 04.0927

exhaust velocity [for a turbine] 排气速度 04.1542

existing railway 既有铁路 01.0032

expanded preliminary design 扩大初步设计 02.0016

expansion bearing 活动支座 02.0516

expansion drum 膨胀水箱 04.1030

expansion joint 伸缩缝 02.0508

expansion rail joint 钢轨伸缩调节器，＊温度调节器 03.0075

expansion ratio 膨胀比 04.1544

expansion ratio of turbine 涡轮膨胀比 04.0736

expansion shoe 滑台 04.1221

expansion stroke 膨胀行程 04.0725

expansion tank 膨胀水箱 04.1030

expansion terminal pressure 膨胀终点压力 04.0757

expansion terminal temperature 膨胀终点温度 04.0756

expert system of tunnel 隧道专家系统 02.0703

exploration 勘探 02.0050

exploration drilling 钻探 02.0052

explosion accident 爆炸事故 06.0636

explosion-proof door of crankcase 曲轴箱防爆门 04.0869

explosive charge 安全炸药 02.1023

express extra ticket 特快加快票 06.0037

express train 旅客特别快车 06.0204

expropriation 征地 02.0832

extension 分机 07.0134

extension of line 展线 02.0163

extension rod 延长杆 05.0662

extrados 拱背线 02.0489

extra passenger train 临时旅客列车 06.0212

extruding concrete tunnel lining 挤压混凝土衬砌 02.0692

extruding swirl 挤压涡流 04.0803

F

fabricated short rail used on curves 缩短轨 03.0058

fabricated universal steel members 万能杆件 02.0955

facsimile apparatus 传真机 07.0723

facsimile receiver 传真接收机 07.0725

facsimile transceiver 传真收发机 07.0726

facsimile transmitter 传真发送机 07.0724

fading 衰落 07.0538

fail-safe 故障－安全 08.0666

failure accumulation 故障积累 08.0704

failure diagnostic 部件故障检测，＊故障诊断 04.0277

failure to the safe side 导向安全 08.0670

falling protector 防坠器 02.1028

false gradient 前后风压差 04.0154

false indication 乱显示 08.0048

false indication of a switch 道岔错误表示 08.0682

false locking 错误锁闭 08.0677

false release 错误解锁 08.0678

false release by itself 自动缓解 08.0683

false stopping of a signal 错误关闭信号 08.0680

fan 风轮 05.0613

far-end crosstalk 远端串音 07.0040

far-end crosstalk attenuation 远端串音衰减 07.0142

far field 远场 07.0460

far field area 远场区 07.0546

fastening-down temperature of rail 锁定轨温 03.0227

fast extra ticket 加快票 06.0036

fast passenger train 旅客快车 06.0203

fatigue residual life of bridge 桥梁疲劳剩余寿命 03.0540

fault isolating switch for phase splitter 劈相机故障隔离开关 04.0571

fault locator for overhead contact line equipment 接触网故障探测装置 04.1417

favorable condition for car rolling 溜车有利条件 06.0472

Fax 传真机 07.0723

feasibility study 可行性研究 02.0009

feed end 送电端 08.0514

feeder 供电线 04.1310

feed holes 输纸孔 07.0719

feeding current 馈电电流 07.0087

feeding section 供电臂 04.1292

feed valve 给风阀 04.0163

feed water heater 给水预热装置 04.1168

feed water rigging 给水装置 04.1165

ferromagnetic attraction force 铁磁吸力 04.1623

ferry boat 渡轮 02.0642

ferry slip 轮渡引线，＊轮渡斜引道 02.0643

ferry station 轮渡站 02.0640

ferry trestle bridge 轮渡栈桥 02.0641

field coat 工地漆 03.0554

field control point of aerophotogrammetry 航测外控点 02.0136

field equipment of CTC 调度集中分机 08.0462

field location 线路点 08.0464

field strength coverage 场强覆盖区 07.0540

field weakening 磁场削弱率 04.0436

field weakening contactor 磁场削弱接触器 04.0551

figurative progress of construction work 施工形象进度 02.0876

filament burn-out 灯丝断丝 08.0689

fill 路堤 02.0238

filler 间隔铁 03.0365

filling behind abutment 台后填方 02.0558

filling limiting valve 充量限制阀 04.1022

filling material of embankment 路堤填料 02.0252

filling pipe end 注水口 05.0665

fill slope 路堤边坡 02.0248

filter 过滤层 03.0475, 滤波装置 04.1304, 过滤器 05.0636, 滤波器 07.0057

filter capacitor 滤波电容器 04.0524

filter circuit 滤波电路 04.0374

filter reactor 滤波电抗器 04.0511

final accounts of completed project 竣工决算 02.0035

final location survey 定测 02.0101

financial management information system 财务管理信息系统 06.0762

financial statements review 财务决算审查 06.0761

fine coal locomotive 煤粉机车 04.1082

fine-grained soil fill 细粒土填料 02.0256

fine-grained soil filler 细粒土填料 02.0256

fire accident 火灾事故 06.0637

fire alarm signal 火警信号 06.0660

firebox 火箱 04.1116

firebox tube sheet 火箱管板 04.1132

fire door 炉门 04.1127

fire escape 安全梯 02.1013

fire-fighting apparatus and materials 消防器材 02.1019

fire-fighting equipment 消防设施 02.1017

fire hole 炉口 04.1126

fireman 司炉 04.0204

fire protection distance 防火净距 02.1021

fire protection gate 防火门 02.1018

fire protection strip 隔火带 02.1020

firewood raft 柴排 02.0320

firing order 发火次序 04.0772

first approach section　第一接近区段　08.0206

first departure section　第一离去区段　08.0209

fish-belly sill　鱼腹梁　05.0422

fish bolt　接头螺栓，＊鱼尾螺栓　03.0083

fish plate　接头夹板，＊鱼尾板　03.0077

fish plate with apron　带裙鱼尾板　03.0081

five-way cock　五通塞门　05.0667

fixed asset investment　固定资产投资　06.0710

fixed bearing　固定支座　02.0515

fixed capital　固定资金　06.0743

fixed-end arch　固端拱，＊无铰拱　02.0397

fixed-point frequency switching　频率定点切换
　07.0607

fixed pulley　定滑轮装置　04.1420

fixed reservation system　固定备用方式　04.1299

fixed seat　固定座椅　05.0771

fixed signal　固定信号　08.0010

fixed zone　固定区　03.0236

flake crack　白点　03.0174

flame cleaning　火焰除锈　03.0549

flame tube　火焰管　04.1556

flange　翼缘　02.0445

flange clearance　轮缘槽　03.0343

flange height　轮缘高度　05.0223

flange lubricator　轮缘喷油器　04.0638

flange plate　翼缘板　02.0446

flange thickness　轮缘厚度　05.0224

flange-way　轮缘槽　03.0343

flanging　车轮贴靠　05.0878

flash butt welding　电阻焊　03.0239

flashing power source　闪光电源　08.0663

flashing signal　闪光信号　08.0015

flat car　平车　05.0798

flat floor　平板地板　05.0402

flat grade　缓和坡度　02.0197

flat gradient for starting　起动缓坡　02.0198

flat joint bar　平型双头夹板　03.0078

flat marshalling yard　平面调车场　06.0508

flat pallet　平托盘　06.0566

flat regulation　平调　07.0311

flat sliding　踏面擦伤　05.0251

flexible arch bridge with rigid tie　朗格尔式桥，＊刚
　性系杆柔性拱桥　02.0404

flexible freight container　集装袋　06.0572

flexible pier　柔性[桥]墩　02.0563

flexible switch　可弯式尖轨转辙器　03.0345

flight strip design　航带设计　02.0132

floating box　浮箱　02.0954

floating bridge　浮桥　02.0395

floating caisson foundation　浮式沉井基础　02.0584

floating charge power supply　浮充供电　08.0633

floating crane　浮式起重机，＊浮吊　02.0898

flood control　防洪　03.0493

flood damage　水害　03.0497

flood frequency　洪水频率　02.0090

flood inundation on tracks　洪水淹没　03.0586

flood mark　洪水标记　03.0584

flood survey　洪水调查　02.0069

floor　底板　02.0680

floor area　地板面积　05.0038

floor beam　横梁　02.0463

floor draining device　地板排水装置　05.0669

floor height　地板面高度　05.0033

floor loading capability　箱底承载能力　06.0559

floor rack　离水格子　05.0525

floor stringer　辅助梁　05.0399

floor system　桥面系　02.0464

flow control　流量控制　07.0867

flow of rail head　轨头肥边　03.0169

flow passage　通流部分　04.1545

flow relay　流速继电器　04.0600

fluctuating coefficient of freight traffic　货运波动系
　数　06.0306

flue [tube]　大烟管　04.1135

flush valve　冲便阀　05.0784

flute or rib on sheathing　墙板压筋　05.0449

flutter　颤振　02.0526

flying arch method　先拱后墙法　02.0751

flyover　跨线桥，＊立交桥　02.0357

fly-shunting　溜放调车　06.0260

flywheel　飞轮　04.0893

folding seat　折叠座椅　05.0774

follower　从板　05.0573

following stretcher bar　道岔连接杆　03.0364

follow-up acceptance　随工验收　02.1052

foot bridge　人行桥　02.0360

foot plank　步行板　02.0506

forced circulation　强迫循环　04.0503

forced concrete mixer　强制式混凝土搅拌机　02.0929

forced guided circulation　强迫导向循环　04.0504

forced releasing　强拆　07.0127

forced-steering radial truck　迫导向径向转向架　05.0070

force ventilated motor　强迫通风式电动机　04.0414

forcing open of the point　挤岔　06.0630

fore and aft motion　伸缩运动　01.0087

forecast of traffic volume　运量预测　06.0688

forepoling　插板支护　02.0774

forestation against flood　防水林　03.0587

forest railway　森林铁路　01.0030

fore work district for power supply　供电领工区　04.1455

fork-lift truck　叉车　06.0600

fork pockets　叉槽　06.0562

form　模板　02.0956

formation　路基面　02.0233，基床　02.0241

formation base layer　基床底层　02.0243

formation level　路肩高程　02.0237

formation subgrade　路基　02.0210

formation top layer　基床表层　02.0242

formation width　路基面宽度　02.0234

formed cable　成端电缆　07.0190

fortuitous distortion　不规则畸变　07.0712

forward channel　正向信道　07.0802

forward propagation　正向传播　07.0535

fouling post　警冲标　06.0514

foundation　地基　02.0577

foundation base　基底　02.0247

foundation brake rigging　基础制动装置　05.0272

foundation examination by drilling　墩台基础钻探　03.0524

foundation examination by excavation　墩台基础挖验　03.0523

foundation soil　地基　02.0577

four-aspect automatic block　四显示自动闭塞　08.0339

four-axle car　四轴车　05.0003

four-frequency group mode　四频组方式　07.0601

four-function public address equipment　四用广播机　05.0693

four stroke diesel engine　四冲程柴油机　04.0692

frame　机座　04.0451，车架　04.1055

framed shed gallery　刚架式棚洞　02.0701

framed shed tunnel　刚架式棚洞　02.0701

frame format　帧格式　07.0897

frame mounted traction motor　架承式牵引电动机　04.0401

frame start delimiter　帧首定界符　07.0899

frazil ice survey　冰凌调查　02.0071

free cantilever erection with segments of precast concrete　悬臂拼装法　02.0537

free cantilever segmental concreting with suspended formwork　悬臂灌注法　02.0536

free height of center plate wearing surface from rail top　心盘面自由高　05.0042

free inscribing　自由内接　03.0137

freezable freight　易冻货物　06.0120

freezing depth　冻结深度　02.0065

freezing-plate refrigerator car　冷冻板冷藏车　05.0816

freight area　货区　06.0088

freight car　货车　05.0791

freight car repairing point　货车站修所，*站修所　05.0906

freight car technical condition handing-over post　货车技术交接所　05.0917

freight car temporary repairing shed　修车库，*修车棚　05.0899

freight car washing point　货车洗刷所　05.0905

freight container　集装箱　06.0538

freight container traffic　集装箱运输　06.0578

freight dispatching telephone　货运调度电话　07.0413

freight flow　货流　06.0143

freight flow diagram　货流图　06.0145

freight flow volume　货流量　06.0144

freight invoice　货票　06.0127

freight label　货物标记　06.0136

freight locomotive　货运机车　04.0015

freight operation at destination station　货物到达作业

06.0134

freight operation at originated station　货物发送作业
06.0133

freight operation en route　货物途中作业　06.0135

freight platform　货物站台　06.0091

freight rate　货物运价　06.0734, 货物运价率
06.0740

freight section　货位　06.0094

freight shed　货棚　06.0092

freight special line　货运专线　01.0040

freight station　货运站　06.0386

freight tariff No.　货物运价号　06.0739

freight traffic accident　货运事故　06.0619

freight traffic only line　货运专线　01.0040

freight traffic plan　货物运输计划　06.0681

freight traffic source　货源　06.0142

freight traffic volume　货物运输量　06.0682

freight train　货物列车　06.0214

freight train formation plan　货物列车编组计划
06.0278

freight train inspection and service point　货物列车检
修所，＊列检所　05.0907

freight train main inspection and service point　货物
列车主要检修所　05.0908

freight train ordinary inspection and service point　货
物列车一般检修所　05.0910

freight transfer point　货物交接所　06.0195

freight transit period　货物运到期限　06.0125

freight turning rack　货物转向架　06.0162

freight yard　货场　06.0082

frequency　频率　07.0007

frequency band　频带　07.0006

frequency channel　频道　07.0005

frequency converter　变频器　07.0053

frequency diversity　频率分集　07.0579

frequency division multiplex　频分复用　07.0011

frequency doubling attenuation　倍频衰减　04.1520

frequency group mode　频组方式　07.0600

frequency hopping communication　跳频通信
07.0675

frequency modulated track circuit　调频轨道电路
08.0499

frequency modulation　调频　07.0023

frequency offset　频[率]偏[差]　07.0284

frequency shift keying　频移键控　07.0732

frequency-shift modulated track circuit　移频轨道电
路　08.0500

frequency tracking switching mode　频率跟踪切换方
式　07.0605

fresh and live freight　鲜活货物　06.0117

Fresnel reflection　菲涅尔反射　07.0497

frictional clutch　摩擦联结器　08.0611

frictional working current　摩擦电流，＊故障电流
08.0612

friction braking　摩擦制动　05.0322

friction clutch　摩擦离合器　04.1049

friction pile　摩擦桩　02.0597

friction plate　摩擦板　05.0161

frog　辙叉　03.0307

frog angle　辙叉角　03.0308

frog heel　辙叉跟端　03.0323

frog number　辙叉号数　03.0309

frog toe　辙叉趾端　03.0322

front bumper　前缓冲铁　04.1218

front coil　前圈　08.0576

front contact　前接点　08.0552

front cylinder head　汽缸前盖　04.1183

front draft lug　前从板座　05.0411

front draft stop　前从板座　05.0411

front-end processor　前置机　07.0891

frontier station　国境站　06.0392

front part theoretical length of turnout　道岔前部理
论长度　03.0329

front rod of a point　尖端杆　08.0303

front slope　仰坡，＊正面坡　02.0665

front wall　前墙　02.0554

front window　前窗　04.1057

frost damage in tunnel　隧道冰害　03.0499

frost heave board　冻害垫板　03.0098

frost heaving　冻害　03.0484

frost heaving force　冻胀力　02.0347

frost shim　冻害垫板　03.0098

frozen freight　冻结货物　06.0174

frozen joint　冻结接头　03.0073

FSK　频移键控　07.0732

fuel consumption　燃油消耗量　04.0767

fuel dipstick 油尺 04.1071

fuel feed pump 燃油输送泵 04.1034

fuel injector 喷油器 04.0954

fuel level gage 油位表 04.1070

fuel precision filter 燃油精滤器 04.1036

fuel prefilter 燃油粗滤器 04.1035

fuel space 煤槽 04.1241

fuel tank 燃油箱 04.1069

full car vibration test rig 全车振动试验台 05.0993

full circle girder erecting crane 全回转式架梁起重机 02.0946

full-face tunnelling method 全断面开挖法 02.0728

full field 满磁场,＊全磁场 04.0432

full-laminated frame 全叠片机座 04.0455

fullness coefficient of indicator diagram 示功图丰满系数 04.0762

full open position of coupler 全开位置 05.0568

full section ballast consolidating machine 全断面道床夯实机 03.0425

full service application 最大常用制动,＊常用全制动 04.0130

full service braking 最大常用制动,＊常用全制动 04.0130

full service reduction 最大常用减压 04.0151

full-web section switch rail 特种断面尖轨 03.0357

fundamental construction clearance 基本建筑限界 01.0056

fundamental interlocking circuit 基本联锁电路 08.0228

fundamental structure gauge 基本建筑限界 01.0056

fundamental technical requirements for repair and inspection 检修基本技术条件 04.0264

fundamental transposition interval 基本交叉间隔 07.0230

fuse 熔断器 08.0664

fuse break alarm 熔断器断丝报警 08.0178

fuse burn-out 熔断器断丝 08.0690

fusible plug 易熔塞,＊铅堵 04.1179

G

gabion 石笼 03.0506

gage correction 改道 03.0202

gage elastically widened 弹性挤开 03.0148

gage line 钢轨工作边 03.0062

gage line shelly cracks 轨头剥离 03.0176

gage rod 轨距杆 03.0105

gage tie 轨距杆 03.0105

gage tie bar 轨距杆 03.0105

gage widening 挤轨 05.0875

gaging of track 改道 03.0202

gain 增益 07.0031

gallery 明洞 02.0694

gallery portal 明洞门 02.0675

galloping 驰振 02.0525

gangway 内走廊,＊通道 04.1063,[客室]通道 05.0788

gangway door 外端门,＊折棚门 05.0487

gangway foot plate 渡板 05.0453

gap in the frog 辙叉有害空间 03.0340

'gap' in the train diagram 运行图'天窗' 06.0344

gas consistency 瓦斯浓度 02.0996

gas control 瓦斯治理 02.1022

gas density 瓦斯浓度 02.0996

gas filled arrester 充气避雷器 07.0206

gas fired locomotive 煤气机车 04.1083

gas generator 燃气发生器 04.1532

gas maintenance type optical fiber cable 充气维护型光缆 07.0487

gas projection 瓦斯突出 02.0995

gas ring 气环 04.0907

gas-sensitive leak detector 气敏查漏仪 07.0218

gas-tight block 气闭头 07.0199

gasturbine locomotive 燃气轮机车 04.0008

gate change-over valve 闸板转换阀 04.1024

gate valve 闸板阀 04.1023

gateway 网关 07.0771

gauge 限界 01.0053

gauge widening 轨距加宽 03.0023

GB freight container 国家标准集装箱 06.0539

gearless drive 直接驱动 04.0357

general calling　全呼　07.0422

general freight station　综合性货运站　06.0077

general goods station　综合性货运站　06.0077

general inspection　一般性检查　04.0656

general purpose container　通用集装箱　06.0541

general-purpose freight car　通用货车　05.0792

general rate　普通运价　06.0735

general scour　一般冲刷　02.0096

general subgrade　一般路基　02.0244

generator car　发电车　05.0741

geodetic control point of portal location of adit　隧道洞口投点　02.0144

geographical circuitry　站场型网路　08.0218

geogrid　土工格栅，＊土工网格　03.0508

geologic survey　地质调查　02.0041

geomorphologic survey　地貌调查　02.0040

geophysical prospecting　物探　02.0053

geotextile　土工织物，＊土工布　03.0507

geothermal gradient　地温梯度　02.0715

Giesl ejector　矩形通风装置，＊扁烟筒　04.1140

girder without ballast and sleeper　无碴无枕梁　02.0377

girder with rolled steel section encased in concrete　型钢混凝土梁，＊劲性骨架混凝土梁　02.0378

global positioning system　全球定位系统　02.0137

glued insulated joint　胶结绝缘接头　03.0071

gondola car　敞车　05.0794

goods area　货区　06.0088

goods category　货物品类　06.0100

goods locomotive　货运机车　04.0015

goods platform　货物站台　06.0091

goods section　货位　06.0094

goods shed　货棚　06.0092

goods train　货物列车　06.0214

goods yard　货场　06.0082

goose neck tunnel　鹅颈槽　06.0564

governor　调速器　04.0960

governor drive gear　调速器驱动齿轮，＊调速器传动齿轮　04.0943

GPS　全球定位系统　02.0137

grab iron　扶手　05.0418

grade　坡度　02.0180

grade compensation　坡度折减　02.0200

grade crossing　道口　03.0273

grade crossing pavement　道口铺面　03.0274

grade crossing watchman　道口看守工　03.0008

graded index multimode optical fiber　多模渐变型光纤　07.0444

grade of station site　站坪坡度　02.0169

grade post　坡度标　03.0251

grader　平地机　02.0886

grade section　坡段　02.0192

grade separation　立体交叉　03.0272

grade separation bridge　跨线桥，＊立交桥　02.0357

gradient　坡度　02.0180

gradient compensation　坡度折减　02.0200

gradient resistance　坡道阻力　04.0100

gradient within the switching area　道岔区坡　06.0497

grading　分品复接　07.0156

graduated application　阶段制动　04.0131

graduated increasing　阶段提升　04.0145

graduated release　阶段缓解　04.0137

graphic logs of strata　地层柱状图　02.0068

grate　炉床　04.1125

grate area　炉床面积　04.1254

grave accident　重大事故　06.0622

gravel ballast　卵石道碴　03.0129

gravel car　砂石车　05.0806

gravitation type relay　重力继电器　08.0587

gravity abutment　重力式桥台　02.0548

gravity pier　重力式桥墩　02.0562

gravity retaining wall　重力式挡土墙　02.0285

gravity retaining wall with relieving platform　衡重式挡土墙　02.0286

gross mass of freight container　集装箱额定质量，＊集装箱总重　06.0554

gross ton-kilometers　总重吨公里　06.0691

gross weight　总重　05.0863

ground cable　地下电缆　07.0179

grounding alarm　接地报警　08.0179

grounding reactor　接地电抗器　04.0508

ground pressure　地层压力　02.0706

ground signal　地面信号　08.0013

group amplifier　群放大器　07.0336

group calling　组呼　07.0423

group demodulator　群解调器　07.0335

group frequency selector　频组选择器　07.0565

grouping contactor　组合接触器　04.0549

group modulator　群调制器　07.0334

group retarder　线束减速器　08.0380

group starting signal　线群出站信号机，＊线群出发信号机　08.0068

group through-connection station　群转接站　07.0320

grouter　注浆机　02.0978

grouting machine　注浆机　02.0978

grouting pressure　灌浆压力　02.0712

grouting pump　注浆泵　02.0979

guard rail　护轨　03.0060

[guard rail] check gage　护轨与心轨的查照间隔　03.0338

guard rail face gage　护背距离　03.0339

guard rail of bridge　桥梁护轨　02.0502

guard's valve　车长阀　05.0317

guard's van　守车　05.0802

guard timber of bridge　桥梁护木　02.0503

guard wall　护墙　02.0293

guidance force　导向力　04.1625

guidance system　导向系统　04.1588

guide　滑板　04.1194

guided modes　传导模　07.0454

guideway suspension interaction　导轨与悬浮系统相互作用　04.1628

guide wheel　导轮　04.1012

guiding design for construction scheme　指导性施工组织设计　02.0854

gushing water　突水　02.0992

gusset plate　节点板　02.0454

gutter　天沟　02.0297

H

hair crack of rail head　轨头发裂　03.0189

half-duplex operation　半双工　07.0706

half-duplex transmission　半双工传输　07.0782

half section effect　半段效应　04.1504

half through bridge　半穿式桥，＊中承式桥　02.0426

half wet shotcreting　潮喷混凝土，＊半湿喷混凝土　02.0781

half wet shotcreting machine　半湿喷混凝土机　02.0973

Hall plate　霍尔片　04.1622

halogen leak detector　卤素查漏仪　07.0215

hand brake　手制动机　05.0265

hand brake bell crank　手制动曲拐　05.0363

hand brake sheave wheel　手制动链轮　05.0364

hand cart　单轨小车　03.0472

hand electric tamper　手提电动捣固机，＊电镐　03.0413

hand hold　扶手　05.0418

hand hole　手孔　07.0202

handle　手柄　08.0267

handling　装卸搬运　06.0579

handling automation　装卸作业自动化　06.0581

handling by shipper-self　自理装卸　06.0592

handling capacity　装卸能力　06.0582

handling mechanization　装卸作业机械化　06.0580

handling quota　装卸定额　06.0585

handling sheet　装卸工作单　06.0586

handling volume　装卸作业量　06.0584

handling volume by machine　装卸机械作业量　06.0589

handrail　栏杆　02.0510

handrailing　栏杆　02.0510

handset　手机　07.0074

hand signal　手信号　08.0014

hand water pump　手动水泵　05.0603

hanger　吊杆　02.0451，吊弦　04.1401

hanger ear　吊弦线夹　04.1402

hanging scaffold　悬空脚手架　02.0624

hanging stage　悬空脚手架　02.0624

hanging up　挂断　07.0117

harbour station　港湾站　06.0391

hard rolling car　难行车　06.0489

hard running track　难行线　06.0429

hard spot　硬点　04.1442

harmful district　有害地段　02.0189

harmless district　无害地段　02.0190

harmonic interference　谐波干扰　08.0674

hatch　装货口，＊吊装口　05.0462

hatch cover seat　顶盖座　05.0464

hauling hook　牵引钩　05.0420

haunch　梁腋　02.0481

head checks　轨头发裂　03.0189

head contact flat joint bar　轨头底面接触夹板，＊楔型接头夹板　03.0079

header　报头　07.0740

head free flat joint bar　轨腹上圆弧接触夹板，＊铰型接头夹板　03.0080

head hardened alloy steel rail　合金淬火轨　03.0053

head hardened rail　淬火轨　03.0051

heading　导坑　02.0733，标题　07.0739

head lamp　前照灯　04.0641

head light　前照灯　04.0641

head set　头戴送受话器　07.0100

headspan suspension　软横跨　04.1406

headspan wire　横承力索　04.1363

headspan wire clamp　横承力索线夹　04.1372

heat balance　热平衡　04.0760

heat balance test　热平衡试验　04.0860

heat dissipation　散热损失　04.1110

heated container　加热集装箱　06.0550

heat engineering laboratory　热工试验室　05.0991

heater branch pipe　暖汽支管　05.0623

heater car　加温车　05.0812

heater hose　暖汽软管　05.0624

heater hose coupler　暖汽软管连接器　05.0625

heater train pipe　暖汽主管　05.0622

heating system　采暖装置　05.0591

heating transport　加温运输　06.0180

heat loss due to combustibles in refuse　机械不完全燃烧热损失　04.1108

heat loss due to exhaust gas　排烟热损失　04.1109

heat loss due to incomplete combustion　不完全燃烧热损失　04.1107

heat pump　热泵　05.0630

heat value　热值，＊发热量　04.0052

heavy duty switch machine　大功率转辙机　08.0606

heavy freight　重质货物　06.0122

heavy haul railway　重载铁路　01.0018

heavy haul train　重载列车　01.0081

heavy motor trolley　重型轨道车　03.0434

heavy permanent way machine　大型线路机械，＊重型线路机械　03.0381

heavy rail motor car　重型轨道车　03.0434

heavy tamping method　重锤夯实法　02.0614

heel end of frog　辙叉跟端　03.0323

heel length of frog　辙叉跟长　03.0334

heel of switch rail　尖轨跟端　03.0318

heel spread of frog　辙叉跟宽　03.0336

height inside car body　车体内高　05.0031

height inside car body from floor to roof center　车体内中心处高度　05.0032

height of car　车辆高度　05.0030

height of coupler center from top of rail　车钩中心线高度　05.0028

height of floor from rail top　地板面高度　05.0033

height of lifting　拔起高度，＊克服高度　02.0161

helical spring　螺旋弹簧　05.0167

herringbone track　箭翎线　06.0445

hidden construction work　隐蔽工程　02.0840

hidden project　隐蔽工程　02.0840

hidden project inspection　隐蔽工程检查　02.1043

hiding-place for personnel　人员掩蔽所　02.1006

hierarchical network　分级网[络]，＊分层网[络]　07.0757

highest water level　最高水位　02.0079

high frequency section　高频转接段　07.0281

high frequency terminal box　高频分线盒　07.0197

high friction composite brake shoe　高摩合成闸瓦　05.0197

high grade project　优质工程，＊优良工程　02.1057

high level bend　高鹅头弯管　05.0369

high phosphor cast iron brake shoe　高磷闸瓦　05.0196

high pressure fuel pipe　高压油管　04.0959

high pressure lubricating oil pump　高压机油泵　04.1040

high pressure safety valve　高压保安阀　04.0171

high pressure tank car 高压罐车 05.0800

high quality project 优质工程，＊优良工程 02.1057

high signal 高柱信号机 08.0061

high speed diesel engine 高速柴油机 04.0695

high speed railway 高速铁路 01.0019

high speed train 高速列车 01.0082

high strength friction grip bolt 摩擦结合式高强度螺栓 02.0366

high tension circuit 高压电路 04.0369

high tension tap changer 高压调压开关 04.0557

high tension winding 高压绕组 04.0496

high tone air horn 高音风喇叭 04.0651

high voltage circuit 高压电路 04.0369

high voltage interference 强电干扰 08.0673

high voltage regulation 高压侧调压 04.0318

high voltage tap changer 高压调压开关 04.0557

high voltage winding 高压绕组 04.0496

highway level crossing announcing device 道口通知设备 08.0438

highway level crossing flashing signal 道口闪光信号 08.0447

highway level crossing obstruction signal 道口遮断信号 08.0443

highway level crossing out door audible device 道口室外音响器 08.0441

highway level crossing signal 道口信号机 08.0444

highway level crossing signal control panel 道口信号控制盘 08.0445

hilly land location 丘陵地段选线 02.0150

hindrance fault 妨害故障 08.0740

hinged cantilever 旋转腕臂 04.1382

historic flood level 历史洪水位 02.0078

holding position 保持位 04.0146

hollow axle 空心车轴 05.0108

hollow axle drive 车轴空心轴驱动 04.0354

hollow pier 空心桥墩 02.0560

hollow shaft motor drive 电机空心轴驱动 04.0355

hollow slab bridge 空心板桥 02.0383

home signal 进站信号机 08.0065

homologation test 鉴定试验 04.0282

hook bolt 钩螺栓 03.0084

hook switch 叉簧 07.0078

hook-type saddle 钩头鞍子 04.1388

hopper 漏斗 05.0467

hopper car 漏斗车 05.0808

hopper-shed support tunnelling method 漏斗棚架法 02.0736

horizontal hole drainage 平孔排水 02.0307

horizontal plate 水平板，＊水平反射板 04.1148

horizontal point of tunnel portal 隧道洞口投点 02.0144

horizontal split of rail head 轨头水平劈裂 03.0186

horizontal wheel 导线平轮 08.0308

horizontal wheel assembly 导线平轮组 08.0311

hose 制动软管 05.0281

hospital car 医疗车 05.0748

hot air heating equipment 热风采暖装置 05.0616

hot box 热轴 05.0956

hot corrosion 热腐蚀 04.1558

hot-running 热滑 04.1441

hot water heating equipment 温水采暖装置 05.0599

hot water pump 热水泵 04.1172

hot water tank 温水箱 05.0661

H-type diesel engine H型柴油机 04.0704

hub bore diameter 轮毂孔直径 05.0228

hub diameter 轮毂直径 05.0226

hub length 轮毂长度 05.0227

hub thickness 轮毂厚度 05.0225

human failure 人为故障 08.0701

hump 驼峰 06.0447

hump crest 峰顶 06.0448

hump height 驼峰高度，＊峰高 06.0493

humping 驼峰调车 06.0261

humping control system 驼峰溜放控制系统 08.0363

humping fast signal 加速推送信号 08.0036

humping signal 驼峰信号 08.0033

humping signal repeater 驼峰复示信号机 08.0080

humping slow signal 减速推送信号 08.0037

hump lead 驼峰溜放线 06.0427

hump mechanics repair room 驼峰机械修理室 08.0721

hump profile 驼峰纵断面 06.0494

hump signal 驼峰信号机 08.0079

hump yard 驼峰调车场 06.0509

hump yard classification throat 驼峰调车场头部 06.0510

hunting 蛇行运动 01.0088

hunting [vibration] 摇头振动 01.0104

HWL 最高水位 02.0079

hybrid coil 混合线圈 07.0347

hydraulic brake 液力制动 04.0116, 液力制动器 04.1007

hydraulic coupler 液力偶合器 04.1006

hydraulic damper 液压减振器, *油压减振器 05.0163

hydraulic draft gear 液压式缓冲器 05.0552

hydraulic drop 跌水 02.0299

hydraulic pile driver 液压打桩机 02.0907

hydraulic pressure engine room 油压动力室 08.0722

hydraulic radius 水力半径 02.0082

hydraulic tamping machine 液压捣固机, *液压捣固车 03.0417

hydraulic tensioning jack 液压式张拉千斤顶 02.0545

hydraulic torque converter 液力变扭器, *液力变矩器 04.1008

hydraulic transmission 液力传动 04.0990

hydraulic transmission gear box 液力传动箱 04.0998

hydraulic transmission system 液力传动系统 04.0997

hydraulic unit 液力循环元件 04.1005

hydrodynamic brake operating valve 液力制动操纵阀 04.1021

hydrodynamic brake valve 液力制动阀 04.1020

hydrodynamic reverser 液力换向传动箱 04.0999

hydrogeologic survey 水文地质调查 02.0043

hydrological survey 水文测量 02.0086

hydrologic cross-section 水文断面 02.0073

hydrologic observation of bridge site 桥址水文观测 03.0516

hydrologic section 水文断面 02.0073

hydrologic sectional drawing 水文断面 02.0073

hydrology of bridge and culvert 桥涵水文 02.0081

hydromechanical drive 液力机械传动 04.0993

hydromechanical drive with direct step 混合液力机械传动 04.0996

hydromechanical drive with inner ramification 单流液力机械传动 04.0994

hydromechanical drive with outer ramification 双流液力机械传动 04.0995

hydropneumatic draft gear 液力气动式缓冲器 05.0553

hydrostatic drive 液压传动 04.0989

hydrostatic motor 静液压马达 04.1052

hydrostatic pressure 静水压力 02.0333

hydrostatic pump 静液压泵 04.1051

I

IAT 实制动时间 05.0345

I-beam 工形梁 02.0385

ice apron 破冰体 02.0576

ice-breaking cutwater 破冰体 02.0576

ice-bunker refrigerator car 车端冰箱冷藏车 05.0814

ice drift 流冰 02.0638

ice floe survey 冰凌调查 02.0071

ice guard 破冰体 02.0576

ice pressure 冰压力 02.0346

idle car 游车 06.0166

idle running pump 惰行泵 04.1028

idle running resistance 惰行阻力 04.0097

idling braking time 空走时间 05.0344

idling stopping distance 空走距离 05.0348

ignition 点火 04.0775

illuminating lamp 照明灯 04.0648

illumination equipment 照明装置 05.0672

illumination voltage stabilizer 照明稳压器 05.0701

immerseable embankment 浸水路堤 02.0313

immersed tunnelling method 沉管法 02.0726

immersion type vibrator for concrete 插入式混凝土

振捣器 02.0940

immunity to interference 抗扰度 04.1509

impact factor 冲击因数 03.0535

impact force of falling stone 落石冲击力 02.0713

impact force of train 列车冲击力，＊冲击荷载 02.0337

impact of rolling stock 机车车辆冲击 01.0106

impact-type drill machine 冲击式钻机 02.0915

impedance bridge 阻抗电桥 07.0221

impedance matching transformer ［阻抗］匹配变压器 07.0198

impedance of traction electric network 牵引网阻抗 04.1335

impedance transformer 扼流变压器 08.0548

impervious embankment 非渗水土路基 02.0213

improved foundation 加固地基 02.0578

improved ground 加固地基 02.0578

impulse 脉冲 07.0025

impulse relay 脉冲继电器 08.0583

in advance of a signal 信号机前方，＊信号机外方 08.0713

incising of wooden tie 木枕刻痕 03.0220

inclined cable 斜缆 02.0493

inclined catenary 斜链形悬挂 04.1355

inclined shaft 斜井 02.0787

incoming equipment 进局设备 07.0294

incoming trunk circuit 入中继电路，＊入中继器 07.0263

independent brake valve 单独制动阀，＊小闸 04.0168

independent wheel 自由车轮 05.0882

index of estimate 估算指标 02.0029

index of locomotive operation 机车运用指标 04.0205

index of photography 镶辑复照图，＊像片索引图 02.0138

indicated object 表示对象 08.0467

indicated power 指示功率 04.0813

indicated specific fuel consumption 指示燃油消耗率 04.0765

indicated thermal efficiency 指示热效率 04.0827

indicated tractive effort 指示牵引力 04.1087

indicated work 指示功 04.0764

indicating panel 表示盘 08.0255

indication 表示 08.0158

indication circuit 表示电路 08.0235

indication cycle 表示周期 08.0482

indication lamp 表示灯 08.0272

indication rod 表示杆 08.0614

indicator 表示器 08.0088

indicator diagram 示功图 04.0761

indicator lamp 指示灯 04.0649

indicator valve 示功阀 04.0924

indirect cost of project 工程间接费 02.0032

indirect expense of project 工程间接费 02.0032

indirect holding fastening 分开式扣件 03.0086

individual approximate estimate 个别概算 02.0020

individual budget 单项预算 02.0852

individual calling 单呼 07.0424

individual drive 独立驱动 04.0358

individual drive locomotive 独立传动机车 04.0362

individual inverter 单管逆变器 05.0699

individual level type all-relay interlocking 单独操纵继电式电气集中联锁 08.0110

individual synchronized mode 独立同步方式 07.0602

indoor test 室内测试 02.0054

inducer 导风轮 04.0975

induction coil 感应线圈 07.0075

inductive accelerometer 电感式加速度计 04.1608

inductive cab signaling 感应式机车信号 08.0415

inductive communication in yard and station 站场感应通信 07.0620

inductive radio communication for railway 铁路感应无线电通信 07.0510

inductive reactor 限流电抗器 04.0513

inductive shunt 分流电抗器，＊感应分路 04.0510

inductive train radio communication 感应式列车无线电通信 07.0581

inductive transmission mode 感应传输方式 07.0548

inductor type alternator 交流感应子发电机 05.0698

industrial and mining locomotive 工矿机车 04.0013

269

industrial interference 工业干扰 07.0534

industrial station 工业站 06.0390

industry railway 工业企业铁路 01.0024

ineffective traffic 无效运输 06.0110

inexhaustibility 制动不衰竭性 05.0355

inflammable freight 易燃货物 06.0119

influence of ground current 地电流影响 04.1467

influencing current 影响电流 04.1468

information 信息 07.0003

information desk 查号台 07.0258

information office 问讯处 06.0022

infrared journal temperature detection point 红外线轴温检测所 05.0958

infrared journal temperature detection system 红外线轴温探测系统 05.0959

inherrent distortion 固有畸变 07.0844

initial speed at brake application 制动初速 05.0349

injection advance angle 喷油提前角 04.0781

injection pump 注浆泵 02.0979, 喷油泵 04.0947

injection pump cam 喷油泵凸轮 04.0940

injection pump drive gear 喷油泵驱动齿轮, *喷油泵传动齿轮 04.0944

injection pump transmission mechanism 喷油泵传动装置 04.0945

injection rate 喷油速率 04.0777

injection timer 自动喷调装置 04.0946

injector filter core 喷油器滤芯 04.0958

injector nozzle matching parts 喷油嘴偶件 04.0955

inlet cam 进气凸轮 04.0938

inlet submerged culvert 半压力式涵洞 02.0633

inlet unsubmerged culvert 无压力涵洞 02.0631

inlet valve 进气门 04.0926

in line type diesel engine 直列式柴油机 04.0701

inner dead center 内止点 04.0714

inner dead point 内止点 04.0714

inner firebox 内火箱 04.1118

input shaft 输入轴 04.1014

inquiry office 问讯处 06.0022

in rear of a signal 信号机后方, *信号机内方 08.0714

inscribed to curves 曲线内接 03.0136

insensitivity 制动稳定性 05.0352

inserting plate support 插板支护 02.0774

inside corridor 内走廊, *通道 04.1063

inside end wall 内端墙 05.0434

inside firebox 内火箱 04.1118

inside-frame truck 内构架转向架 05.0073

inside reinforcement of side post 侧柱内补强 05.0428

in situ test 原位测试 02.0055

inspect by instrument 仪器检查 02.1047

inspection and measurement 检测 08.0735

inspection and repair of buildings and structures 房建检修 03.0591

inspection and service point for car before loading or after unloading 装卸检修所 05.0911

inspection and test of locomotive after completion of construction 机车组装后的检查与试验 04.0655

inspection of component failure 部件故障检测, *故障诊断 04.0277

inspection of engineering quality 工程质量检验 02.1042

inspection well 检查井 02.0309

installation vehicle for contact wire 接触网架线车 02.0983

instantaneous application time 实制动时间 05.0345

instantaneous loss of shunting 瞬时分路不良 08.0671

instantaneous maximum current of feeding section 供电臂瞬时最大电流 04.1328

instantaneous shunt 瞬时分路 08.0528

instantaneous speed change rate 瞬时调速率 04.0832

instructions for train operation at station 车站行车工作细则 06.0198

instrument circuit 仪表电路 04.0382

insulant shoe 绝缘鞋 02.1027

insulated box car 隔热棚车 05.0811

insulated container 绝热集装箱 06.0548

insulated gage rod 绝缘轨距杆 03.0106

insulated joint 绝缘接头 03.0070

insulated joint located within the clearance limit　侵入限界绝缘　08.0537

insulated joint within a turnout　岔中绝缘　08.0536

insulated ladder trolley　绝缘梯车　04.1463

insulated overlap　绝缘锚段关节　04.1430

insulated switch section　道岔绝缘段　06.0435

insulated transition mast　绝缘转换柱　04.1409

insulated transport　保温运输　06.0178

insulating course　隔断层　02.0311

insulating layer　隔断层　02.0311

insulation gap　绝缘间隙　04.1438

insulation resistance　绝缘电阻　07.0224

insulator　绝缘子　07.0162

insulator cleaning car　绝缘子清洗车　04.1462

insurance charge　保险费　06.0733

insured rail traffic　铁路保险运输　06.0010

intake duct　进气道　04.1536

intake pressure　进气压力　04.0732

intake stroke　进气行程　04.0723

intake swirl　进气涡流　04.0804

intake temperature　进气温度　04.0731

intangible wear　无形损耗　06.0700

integral lining　整体式衬砌　02.0681

integral loadcarrying structure　整体承载结构　05.0384

integral train　整体列车　05.0841

integrated ballast bed　整体道床　03.0121

intelligibility　可懂度　07.0148

intelligible crosstalk　可懂串音　07.0138

interactive protocol　交互式协议　07.0882

intercepting ditch　天沟　02.0297，截水沟　02.0300

interchange　互换　07.0744

inter-channel crosstalk　路际串音　07.0362

inter-continental railway　大陆桥，＊洲际铁路　01.0122

intercooler　中冷器　04.0980

interference　干扰　07.0042

interference field　干涉场　07.0539

interference field strength　干扰场强　07.0522

interference from neighboring line　邻线干扰　08.0672

interference source　干扰源　04.1508

interference voltage　干扰电压　07.0521

interlocked switch　联锁道岔　08.0196

interlocking　联锁　08.0103

interlocking area　联锁区　08.0200

interlocking by electric locks　电锁器联锁　08.0121

interlocking by electric locks with color light-signals　色灯电锁器联锁　08.0122

interlocking by electric locks with electric semaphore　电动臂板电锁器联锁　08.0124

interlocking by electric locks with semaphore　臂板电锁器联锁　08.0123

interlocking by point detector　联锁箱联锁　08.0120

interlocking chart and table　联锁图表　08.0211

interlocking circuit　联锁电路　04.0383

interlocking equipment　联锁设备　08.0104

interlocking for shunting area　调车区电气集中联锁　08.0116

interlocking table　联锁表　08.0212

intermediate distributing frame　中间配线架　07.0129

intermediate acceptance　中间验收　02.1053

intermediate buffer　中间缓冲装置　04.1245

intermediate cable terminal box　中间电缆盒　08.0288

intermediate coat　中间漆　03.0552

intermediate draw gear　中间牵引装置　04.1244

intermediate equipment　中间设备　07.0878

intermediate gear box　中间齿轮箱　04.1001

intermediate grade　中间坡　06.0496

intermediate inspection at the technological process　中间工艺检验　04.0272

intermediate line filter　中间线路滤波器　07.0291

intermediate relay　中间继电器　04.0593

intermediate repair　架修　04.0242

intermediate repair of track　线路中修　03.0195

intermediate station　中间站　06.0365

intermediate straight line　夹直线　02.0179

intermediate test　中间试验　05.0987

intermittent type cab signaling　点式机车信号　08.0410

intermode transport　综合运输　01.0120

intermodulation interference　互调干扰　07.0527

international express train 国际联运旅客特别快车 06.0206

international railway through traffic 国际铁路联运 01.0121

international through freight shipping documents 国际联运货物票据 06.0190

international through freight traffic 国际货物联运 06.0183

international through traffic station 国际联运站 06.0393

inter office through trunk 局间直通中继方式 07.0406

inter operation 网络互连 07.0768

interpole coil 换向极线圈 04.0449

interpole core 换向极铁心 04.0448

interposition trunk 台间联络线，＊座席间中继线 07.0405

intersection point 交点 02.0120

interstation telephone 中间站公务电话，＊各站电话 07.0411

interstation train operation telephone 站间行车电话，＊闭塞电话 07.0415

interstitial wire 填充线 07.0485

inter-system crosstalk 制际串音 07.0361

intrados 拱腹线 02.0488

in-train repair 不摘车修 05.0932

intunnel control survey 隧道洞内控制测量，＊贯通测量 02.0143

inventory stock rails 周转轨 03.0241

invert 仰拱 02.0679

inverted arch 仰拱 02.0679

inverted filter 反滤层 02.0308

inverted siphon 倒虹吸管 02.0635

inverter 逆变器 07.0049

investigation gallery 调查坑道 02.0791

investigation survey 调查测绘 02.0038

investigation surveying and sketching 调查测绘 02.0038

investigation test 研究性试验 04.0296

investment estimate 投资估算 02.0028

irregularity without load 静态不平顺 03.0143

ISO freight container 国际标准集装箱 06.0540

isolating switch for control supply 控制电源隔离开关 04.0572

isolating switch for main silicon rectifier cubicle 主整流柜隔离开关 04.0570

isolating transformer 隔离变压器，＊绝缘变压器 04.0490

isolation layer 隔离层 02.0689

item project 分项工程 02.0006

J

jack 千斤顶 02.0903，插孔，＊插座 07.0102

jack hammer 手持式凿岩机 02.0976

jacking floor beam 起重横梁 02.0466

jacking pad 顶车座 05.0425

jack-in method 顶进法 02.0543

jack strip 插孔排 07.0103

jack up car body 架车 05.0935

jelly filled type optical fiber cable 充油型光缆 07.0486

jerking 溜放调车 06.0260

jitter 抖动 07.0869

joint bar 接头夹板，＊鱼尾板 03.0077

jointed track 有缝线路 03.0016

joint gap 轨缝，＊接头缝 03.0063

joint investment railway 合资铁路 01.0010

jointless track 无缝线路 03.0017

jointless track circuit 无绝缘轨道电路 08.0498

jointly owned railway 合资铁路 01.0010

joint resistance 接头阻力 03.0043

joint responsibility of railway 铁路的连带责任 06.0184

journal 轴颈 05.0110

journal back fillet 轴颈后肩 05.0111

journal bearing babbit metal lining room 挂瓦室 05.0896

journal bearing wedge 轴瓦垫 05.0134

journal box 轴箱 05.0127

journal box body 轴箱体 05.0132

journal box front cover 轴箱前盖 05.0130

journal box lid 轴箱盖 05.0129

journal box packing room　油线室　05.0895

journal box rear cover　轴箱后盖　05.0131

journal box spring　轴箱弹簧　05.0165

journal box spring guide post　轴箱弹簧支柱
　05.0155

journal temperature　轴温　05.0957

jumper　跳线　08.0539

jump on rail　跳轨　05.0877

junction box　车电分线盒　05.0704

junction station　联轨站　06.0394

junction terminal transfer train　枢纽小运转列车
　06.0226

K

key　电键　07.0691

keyboard　键盘　07.0694

key pad　按键式拨号盘　07.0085

key pad telephone set　按键电话机　07.0095

key project　关键工程　02.0861

key switch　按键开关　04.0576

kilometer post　公里标　03.0249

kinds of traction　牵引种类　01.0070

king pin　中心销　05.0211

kitchen　厨房　05.0757

kitchen car　厨房车　05.0735

knob　手柄　08.0267

knock　敲缸　04.0808

knuckle pin　肘销　04.1198

knuckle pivot pin　钩舌销　05.0546

L

labels for packages of dangerous goods　危险货物包
　装标志　06.0574

laboratory test　室内测试　02.0054

labor norm　劳动定额　06.0717

labor productivity of railway transport　铁路运输全员
　劳动生产率　06.0716

labor ratings　劳动定额　06.0717

labyrinth seal　迷宫式密封　05.0144

lacing bar　缀条　02.0456

ladder　梯　05.0419

ladder track　梯线　03.0299

lagging phase feeding section　滞后相供电臂
　04.1306

laminated rubber bearing　板式橡胶支座　02.0521

laminated spring　叠板弹簧　05.0168

LAN　局域网　07.0756

land-railway　大陆桥，＊洲际铁路　01.0122

landslide　滑坡　03.0477

land slide warning device　塌方落石报警器
　03.0456

landslip　滑坡　03.0477

Langer bridge　朗格尔式桥，＊刚性系杆柔性拱桥
　02.0404

large ballast undercutting cleaners　大型全断面清筛
　机　03.0429

large permanent way machine　大型线路机械，＊重
　型线路机械　03.0381

large scale ore car　大型矿车　02.0969

large-scale temporary project　大型临时工程
　02.0857

laser［diode］　激光器　07.0470

lateral bracing　侧向水平联结　02.0458

lateral connecting rod　横向拉杆　05.0213

lateral direction of car　车辆横向　05.0007

lateral elasticity　横向弹性　04.1444

lateral force　横向力　05.0886

lateral impact　横向冲击　01.0108

lateral play of wheel set　轮对横动量　04.0050

lateral resilience　横向弹性　04.1444

lateral rigidity of bridge　桥梁横向刚度　03.0542

lateral shift for center of gravity of goods　货物重心
　的横向位移　06.0167

lateral stability　横向稳定性　05.0850

lateral swaying force of train　列车横向摇摆力
　02.0344

lateral turnout　单开道岔　03.0282

lateral vibration 横向振动 01.0097

lattice arch 格构拱, * 钢花拱, * 格栅拱 02.0777

launching method 拖拉架设法 02.0533

lavatory 厕所 05.0777

law of heat release 放热规律 04.0759

law of injection 喷油规律 04.0778

lay out of route 放线 02.0119

LD 激光器 07.0470

leaching well 渗井 02.0305

lead 导程 04.1257

lead curve 导曲线 03.0303

leading line 导向线 02.0160

leading locomotive 主机, * 本务机车 04.0226

leading locomotive running kilometers 本务走行公里 04.0215

leading phase feeding section 引前相供电臂 04.1307

leading truck 导轮转向架 04.1231

leading truck wheel set 导轮对 04.1232

lead-in rack 引入架 07.0351

lead-in test rack 引入试验架 07.0353

lead track 牵出线 06.0410

leakage 漏泄 04.0139

leakage current 漏泄电流, * 迷流 04.1470

leak tunnel 渗水隧洞 02.0304

leaky coaxial cable 漏泄同轴电缆 07.0173

leaky modes 漏泄模 07.0456

leaky pipe 渗管 02.0306

leased circuit 租用电路 07.0111

leased locomotive 机车出租 04.0028

leaving velocity loss 余速损失 04.1555

LED 发光二极管 07.0471

left-handed machine 反装 08.0617

left hand turnout 左开道岔 03.0288

left side of car 一位侧 05.0010

length between truck centers 车辆定距 05.0034

length inside car body 车体内长 05.0023

length of car body 车体长度 05.0020

length of connecting rod 连杆长度 04.0710

length of grade section 坡段长度 02.0193

length of repair position 修车台位长度 05.0903

length of station site 站坪长度 02.0168

length of switch rail 尖轨长度 03.0332

length of tank 罐体长度 05.0022

length of underframe 底架长度 05.0021

length over end sills 底架长度 05.0021

length over ends of body 车体长度 05.0020

length over pulling faces of couplers 车辆全长, * 车钩连结线间长度 05.0019

less-than-carload freight 零担货物 06.0112

less-than-carload freight transhipment station 零担货物中转站 06.0079

lettering and marking of car 车辆标记 05.0014

level 电平 07.0030

level crossing 平面交叉 03.0271, 道口 03.0273

level crossing watchman 道口看守工 03.0008

leveling valve 高度调整阀 05.0181

level meter 电平表 07.0356

level stretch between opposite sign gradient 分坡平段 02.0196

level stretch of grade crossing 道口平台 03.0275

leverage ratio 制动倍率 05.0338

lever connecting rod 杠杆上拉杆 05.0188

lever connection 制动拉杆 05.0187

liaison circuit 联系电路 08.0236

liaison circuit between stations 站间联系电路 08.0240

liaison circuit between yards 场间联系电路 08.0241

liaison circuit with a locodepot 机务段联系电路 08.0242

liaison circuit with automatic blocks 自动闭塞联系电路 08.0238

liaison circuit with block signaling 区间联系电路 08.0237

liaison circuit with highway crossings within the station 站内道口联系电路 08.0243

liaison circuit with semi-automatic blocks 半自动闭塞联系电路 08.0239

life guard 排障器 04.1065

lifesaving appliance 救生设施 02.1031

lift bridge 升降桥 02.0411

lift force 悬浮力 04.1597

lifting height 拔起高度, * 克服高度 02.0161

lift one end of car 顶车 05.0936

light and bulk freight 轻浮货物 06.0121

light-emitting diode 发光二极管 07.0471

light hydraulic tamping machine 小型液压捣固机 03.0416

lighting circuit 照明电路 04.0381

light locomotive running 单机运行 04.0223

light locomotive running kilometers 单机走行公里 04.0216

light motor trolley 轻型轨道车 03.0433

lightning arrester 避雷器 08.0665

lightning conductor 避雷线 07.0211

lightning interference 雷电干扰 08.0676

lightning protection device 防雷装置 02.1032

light permanent way machine 小型线路机械，＊轻型线路机械 03.0382

light rail 轻轨铁路 01.0026

light rail motor car 轻型轨道车 03.0433

light railway 轻轨铁路 01.0026

light repair 定修 04.0243

light strip 光带 08.0274

light weight 自重 05.0861

lime sand pile 石灰砂桩 02.0266

limited speed 限制速度 01.0076

limited subscriber 限制用户 07.0402

limiting error 最大误差，＊极限误差 02.0111

limiting grade 限制坡度 02.0182

limiting valve 限压阀 05.0305

limit of smoke 烟度 04.0837

limit of wear 磨耗限度 04.0267

limit state design method 极限状态设计法 02.0207

line air-blast circuit-breaker 空气主断器 04.0531

line amplifier 线路放大器 07.0337

linear asynchronous motor 直线异步电动机，＊直线感应电动机 04.0417

linear induction motor 直线异步电动机，＊直线感应电动机 04.0417

linear motor 直线电动机，＊线性电动机 04.0415

linear synchronous motor 直线同步电动机 04.0416

line blockade due to flood 水害断道 03.0577

line circuit-breaker 主断路器 04.0529

line contactor 线路接触器 04.0550

line development 展线 02.0163

line efficiency 线路利用率 07.0154

line feed 换行 07.0715

line feeder 加强线 04.1367

line filter 线路滤波器 07.0349

line interruption 线路中断 03.0257

line noise 线路杂音 07.0360

line occupation for works 施工封闭线路 03.0258

line plan 线路平面图 02.0166

line profile 线路纵断面图 02.0167

line vacuum circuit-breaker 真空主断路器 04.0530

lining 衬砌 02.0676，[货车]内墙板 05.0452

lining bar 撬棍 03.0469

lining corrosion 衬砌腐蚀 03.0528

lining cracking 衬砌裂损 03.0527

lining deformation 衬砌变形 03.0529

lining of tunnel portal section 洞口段衬砌 02.0684

link cable 链接电缆 07.0189

link control procedure 链路控制规程 07.0886

link management 链路管理 07.0806

link protocol 链路协议 07.0845

lipping of rail head 轨头肥边 03.0169

liquid fund 流动资金 06.0748

liquid level indicating plate 液面高度指示牌 05.0474

liquid receiver 贮液桶 05.0633

liquid trap 聚液窝 05.0477

live fish car 活鱼车 05.0822

live load of special type wagon 特种车辆活载 03.0522

live load of train 列车活载 02.0335

load characteristic 负荷特性，＊负载特性 04.0849

loaded cable 加感电缆 07.0192

loaded car 重车 05.0013

loaded cars to be delivered at junction stations 移交车 06.0321

loaded test 负载试验 04.0286

loader 装载机 02.0884

load factor rating 承载系数检定 03.0519

loading and unloading accident　装卸事故　06.0648

loading and unloading track　装卸线　06.0086

loading capacity　载重　05.0862

loading clearance limit　装载限界　01.0062

loading coefficient　装载系数　01.0073

loading coil box　加感箱　07.0194

loading coil spacing　加感节距　07.0235

loading combination　荷载组合　02.0354

loading gauge　装载限界　01.0062

loading machine with chain buckets　链斗装车机　06.0597

load on axle journals　轴载荷　05.0231

load on center plate　心盘载荷　05.0233

load on side bearing　旁承载荷　05.0234

load per meter of track　每延米重量　05.0860

load proportional valve　载荷比例阀　05.0303

load sensor valve　载荷传感阀　05.0302

load test　负载试验　04.0286

local area network　局域网　07.0756

local battery　本地电池　07.0089

local car　本站作业车　06.0235

local cars to be unloaded　管内工作车　06.0320

local control　局部控制　08.0185

local control circuit　局部控制电路　08.0245

local control panel　局部控制盘　08.0284

local loading and unloading rate　管内装卸率　06.0325

locally operated switch　非集中道岔　08.0199

locally supplied power source　局部电源　08.0646

local marshalling station　地方性编组站　06.0372

local passenger flow　管内客流　06.0025

local passenger train　管内旅客列车　06.0208

local railway　地方铁路　01.0008

local reduction　局部减压　04.0149

local scour　局部冲刷　02.0097

local telephone network　地区电话网　07.0381

local telephone switching system　地区电话交换机　07.0390

local terminal　本地终端　07.0907

location in plain region　平原地区选线　02.0147

location of line in mountain and valley region　山区河谷选线　02.0149

location of line in mountain region　越岭选线　02.0148

location of line on hilly land　丘陵地段选线　02.0150

location of mountain line　越岭选线　02.0148

location of railway route selection　铁路选线　02.0146

location over mountain　越岭选线　02.0148

location survey　定测　02.0101

location with cross-section method　横断面选线　02.0162

locked position of coupler　闭锁位置　05.0566

locking　锁闭　08.0126

locking center pin　锁紧中心销　05.0212

locking circuit　锁闭电路　08.0230

locking force　锁闭力　08.0624

locking rod　锁闭杆　08.0615

locking system　锁闭系统　08.0609

lockset position of coupler　开锁位置　05.0567

locomotive　机车　04.0002

locomotive adhesive weight　机车粘着重量　04.0046

locomotive boiler　机车锅炉　04.1115

locomotive brake gear　机车制动机　04.0160

locomotive braking distance　机车制动距离　04.0126

locomotive braking period　机车制动周期　04.0125

locomotive caboose crew system　机车随乘制　04.0199

locomotive centralized power supply　机车集中供电　04.0365

locomotive crew　机车乘务组　04.0200

locomotive crew pooling system　机车轮乘制　04.0198

locomotive crew working system　机车乘务制度　04.0196

locomotive depot　机务段　04.0067

locomotive dispatching order　机车调度命令　04.0230

locomotive distributing valve　机车分配阀　04.0165

locomotive emergency vent valve　机车紧急放风阀　04.0170

locomotive energy consumption　机车能耗　06.0728

locomotive equipment　机车设备　08.0419

locomotive failure 机车故障 04.0073

locomotive for district transfer 小运转机车 04.0017

locomotive general overhaul 机车大修 04.0241

locomotive height 机车高度 04.0042

locomotive inductor 机车感应器 08.0421

locomotive in operation 运用机车 04.0023

locomotive in reserve 备用机车 04.0025

locomotive inspection and repair 机车检修 04.0239

locomotive kilometers 机车公里 06.0692

locomotive logbook 机车履历簿 04.0299

locomotive maintenance 机车保养 04.0238

locomotive noise 机车噪声 04.0051

locomotive of service train 路用机车 04.0018

locomotive overall length 机车长度 04.0040

locomotive overhaul [repair] 机车大修 04.0241

locomotive periodical repair 机车定期修 04.0247

locomotive power 机车功率 04.0082

locomotive power curve at wheel rim 机车轮周功率曲线 04.0079

locomotive radio-control 机车无线电遥控,＊机车无线电操纵 07.0679

locomotive ratio 机车比率,＊机车比值 04.0004

locomotive repair depot 机车检修段 04.0069

locomotive repair in depot 机车段修 04.0246

locomotive repair in works 机车厂修 04.0245

locomotive repair limit 检修限度 04.0268

locomotive reservation 机车储备 04.0030

locomotive retirement 机车报废 04.0029

locomotive rigid wheel base 机车固定轴距 04.0039

locomotive routing 机车交路 04.0187

locomotive running depot 机车运用段 04.0068

locomotive running kilometers 机车走行公里 04.0211

locomotive running preparation 机车整备 04.0031

locomotive running track 机车走行线 06.0412

locomotive service capacity 机车整备能力 04.0032

locomotive service weight 机车整备重量 04.0044

locomotive servicing 机车整备 04.0031

locomotive shed 机车库 06.0537

locomotive speedmeter 机车速度表 04.0626

locomotive station 机车电台 07.0556

locomotive storage 机车储备 04.0030

locomotive stored up 封存机车 04.0026

locomotive supervise and record apparatus 机车监控记录装置 04.0072

locomotive technical specification 机车技术规范 04.0033

locomotive temporary repair 机车临修 04.0250

locomotive total wheel base 机车全轴距 04.0037

locomotive [traction] power test 机车功率试验 04.0675

locomotive tractive characteristic 机车牵引特性 04.0074

locomotive tractive characteristic curve 机车牵引特性曲线 04.0078

locomotive tractive district 机车牵引区段 04.0186

locomotive tractive effort 机车牵引力 04.0089

locomotive tractive effort curve 机车牵引力曲线 04.0080

locomotive trial run 机车试运转 04.0279

locomotive turnaround depot 机务折返段 04.0070

locomotive under repairing 检修机车 04.0024

locomotive waiting for repair 待修机车 04.0027

locomotive waiting track 机待线 06.0413

locomotive weight 机车重量 04.0045

locomotive wheel base of bogie 机车转向架轴距 04.0038

locomotive width 机车宽度 04.0041

locomotive working diagram 机车周转图 06.0288

logical channel 逻辑信道 07.0879

Lohse bridge 洛泽式桥,＊直悬杆式刚性拱刚性梁桥 02.0405

long heavy grade 长大坡道 02.0186

longitudinal arrangement type junction terminal 顺列式枢纽 06.0399

longitudinal choke coil 纵向扼流线圈 07.0209

longitudinal dam 顺坝 02.0619

longitudinal direction of car 车辆纵向 05.0006

longitudinal dynamic force of train 列车纵向动力 05.0586

longitudinal electro-motive force 纵电动势 04.1471

longitudinal force 纵向力 05.0887

longitudinal impact 纵向冲击 01.0107

longitudinal level of rail 轨道前后高低，＊前后高低 03.0151

longitudinal reinforcement 纵向钢筋 02.0475

longitudinal shift for center of gravity of goods 货物重心的纵向位移 06.0168

longitudinal survey 线路测量 02.0106

longitudinal suspension 链形悬挂 04.1352

longitudinal tie 纵向轨枕 03.0110

longitudinal type district station 纵列式区段站 06.0368

longitudinal vibration 纵向振动 01.0096

long rail string 长轨条 03.0021

long routing 长交路 04.0194

long stator 长定子 04.1585

long steep grade 长大坡道 02.0186

long-term railway plan 铁路长期计划 06.0674

long tunnel 长隧道 02.0657

long wave undulation of rail head 轨头长波浪磨损 03.0167

long welded rail track 长轨线路 03.0018

long welded rail transporting and working train 长钢轨运输作业列车 03.0438

loop 环线 06.0421

looply connected power supply 环状供电 08.0638

loop resistance 环路电阻 07.0223

loop routing 循环交路 04.0191

loop-type junction terminal 环形枢纽 06.0401

loose heel switch 间隔铁式尖轨转辙器 03.0346

loose tie 轨道暗坑，＊空吊板 03.0154

Los Angeles abrasion test 洛杉矶磨损试验 03.0126

loss 损耗 07.0033

loss accident 丢失事故 06.0639

loss of indication of a switch 道岔失去表示 08.0681

loss of interlocking 失去联锁 08.0686

loudness 响度 07.0067

loud-speaker 扬声器 07.0073

loud speaking telephone set 扬声电话机 07.0099

louver 百叶窗 05.0508

low end 矮端板 05.0498

lower dead center 下止点 04.0713

lowered draft sill 刀把梁 05.0423

lowering of track 落道 03.0200

lower water tank 车下水箱 05.0659

low friction composite brake shoe 低摩合成闸瓦 05.0199

low pressure safety valve 低压保安阀 04.0172

low side 矮侧板 05.0497

low speed diesel engine 低速柴油机 04.0693

low tension circuit 低压电路 04.0370

low tension tap changer 低压调压开关 04.0556

low tension winding 低压绕组 04.0497

low tone air horn 低音风喇叭 04.0652

low value and easily wornout articles 低值易耗品 06.0698

low voltage circuit 低压电路 04.0370

low voltage regulation 低压侧调压 04.0319

low voltage tap changer 低压调压开关 04.0556

low voltage winding 低压绕组 04.0497

lubricant for locomotive 机车用润滑剂 04.0059

lubricating oil carry-over 窜机油 04.0840

lubricating oil filter 机油滤清器 04.1041

lubricating oil heat exchanger 机油热交换器 04.1045

lubricating oil pump 机油泵 04.1037

lubricating prefilter 机油粗滤器 04.1042

lubricating roll 油卷 05.0138

luggage 行李 06.0015

luggage and parcel traffic accident 行包事故 06.0621

luggage compartment 行李室 05.0762

luggage office 行李房 06.0018

luggage rack 行李架 05.0768

luggage ticket 行李票 06.0049

luggage van 行李车 05.0736

lumped loading 集总加感 07.0234

L/V ratio L/V 比值 04.0184

M

machine for roadway work 路基机械 03.0370

machine operator 养路工 03.0006

maglev 磁浮铁路 01.0029, 磁[悬]浮 04.1578

[maglev] lateral force 侧力 04.1598

maglev vehicle 磁[悬]浮车辆 04.1584

magnaflux inspection for axle 车轴电磁探伤 05.0947

magnet driver 电磁铁驱动器, ＊斩波器 04.1593

magnetic circuit 磁路[系统] 08.0574

magnetic induction 磁感应 04.1465

magnetic levitation 磁[悬]浮 04.1578

magnetic levitation railway 磁浮铁路 01.0029

magnetic particle inspection for axle 车轴电磁探伤 05.0947

magneto 磁石发电机 07.0080

magneto telephone set 磁石电话机 07.0091

magneto telephone switch board 磁石电话交换机 07.0391

magnet wheel 磁轮 04.1630

mail sorting room 分检室 05.0761

mail van 邮政车 05.0737

main air reservoir 总风缸 04.0175

main alternator 交流主发电机 04.0464

main bearing 主轴承 04.0878

main bearing cap 主轴承盖 04.0880

main bearing housing 主轴承座 04.0879

main bearing nut 主轴承螺母 04.0886

main bearing seat 主轴承座 04.0879

main bearing shell 主轴瓦 04.0882

main bearing stud 主轴承螺栓 04.0885

main bridge 正桥, ＊主桥 02.0423

main button switch group 主按键开关组 04.0579

main cable 主干电缆 07.0187

main circuit 主电路 04.0368

main circuit socket for shed supply 主电路库用插座 04.0605

main circuit transfer switch for shed supply 主电路库用转换开关 04.0564

main connecting rod 主连杆 04.0917

main crank pin 主曲拐销 04.1228

main distribution frame 总配线架 07.0130

main driving wheel set 主动轮对 04.1227

main elastic coupling 主弹性联轴节 04.1000

main frame 机体 04.0863, 主车架片 04.1215

main generator 主发电机 04.0462

main journal 主轴颈 04.0888

main key switch set 主按键开关组 04.0579

main line 干线 01.0037

main linely connected power supply 干线供电 08.0637

main line of turnout 道岔主线 03.0301

main line railway 干线铁路 01.0021

main marshalling station 主要编组站 06.0379

mainpole coil 主极线圈 04.0447

mainpole core 主极铁心 04.0446

main power source 主电源 08.0634

main river channel 主河槽 02.0074

main rod 摇杆 04.1196

main signal 主体信号机 08.0074

maintaining position 保持位 04.0146

main technical requirement of railway 铁路主要技术条件 02.0003

main technical standard of railway 铁路主要技术条件 02.0003

maintenance car 维修车 05.0746

maintenance gang for catenary 接触网工区 04.1456

maintenance of bridge and tunnel 桥隧养护 03.0564

maintenance of track 线路维修 03.0196

maintenance of way expenditures 养路费用 03.0265

maintenance-of-way report and forms 养路表报 03.0261

main track of turnout 道岔主线 03.0301

main valve 主阀 04.1161

major bridge 大桥 02.0417

major repair of bridge and tunnel 桥隧大修 03.0545

major repair of buildings and structures 房建大修 03.0590

major repair of track 线路大修 03.0194

make-up capacity 编组能力 06.0507

make-up of trains 编组调车 06.0256

Manchester encoding 曼彻斯特编码 07.0734

manhole 检查井 02.0309, 人孔 05.0472

manhole cover 人孔盖 05.0473

manhole ladder 内梯 05.0475

manometer regulator 风压调整器 08.0403

manual non-time release 不限时人工解锁 08.0148

manual operation 单独操纵作业，＊手动作业 08.0375

manual release 人工解锁 08.0144

manual release button panel 解锁按钮盘 08.0257

manual release indication 人工解锁表示 08.0166

manual release of a locked switch 道岔人工解锁 08.0146

manual route release 进路人工解锁 08.0145

manual shunt 人工分路 08.0527

manual telegraph set 人工电报机 07.0690

manual telephone office 人工电话所 07.0389

manual time release 限时人工解锁 08.0147

manual toll switching board 长途接续台 07.0252

manual voltage regulation 手动调压 08.0657

manway 人孔 05.0472

marked loading capacity of car 货车标记载重量 06.0152

marker 标志 08.0093, 标志器 07.0400

marker lamp 标志灯 04.0643

marshalling-departure track 编发线 06.0408

marshalling facilities 调车设备 06.0446

marshalling station 编组站 06.0369

marshalling yard 调车场 06.0443, 编组站 06.0369

masonry bridge 圬工桥 02.0361

mass freight 大宗货物 06.0113

master communication center of railway whole administration 通信总枢纽 07.0243

master group 主群 07.0310

master oscillator 主振器 07.0343

master retarder 峰下减速器 08.0379

master station 主站 07.0847

mast gauge 支柱侧面限界 04.1439

material consumption norm 材料消耗定额 06.0718

material consumption ratings 材料消耗定额 06.0718

material requisition plan 材料申请计划 06.0720

material supply plan 材料供应计划 06.0719

mattress 柴排 02.0320

maximum allowable axle load 最大轴重 01.0051

maximum combustion temperature 最高燃烧温度 04.0753

maximum error 最大误差，＊极限误差 02.0111

maximum explosive pressure 最高爆发压力 04.0754

maximum field 最大磁场 04.0434

maximum firing pressure 最高爆发压力 04.0754

maximum grade 最大坡度 02.0184

maximum height of car 车辆最大高度 05.0029

maximum horizontal displacement of contact wire 接触线最大水平偏移值 04.1452

maximum inclining position 最大倾斜位置 05.0848

maximum joint gap structurally obtainable 构造轨缝 03.0064

maximum lateral amplitude of bridge 桥梁最大横向振幅 03.0543

maximum load current of feeding section 供电臂最大负荷电流 04.1329

maximum number of passengers in peak hours 旅客最高聚集人数 06.0033

maximum output power 最大输出功率 04.0425

maximum outward position 最大外移位置 05.0847

maximum permissible speed of car 车辆最大容许速度 05.0843

maximum power 最大功率 04.0816

maximum service output power 最大运用功率，

＊装车功率　04.0084

maximum speed　最高速度　01.0079

maximum stroke volume　气缸最大容积　04.0716

maximum voltage between segments　片间最高电压
04.0431

maximum voltage of overhead contact line　接触网最
高电压　04.1331

maximum width of car　车辆最大宽度　05.0024

mean effective pressure　平均有效压力　04.0824

mean indicated pressure　平均指示压力　04.0823

mean mechanical loss pressure　平均机械损失压力
04.0825

mean operating air gap　平均工作气隙　04.1631

mean piston speed　活塞平均速度　04.0833

mean square error　均方[误]差，＊中误差
02.0110

mean voltage between segments　片间平均电压
04.0430

measure　丈量　02.1046

measured shovel packing　垫砂起道　03.0201

measurement of efficiency of locomotive　机车效率测
定　04.0677

measurement of percentage of harmonic current　谐波
电流百分比测定　04.0678

measurement of power factor　机车功率因数测定
04.0676

measurements of vibration parameters　机车振动参数
测试　04.0668

measures against tunnel fire　隧道防火措施
03.0560

measuring car for overhead contact line equipment　接
触网测试车　04.1459

mechanical efficiency　机械效率　04.0826

mechanical interlocking　机械集中联锁　08.0106

mechanical load　机械负荷　04.0835

mechanical locking　机械锁闭　08.0136

mechanical loss power　机械损失功率　04.0815

mechanical lubricator　压油机　04.1248

mechanically operated semaphore signal　机械臂板信
号机　08.0057

mechanical refrigerator car　机械冷藏，＊单节机
械冷藏车　05.0817

mechanical refrigerator car depot　机械保温车辆段

05.0898

mechanical refrigerator car group　机械冷藏车组
05.0818

mechanical ventilation　机械通风　02.0804

mechanical ventilation equipment　机械通风装置
05.0618

mechanized hump　机械化驼峰　06.0456

mechanized hump yard equipment　机械化驼峰设备
08.0366

median of field strength　场强中值　07.0520

medium　工质　04.0721

medium ballast undercutting cleaners　中型清筛机
03.0430

medium bridge　中桥　02.0418

medium phosphor cast iron brake shoe　中磷闸瓦
05.0195

medium speed diesel engine　中速柴油机　04.0694

medium tunnel　中长隧道　02.0658

megger　兆欧表　07.0219

member　杆件　05.0386

member support　构件支撑　02.0766

mesh network　网形网[络]　07.0759

message　报文　07.0738

message register　通话计数器，＊电磁计数器
07.0105

message switching　电文交换　07.0722

meteorological data　气象资料　02.0064

meter-gage railway　米轨铁路　01.0013

method of lime-soil replacement and tamping　灰土
换填夯实法　02.0615

metro　地下铁道，＊地铁　01.0023

metro engineering　地铁工程　02.0661

metro station　地铁车站　02.0655

metro tunnel　地铁隧道　02.0648

microcomputer interlocking　微机联锁　08.0113

microcomputer-relay interlocking　微机－继电式电气
集中联锁　08.0112

microwave communication　微波通信　07.0666

microwave communication vehicle　微波通信车
07.0668

microwave relay communication for railway　铁路微
波中继通信　07.0663

middle rolling car　中行车　06.0490

midheight deck bridge 半穿式桥，* 中承式桥 02.0426

mid-point anchor 中心锚结 04.1425

mid-point anchor clamp 中心锚结线夹 04.1426

midway between tracks 线间距[离] 02.0152

midway return operation 中途返回，* 中途折返 08.0193

military train 军用列车 06.0213

mine railway 矿山铁路 01.0025

mine tunnelling method 矿山法 02.0720

minimum available receiving level 最小可用接收电平 07.0518

minimum field 最小磁场 04.0435

minimum fill height of subgrade 最小填筑高度 02.0245

minimum height of fill 最小填筑高度 02.0245

minimum idling stabilized speed 最低空载稳定转速 04.0830

minimum no-load speed test 最低空载转速试验 04.0859

minimum radius of curvature negotiable 通过最小曲线半径 05.0844

minimum radius of curve 最小曲线半径 02.0171

minimum receiving field strength 最小接收场强 07.0519

minimum steady speed test 最低工作稳定转速试验 04.0858

minimum stroke volume 气缸最小容积 04.0717

minimum voltage of overhead contact line 接触网最低电压 04.1333

minimum voltage of overhead contact line at accident condition 事故状态的接触网最低电压 04.1334

mining method 矿山法 02.0720

mining subsidence 煤矿沉陷 03.0491

minor bridge 小桥 02.0419

minor telephone office 支电话所 07.0388

miscellaneous fees of goods traffic 货运杂费 06.0741

miseroute 错溜 08.0368

missing a train 漏乘 06.0057

missing locking 漏锁闭 08.0684

missing release 漏解锁 08.0685

mistake rate of goods 货差率 06.0647

misuse fault 误用故障 08.0739

mixed coal 混煤 04.1102

mixed feed water heater 混合式给水预热装置 04.1169

mixed gage turnout 套线道岔 03.0300

mixed holding fastening 半分开式扣件 03.0088

mixed passenger and freight station 客货运站 06.0385

mixed train 混合列车 06.0210

mixed-type freight yard 混合式货场 06.0085

mixer 混频器 07.0054

mobile communication 移动通信 07.0676

mobile microwave communication 移动微波通信 07.0667

mobile station 移动电台 07.0554

mode field diameter 模场直径 07.0450

model of car 车型 05.0016

modem 调制解调器 07.0849

mode of axle drive 车轴驱动方式 04.0348

mode of electric drive 电力传动方式 04.0340

mode of rectification 整流方式 04.0325

mode of speed control 调速方式 04.0344

mode of traction 牵引方式 01.0071

modular block 组匣 08.0260

modular block rack 组合柜 08.0263，组匣柜 08.0264

modular type relay interlocking 组匣式电气集中联锁 08.0118

modular type signal mechanism 组合式信号机构 08.0087

modulation 调制 07.0018

modulus of elasticity of rail support 钢轨支点弹性模量 03.0040

modulus of tractive effort 模数牵引力 04.1086

momentum grade 动力坡度 02.0187

monitoring 监听 07.0126

monitor signal 监测信号 07.0609

monitor TV for freight yard 货场监视电视 07.0656

monitor TV for highway crossing 道口监视电视 07.0660

monitor TV for passenger service 客运监视电视

07.0652

monitor TV for ticket check 检票监视电视 07.0654

monitor TV for yard and station 站场监视电视 07.0655

monitor TV under water 水下监视电视 07.0662

mono-bloc wheel 整体车轮 05.0096

monocoque structure 整体承载结构 05.0384

monolithic concrete bed 整体道床 03.0121

monomotor drive 单电动机驱动 04.0353

monorail 单轨铁路，＊独轨铁路 01.0028

monorail railway 单轨铁路，＊独轨铁路 01.0028

monthly freight traffic plan 月度货物运输计划 06.0096

monthly inspection 月检 04.0257

more favorable signal 较大允许信号 08.0044

more restrictive signal 较大限制信号 08.0042

Morse code 莫尔斯电码 07.0702

most favorable signal 最大允许信号 08.0043

most restrictive signal 最大限制信号 08.0041

motor car 动车 04.0010

motor characteristic 电动机特性 04.0420

motor coach set 电动车组 04.0309

motor individual power supply 电动机独立供电 04.0367

motor speed 电动机转速 04.0418

motor torque 电动机转矩，＊电动机扭矩 04.0421

motor train set 动车组 04.0009

motor train unit 动车组 04.0009

moulded coal 型煤 04.1103

moulded lining 模筑衬砌 02.0683

mountain and valley region location 山区河谷选线 02.0149

mountain railway 山区铁路 01.0031

mountain tunnel 山岭隧道 02.0645

movable bearing 活动支座 02.0516

movable bridge 活动桥 02.0408

movable contact 动接点 08.0551

movable plate 动板 05.0562

movable-point frog 可动心轨辙叉 03.0349

movable reservation system 移动备用方式 04.1298

movable signal 移动信号 08.0011

movable traction substation 移动牵引变电所 04.1346

movable traction transformer 移动牵引变压器 04.1345

movable-wing frog 可动翼轨辙叉 03.0350

moving blade loss 动叶损失 04.1554

mucking and removing 出碴 02.0756

mud and rock flow 泥石流 03.0483

mud-pumping 翻浆冒泥 03.0155

mud ring 底圈 04.1124

muffler 消声器 04.0987

multi-axle car 多轴车 05.0004

multi-axle truck 多轴转向架 05.0061

multi-circuit hydraulic transmission 多循环液力传动 04.0992

multi-lenses signal 透镜式色灯信号机，＊多灯信号机 08.0054

multilevel encoding 多电平编码 07.0737

multilink 多链路 07.0807

multi-locomotive electric coupler 机车重联电连接器 04.0393

multi-locomotive running kilometers 重联机车走行公里 04.0217

multi-locomotive traction 多机牵引 04.0225

multimagnet system 多磁铁系统 04.1600

multi-mode transport 综合运输 01.0120

multi-office system 多局制 07.0384

multipath propagation 多径传播 07.0537

multiple bearing type guide 多滑面滑板 04.1195

multiple system electric locomotive 多电流制电力机车 04.0305

multiple track railway 多线铁路 01.0017

multiple track tunnel 多线隧道 02.0653

multiple unit socket 重联插座 04.0608

multiple wear wheel 多次磨耗车轮 05.0104

multiplex 多路复用 07.0009

multiply connected track circuit 并联式轨道电路 08.0523

multipoint connection 多点连接 07.0800

multi rectifier bridge 多段桥[联结] 04.0336

multi-track bridge 多线桥 02.0422

mushroom-type tunnelling method 蘑菇形开挖法

02.0737

mutual inductance coefficient 互感系数，＊磁耦合系数 04.1473

mutual inductive impedance 互感阻抗 04.1474

N

nailable floor 可钉地板 05.0404

NAK 否认 07.0809

narrowband interference 窄带干扰 04.1507

narrow-gage railway 窄轨铁路 01.0012

national railway 国有铁路 01.0007

NATM 新奥法 02.0722

natural circulation 自然循环 04.0502

natural foundation 天然地基 02.0579

natural frequency of bridge span 桥梁自振频率 03.0544

natural ground 天然地基 02.0579

natural ventilation 自然通风 02.0803

natural ventilation equipment 自然通风装置 05.0617

natural vibration period of bridge 桥梁自振周期 02.0529

navigation water level 通航水位 02.0080

2nd limit 第二限度 04.0270

nearby protection 近体防护 02.1002

near-end crosstalk 近端串音 07.0039

near-end crosstalk attenuation 近端串音衰减 07.0141

near field 近场 07.0459

near field area 近场区 07.0545

negative acknowledge 否认 07.0809

negative benching tunnelling method 反台阶法 02.0732

negative superelevation 反超高 03.0029

net boiler efficiency 锅炉净效率 04.1112

net braking ratio 净制动率 05.0341

net weight of diesel engine 柴油机净重 04.0838

network 网络 07.0029

network management 网络管理 07.0769

network marshalling station 路网性编组站 06.0370

network planning technique 网络计划技术 02.0869

neutral relay 无极继电器 08.0591

neutral section 分相装置 04.1308

neutral section insulator 分相绝缘器 04.1416

neutral temperature 中和轨温 03.0229

neutralizing transformer 中和变压器 04.1495

New Austrian Tunnelling Method 新奥法 02.0722

newly-built railway 新建铁路 01.0033

newly-built railway construction 铁路新线建设 02.0001

Nielsen system bridge 尼尔森体系桥 02.0406

Nielsen type Lohse bridge 尼尔森式洛泽梁桥，＊斜悬杆式刚性拱刚性梁桥 02.0407

night signal 夜间信号 08.0008

nodding [vibration] 点头振动 01.0103

node 结点 07.0767

no-delay [demand] working 立接制 07.0274

no-humping car storage 禁溜车停留线 06.0437

noise 噪声 07.0037

noise silencer 消声器 04.0987

noise suppression coil 杂音抑制线圈 04.1498

noise suppressor 杂音抑制器 04.1499

no-load characteristic 空载特性 04.0853

no-load starting electro pneumatic valve 无载起动电空阀 04.0173

nominal power 标称功率，＊额定功率 04.0083

nominal speed 标称转速 04.0829

nominal voltage 标称电压 04.0428

nominal voltage at pantograph 受电弓标称电压 04.0316

nominal voltage of overhead contact line 接触网标称电压 04.1332

nominal voltage of traction substation 牵引变电所标称电压 04.1330

non-adhesion braking 非粘着制动 05.0324

non-attraction injector 非吸上式注水器 04.1167

nonbreathing zone 固定区 03.0236

non-centralized interlocking 非集中联锁

08.0119

non-circularity of core [cladding] 芯[包层表面]不圆度 07.0452

noncontacting proximity sensor 非接触式传感器 04.1601

non-interlocked switch 非联锁道岔 08.0197

non-interlocking area 非联锁区 08.0201

non-linear resistor 非线性电阻器 04.0523

non-mechanized hump 非机械化驼峰 06.0455

non-operating outlay 营业外支出 06.0729

non-operating station 非营业站 06.0081

non-parallel train diagram 非平行运行图 06.0290

non-pedestal truck 无导框式转向架 05.0067

non-permeable soil subgrade 非渗水土路基 02.0213

nonresponsible accident 非责任事故 06.0617

non-self-reset push-key switch 非自复式按键开关 04.0578

nonseparated rail fastening 不分开式扣件 03.0087

non-serviceable car 非运用车 06.0272

nonstick button 自复式按钮 08.0270

non-suspended weight 簧下重量 05.0866

non-vital circuit 非安全电路 08.0215

normal acting relay 正常动作继电器 08.0601

normal and reverse locking 定反位锁闭 08.0139

normal contact 定位接点 08.0554

normal locking 定位锁闭 08.0137

nose rail 心轨 03.0359

nose suspension traction motor 抱轴式牵引电动机 04.0400

nosing 蛇行运动 01.0088

notch indicator 位置指示器 04.0612

not on-duty casualty 路外人员伤亡 06.0651

not well inspected and repaired 检修不良 08.0694

not well maintained 维修不良 08.0693

nozzle needle valve 喷油嘴针阀 04.0956

nozzle needle valve body 喷油嘴针阀体 04.0957

nozzle ring 喷嘴环 04.0978

nucleus flaw 核伤 03.0175

number of allocated passenger cars 客车配属辆数 05.0978

number of allocated passenger trains 车底数 05.0979

number of car 车号 05.0017

number of car loadings 装车数 06.0313

number of cars in a train 列车编成辆数 06.0281

number of car unloadings 卸车数 06.0314

number of empty cars delivered 交出空车数 06.0318

number of empty cars received 接入空车数 06.0317

number of freight cars on hand 货车保有量 05.0980

number of inbound and outbound car handled at station 车站办理车数 06.0268

number of loaded cars delivered 交出重车数 06.0316

number of loaded cars received 接运重车数 06.0315

number of locomotives in service 运用机车台数 04.0221

number of passenger cars on hand 客车保有量 05.0977

number of passengers arrived 旅客到达人数 06.0031

number of passengers despatched 旅客发送人数 06.0030

number of passengers originated 旅客发送人数 06.0030

number of passengers transported 旅客运送人数 06.0032

number of passenger train set required 列车车底需要数 06.0070

number of railway staff and workers 铁路职工数 06.0715

number of serviceable cars held kept 运用车保有量 06.0327

number of serviceable cars turnround 运用车工作量 06.0319

numerical aperture 数值孔径 07.0449

NWL 通航水位 02.0080

O

oblique exposure 斜接近 04.1487

oblique exposure length 斜接近段长度 04.1488

oblong ejector 矩形通风装置，＊扁烟筒 04.1140

observation car 了望车 05.0739

observation window 了望窗 05.0507

obstruction signal 遮断信号 08.0032，遮断信号机 08.0081

obstruction signal button 遮断信号按钮 08.0453

obtuse frog 钝角辙叉 03.0351

occupancy indication 占线表示 08.0485

odd-even check code 奇偶校验码 07.0015

off-hook 摘机状态 07.0118

office car for peddler train 岔担办公车 05.0832

office equipment for radio train-dispatching 列车无线电调度总机 07.0589

officer's car 公务车 05.0742

off line maintenance 脱线修 08.0730

offset of lead curve 导曲线支距 03.0305

offset stake 外移桩 02.0115

off tread braking 非踏面制动 05.0326

oil bath air filter 油浴式空气滤清器 04.1046

oil burning heater 燃油温水锅炉 05.0605

oil-burning [heater type] hot water heating equipment 燃油独立温水采暖装置 05.0601

oil consumption 机油消耗量 04.0769

oil cylinder of shutter 百叶窗油缸 04.1054

oil damper 液压减振器，＊油压减振器 05.0163

oil extractor 油分离器 05.0637

oil feed pump 供油泵 04.1018

oil guide cylinder type journal box positioning device 油导筒式轴箱定位装置 05.0148

oil-immersed type traction transformer 油浸式牵引变压器 04.0487

oil pan 油底壳 04.0868

oil precision filter for turbocharger 增压器机油滤清器 04.1044

oil-pressure relay 油压继电器 04.0603

oil ring 油环 04.0908

oil separator 油气分离器 04.0870

oil sump 油底壳 04.0868

oil-water separator 油水分离器 04.0174

omnidirectional antenna 全向天线 07.0567

on-board rail lubricator 车载钢轨涂油器 03.0404

on-duty casualty 路内人员伤亡 06.0650

one-phase design 一阶段设计 02.0013

one-piece piston 整体活塞 04.0900

one-step design 一阶段设计 02.0013

one-way communication 单向通信 07.0777

one way feeding 单边供电 04.1293

one-wear wheel 一次磨耗车轮 05.0103

on-hook 挂机状态 07.0119

on line maintenance 在线修 08.0729

on-track full section undercutting cleaners 大型全断面清筛机 03.0429

on-track rail lubricator 地面钢轨涂油器 03.0403

open caisson foundation 沉井基础 02.0583

open channel 明渠 02.0634

open-cut foundation 明挖基础 02.0582

open-cut-tunnel portal 明洞门 02.0675

open deck 明桥面 02.0499

open ditch 明渠 02.0634

open drain 明渠 02.0634

open excavation foundation 明挖基础 02.0582

open floor 明桥面 02.0499

open goods wagon 敞车 05.0794

opening enlargement of bridge and culvert 桥涵扩孔 03.0568

open regenerative cycle gas turbine 开式回热循环燃气轮机 04.1525

open simple cycle gas turbine 开式简单循环燃气轮机 04.1524

open-spandrel arch 空腹拱 02.0401

open system interconnection 开放系统互连 07.0770

open throat 辙叉有害空间 03.0340

open top container 敞顶集装箱 06.0544

open type sleeping car 开敞式卧车 05.0729

open type track circuit 开路式轨道电路 08.0504

open wire 架空明线 07.0157

open-wire communication 明线通信 07.0316

open-cut method 明挖法 02.0719

open-cut tunnel 明洞 02.0694

operating console 操纵台 08.0254

operating distance 铁路运营长度，* 运营里程 01.0043

operating length of railway 铁路运营长度，* 运营里程 01.0043

operating railway 运营铁路 01.0035

operating rod for driving a switch 动作连接杆 08.0626

operating station 营业站 06.0080

operating wavelength 工作波长 07.0467

operation limit 使用限度 04.0271

operation office for train receiving-departure 运转室 06.0529

operation test 运用试验 04.0291

operation vehicle for contact wire 接触网作业车 02.0984

operative construction organization design 实施性施工组织设计 02.0855

operator's circuit 座席电路 07.0260

opposed-piston type diesel engine 对活塞式柴油机 04.0702

opposite joint 相对式接头 03.0066

optical adapter 光适配器 07.0474

optical attenuator 光衰减器 07.0475

optical connector 光连接器 07.0473

optical detector 光探测器 07.0472

optical digital section 光数字段 07.0503

optical fiber 光纤 07.0442

optical fiber cable 光缆 07.0480

optical fiber cable joint closure 光缆接头 07.0491

optical fiber communication system 光纤通信系统 07.0500

optical fiber cutter 光纤切断器 07.0489

optical fiber digital line system 光纤数字线路系统 07.0501

optical fiber distribution frame 光纤分配架 07.0479

optical fiber fusion splicing machine 光纤熔接机 07.0488

optical fiber splice loss 光纤接续损耗 07.0499

optical fiber stripper 光纤剥除器 07.0490

optical interface 光接口 07.0478

optical line protection switching equipment 光线路保护切换设备 07.0504

optical line terminal equipment 光线路终端设备 07.0463

optical power meter 光功率计 07.0493

optical [regenerative] repeater 光[再生]中继器 07.0477

optical regenerator section 光再生[中继]段 07.0502

optical source 光源 07.0469

optical time domain reflectometer 光时域反射仪 07.0495

optical transmission mode 光传输模式 07.0468

optimum density 最佳密度 02.0261

optimum handling 优化操纵 04.0236

optimum moisture content 最佳含水量 02.0260

optimum operation 优化操纵 04.0236

option 选件，* 选项 07.0750

ordinary accident 一般事故 06.0625

ordinary subgrade 一般路基 02.0244

ore car 矿石车 05.0807

organization of car flow 车流组织 06.0277

organization of handling 装卸作业组织 06.0583

organization of station operation 车站工作组织 06.0230

organization of train operation 铁路行车组织 06.0196

original value of fixed assets 固定资产原价 06.0744

originating railway 发送路 06.0185

originator 始发者 07.0905

orthonormal tunnel portal 正洞门 02.0673

orthotropic plate 正交异性板 02.0509

oscillator 振荡器 07.0055

OSI 开放系统互连 07.0770

outer dead center 外止点 04.0715

outer dead point 外止点 04.0715

outer firebox 外火箱 04.1117

outgoing trunk circuit 出中继电路，＊出中继器 07.0262

outlet submerged culvert 压力式涵洞 02.0632

out-of-face surfacing 全面起道捣固 03.0198

out-of-gauge freight 超限货物 06.0157

out-of-plan freight traffic 计划外运输 06.0102

output power at wheel rim 轮周功率 04.0085

output shaft 输出轴 04.1016

outside corridor 外走廊 04.1064

outside end wall 外端墙 05.0433

outside firebox 外火箱 04.1117

outside tunnel control survey 隧道洞外控制测量 02.0142

oval flaw 核伤 03.0175

overall acceptance 全验 02.1048

overall length of bridge 桥梁全长 02.0433

overbreak 超挖 02.0753

over-bridge 天桥 06.0521

overcharge 过量充风，＊过量充气 04.0144

overcharging 过充风，＊过充气 04.0142

over-current relay 过电流继电器 04.0598

overflow call 溢呼 07.0264

overhang 中梁悬臂部 05.0393

overhaul of track 线路大修 03.0194

overhead brine tank refrigerator car 车顶冰箱冷藏车 05.0815

overhead catenary system gauge 接触网限界 01.0064

overhead contact line 接触网悬挂 04.1351

overhead contact line equipment 接触网 04.1350

overhead contact line with catenary 链形悬挂 04.1352

overhead crossing 线岔 04.1404

overhead ditch 天沟 02.0297

overhead earth wire 接触网架空地线 04.1377

overhead side voltmeter 网侧电压表 04.0622

overhead waiting hall 高架候车厅 06.0021

overlap 锚段关节 04.1429

overlapping line 套线 02.0165

overlap protection block section 保护区段 08.0342

overlap section 重叠区段 08.0341

overlap track circuit 叠加轨道电路 08.0496

over-load test 超负荷试验 04.0861

over mountain line tunnel 越岭隧道 02.0646

overpass bridge 跨线桥，＊立交桥 02.0357

over reduction 过量减压 04.0148

overrunning of signal 冒进信号 06.0631

overshooting interference 越区干扰 07.0532

oversize commodity car 长大货物车 05.0833

overspeed trip 超速停车装置 04.0968

overtaking station 越行站 06.0364

overtaking the station 越站乘车 06.0058

over-zone feeding 越区供电 04.1295

owner of freight 货主 06.0132

oxyacetylene pressure welding 气压焊 03.0238

P

package 包，＊分组 07.0742

packed freight 成件包装货物 06.0116

packer 捣镐，＊道镐 03.0467

packet 包，＊分组 07.0742

packet assembler/disassembler 包装/拆器，＊分组装拆器 07.0895

packet mode terminal 包式终端，＊分组式终端 07.0896

packet switching 包交换，＊分组交换 07.0850

packet switching network 包交换网，＊分组交换网 07.0766

packing 填料 04.1188

packing sedimentation by throwing stones 抛石挤淤 02.0269

packing up sedimentation by dumping stones 抛石挤淤 02.0269

pad 衬垫 03.0097

PAD 包装/拆器，＊分组装拆器 07.0895

pallet 托盘 06.0565

pallet traffic 托盘运输 06.0577

panel 节间 02.0440，[客车]内墙板 05.0451

panel length 节间长度 02.0436

panel point 节点 02.0453

pantograph 受电弓 04.0534

pantograph bow 弓头 04.0537

pantograph-contact line relation 弓网关系 04.0185

pantograph cylinder 受电弓气缸 04.0540

pantograph horn 弓角 04.0539

pantograph isolating switch 受电弓隔离开关 04.0569

pantograph pan 受电弓滑板 04.0538

pantograph static contact force 受电弓上抬力 04.1445

pantograph test 受电弓试验 04.0660

pantograph upthrust 受电弓上抬力 04.1445

pantograph valve 受电弓电空阀 04.0585

paper location of line 纸上定线 02.0156

parallel arrangement type junction 并列式枢纽 06.0400

parallel crossover 平行渡线 03.0298

parallel heading 平行导坑 02.0789

parallelism approach 平行接近 04.1486

parallel relay network 继电并联网路 08.0219

parallel route 平行进路 06.0250

parallel train diagram 平行运行图 06.0289

parallel transmission 并行传输 07.0784

parcel 包裹 06.0016

parity check 奇偶检验 07.0798

parking braking 停驻制动 05.0365

part-cut and part-fill section 半堤半堑 02.0240

part cut-type gallery 半路堑式明洞 02.0698

part cut-type open cut tunnel 半路堑式明洞 02.0698

part cut-type tunnel without cover 半路堑式明洞 02.0698

partial excavation method 分部开挖法 02.0729

partially prestressed concrete bridge 部分预应力混凝土桥 02.0374

partial scour 局部冲刷 02.0097

particular load 特殊荷载 02.0348

partition 间壁 05.0790

partition door [客车]隔门 05.0489

part-load transhipment station 零担货物中转站 06.0079

part project 分部工程 02.0005

party-line automatic telephone 共线自动电话 07.0420

party-line telephone control board with VF selective calling 音选同线电话总机 07.0429

party-line telephone distributor for VF selective calling 音选同线电话分配器 07.0430

party-line telephone subset with VF selective calling 音选同线电话分机 07.0431

passenger building 旅客站舍 06.0519

passenger car 客车 05.0718

passenger car journal temperature warning 客车轴温报警 05.0960

passenger car servicing depot 客车整备所 06.0525

passenger casualty 旅客伤亡 06.0652

passenger compartment 客室 05.0752

passenger fare 旅客票价 06.0731

passenger fatigue time 旅客疲劳时间 05.0853

passenger flow 客流 06.0023

passenger flow diagram 客流图 06.0029

passenger flow investigation 客流调查 06.0028

passenger flow volume 客流量 06.0027

passenger foot-bridge 天桥 06.0521

passenger locomotive 客运机车 04.0014

passenger radiotelephone on train 列车旅客无线电话 07.0588

passenger room 客室 05.0752

passenger special line 客运专线 01.0041

passenger station 客运站 06.0382

passenger stop point 旅客乘降所 06.0526

passenger ticket 客票 06.0035

passenger ticket for international through traffic 国际联运旅客车票 06.0044

passenger traffic accident 客运事故 06.0620

passenger traffic density 客运密度 06.0067

passenger traffic only line 客运专线 01.0041

passenger traffic plan 旅客运输计划 06.0676

passenger train 旅客列车 06.0202

passenger train crew 旅客列车乘务组 06.0059

passenger train formation 旅客列车编组 06.0201

passenger train inspection and service point 旅客列车检修所 05.0913

passenger train technical servicing point 客车技术

整备所　05.0916

passenger transference　旅客换乘　06.0054

passing station　会让站　06.0363

path　通路　07.0004

path free signal　频道空闲信号　07.0610

pattern generator　码型发生器　07.0371

payload of freight container　集装箱载重　06.0556

PCM terminal　脉码调制终端机　07.0367

peak current　峰值电流　04.0427

peak detector　峰值检波器　07.0570

peak pressure　最高爆发压力　04.0754

peak speech power　峰值话音功率　07.0068

peak torque　峰值转矩　04.0423

pedestal brace　轴箱托板　04.1217

pedestal truck　导框式转向架　05.0065

pedestrian bridge　人行桥　02.0360

pendulum damper　摆式减振器　04.0897

pendulum type car body　摆式车体　05.0376

percentage of empty to loaded car kilometers　空车走行率　06.0322

percentage of machine handling in good condition　装卸机械完好率　06.0590

percentage of passenger seats utilization per car　客车客座利用率　06.0074

percentage of passenger seats utilization per train　列车客座利用率　06.0073

percentage of punctuality of trains despatched to total trains　列车出发正点率　06.0332

percentage of punctuality of trains running to total trains　列车运行正点率　06.0333

percent of call completed　接通率　07.0155

percent of call loss　呼损率　07.0153

percussion type drill machine　冲击式钻机　02.0915

performance test　性能试验　04.0281

periodical maintenance　定期维修　08.0724

periodical repair　定修　04.0243

periodical ticket　定期票　06.0047

periodic repair　定期修　05.0918

period in the train diagram　运行图周期　06.0309

period of inspection and repair　检修周期　04.0256

period of one complete turnround of locomotive　机车全周转时间　04.0210

perishable freight　易腐货物　06.0173

permanent bridge　永久性桥　02.0429

permanent magnet suspension system　永磁悬浮系统　04.1582

permanent ventilation of tunnel　隧道运营通风　02.0802

permanent virtual circuit　永久虚[拟]电路　07.0851

permanent way　线路　03.0015

permanent way clearing machine　线路清理机械　03.0379

permanent way gang　养路工区　03.0003

permanent way machine　线路机械　03.0368

permanent way signs　线路标志　03.0248

permanent way tool　线路机具　03.0369

permeable embankment　透水路堤　02.0312

permeable soil subgrade　渗水土路基　02.0212

permeable tunnel　渗水隧洞　02.0304

permissive period of transport　容许运输期限　06.0182

permissive prehumping signal　允许预推信号　08.0035

permissive signal　容许信号　08.0025

personal communication　个人通信　07.0677

pervious embankment　渗水土路基　02.0212, 透水路堤　02.0312

phantom resonant transformer　幻通谐振变压器　04.1496

phase alternating connection　换相联接　04.1283

phase control　相控调压　04.0322

phase detecting track circuit　相敏轨道电路　08.0503

phase jitter tester　相位抖动测量仪　07.0370

phase modulation　调相　07.0024

phase shift keying　相移键控　07.0733

phase splitter　劈相机　04.0470

photoelectric traverse　光电导线　02.0103

photographic facsimile apparatus　相片传真机　07.0727

physical and mechanical properties of soil and rock　土石物理力学性质　02.0045

pick-up and drop train　摘挂列车　06.0224

pick-up time　吸起时间　08.0560

pick-up value　吸起值　08.0566

pictorial markings for handling of packages　包装储运图示标志　06.0573

pier　桥墩　02.0559

pier body　墩身　02.0572

pier constructed with precast units　拼装式桥墩　02.0564

pier coping　墩帽　02.0573

pier protection against collision　墩台防撞　03.0573

pier shaft　墩身　02.0572

pier sheathing　窗间板　05.0448

piezoelectric accelerometer　压电式加速度计　04.1611

piggyback car　背负运输车　05.0827

pile bent pier　排架式桥墩　02.0571

pile casing　护筒　02.0613

pile driver　打桩机　02.0904

pile follower　送桩　02.0610

pile foundation　桩基础　02.0593

pillow shaped wall lamp　枕形壁灯　05.0689

pilot　排障器　04.1065

pilot amplifier　导频放大器　07.0339

pilot audio fequency signal　导音频信号　07.0612

pilot freqency　导频　07.0305

pilot unattended repeater　导频无人增音机　07.0288

pilot valve　先开阀　04.1160

pinch device　压紧装置　05.0521

pin-point blasting　抛掷爆破　02.0276

pipe bracket　中间体　05.0296

pipe çompensator　导管调整器　08.0294

pipe culvert　管涵　02.0627

piped rail　轨腰劈裂　03.0187

pipe installation　导管装置　08.0293

pipe jaw　接杆　08.0295

pipe link　连接杆　08.0296

pipe nipple　端接管　05.0276

pipe roofing support　管棚支护　02.0776

pipe-shed support　管棚支护　02.0776

piston　活塞，＊勾贝　04.0899

piston action of train　列车活塞作用　02.0807

piston area　活塞面积　04.0819

piston body　活塞体　04.0902

piston bush　活塞衬套　04.0905

piston cooling nozzle　活塞冷却喷嘴　04.0871

piston head　活塞头　04.0903

piston pin　活塞销　04.0909

piston ring　活塞环　04.0906

piston ring type seal　涨圈式密封　05.0145

piston rod　活塞杆，＊勾贝杆　04.1190

piston scraping　拉缸　04.0842

piston seizure　抱缸　04.0843

piston skirt　活塞裙　04.0904

piston stroke　活塞行程　04.0708

piston travel　制动缸活塞行程，＊勾贝行程　05.0342

piston unit area power　单位活塞面积功率　04.0820

piston valve　汽阀　04.1201

pitching　护坡　02.0294

pitching [vibration]　点头振动　01.0103

pivot pin　中心销　05.0211

plain bearing　滑动轴承　05.0128

plain journal bearing　轴瓦　05.0133

plain location　平原地区选线　02.0147

plane-coordinate azimuth　坐标方位角　02.0124

plane rectangular coordinate　经纬距　02.0126

planned freight traffic　计划内运输　06.0101

planned maintenance　计划修　08.0728

planned management of construction　施工计划管理　02.0868

plan of capital construction　基本建设计划　06.0708

plan of capital repair　大修计划　06.0702

plan of labor and wages　劳动工资计划　06.0714

plan of renewal and upgrading　更新改造计划　06.0704

plan of technical indices for freight traffic　运输工作技术计划　06.0312

plastic stage design method　破损阶段设计法　02.0206

plastic wearing end for plain bearing　塑料轴瓦头　05.0135

plate　辐板　05.0121

plate bearing　平板支座　02.0517

plate frame　板式车架　04.1213

plate girder　板梁　02.0384

plate girder type steel structure 板梁式钢结构 05.0381

plate hole 辐板孔 05.0126

plate vibrator 平板式混凝土振捣器 02.0942

platform 站台 06.0520

platform based container 台架式集装箱 06.0545

platform container 平台集装箱 06.0546

platform of cutting 路堑平台 02.0282

platform of hump crest 峰顶平台 06.0452

platform ticket 站台票 06.0039

platform trap door 翻板 05.0466

plug 插塞 07.0101

plug bond 塞钉式钢轨接续线 08.0541

plug door 塞入门 05.0499

plug-in trouble 插接不良 08.0691

plug-in type relay 插入式继电器 08.0586

plunger 柱塞 04.0949

plunger lock 转辙锁闭器 08.0301

plunger matching parts 柱塞偶件 04.0948

plunger sleeve 柱塞套 04.0950

plus stake 加桩 02.0114

PMM 永磁悬浮系统 04.1582

pneumatic ballast hopper car 风动石碴[漏斗]车 05.0830

pneumatic caisson foundation 沉箱基础 02.0591

pneumatic control device 空气控制装置 05.0514

pneumatic grate shaking rigging 风动摇炉装置 04.1250

pneumatic/manual hopper door operating device 漏斗门风手动传动装置 05.0513

pneumatic pick 风镐 02.0961

pneumatic rock drill 风动凿岩机 02.0975

point-continued type speed control system 点连式调速系统 06.0488

point detector 联锁箱 08.0290

point of gradient change 变坡点 02.0191

point rail 心轨 03.0359

point to point connection 点对点连接 07.0801

point to point toll automatic dialling 点对点长途自动接续 07.0268

point type speed control system 点式调速系统 06.0486

poison car 毒品车 05.0831

poisonous goods wagon 毒品车 05.0831

poking fork 拨叉 05.0524

polar biased relay 偏极继电器 08.0593

polar-frequency pulse track circuit 极频轨道电路 08.0497

polarity checking circuit 极性检查电路 08.0225

polarized relay 有极继电器 08.0592

polar transposition 极性交叉 08.0532

pole 电杆 07.0164

pole balcony 杆上工作台 07.0213

pole changer operated by semaphore 臂板转极器 08.0316

pole changing speed control 变极调速 04.0347

pole-changing time 转极时间 08.0563

pole-changing value 转极值 08.0570

polling 轮询，＊探询 07.0852

polygonal catenary 直链形悬挂 04.1356

poly-tetrafluoroethylene bearing 聚四氟乙烯支座 02.0523

polytropic index of compression 压缩多变指数 04.0750

polytropic index of expansion 膨胀多变指数 04.0758

pontoon 浮箱 02.0954

pontoon bridge 浮桥 02.0395

poppet valve 提升阀 05.0300

portable electric tamper 手提电动捣固机，＊电镐 03.0413

portable gasoline-powered tamper 手提内燃捣固机 03.0415

portable pneumatic tamper 手提风动捣固机 03.0414

portable radio set 便携电台 07.0555

portable rock drill 手持式凿岩机 02.0976

portable telephone set 携带电话机 07.0097

portal frame 桥门架 02.0461

portal structure 硬横跨 04.1405

position changeover switch 位置转换开关 04.0558

positive benching tunnelling method 正台阶法 02.0731

postal car 邮政车 05.0737

post office compartment 邮政间 05.0760

post pallet 立柱式托盘 06.0570

post-tentioned prestressed concrete girder　后张法预应力梁　02.0373

post tunnel portal　柱式洞门　02.0669

pot hole blasting　药壶爆破　02.0278

pot rubber bearing　盆式橡胶支座　02.0522

poultry car　家禽车　05.0821

powdered goods car　粉末货物车　05.0801

power/brake changeover switch　牵引－制动位转换开关　04.0561

power circuit　主电路　04.0368

power dispatching telephone　电力调度电话　07.0414

power divider in tunnel　隧道功率分配器　07.0591

power factor compensation device　功率因数补偿装置　04.0589

power failure　停电　08.0687

power generating rail car　发电轨道车　03.0435

power input to compressor　压气机耗功　04.1561

power off contact　安全接点　08.0628

power per cylinder　单缸功率　04.0817

power plant car train-set　电站列车[车组]　05.0839

power plant compartment　发电室　05.0765

power regulating system　功率调节系统　04.0963

power reverse gear　动力回动机　04.1210

power room　动力室　04.1060

power shovel　单斗挖土机　02.0878

power source for indication lamp　表示灯电源　08.0662

power source for relay control　继电器控制电源　08.0661

power source for switch control　道岔控制电源　08.0658

power source for switch indication　道岔表示电源　08.0660

power source for switch operation　道岔动作电源　08.0659

power supply cable and coupling　电力电缆及连接器　05.0675

power supply circuit for train　列车供电电路　04.0376

power supply panel　电源屏　08.0650

power supply room　电源室　08.0716

power supply system of electric traction　电力牵引供电系统　04.1270

power switching over panel　电源转换屏　08.0654

power turbine　动力涡轮　04.1539

power/weight ratio　比功率　04.0086

practical design for construction scheme　实施性施工组织设计　02.0855

preboring of spike holes　木枕预钻孔　03.0219

preboring of wooden tie　木枕预钻孔　03.0219

precast concrete members　预制混凝土构件　02.0538

precast concrete units　预制混凝土构件　02.0538

precast lining　装配式衬砌　02.0682

precast pile　预制桩　02.0594

precautionary work against flood　防洪预抢工程　03.0576

pre-chamber　预燃室式燃烧室　04.0799

precise traverse survey　精密导线测量　02.0140

precision of survey　测量精度　02.0109

precombustion chamber　预燃室式燃烧室　04.0799，预燃室　04.0922

precombustion chamber nozzle　预燃室喷嘴　04.0923

precutting trough　预切槽　02.0784

predelivery test　出厂试验　04.0283

predetermined tractive characteristic curve of locomotive　机车预期牵引特性曲线　04.0081

prefabricated lining　装配式衬砌　02.0682

prefabricated steel shell bored foundation　预制钢壳钻孔基础　02.0588

pre-feasibility study　预可行性研究　02.0007

prefix number for railway toll call　铁路长途字冠　07.0250

pregroup　前群　07.0307

pre-grouting with small duct　小导管预注浆　02.0775

preheating boiler　预热锅炉　04.1033

preliminary design　初步设计　02.0014

preliminary survey　初测　02.0100

preparation of locomotive for leaving and arriving at depot　机车出入段作业　04.0206

preparation of the route　准备进路　06.0244

presence monitor　车辆存在监测器　08.0408

preservation of wooden tie 木枕防腐 03.0218

presetting of a route 预排进路 08.0188

presplit blasting 预裂爆破 02.0749

pressing block 压块 05.0560

pressure grouting of tunnel 隧道压浆 02.0798

pressure of surrounding rock 围岩压力 02.0705

pressure of water flow 流水压力 02.0345

pressure pipe 压力主管 05.0278

pressure reducing reservoir 降压气室 05.0310

[pressure] reducing valve [给水]减压阀 05.0663

pressure regulating valve 给气调整阀 05.0517

pressure regulator 调压器 04.0162

pressure relay 压力继电器 04.0596

pressure reservoir 压力风缸 05.0314

pressure sensor 压力传感器 04.0632

pressure step-up ratio 压力升高比 04.0755

prestressed concrete bridge 预应力混凝土桥
  02.0371

pretensioned prestressed concrete girder 先张法预应
  力梁 02.0372

prevention for repetitive clear of a signal 防止重复
  08.0134

prevention of accident 事故预防 06.0655

preventive maintenance 预防维修 08.0726

preventive maintenance system 预防维修制
  04.0255

preworking a block 预办闭塞 08.0348

primary air 一次空气 04.1568

primary cell power supply 一次电池供电 08.0639

primary group 一次群 07.0376

primary parameter 一次参数 08.0533

primary station 主控站 07.0854

primary support 初期支护 02.0687

primary suspension 第一系悬挂 05.0178

primary train diagram 基本运行图 06.0296

prime coat 底漆 03.0551

priming 汽水共腾 04.1265

principal load 主力 02.0330

printed circuit board socket 印制电路板插座
  04.0609

private car 企业自备车 06.0095

private circuit 专用电路 07.0112

private communication network 专用通信网
  07.0062

private data network 专用数据网 07.0754

private railway 私有铁路 01.0009

private siding 专用线 06.0417

private telephone network 专用电话网 07.0108

probabilitic limit state design method 概率极限状态
  设计法，＊可靠度设计法 02.0208

procedure 规程 07.0846

proceed signal 进行信号，＊允许信号 08.0020

profile survey 线路测量 02.0106

progression of failure 故障升级 08.0705

project appraisal 项目评估 06.0713

projecting mud soil 突泥 02.0991

projection of broken chain 投影断链 02.0128

project management 项目管理 06.0712

project quoted price 工程报价 02.0847

propelling trolley 推送小车辆 08.0399

proposed task of project 项目建议书 02.0008

propulsion force 推力 04.1599

propulsion system 推进系统 04.1587

prospecting 勘探 02.0050

protected 5-unit numerical code 五单位数字保护电
  码 07.0703

protected section 防护区段 08.0343

protecting acoustic signal 防护音响信号 02.1000

protection 防护 04.1503

protection circuit for approaching heavy down grade
  下坡道防护电路 08.0246

protection circuit with switch lying in receiving-departure track 到发线出岔电路 08.0247

protection distance 防护间距 04.1522

protection filter 防护滤波器 04.1500

protection railing 防护栏杆 02.1014

protection ratio 防护率 04.1510

protection signal 防护信号 06.0656

protection sleeve 保护套管 07.0492

protective capacitor 保护电容器 04.0526

protective circuit 保护电路 04.0386

protective coating of steel bridge 钢梁油漆
  03.0547

protective earth wire 保护地线 02.1033

protective filter 反滤层 02.0308

protective grounding 保护接地 02.1034

protective stake at grade crossing　道口护桩 03.0279

protective transformer　防护变压器　08.0547

protective turnout　防护道岔　08.0194

protective valve　保护阀　04.0584

protective wire　保护线　04.1316

protector　保安器　07.0208

protocol　协议　07.0881

protocol of international through traffic　国际联运议 定书　01.0124

protocol specification　协议规范　07.0885

PSK　相移键控　07.0733

psophometer　杂音测试器　07.0354

PTFE bearing　聚四氟乙烯支座　02.0523

public address coupling　播音连接器　05.0673

public address room　播音室　05.0767

public address system　播音装置　05.0692

public data network　公用数据网　07.0753

public hall　广厅　06.0017

public switched telephone network　公用电话交换网 07.0109

public telephone network　公用电话网　07.0107

pull　角深　07.0228

pull-off arm　软定位器　04.1393

pull-off mode　正定位　04.1396

pull-out button　拉钮　08.0268

pull trough　拉槽　02.0745

pulp car　纸浆车　05.0824

pulsating current traction motor　脉流牵引电动机 04.0396

pulse　脉冲　07.0025

pulse charging　脉冲增压　04.0791

pulse code modulation　脉冲编码调制　07.0027

pulse echo fault locator　线障脉冲测试器　07.0214

pulse track circuit　脉冲式轨道电路　08.0494

pulse transformer　脉冲变压器　04.0492

pulverized coal locomotive　煤粉机车　04.1082

pump impeller　泵轮　04.1010

pump motor　泵电动机　04.0476

pump supporting box　泵支承箱　04.0877

purline　车顶纵梁　05.0458

pusher grade　加力牵引坡度　02.0183

push-in button　按钮　08.0269

pushing section of hump　驼峰推送部分　06.0450

pushing speed　推送速度　06.0481

pushing track of hump　驼峰推送线　06.0426

push-off mode　反定位　04.1397

push-pull shunting　推送调车　06.0259

put into operation　开通　08.0710

pylon　桥塔　02.0490

Q

quadrant　扇形齿板　04.1162

quadrantal angle　象限角　02.0125

quadrant link　月牙板　04.1205

quadrant revetment　锥体护坡　02.0557

qualified project　合格工程　02.1055

quality assurance　质量保证　02.1038

quality control　质量控制　02.1039

quality control of railway transportation　铁路运输 质量管理　06.0005

quality management　质量管理　02.1037

quality superintendence　质量监督　02.1041

quality surveillance　质量监督　02.1041

quality system　质量体系　02.1040

quantizing　量化　07.0373

quasi-peak detector　准峰值检波器　07.0571

quenched alloy steel rail　合金淬火轨　03.0053

quenched rail　淬火轨　03.0051

quenched switch rail　淬火尖轨　03.0356

quick-acting relay　快动继电器　08.0599

quick-acting switch machine　快速转辙机　08.0607

quick action　紧急局减　04.0153

quick coupling　快速接头　05.0485

quick pick-up relay　快吸继电器　08.0600

quick sand　流砂　02.0609

quick service　常用局减　04.0152

quill drive　车轴空心轴驱动　04.0354

R

rack locomotive 齿轨[传动]机车，＊齿条[传动]机车 04.0363

radar speedometer 雷达测速器 08.0396

radial crack in plate 辐板径向裂纹 05.0259

radiated interference 辐射干扰 07.0526

radiation modes 辐射模 07.0455

radiator 散热器 04.1031，散热管 05.0607

radio communication for emergency purpose 应急救灾无线电通信 07.0644

radio communication for engineering construction 工程施工无线电通信 07.0639

radio communication for highway crossing protection 道口防护无线电通信 07.0632

radio communication for maintenance of locomotive 机务维修无线电通信 07.0637

radio communication for maintenance of signal and communication equipment 电务维修无线电通信 07.0638

radio communication for multiple-operated locomotive units 多机牵引无线电通信 07.0587

radio communication for number taker 车号员无线电通信 07.0624

radio communication for passenger service 客运业务无线电通信 07.0621

radio communication for protection of construction 施工防护无线电通信 07.0631

radio communication for railway protection 铁路防护无线电通信 07.0630

radio communication for railway public security 铁路公安无线电通信 07.0643

radio communication for railway warning 铁路告警无线电通信 07.0629

radio communication for shunting 调车无线电通信 07.0622

radio communication for special train 专运列车无线电通信 07.0646

radio communication for survey and design 勘测设计无线电通信 07.0640

radio communication for track maintenance 工务维修无线电通信 07.0636

radio communication for track walker 巡道工无线电通信 07.0641

radio communication for train inspection 列检无线电通信 07.0623

radio communication for train protection 列车防护无线电通信 07.0633

radio communication for train reception and starting 接发车无线电通信 07.0584

radio communication for train relieving 救援列车无线电通信 07.0645

radio communication for train service 列车业务无线电通信 07.0585

radio communication for warning of train approaching 列车接近告警无线电通信 07.0635

radio dispatching communication for train 列车无线电调度通信 07.0583

radio distance-measurement 无线电测距 07.0681

radio disturbance 无线电干扰 07.0524

radio frequency 射频 07.0014

radio interference 无线电干扰 07.0524

[radio] interference meter [无线电]干扰测量仪 04.1511

radio interference meter 无线电干扰测量仪 07.0574

radio locomotive signal 无线电机车信号 07.0682

radio noise 无线电噪声 04.1505

radio operated signal for level shunting 平面无线电调车信号 07.0685

radio paging set 无线寻呼机 07.0561

radio relay set in yard and station 站场无线电中继转发台 07.0625

radio repeating set 差转电台 07.0562

radio set in stand-by state 无线电台守候状态 07.0576

radio shunting signal 无线电调车信号 07.0683

radio shunting signal at hump 驼峰无线电调车信号

07.0684

radio station 无线电台 07.0552

radio telecontrol for locomotive 机车无线电遥控，
* 机车无线电操纵 07.0679

radio telecontrol for railway 铁路无线电遥控
07.0678

radio telecontrol of locomotive for shunting at hump
驼峰调车机车无线电遥控 07.0680

radius bar 半径杆 04.1206

radius of lead curve 导曲线半径 03.0304

radius rod 半径杆 04.1206

rail 钢轨 03.0047，横带 05.0437

rail anchor 防爬器 03.0100

rail base 轨底 03.0050

rail bender 平轨器 03.0398

rail bending tool 平轨器 03.0398

rail bond 钢轨接续线 08.0540

rail bottom 轨底 03.0050

rail brace 轨撑 03.0104

rail cant 轨底坡 03.0061

rail corrosion 钢轨锈蚀 03.0173

rail corrugation 轨头波纹磨损 03.0165

rail cracks 钢轨裂纹 03.0188

rail crane 轨道起重机 02.0897

rail creep indication posts 钢轨位移观测桩
03.0255

rail cutting machine 锯轨机 03.0394

rail defects and failures 钢轨伤损 03.0171

rail drilling machine 钢轨钻孔机 03.0389

rail drilling tool 钢轨钻孔器 03.0390

rail end batter 轨端马鞍形磨损 03.0168

rail end breakage 轨端崩裂 03.0181

rail end chamfering 轨端削角，* 轨端倒棱
03.0183

rail ends unevenness in line or surface 错牙接头
03.0170

rail expansion device 钢轨伸缩调节器，* 温度调节
器 03.0075

rail fastening 扣件，* 中间联结零件 03.0085

rail flat car 轨道平车 03.0437

rail flaw detection car 钢轨探伤车 03.0449

rail flaw detector 钢轨探伤仪 03.0450

rail gage 轨距 01.0048

rail gap 轨缝，* 接头缝 03.0063

rail gap adjuster 轨缝调整器 03.0395

rail gauge 轨距 01.0048

rail grinding 打磨钢轨 03.0212

rail grinding car 磨轨车 03.0392

rail grinding machine 磨轨机 03.0391

rail grinding train 磨轨列车 03.0393

rail head 轨头 03.0048

rail-head edges planing machine 钢轨刨边机
03.0399

rail head reprofiling 轨头整形 03.0213

rail impedance 钢轨阻抗 08.0518

railing 栏杆 02.0510

railing around coping of pier or abutment 围栏
02.0574

rail insulation 钢轨绝缘 08.0535

rail jack 起道器 03.0386

rail joint 钢轨接头 03.0065

rail joint accessories 接头联结零件 03.0076

rail joint bond 钢轨连接器 04.1339

rail joint fastenings 接头联结零件 03.0076

rail laying machine with cantilever 悬臂式铺轨机
03.0442

rail link 轨节 03.0020

rail-mounted crane 轨行式起重机 03.0440

rail-mounted handling and transportation machine 轨
行式装运机械 03.0376

rail potential to ground 钢轨对地电位 04.1338

rail profile gauge 钢轨磨损检查仪 03.0452

rail profile measuring car 钢轨磨损检查车
03.0451

rail puller 轨缝调整器 03.0395

railroad 铁道，* 铁路 01.0001

railroad break down due to flood 水害断道
03.0577

railroad communication 铁道通信，* 铁路通信
07.0001

railroad line 铁路线 01.0002

railroad network 铁路网 01.0003

rail/road permanent way machine 公铁两用线路机
械 03.0380

railroad signaling 铁道信号，* 铁路信号 08.0001

rail sawing machine 锯轨机 03.0394

rail screw-bolt power wrench 机动螺钉-螺栓搬手 03.0402

rail shearing device 钢轨推凸器 03.0401

rail spike 道钉, *狗头钉, *钩头钉 03.0090

rail straightener 直轨器 03.0397

rail straightening tool 直轨器 03.0397

rail supporting modulus 钢轨基础模量 03.0039

rail tensor 钢轨拉伸器 03.0396

railway 铁道, *铁路 01.0001

Railway Act 铁道法规 01.0117

railway administration toll communication network 局线长途通信网 07.0246

railway aerial photogrammetry 铁路航空摄影测量, *铁路航测 02.0130

railway aerial surveying 铁路航空勘测 02.0131

railway bridge 铁路桥 02.0355

railway car 铁道车辆 05.0001

railway car ferries 铁路轮渡 02.0639

railway classification 铁路等级 01.0006

railway code 铁路条例 01.0118

railway communication 铁道通信, *铁路通信 07.0001

railway construction clearance 铁路建筑限界 01.0055

railway construction fund 铁路建设基金 06.0754

railway culvert 铁路涵洞 02.0625

railway data exchange system 铁路数据交换系统 01.0114

railway finance 铁路财务 06.0742

railway financial condition 铁路财务状况 06.0759

railway financial result 铁路财务成果 06.0758

railway fixed assets 铁路固定资产 06.0697

railway freight traffic 铁路货物运输 06.0075

railway freight traffic organization 铁路货运组织 06.0076

railway heavy haul traffic 铁路重载运输 06.0008

railway high speed traffic 铁路高速运输 06.0009

railway in operation 运营铁路 01.0035

railway junction terminal 铁路枢纽 06.0396

Railway Law 铁路法 01.0116

railway line 铁路线 01.0002

railway line for freight traffic 货运专线 01.0040

railway line for mixed passenger and freight traffic 客货运混合线路 01.0042

railway line for passenger traffic 客运专线 01.0041

railway local telephone 铁路地区电话 07.0382

railway location 铁路选线 02.0146

railway long distance telephone network 铁路长途电话网 07.0249

railway military service 铁路军事运输 06.0012

railway mobile communication 铁路移动无线电通信 07.0508

railway motive power 铁道牵引动力 04.0001

railway network 铁路网 01.0003

railway operation 铁路运营 06.0003

railway operation information system 铁路运营信息系统 01.0113

railway passenger traffic 铁路旅客运输 06.0013

railway passenger traffic organization 铁路客运组织 06.0014

railway plan 铁路计划 06.0673

railway radio communication 铁路无线电通信 07.0507

railway rate 铁路运价 06.0730

railway reconnaissance 铁路勘测 02.0037

railway science 铁道科学 01.0004

railway service train 路用列车 06.0227

railway signaling 铁道信号, *铁路信号 08.0001

railway special line 铁路专用线 01.0039

railway station telephone 铁路站内电话 07.0437

railway structure gauge 铁路建筑限界 01.0055

railway survey 铁路测量 02.0098

railway tariff 铁路运价 06.0730

railway technology 铁路技术 01.0005

railway traction power 铁道牵引动力 04.0001

railway traffic 铁路运输 06.0001

railway traffic control 铁路运输调度 06.0334

railway traffic dispatching 铁路运输调度 06.0334

railway traffic organization 铁路运输组织 06.0004

railway traffic profit 铁路运输利润 06.0760

railway traffic turnover fund 铁路运输周转基金

06.0755

railway train load norm 列车重量标准 06.0228

railway transport administration 铁路运输管理 06.0002

railway transportation 铁路运输 06.0001

railway transport economy 铁路运输经济 06.0672

railway trunk line locomotive 铁路干线机车 04.0012

railway tunnel 铁路隧道 02.0644

railway TV 铁路电视 07.0651

railway vehicle 铁道车辆 05.0001

rail web 轨腰 03.0049

rail welding machine 焊轨机 03.0405

rail weld seam shearing machine 钢轨推凸机 03.0400

rail/wheel contact stress 轮/轨接触应力 03.0046

rain gage 雨量计 03.0583

raining season construction 雨季施工 02.0844

rainy season construction 雨季施工 02.0844

raising of track 起道 03.0199

raking pile 斜桩 02.0612

range 炉灶 05.0756

range of a signal 显示距离 08.0049

rated load-bearing coefficient for bridge 桥梁检定承载系数 03.0521

rated load-bearing coefficient for bridge as compared with standard live loading 桥梁检定承载系数 03.0521

rated value 额定值 08.0568

rate of accumulated freight damage 货物积累损伤指数 05.0852

rate of bad order cars 残车率 05.0975

rate of cars under repair 车辆检修率 05.0971

rate of combustion 燃烧率 04.1100

rate of easement curvature 缓和曲线半径变更率 02.0175

rate of energy absorbed by draft gear 缓冲器能量吸收率 05.0583

rate of evaporation for engine 供汽率 04.1098

rate of firing 燃烧率 04.1100

rate of fixed assets renewal 固定资产更新率 06.0747

rate of fixed assets retirement 固定资产退废率 06.0746

rate of handling charge 装卸费率 06.0593

rate of heat release 放热率 04.1101

rate of pressure rise 压力升高比 04.0755

rate of speed fluctuation 转速波动率 04.0831

rate of traffic revenue 运输收入率 06.0751

rate of transition curve 缓和曲线半径变更率 02.0175

rate of utilization of repair positions in car depot 车辆段检修台位利用率 05.0974

rate of wheel load reduction 轮重减载率 05.0858

rating form for budget 预算定额 02.0026

rating form for estimate 概算定额 02.0027

rating of approximate estimate 概算定额 02.0027

rating of budget 预算定额 02.0026

rating of freight container 集装箱额定质量，＊集装箱总重 06.0554

rating per machine per team 机械台班定额 02.0030

rating per machine-team 机械台班定额 02.0030

rational traffic 合理运输 06.0105

ratio of car body length to length between truck centers 车辆长距比 05.0035

ratio of construction machinery in good condition 施工机械完好率 02.0875

ratio of freight cars under repair 货车检修率 06.0694

ratio of light weight to loading capacity 自重系数 05.0859

ratio of locomotives under repair 机车检修率 06.0696

ratio of spring deflections of primary and secondary suspension 第一第二系弹簧挠度比 05.0240

Rayleigh scattering 瑞利散射 07.0496

reactance ratio of compensator 补偿装置的电抗比 04.1303

reaction plate 反应板 04.1592

reaction rail 反应轨 04.1603

reading lamp 阅读灯 05.0684

rear bumper 后缓冲铁 04.1219

rear draft check casting　一体从板座　05.0413

rear draft lug　后从板座　05.0412

rear draft stop　后从板座　05.0412

rear expansion sheet　后膨胀板　04.1222

rear frame　后车架片　04.1216

rear part actual length of turnout　道岔后部实际长
　　度　03.0331

rear part theoretical length of turnout　道岔后部理论
　　长度　03.0330

rebound of shotcrete　喷射混凝土回弹　02.0761

receive channel　接收信道　07.0866

receiver　受话器　07.0071

receiver dynamic range　接收机动态范围　07.0466

receiver sensitivity　接收机灵敏度　07.0465

receiving amplifier　收信放大器　07.0342

receiving coil　接收线圈　08.0420

receiving-departure yard　到发场　06.0441

receiving end　受电端　08.0515

receiving route　接车进路　06.0248

receiving signal　接车信号　08.0026

receiving track　到达线　06.0406

receiving yard　到达场　06.0439

receptor　接受者　07.0906

recharging　再充风，＊再充气　04.0143

reciprocating compressor　往复式空气压缩机
　　02.0921

reciprocating type refrigeration compressor　往复式制
　　冷压缩机　05.0628

recirculating zone　回流区　04.1571

reclining seat　可躺座椅　05.0773

reconstructed railway　改建铁路　01.0034

recording desk　记录台　07.0256

recording phone set　录音电话机　07.0098

record [demand] working　记录制　07.0273

recovery procedure　恢复规程　07.0855

rectangular crossing　直角交叉　03.0292

rectangular pier　矩形桥墩　02.0570

rectifier　整流器　07.0048

rectifier power supply　整流器供电　08.0641

rectifier relay　整流继电器　08.0590

rectifier transformer　整流变压器　04.0488

red board　停车牌　03.0246

redirection　重定向　07.0880

reduced-fare ticket　减价票　06.0040

reducing valve　减压阀　04.0164

reduction relay valve　减压中继阀　05.0318

reduction transformer　屏蔽变压器　04.1497

reference equivalent　参考当量　07.0237

reference line of turnout　道岔基线　03.0324

refractive-index profile　折射率分布　07.0448

refrigerated container　冷藏集装箱　06.0549

refrigerated transport　冷藏运输　06.0179

refrigeration and heating equipment　制冷加温装置
　　05.0647

refrigerator and heater car　冷藏加温车　05.0813

refuge hole　避车洞　02.0815

refuge niche　避车洞　02.0815

refuge place　待避所　02.1007

refuge platform　避车台　02.0507

refuge recess　避车洞　02.0815

refuge siding　避难线　06.0415

regeneration and repetition　再生中继　07.0379

regenerative brake　再生制动　04.0118

regenerator　回热器　04.1533

regenerator effectiveness　回热度　04.1574

regional geology　区域地质　02.0059

regional marshalling station　区域性编组站
　　06.0371

regional railway　地方铁路　01.0008

register　记发器　07.0397

registration arm　定位管　04.1391

registration device　定位装置　04.1390

registration mast　定位柱　04.1414

registration wire　定位索　04.1364

regular maintenance of bridge and tunnel　桥隧经常
　　保养　03.0565

regular maintenance of buildings and structures　房建
　　维修　03.0589

regulated state of a track circuit　轨道电路调整状态
　　08.0513

regulating amplifier　调节放大器　07.0340

regulating plate　调节板　04.1386

regulating resistor　调节电阻器　04.0520

regulating structure　导流建筑物　02.0617

regulating traction transformer　调压牵引变压器
　　04.0485

regulating winding 调压绕组 04.0498

regulations for passenger/freight car repair in factory/depot 客货车厂段修规程 05.0985

regulations for railway freight traffic 铁路货物运输规程 06.0007

regulations for railway passenger traffic 铁路旅客运输规程 06.0006

regulations of railway technical operation 铁路技术管理规程 01.0119

regulator valve 调整阀，*汽门 04.1159

re-icing point 加冰所 06.0176

reinforced concrete bridge 钢筋混凝土桥 02.0370

reinforced concrete pile 钢筋混凝土桩 02.0606

reinforced concrete pole 钢筋混凝土电杆 07.0165

reinforced earth retaining wall 加筋土挡土墙 02.0288

reinforced soil retaining wall 加筋土挡土墙 02.0288

reinforcement 钢筋 02.0473

reinforcing bar 尖轨补强板 03.0366

relative density 相对密度 02.0258

relative friction coefficient 相对摩擦系数 05.0243

relative humidity 相对湿度 04.0847

relaxation pressure 松弛压力 02.0707

relay 继电器 08.0550

relay case 继电器箱 08.0281

relayed surveillanced network 远程网路 08.0478

relayed surveillanced subsection 远程监督分区，*远程遥信分区 08.0476

relay energized 继电器吸起，*继电器励磁 08.0558

relaying rail 再用轨 03.0243

relay interlocking for large station 大站电气集中联锁 08.0114

relay interlocking for small station 小站电气集中联锁 08.0115

relay radio communication for railway 铁路无线电中继通信 07.0509

relay released 继电器释放，*继电器失磁 08.0559

relay room 继电器室 08.0715

relay sensitivity 继电器灵敏度 08.0572

relay transmission mode 中继传输方式 07.0551

release 缓解 04.0135，解锁 08.0140

release circuit 解锁电路 08.0231

released by checking four sections 四点检查 08.0152

released by checking three sections 三点检查 08.0151

released by checking two sections 两点检查 08.0150

released route 解锁进路 08.0190

release factor 返还系数 08.0571

release portion 缓解部 05.0294

release propagation rate 缓解波速 04.0158

release speed at retarder 减速器出口速度 06.0484

release value 释放值，*落下值 08.0569

release valve 缓解阀 05.0315

releasing force 解锁力 08.0625

reliability 可靠性 08.0667

reliability test 可靠性试验 04.0293

remote control 遥控 08.0454

remote control for a section 区段遥控 08.0458

remote control of a junction terminal 枢纽遥控 08.0459

remote control of hump engines 驼峰机车遥控 08.0391

remotely controlled section 遥控区段 08.0465

remotely surveillanced section 遥信区段 08.0472

remote power feeding system 远供系统 07.0329

remote regulation 遥调 08.0457

remote sensing of railway engineering geology 铁路工程地质遥感 02.0133

remote surveillance 遥信 08.0455

remote surveillance and telemetering for highway level crossing 道口遥信遥测设备 08.0448

remote terminal 远程终端 07.0908

removable hatch cover 活顶盖 05.0463

removing 拆迁 02.0833

renewal and reconstruction of fixed assets 固定资产更新改造 06.0703

renewal and upgrading of fixed assets 固定资产更新改造 06.0703

rented land 租地 02.0834

repair according to condition　状态修　05.0919

repair based on condition of component　状态检修　04.0252

repair based on time or running kilometers　定期检修　04.0251

repair beyond the scope of repairing course　超范围修理　04.0261

repair bridge and tunnel fault　桥隧病害整治　03.0567

repairing car for overhead contact line equipment　接触网检修车　04.1460

repair procedure　检修作业程序　04.0266

repair siding　站修线　05.0938, 修车线　05.0941

repair track in station　站修线　05.0938

repair track maintenance　站修　05.0929

repair with interchangeable component　配件互换修　04.0254

repayment period of capital cost　投资回收期　02.0155

repeated checking　重复检查　08.0224

repeated traffic　重复运输　06.0108

repeater station　增音站　07.0319, 中继站　08.0463

repeating signal　复示信号　08.0031, 复示信号机　08.0083

repetition survey of existing railway　线路复测　02.0108

repetive survey　施工复测　02.0836

report on starting of construction work　开工报告　02.0866

repulsion force　排斥力　04.1596

request data transfer　请求数据传送　07.0856

re-railer　复轨器　06.0669

rerailing device　复轨器　06.0669

re-ringing　再振铃　07.0123

rescue train　救援列车　06.0664

reserved cars　备用货车　06.0341

reserve fund of project　工程预备费　02.0033

reserve power　储备功率　04.0821

reserve settlement　预留沉落量　02.0264

residual error-rate　差错漏检率　07.0857

residual gas　残余废气　04.0739

resilient gear drive　弹性齿轮驱动　04.0349

resonance of rolling stock　机车车辆共振　01.0105

resorting　调车　06.0254

resorting capacity of lead track　牵出线改编能力　06.0503

responsibility crew system of passenger train　旅客列车包车制　06.0063

responsible accident　责任事故　06.0616

restoration after a failure　故障复原　08.0703

restoration of flood damaged structures　水害复旧　03.0581

restoration work for flood damage　水害复旧　03.0581

restoring traffic　恢复通车　06.0671

restraint capability　栓固能力　06.0558

restriction section of carrying capacity　通过能力限制区间　06.0310

restriction signal　减速信号　08.0022

resurfacing of rail　焊修钢轨, ＊堆焊钢轨　03.0211

resurvey of existing railway　线路复测　02.0108

retaining ring [of tire]　车轮扣环　05.0100

retaining valve　[制动缸压力]保持阀　05.0319

retaining wall　挡土墙　02.0284

retarder　减速器　06.0460

retarder in closed state　减速器制动状态　08.0381

retarder in working state　减速器工作状态　08.0382

retarder location　制动位　06.0469

retarder released　减速器缓解状态　08.0383

retransmission　重传　07.0889

return wire　回流线　04.1314

revenue length　铁路运营长度, ＊运营里程　01.0043

reverse contact　反位接点　08.0555

reverse curve　反向曲线　02.0178

reverse filtration layer　反滤层　02.0308

reverse lever　回动手把　04.1211

reverse link　月牙板　04.1205

reverse link block　月牙板滑块　04.1209

reverse locking　反位锁闭　08.0138

reverse propagation　反向传播　07.0536

reverse pull rod 回动拉杆 04.1212

reverser 反向器 04.0562

reverse superelevation 反超高 03.0029

reversing shaft 换向轴 04.1017

revetment 护坡 02.0294，护岸 02.0316

revised general estimate 修正总概算 02.0023

rheostatic brake 电阻制动 04.0117

rheostatic control 变阻调压 04.0324

ride comfort 乘座舒适度 05.0849

riding comfortableness 乘座舒适度 05.0849

riding index 平稳性指标 05.0851

riding quality 运行品质 05.0888

right angle crank 直角拐肘 08.0298

right bridge 正交桥 02.0412

right-handed machine 正装 08.0616

right hand turnout 右开道岔 03.0289

right-of-way 铁路用地 02.0036

right side of car 二位侧 05.0011

rigid arch bridge with rigid tie and inclined suspenders
尼尔森式洛泽梁桥，* 斜悬杆式刚性拱刚性桥梁
02.0407

rigid arch bridge with rigid tie and vertical suspenders
洛泽式桥，* 直悬杆式刚性拱刚性梁桥
02.0405

rigid frame bridge 刚架桥，* 刚构桥 02.0390

rigid frame truck 一体构架转向架 05.0063

rigidity of track panel 轨道框架刚度 03.0037

rigidly positioned wheelset 刚性定位轮对
05.0881

rigid piston valve 固定式汽阀 04.1202

rigid wheelbase 固定轴距 05.0045

rim width 轮辋宽度 05.0222

ring back tone 回铃音 07.0278

ring current 铃流 07.0124

ringing 振铃 07.0122

ringing signal oscillator 振铃信号振荡器
07.0344

ring network 环形网[络] 07.0761

riprap protection 抛石防护 03.0579

rise of arch 拱矢 02.0485

river bank protection out of repair 河调－防护失修
03.0515

river bed paving 河床铺砌 02.0620

river course survey 河道调查 02.0070

riveted steel bridge 铆接钢桥 02.0363

r.m.s current of feeding section 供电臂有效电流
04.1325

r.m.s current of train 列车有效电流 04.1322

road bed 路基 02.0210

road crown 路拱 02.0235

road cutting 路堑 02.0239

road-railway repairing vehicle 公铁两用检修车
04.1458

[road] shoulder 路肩 02.0236

road test 线路试验 05.0867

road way signs 线路标志 03.0248

robbery accident 被盗事故 06.0638

rock blasting in cut 路堑石方爆破 02.0270

rock block filler 岩块填料 02.0254

rock burst 岩爆 02.0994

rock cutting blasting 路堑石方爆破 02.0270

rock drill 凿岩机 02.0890

rock drilling jumbo 凿岩台车，* 钻孔台车
02.0962

rocker bearing 摇轴支座 02.0518

rocker type journal box positioning device 转臂式轴
箱定位装置 05.0152

rock fall 落石 03.0482

rock fill 岩块填料 02.0254

rock filler 岩块填料 02.0254

rock loader 装岩机 02.0970

rock-roll vibration 摆滚振动 01.0099

rock subgrade 岩石路基 02.0211

rod drive 连杆驱动 04.0360

rod insulator 棒式绝缘子 04.1379

rollability measurement 测阻 08.0389

rolled steel wheel 辗钢车轮 05.0102

roller 压路机 02.0888

roller bearing 滚轴支座 02.0519

roller bearing end cap 轴承端盖 05.0143

roller bearing shop 滚动轴承间 05.0894

roller bearing test stand 滚柱轴承试验台
05.0994

roller rocker 滚轮摇臂 04.0935

rolling car resistance due to wind effects 货车溜放风
阻力 06.0475

rolling direction of hump　驼峰溜车方向　06.0453

rolling rig　滚动试验台　05.0990

rolling ring　滚圈　05.0367

rolling section of hump　驼峰溜放部分　06.0451

rolling speed　溜放速度　06.0498

rolling stock clearance [for railway]　铁路机车车辆限界　01.0059

rolling stock damage　机车车辆破损　06.0632

rolling stock utilization plan　机车车辆运用计划　06.0689

rolling track of hump　驼峰溜放线　06.0427

rolling [vibration]　侧滚振动　01.0101

roof　车顶　05.0454

roof cant rail　车顶侧梁　05.0459

roof cross beam　车顶横梁　05.0461

roof end rail　车顶端横梁　05.0460

roof fall　冒顶　02.0993

roof sheet　外火箱顶板　04.1119,外顶板　05.0455

roof water tank　车上水箱　05.0656

room thermostat　室温控制器,＊恒温控制器　05.0645

root mean square detector　均方根值检波器　07.0573

rope lug　绳栓　05.0416

rotary dial　旋转式拨号盘　07.0084

rotary dial telephone set　旋转号盘电话机　07.0094

rotary drill machine　旋转式钻机　02.0912

rotary dump coupler　转动式车钩　05.0529

rotary dumping coal car　转动车钩煤车　05.0804

rotary operation anticreep ledge　下防跳台　05.0542

rotary type barring mechanism　回转式盘车机构　04.0972

rotating seat　转动座椅　05.0772

rotor　转子　04.0445

rough-terrain crane　轮胎式起重机　02.0895

round about line　迂回线　06.0420

round about line of hump　驼峰迂回线　06.0428

round about traffic　迂回运输　06.0109

round-ended pier　圆端形桥墩　02.0568

round section　绕行地段　02.0203

route　进路　06.0243

route alternative　方案比选　02.0154

route diversion　变更径路　06.0055

route indicator　进路表示器　08.0089

route indicator circuit　进路表示器电路　08.0248

route locking　进路锁闭　08.0127

route locking indication　进路锁闭表示　08.0165

route reconnaissance　线路踏勘,＊草测　02.0099

route release　进路解锁　08.0141

route release at once　进路一次解锁　08.0142

route selecting circuit　进路电路　08.0229

route selection　选路　08.0191

route setting　排列进路　08.0187

route sheet　进路表　08.0213

route signal　进路信号机　08.0069

route signal for departure　发车进路信号机　08.0071

route signal for receiving　接车进路信号机　08.0070

route signal for receiving-departure　接发车进路信号机　08.0072

route signaling　选路制信号　08.0017

route storaging devices　进路储存器　08.0405

route survey　线路测量　02.0106

route type all-relay interlocking　进路继电式电气集中联锁　08.0111

route with overlapped section in the opposite direction　对向重叠进路　08.0155

route with overlapped section in the same direction　顺向重叠进路　08.0156

routine inspection　日常检查　04.0248

routine test　例行试验　04.0294

rubber draft gear　橡胶缓冲器　05.0549

rubber elastic guide post type journal box positioning device　橡胶弹性导柱式轴箱定位装置　05.0150

rubber friction draft gear　橡胶摩擦式缓冲器　05.0550

rubber-metal pad　橡胶堆　05.0174

rubber pad　橡胶垫　05.0175

rubber spring　橡胶弹簧　05.0173

rubber tie plate　橡胶垫板,＊弹性垫板　03.0096

rubber tired crane　轮胎式起重机　02.0895

rules for organization of train operation 铁路行车组织规则 06.0197

rules of maintenance of way 线路维修规则 03.0266

ruling grade 限制坡度 02.0182

run away 飞车 04.0844

runaway of locomotive or car 机车车辆溜逸 06.0634

runaway speed 电动机超速 04.0419

running at dropping water level 锅炉借水 04.1263

running board 外走廊 04.1064, 走板 05.0417

running gear 走行装置, ＊走行部 05.0056

running-in 磨合 04.0855

running inspection 运行检查, ＊行修 04.0249

running kilometers between predetermined repairs 定检公里 04.0258

running kilometers on the road 沿线走行公里 04.0213

running of extra train 列车加开 06.0357

running resistance 运行阻力 04.0096

running service-bulletin of depot 机务段运行揭示 04.0231

running test 线路试验 05.0867

running token 行车凭证 06.0241

running torque converter 运转变扭器, ＊运转变矩器 04.1004

running without water in gage 白水表行车 04.1264

run-off elevation 顺坡 03.0244

rush repair of flood damage to open for traffic 水害抢修 03.0578

S

saddle wear of rail end 轨端马鞍形磨损 03.0168

safety 安全性 08.0668

safety belt 保险带 02.1012

safety cap 安全帽 02.1009

safety circuit 安全电路 08.0214

safety contact 安全接点 08.0628

safety control system of traffic 运输安全控制系统 06.0608

safety device 安全装置 02.1030

safety distance 安全距离 02.1005

safety education 安全教育 02.0988

safety equipment 安全装置 02.1030

safety evaluation of traffic 运输安全评估 06.0606

safety explosion 安全炸药 02.1023

safety explosive 安全炸药 02.1023

safety factor of stability 稳定性安全系数 05.0889

safety helmet 安全帽 02.1009

safety indicator 安全标志 02.1003

safety inspection of traffic 运输安全检查 06.0605

safety in utilizing electric energy 安全用电 02.1015

safety management of traffic 运输安全管理 06.0604

safety mark 安全标志 02.1003

safety net 安全网 02.1011

safety of railway traffic 铁路运输安全 06.0602

safety place 安全地点 02.1008

safety protection equipment 安全防护设备 03.0378

safety rope 安全绳 02.1010

safety strap 安全绳 02.1010

safety supervision of traffic 运输安全监察 06.0607

safety symbol 安全标志 02.1003

safety system engineering of traffic 运输安全系统工程 06.0603

safety technical measures 安全技术措施 02.0987

safety valve 安全阀 04.1053

safety voltage 安全电压 02.1016

sag 垂度 07.0227

saloon car 客厅车 05.0738

same frequency interference 同频干扰 07.0529

same frequency interference area 同频干扰区

07.0544

same-frequency simplex radio communication 同频单工无线电通信 07.0513

same-side anchor 同侧下锚 04.1432

sample 抽样 07.0372

sample project 样板工程 02.1056

sampling inspection 取样试验检查 02.1044

sampling test 抽样试验 04.0297

sand ballast 砂道碴 03.0130

sand blasting 喷砂除锈 03.0548 .

sand blockade 砂害 03.0498

sand box 砂箱 04.0178

sand compaction pile 挤密砂桩 02.0608

sand drain 砂井，＊排水砂井 02.0310

sand drift 砂害 03.0498

sand-drift control 防砂 03.0494

sand filled drainage layer 排水砂垫层 02.0314

sanding device 撒砂装置 04.0177

sanding sprayer 撒砂器 04.0180

sanding valve 撒砂阀 04.0179

sand liquefaction 砂土液化 02.0324

sand pile 砂桩 02.0607

sand protection 防砂 03.0494

sand removing machine 除砂机 03.0459

satellite communication 卫星通信 07.0669

saturated chamber 饱和蒸汽室 04.1156

scaffold 脚手架 02.0623

scattered freight storehouse 零担仓库 06.0528

scavenging box 扫气箱 04.0875

scavenging efficiency 扫气效率 04.0745

scavenging period 换气过程 04.0728

scavenging pressure 扫气压力 04.0730

scavenging pump 扫气泵 04.0988

scheduled train number 核心车次 06.0287

scheme comparison 方案比选 02.0154

schnabel 钳形梁 05.0424

schnabel car 钳夹车 05.0836

scissors crossing 交叉渡线 03.0297

scope of repair 检修范围 04.0260

scope of repairing course 检修范围 04.0260

scoper 铲运机 02.0883

scouring 冲刷 03:0481

scouring and depositing around pier 墩周冲淤 03.0570

scraper 铲运机 02.0883

scrap value of fixed assets 固定资产残值 06.0745

screw compressor 螺杆式空气压缩机 02.0922

screw coupling 链子钩 05.0556

screw spike 螺纹道钉 03.0092

screw type refrigeration compressor 螺杆式制冷压缩机 05.0629

seal 墙顶封口 02.0760

seal at the top of wall 墙顶封口 02.0760

sealing 加封 08.0711

seal ring 密封圈 05.0146

searchlight signal 探照式色灯信号机 08.0055

seating capacity 定员 05.0018

seat-type water closet 坐式便器 05.0783

second approach section 第二接近区段 08.0207

secondary air 二次空气 04.1569

secondary flow loss 二次流损失 04.1552

secondary group 二次群 07.0377

secondary key switch set 副按键开关组 04.0580

secondary lining 二次衬砌 02.0688

secondary load 附加力 02.0340

secondary parameter 二次参数 08.0534

secondary push-key switch group 副按键开关组 04.0580

secondary station 受控站 07.0858

secondary stress 次应力 03.0538

secondary suspension 第二系悬挂 05.0179

second brake pipe 压力主管 05.0278

second departure section 第二离去区段 08.0210

second hand rail 再用轨 03.0243

section 区间 01.0046

sectional communication center of railway branch administration 分通信枢纽 07.0242

sectional release of a locked route 进路分段解锁 08.0143

section blocked 区间闭塞 08.0318

section cleared 区间空闲 08.0319

section closed up 区间封锁 08.0707

section for power supply 供电段 04.1454

sectioning 电分段 04.1336

section insulator 分段绝缘器 04.1415

section locking 区段锁闭 08.0130

section occupancy indication 区段占用表示
　08.0171

section occupied 区间占用 08.0320

section of easy grade 缓坡地段 02.0157

section of gentle slope 缓坡地段 02.0157

section of insufficient grade 非紧坡地段 02.0159

section of sufficient grade 紧坡地段 02.0158

section of unsufficient grade 非紧坡地段 02.0159

section point 电分段装置 04.1309

section post 分区所 04.1349

section sign 管界标 03.0254

section with a switch or switches 道岔区段
　08.0202

section without a switch 无岔区段 08.0203

seepage well 渗井 02.0305

segmental span-by-span construction using form tra-
　veller 活动模架逐跨施工法 02.0539

segment of survey 测段 02.0134

seismic basic intensity 地震基本烈度 02.0066

seismic coefficient method 地震系数法 02.0209

seismic force 地震力 02.0350

selecting 选择 07.0853

selective acceptance 抽验 02.1049

selective call 选呼 07.0616

selective calling and talking box for dispatching office
　调度所选叫通话箱 07.0428

selective call with audio frequency coding 音频组合
　选呼 07.0617

selective call with digital pulse coding 数字编码选
　呼 07.0618

selective level meter 选频电平表 07.0357

selectivity 选择性 07.0035

selectivity signal 选控信号 07.0613

self-excited vibrational frequency of bridge span 桥梁
　自振频率 03.0544

self-folding seat 活动座椅 05.0775

self-loop test 自环检测 07.0598

self-propelled power generating car 发电走行两用车
　03.0436

self-reset push-key switch 自复式按键开关
　04.0577

self-stabilization time of tunnel 坑道自稳时间
　02.0710

self-steering radial truck 自导向径向转向架
　05.0069

self-stick circuit 自闭电路，＊自保电路
　08.0217

self-supporting capacity of surrounding rock 围岩自
　承能力 02.0709

self-ventilated motor 自通风式电动机 04.0413

semaphore signal 臂板信号机 08.0056

semi-automatic block machine 半自动闭塞机
　08.0355

semi-automatic block system 半自动闭塞 08.0324

semi-automatic hump 半自动化驼峰 06.0457

semi-automatic hump yard system 半自动化驼峰系
　统 08.0365

semi-automatic operation by route 进路操纵作业，
　＊半自动作业 08.0376

semi-auto-tensioned catenary equipment 半补偿链形
　悬挂 04.1359

semi-cushioned berth sleeping car 硬卧车 05.0727

semi-cushioned couchette 硬卧车 05.0727

semi-cushioned seat coach 硬座车 05.0725

semi-duplex radio communication 半双工无线电通
　信 07.0516

semi-laminated frame 半叠片机座 04.0454

semi-loop routing 半循环交路 04.0190

semi-separated rail fastening 半分开式扣件
　03.0088

semi-trailer 半拖车 05.0828

sending and receiving separated circuit 送受分开电
　路 08.0226

sensible combustion period 显燃期 04.0783

sensitivity 灵敏度 07.0034，制动灵敏度
　05.0351

sensitivity coefficient 敏感系数 04.1494

sensor 传感器 04.0629

separate control lift/guidance 悬浮导向分别控制
　04.1632

separated rail fastening 分开式扣件 03.0086

separately excited motor 他励电动机 04.0407

separate power supply [system] for car 车辆分散供
　电 05.0716

separation distance 接近距离 04.1485

separation of waybill from shipment 票货分离 06.0644

separation point 脱钩点 06.0499

sequential starting of switches 道岔顺序启动 08.0183

sequential transiting of switches 道岔顺序转换 08.0184

serially connected track circuit 串联式轨道电路 08.0522

serial transmission 串行传输 07.0785

series excited motor 串励电动机 04.0404

series relay network 继电串联网路 08.0220

serious accident 大事故 06.0623

serviceable car 运用车 06.0271

serviceable work-done per car day 货车日产量 06.0156

service application 常用制动 04.0129

service braking 常用制动 04.0129

service car 公务车 05.0742

service communication system 业务通信系统 07.0330

service gallery 辅助坑道 02.0785

service locomotives 路用机车 04.0018

service pass 公用乘车证 06.0048

service portion 作用部 05.0292

service stability 制动安定性 05.0353

service test 运用考验 05.0868, 运用试验 04.0291

servicing siding 整备线 06.0424

servicing weight 整备品重量 05.0864

servo accelerometer 伺服型加速度计 04.1610

servo fatigue testing machine 伺服疲劳试验机 05.0992

servomotor 伺服电机, * 伺服马达 04.0478

setting-out of route 放线 02.0119

settlement allowance 预留沉落量 02.0264

settlement of accident 事故处理 06.0609

sever 分隔 07.0125

severe hot box 燃轴 05.0955

shaft 竖井 02.0786, 转轴 04.0443

shaft connection survey 竖井联系测量 02.0792

shallow burying tunnel 浅埋隧道 02.0649

shallow-depth tunnel 浅埋隧道 02.0649

shallow foundation of bridge 桥梁浅基 03.0513

shallow girder 低高度梁 02.0376

shallow hole blasting 浅孔爆破 02.0272

shallow tunnel 浅埋隧道 02.0649

sharing reactor 均流电抗器 04.0512

shatter crack 白点 03.0174

shear connector 抗剪连接件, * 抗剪结合件 02.0513

sheath 套管 02.0480, 护套 07.0483

shed gallery 棚洞 02.0696

shed tunnel 棚洞 02.0696

sheet pile 板桩 02.0604

shelf-type relay 座式继电器 08.0585

shelled tread 踏面剥离 05.0252

shell-type traction transformer 壳式牵引变压器 04.0483

shield 盾构 02.0982, 屏蔽体 04.1512

shielded cable 屏蔽电缆 07.0181

shielding 屏蔽 04.1513

shielding factor 屏蔽系数 04.1477

shielding factor of aerial earth wire 架空地线屏蔽系数 04.1482

shielding factor of bridge and tunnel 桥隧屏蔽系数 04.1480

shielding factor of cable 电缆屏蔽系数 04.1481

shielding factor of track 钢轨屏蔽系数 04.1479

shield method 盾构法 02.0725

shift out stake 外移桩 02.0115

shock absorber 减振器 05.0156

shock absorber base for relays 继电器防震架 08.0282

shoegear 受电靴装置 04.0543

shoofly 临时便线 03.0575

shop coat 工厂漆 03.0555

shop program 检修作业程序 04.0266

shore protection 护岸 02.0316

short-circuit current of feeding section 供电臂短路电流 04.1327

short-circuiting device 短路器 04.0566

shortening device 缩短装置 05.0563

short rail 短轨 03.0057

short routing 短交路 04.0193

short stator 短定子 04.1586

short tie　短枕　03.0118

short tunnel　短隧道　02.0659

short wave communication for railway　铁路短波通信　07.0647

short wave communication for railway emergency　铁路应急短波通信　07.0648

short wave radio communication vehicle　短波通信车　07.0650

short wave undulation of rail head　轨头短波浪磨损　03.0166

shotcrete　喷射混凝土　02.0778

shotcrete and rock bolt support　喷锚支护，* 锚喷支护　02.0770

shotcrete-bolt construction method　喷锚构筑法　02.0723

shotcrete bolt lining　喷锚衬砌　02.0690

shotcrete manipulator　喷射混凝土机械手　02.0974

shotcrete repair　喷射混凝土修理　03.0563

shotcrete support　喷射混凝土支护　02.0768

shoulder level　路肩高程　02.0237

shoulder of ballast bed　道床碴肩　03.0134

shunt　分路　08.0526

shunted state of a track circuit　轨道电路分路状态　08.0512

shunter's cabin at hump crest　峰顶调车员室　06.0531

shunt excited motor　并励电动机　04.0405

shunting　调车　06.0254

shunting calling tone　调车呼叫信号音　07.0628

shunting controller　调车控制器　04.0574

shunting effect　分路效应　08.0511

shunting indicator　调车表示器　08.0091

shunting indicator circuit　调车表示器电路　08.0250

shunting locomotive　调车机车　04.0016

shunting neck　牵出线　06.0410

shunting operation plan　调车作业计划　06.0348

shunting resistor　分流电阻器　04.0518

shunting route　调车进路　06.0246

shunting sensitivity　分路灵敏度，* 分流感度　08.0519

shunting signal　调车信号　08.0019，调车信号机　08.0078

shunting signal to prohibitive humping line　去禁溜线信号　08.0040

shunting tone　调车信号音　07.0626

shunting track　调车线　06.0409

shunting yard　调车场　06.0443

shuttle car　梭式矿车　02.0965

shuttled block train　循环直达列车　06.0218

side barrier at grade crossing　道口栅栏　03.0278

side bearing　旁承　05.0182

side bearing clearance　旁承间隙　05.0036

side bearing loading　旁承承载　05.0055

side bearing truck　旁承支重转向架　05.0072

side buffer　盘形缓冲器　05.0554

side ditch　边沟，* 侧沟　02.0296

side door　侧门　05.0496

side frame bottom chord　侧架下弦杆　05.0082

side frame bottom oblique chord　侧架下斜弦杆　05.0084

side frame column　侧架立柱　05.0086

side frame pedestal　轴箱导框　05.0087

[side frame] pedestal bearing boss　轴箱承台　05.0089

side frame spring seat　侧架弹簧承台　05.0085

side frame top chord　侧架上弦杆　05.0081

side frame top oblique chord　侧架上斜弦杆　05.0083

side heading method　侧壁导坑法　02.0738

side lamp　侧灯　05.0679

side number plate lamp　车号灯　04.0645

side oil outlet valve　侧排油阀　05.0484

side post　侧柱　05.0430

side post connecting rail　侧柱连铁　05.0427

side sheathing　侧墙包板，* 包板　05.0440

side sheet　侧板　04.1123

side sill　侧梁　05.0394

side slope of cut　路堑边坡　02.0280

side slope of embankment　路堤边坡　02.0248

side tone　侧音　07.0144

sidewalk loading　人行道荷载　02.0339

sidewalk on bridge　桥上人行道　02.0505

side wall　边墙　02.0678，侧墙　05.0426

side wall first lining method　先墙后拱法　02.0750

side wear of rail head 轨头侧面磨损，* 轨头侧面磨耗 03.0163

side window 侧窗 04.1058

sidewise bending 旁弯 05.0443

siding 站线 06.0404

sign 标志 08.0093

signal 信号 08.0004，信号机 08.0052

signal aspect and indication 信号显示 08.0046

signal at clear 信号开放 08.0050

signal at stop 信号关闭 08.0051

signal bracket 信号托架 08.0063

signal bridge 信号桥 08.0064

signal centralized maintenance 信号集中修 08.0732

signal check-out maintenance 信号检修 08.0733

signal circuit 信号电路 04.0384

signal control circuit 信号控制电路 08.0232

signal fault 信号故障 08.0738

signal for day and night 昼夜通用信号 08.0009

signal for shunting forward and backward 双面调车信号机 08.0084

signal for stub-end track 尽头信号机 08.0086

signal generator 信号发生器 07.0079

signal indication 信号表示 08.0162

signaling at stations 车站信号 08.0003

signaling circuit 信令电路 07.0077

signaling current 信令电流 07.0086

signaling equipment 振铃器 07.0333

signal inspection 信号检修 08.0733

signal intermediate repair 信号中修 08.0736

signal in throat section 咽喉信号机 08.0085

signal lamp 信号灯 07.0104

signal lever 信号握柄 08.0292

signal lighting circuit 信号机点灯电路 08.0233

signal lighting power source 信号机点灯电源 08.0647

signal location 信号点 08.0345

signal maintenance 信号维修 08.0723

signal major repair 信号大修 08.0737

signal out of order sign 信号无效标 08.0102

signal overhaul repair 信号大修 08.0737

signal quality detection 信号质量检测 07.0859

signal renovation 信号整治 08.0734

signal repeater 信号复示器 08.0273

signal slot 信号选别器 08.0312

signal to crosstalk ratio 信串比，* 串音防卫度 07.0140

signal to low frequency interference rate 低频干扰防卫度 07.0283

signal to noise ratio 信噪比 07.0043

signal tower 信号楼 06.0536

signal transformer 信号变压器 04.0494

silicon oil damper 硅油减振器 04.0895

silicon oil spring damper 硅油弹簧减振器 04.0896

silicon rectifier device 硅整流装置 04.0337

silicon rectifier electric locomotive 硅整流器电力机车 04.0306

sill step [货车]脚蹬 05.0406

silt arresting by explosion 爆破排淤 02.0268

siltation 淤积 02.0637

silting 淤积 02.0637

simple concrete mixing plant 简易混凝土搅拌站 02.0931

simple curve 单曲线 02.0173

simple turnout 单开道岔 03.0282

simplex operation 单工 07.0704

simplex radio communication 单工无线电通信 07.0512

simplex transmission 单工传输 07.0783

simplified hump 简易驼峰 06.0454

simply supported beam bridge 简支梁桥 02.0379

simulation of operation system 运营系统模拟 01.0115

simulator for driver train-handling 司机模拟操纵装置 04.0235

single arm pantograph 单臂受电弓 04.0535

single-arm routing 单肩回交路 04.0188

single-axle truck 单轴转向架 05.0058

single beam girder-erecting machine 单梁式架桥机 02.0948

single break 单断 08.0222

single car test 单车试验 05.0953

single car testing device 单车试验器 05.0371

single-circuit hydraulic transmission 单循环液力传动 04.0991

single circuit power supply 单回路供电 04.1279

single current 单流 07.0707

single cylindrical shade wall lamp 单筒壁灯 05.0690

single-deck pallet 单面托盘 06.0567

single-directional running automatic block 单向自动闭塞 08.0329

single expansion steam locomotive 单胀式蒸汽机车 04.1084

single feeding and multiple receiving track circuit 一送多受 08.0516

single frequency inductor 单频感应器 08.0428

single-mode optical fiber 单模光纤 07.0443

single-office system 单局制 07.0383

single-phase AC electric locomotive 单相交流电力机车 04.0301

single-phase AC traction motor 单相交流牵引电动机 04.0398

single-phase bridge rectifier 单相桥式整流器 04.0331

single-phase industrial frequency AC electric locomotive 单相工频交流电力机车 04.0302

single-phase industrial frequency AC electric traction system 单相工频交流电力牵引制 04.1268

single-phase industrial frequency AC motor train unit 单相交流电动车组 04.0310

single-phase industrial frequency AC system 单相工频交流制 04.0314

single-phase low frequency AC system 单相低频交流制 04.0315

single-phase wat-hour meter 单相电度表 04.0627

single rail track circuit 单轨条式轨道电路 08.0506

single reduction gear drive 单侧减速齿轮驱动 04.0351

single rolling on double pushing track 双推单溜 06.0478

single rotary compressor 单转子滑片式空压机 02.0923

single-shaft gas turbine 单轴燃气轮机 04.1526

single shoe brake 单侧[踏面]制动 05.0330

single side band communication 单边带通信 07.0511

single side-band short wave station 短波单边带无线电台 07.0649

single sided linear induction motor 单边型直线感应电动机 04.1604

single side heading method 单侧导坑法 02.0739

single signal location 单置信号点 08.0346

single slip switches 单式交分道岔 03.0294

single stage suspension 一系悬挂 05.0176

single suspension mast 中间柱 04.1407

single track all-relay semi-automatic block system 单线继电半自动闭塞 08.0326

single track bridge 单线桥 02.0420

single track railway 单线铁路 · 01.0015

single track tunnel 单线隧道 02.0651

single tube fluorescent lamp 单管荧光灯 05.0687

single wire semaphore signal 单线臂板信号机 08.0059

single wire system 单线制 05.0713

sink 餐车洗池 05.0776

siren 警笛 04.0653

skate 制动铁鞋 06.0465

skate throw-off device 脱鞋器 06.0466

skeleton reinforced concrete girder 型钢混凝土梁，* 劲性骨架混凝土梁 02.0378

skew bridge 斜交桥 02.0413

skew tunnel portal 斜洞门 02.0674

skylight 天窗 05.0506

'sky-light' in the train diagram 运行图'天窗' 06.0344

slab bridge 板桥 02.0382

slab culvert 盖板涵 02.0630

slab shed tunnel 盖板式棚洞 02.0700

slab-track 板式轨道 03.0111

slack action 间隙效应 04.0183

slackless drawbar 无间隙牵引杆 05.0557

slad shed gallery 盖板式棚洞 02.0700

slag ballast 矿渣道碴 03.0128

slant-legged rigid frame bridge 斜腿刚架桥，* 斜腿刚构桥 02.0391

slave station 从站 07.0848

sleeper 轨枕 03.0108

sleeping compartment 卧室 05.0753

sleeve with clevis and ring 套管铰环 04.1387

slide 坍方 03.0478

slide bar 滑板 04.1194

slide plate 滑床板 03.0367

slide valve 滑阀 05.0297

sliding door 拉门，＊滑门 05.0492

sliding roof box car 活顶棚车 05.0796

sliding roof goods van 活顶棚车 05.0796

sliding side box car 活墙棚车 05.0797

sliding type barring mechanism 滑动式盘车机构
　　04.0971

slight grade 缓和坡度 02.0197

slip 坍方 03.0478

slip form 滑动模板 02.0958

slip sheet 滑板 04.1194

slip switch 交分道岔 03.0293

slope 坡度 02.0180

slope cutting 刷坡 03.0504

slope of river 河流比降 02.0077

slope of water surface 水面坡度 02.0085

slope protection 护坡 02.0294，坡面防护
　　02.0315

slope regulation 斜调 07.0312

slotting 拉沟 03.0503

slow-acting relay 缓动继电器 08.0596

slow pick-up relay 缓吸继电器 08.0597

slow pick-up time 缓吸时间 08.0564

slow release relay 缓放继电器 08.0598

slow release time 缓放时间 08.0565

slurry jacket method for sinking caisson 泥浆套沉
　　井法 02.0589

small ballast undercutting cleaners 小型枕底清筛机
　　03.0432

small nucleus fissure 白点 03.0174

small permanent way machine 小型线路机械，＊轻
　　型线路机械 03.0382

small-scale temporary project 小型临时工程
　　02.0858

smokebox 烟箱 04.1137

smokebox door 烟箱小门 04.1152

smokebox front 烟箱大门 04.1151

smokebox tube sheet 烟箱管板 04.1138

smoke deflector 挡烟板 04.1153

smokestack 外烟筒 04.1143

[smoke] tube 小烟管 04.1136

smooth blasting 光面爆破 02.0748

smoothing reactor 平波电抗器 04.0506

snack counter 小卖部 05.0759

snow avalanche 雪崩 03.0485

snow blockade 雪害 03.0496

snow drift 雪害 03.0496

snow-drift control 防雪 03.0495

snow fence 防雪栅 03.0500

snow guard 防雪栅 03.0500

snow plough 除雪车 05.0840

snow plow 除雪车 05.0840

snow protection 防雪 03.0495

snow protection bank 防雪障 03.0501

snow protection hedge 防雪树篱 03.0502

snow remover 除雪机 03.0457

snow removing machine 除雪机 03.0457

snow slide 雪崩 03.0485

snow slip 雪崩 03.0485

snubber 摩擦式减振器 05.0157

socket 锚座 02.0497

socket-type saddle 杵座鞍子 04.1389

sodding 铺草皮 03.0509

soffit 拱腹 02.0487

soffit scaffolding 砌拱支架，＊碗扣式脚手架
　　02.0960

soft spots of road bed 路基松软 03.0476

soil replacement 换土 02.0267

solid bed 整体道床 03.0121

solid gear drive 刚性齿轮驱动 04.0350

solid manganese steel frog 高锰钢整铸辙叉
　　03.0347

solid pier 实体桥墩 02.0561

solid-spandrel arch 实腹拱 02.0400

solid wheel 整体车轮 05.0096

source address 源点地址 07.0901

SP 分区所 04.1349

space 间隔 07.0717

space diversity 空间分集 07.0578

space-interval method 空间间隔法 06.0238

spacer block 间隔铁 03.0365

space-wave propagation mode 空间[波]传播方式
　　07.0547

spacing braking 间隔制动 06.0470

spalling of rail head 轨头掉块 03.0177

span 跨径，＊跨度 02.0431

spandrel-filled arch 实腹拱 02.0400

spark arrester netting 火星网 04.1150

spark extinguishing circuit 灭火花电路 08.0227

spark gap 火花间隙 04.1340

spark lighter 点火器 05.0610

speaking post in yard 扩音柱 07.0441

special equipment for locomotive operation 机车专用设备 04.0276

special fund 专用基金 06.0753

special geology 特殊地质 02.0062

special heavy section switch rail 特种断面尖轨 03.0357

specialized freight station 专业性货运站 06.0078

special live load 特种车辆活载 03.0522

special-purpose freight car 专用货车 05.0793

special purpose railway 专用铁路 01.0036

special rate 特定运价 06.0736

special subscriber 特种用户 07.0403

special test 特殊试验 04.0295

specific air consumption 空气消耗率 04.0771

specific apparent energy consumption 视在单位能耗 04.1273

specifications for design of cars 车辆设计规范 05.0986

specific floor area 比面积 05.0054

specific oil consumption 机油消耗率 04.0768

specific purpose container 专用集装箱 06.0542

specific resistance 单位阻力 04.0103

specific volume 比容[积] 05.0053

specific volume power 单位体积功率 04.0088

specified conditions for detaining cars 扣车条件 05.0964

spectacles type tunnelling method 眼镜式开挖法 02.0742

spectrum character 频谱特性 04.1517

speech side tone 话音侧音 07.0145

speed 速度 01.0074

speed characteristic 速度特性 04.0850

speed control device 调速设备 06.0459

speed control system 速度控制系统 04.0105

speed-governing servomechanism 调速伺服机构 04.0962

speed-governing system 调速系统 04.0961

speed indicator 慢行牌 03.0247

speed measurement 测速 08.0386

speed ratio on constant power 恒功调速比 04.0438

speed regulation characteristics 调速特性，＊外特性 04.0852

speed relay 速度继电器 04.0594

speed restriction 限制速度 01.0076

speed sensor 速度传感器 04.0633

speed signaling 速差制信号 08.0016

spherical bearing 球面支座 02.0520

spherical center plate 球形心盘 05.0408

spherical combustion chamber 球型燃烧室 04.0800

spherical roller bearing 球形滚子轴承 05.0142

spike driver 打道钉机 03.0408

spike hammer 道钉锤 03.0468

spike puller 起道钉机 03.0409

spiral curve 螺旋曲线 03.0024

spiral transition curve 缓和曲线 02.0174

spiral unloading machine 螺旋卸车机 06.0599

s-plate s形辐板 05.0122

splice bar 接头夹板，＊鱼尾板 03.0077

splice plate 拼接板 02.0455

split of rail web 轨腰劈裂 03.0187

split reduction 分段减压 04.0150

split return wire 分相回流线，＊裂相回流线 04.1315

split-shaft gas turbine 分轴燃气轮机 04.1527

spoil bank 弃土堆 02.0283

spoiler 扰流板 02.0527

spool [制动]柱塞 05.0298

spot surfacing 找小坑 03.0197

spotted indication light 光点式表示 08.0173

spray concrete 喷射混凝土 02.0778

spread foundation 扩大基础 02.0581

spread rim 轮辋辗出 05.0254

spread spectrum communication 扩频通信 07.0674

spring butt rocking door [客车]摆门 05.0491

spring damper 弹簧式减振器 04.0898

spring flexibility 弹簧柔度 05.0236

spring friction draft gear 弹簧摩擦式缓冲器 05.0551

springing 起拱点 02.0486

spring plank 弹簧托板 05.0209

spring plank carrier 弹簧托梁 05.0208

spring rail anchor 弹簧防爬器 03.0102

spring stiffness 弹簧刚度 05.0237

spring suspension 弹簧悬挂装置 05.0164

spring tensioner 弹簧补偿器 04.1423

spring-type relay 弹力继电器 08.0588

spring washer 弹簧垫圈 03.0099

sprung weight 簧上重量 05.0865

spur dike 丁坝，* 挑水坝 02.0618

spur pile 斜桩 02.0612

square crossing 直角交叉 03.0292

square joint 相对式接头 03.0066

squaring of ties 方正轨枕 03.0215

squat-across type water closet 蹲式便器 05.0782

srew catenary 斜链形悬挂 04.1355

SS 牵引变电所 04.1342

SSP 开闭所 04.1347

stability against derailment 抗脱轨稳定性 05.0855

stability against forcing out during train buckling 抗挤出稳定性 05.0856

stability against overturning 抗倾覆稳定性 05.0854

stability of track 轨道稳定性 03.0042

stabilization for sands by afforestation 固沙造林 02.0321

stabilized light source 稳定光源 07.0494

stabilizing resistor 稳定电阻器 04.0519

stack extension 内烟筒 04.1144

stacking capability 堆码能力 06.0557

stacking operation 堆码作业 06.0594

stack-loading freight 堆装货物 06.0115

stack of freight 货垛 06.0595

staff exchanger 路签授受器 04.1066

staff pouch 路签携带器 08.0358

staff with a key 钥匙路签 08.0362

stagger 拉出值 04.1400

stagger angle 安装角 04.1547

staggered joint 相错式接头，* 相互式接头 03.0067

staggered magnet configuration 交错排列的磁铁布置 04.1633

stake outward 外移桩 02.0115

stake pocket 柱插 05.0415

staking out in survey 测量放样 02.0837

stall 失速 04.1562

standard atmospheric condition 标准大气状况 04.0845

standard coal 标准煤 04.0058

standard curtailed rail 缩短轨 03.0058

standard-gage railway 准轨铁路 01.0011

standard length rail 标准长度钢轨 03.0056

standard live load for bridge 桥梁标准活载 02.0326

standard penetration test 标准贯入试验 02.0058

standard shortened rail 缩短轨 03.0058

standard shunting sensitivity 标准分路灵敏度， * 标准分流感度 08.0520

standards of track maintenance 轨道养护标准 03.0267

stand-by power source 备电源 08.0635

standing time under repair 检修停时 04.0262

standing train protection 停车防护 02.1004

standstill detector valve 换向限止阀 04.1025

stand test 台架试验 04.0854

star network 星形网[络] 07.0760

start humping signal 推送信号 08.0034

starting accelerator 起动加速器 04.0966

starting and acceleration test 起动加速试验 04.0679

starting capacitor 起动电容器 04.0527

starting circuit 起动电路 04.0385

starting current 起动电流 04.0426

starting lubricating oil pump 起动机油泵 04.1038

starting relay 起动继电器 04.0604

starting resistance 起动阻力 04.0098

starting resistor 起动电阻器 04.0515

starting signal 出站信号机，* 出发信号机 08.0066

starting torque　起动转矩　04.0422

starting torque converter　起动变扭器，＊起动变矩器　04.1003

starting tractive effort　起动牵引力　04.0092

start-of-block signal　码组起始信号　07.0860

start-of-heading signal　报头开始信号　07.0861

start-of-text signal　正文开始信号　07.0862

start-stop signal　起止信号　07.0863

start-stop signal distortion tester　起止信号畸变测试器　07.0700

start-stop signal generator　起止信号发生器　07.0699

start-stop type　起止式　07.0713

state railway　国有铁路　01.0007

state transition diagram　状态转移图　07.0748

static inscribing　静力内接　03.0140

static load of car　货车静载重　06.0153

static pressure pile drawing machine　静压力拔桩机　02.0910

static pressure pile extractor　静压力拔桩机　02.0910

static probing　静力触探　02.0056

static sounding　静力触探　02.0056

static spring deflection　弹簧静挠度　05.0238

static strength test rack　静强度试验台　05.0989

static test　静载试验　03.0533

static track irregularity　静态不平顺　03.0143

station　车站　06.0362

station and freight yard radio communication　客站货场无线电通信　07.0642

stationary blade loss　静叶损失　04.1553

stationary test　定置试验　04.0288

station limit　站界　06.0231

station limit sign　站界标　08.0100

station master control　车站控制　08.0470

station operating plan　车站作业计划　06.0345

station radio set　车站电台　07.0558

station shift operating plan　车站班计划　06.0346

station site　站坪　06.0516

station square　站前广场　06.0524

station stage operating plan　车站阶段计划　06.0347

station technical working diagram　车站技术作业表　06.0269

station throat　车站咽喉　06.0506

station track　站线　06.0404

station tunnel　车站隧道　02.0654

station-yard communication　站场通信　07.0436

station-yard radio communication　站场无线电通信　07.0619

stator　定子　04.0444

stay　炉撑　04.1130，拉线　07.0170

stay cable　斜缆　02.0493

stay plate　缀板　02.0457

steady arm　定位器　04.1392

steady ear　定位线夹　04.1395

steady ring　定位环　04.1394

steady running condition　稳定运行工况　05.0051

steam chest　汽室　04.1186

steam chest bushing　汽室套　04.1187

steam heating equipment　蒸汽采暖装置　05.0592

steam jacket　加温套　05.0482

steam lap　进汽余面　04.1258

steam locomotive　蒸汽机车　04.0005

steam locomotive boiler washout repair　蒸汽机车洗修　04.0244

steam locomotive side rod　蒸汽机车连杆　04.1197

steam pile driver　蒸汽打桩机　02.0905

steam pile hammer　蒸汽打桩机　02.0905

steam pipe　主蒸汽管　04.1158

steel arch support　钢拱支撑　02.0767

steel bar　钢筋　02.0473

steel bridge　钢桥　02.0362

steel bridge fatigue damage　钢梁疲劳损伤　03.0539

steel-cored aluminum stranded wire　钢芯铝绞线　07.0161

steel fiber reinforced shotcrete　喷射钢纤维混凝土　02.0782

steel fiber shotcrete　喷射钢纤维混凝土　02.0782

steel pile　钢桩　02.0601

steel pipe pile　钢管桩　02.0602

steel sheet mesh type air filter　钢板网式空气滤清器　04.1047

steel sheet pile 钢板桩 02.0603

steel strand 钢绞线 02.0472

steel structure 钢结构 05.0379

steel tie 钢枕 03.0115

steel wire 钢丝 02.0470, 钢线, *铁线 07.0159

step-by-step telephone switching system 步进制电话交换机 07.0393

step down resistance 降压电阻 05.0609

stepless voltage regulation 无级调压 04.0321

stepped voltage regulation 分级调压 04.0320

stick button 非自复式按钮 08.0271

stiffener 加劲杆 02.0452

still picture videophone 静止图象可视电话 07.0066

stirrup 箍筋 02.0474

stitched catenary equipment 弹性链形悬挂 04.1357

stitched tramway type suspension equipment 弹性简单悬挂 04.1354

1st limit 第一限度 04.0269

stock car 家畜车 05.0820

stock rail 基本轨 03.0354

stock rails per kilometer of track 备用轨 03.0242

stoker 加煤机 04.1246

stone ballast 碎石道碴 03.0127

stone cut off wall 拦石墙 02.0318

stone falling channel 落石槽 02.0319

stone falling wall 拦石墙 02.0318

stop buffer 挡车器 06.0468

stop device 阻车器 02.1029

stop indicator 停车牌 03.0246

stopping at maintaining position 保压停车 04.0156

stopping a train at a target point 定点停车 08.0434

stopping at release 缓解停车 04.0157

stopping device 停车器 06.0467

stopping distance 制动距离 05.0346

stopping train protection 停车防护 02.1004

stop short 空档 06.0485

stop signal 停车信号 08.0023

stop signal indication 信号关闭表示 08.0164

storage battery 蓄电池 05.0707

storage battery box 蓄电池箱 05.0708

storage battery car 电瓶车 02.0967

storage battery power supply 蓄电池供电 08.0640

storage room 储藏室 05.0763

storage siding 存车线 06.0411

storage yard 堆货场 06.0090

storage yard and warehouse 场库 06.0089

store and forward 存储转发 07.0864

stored program controlled telephone switching system 程控电话交换机 07.0396

straddle 跨装 06.0164

straight air brake equipment 直通空气制动装置 05.0271

straightening of kinked rail 矫直钢轨 03.0210

straight-flow combustion 直流燃烧 04.1564

straight switch 直线尖轨 03.0314

straight tunnel portal 正洞门 02.0673

straight type diesel engine 直列式柴油机 04.0701

straight wheel plate 直辐板 05.0125

strain ageing 应变时效 02.0468

strainer check valve 滤尘止回阀 04.0169

strain gage 应变仪, *应变计 03.0536

strainometer 应变仪, *应变计 03.0536

stratum pressure 地层压力 02.0706

streamlined car body 流线型车体 05.0375

strengthening of steel bridge 钢梁加固 03.0574

strength test 强度试验 04.0290

stress corrosion cracking of steel bridge 钢梁应力腐蚀裂纹 03.0559

stressed ribbon bridge 悬板桥, *悬带桥 02.0392

stress free rail temperature 零应力轨温 03.0228

stress liberation 放散温度力 03.0230

stretcher bar 道岔拉杆 03.0363

stretching bed for longline production 长线张拉台座 02.0944

striker 冲击座 05.0414

striking casting 冲击座 05.0414

stringer 纵梁 02.0462

string-line calculator 绳度整正曲线计算器 03.0207

string lining computer 绳度整正曲线计算器 03.0207

string lining of curve 绳正法整正曲线 03.0206

stript indication light 光带式表示 08.0174

stroke 动程 08.0618

stroke/bore ratio 行程缸径比 04.0709

stroke volume 气缸工作容积 04.0718

structural joint gap 构造轨缝 03.0064

structural member 构件 05.0385

structure clearance for railway 铁路建筑限界 01.0055

strut insulator 棒式绝缘子 04.1379

strutted beam bridge 斜腿刚架桥，*斜腿刚构桥 02.0391

stub cable 尾巴电缆 07.0191

stub-end freight station 尽头式货运站 06.0387

stub-end passenger station 尽头式客运站 06.0384

stub-end siding 尽头线 06.0416

stub-end type freight yard 尽头式货场 06.0083

stub-end type junction terminal 尽端式枢纽 06.0403

student ticket 学生票 06.0041

s-type wheel plate s形辐板 05.0122

subaqueous cable 水底电缆 07.0185

subaqueous tunnel 水下隧道，*水底隧道 02.0647

subballast 底碴 03.0131

subballast consolidating machine 道床底碴夯实机 03.0426

subgrade 路基 02.0210，地基 02.0577

subgrade bed 基床 02.0241

subgrade bulge 路基挤起 03.0492

subgrade cross-section 路基横断面 02.0232

[subgrade] crown 路拱 02.0235

subgrade in bog [soil] zone 泥沼地区路基 02.0216

subgrade in cavern zone 洞穴地段路基 02.0228

subgrade in cavity zone 洞穴地段路基 02.0228

subgrade in collapse zone 崩塌地段路基 02.0224

subgrade in debris flow zone 泥石流地段路基 02.0231

subgrade in desert 风沙地段路基 02.0229

subgrade in expansive soil region 膨胀土地区路基，*裂土地区路基 02.0217

subgrade in karst zone 喀斯特地段路基，*岩溶地段路基 02.0227

subgrade in morass region 泥沼地区路基 02.0216

subgrade in permafrost soil zone 多年冻土路基 02.0219

subgrade in reservoir 水库路基 02.0223

subgrade in rock deposit zone 岩堆地段路基 02.0225

subgrade in rock fall district 崩塌地段路基 02.0224

subgrade in saline soil region 盐渍土地区路基 02.0218

subgrade in salty soil zone 盐渍土地区路基 02.0218

subgrade in scree zone 岩堆地段路基 02.0225

subgrade in slide 滑坡地段路基 02.0226

subgrade in snow damage zone 雪害地段路基 02.0230

subgrade in snow disaster zone 雪害地段路基 02.0230

subgrade in soft clay region 软土地区路基 02.0215

subgrade in soft soil zone 软土地区路基 02.0215

subgrade in swampland 泥沼地区路基 02.0216

subgrade in swelling soil zone 膨胀土地区路基，*裂土地区路基 02.0217

subgrade in talus zone 岩堆地段路基 02.0225

subgrade in windy and sandy zone 风沙地段路基 02.0229

subgrade machine 路基机械 03.0370

subgrade of special soil 特殊土路基 02.0214

subgrade settlement 路基下沉 03.0490

subgrade shoulder 路肩 02.0236

subgrade squeeze-out 路基挤起 03.0492

subgrade surface 路基面 02.0233

subgrade under special condition 特殊条件下的路基 02.0220

subhead lamp 副前照灯 04.0642

submarine cable 海底电缆 07.0180

sub-rail foundation 轨下基础 03.0107

sub-rail track bed 轨下基础 03.0107

subscriber's lead-in 用户引入线 07.0132

subscriber's main station 用户主机 07.0133

sub-section post 开闭所 04.1347

subsidiary load 附加力 02.0340

substation testing car 变电所测试车 04.1461

substituting ticket 代用票 06.0046

substructure 桥梁下部结构 02.0546

subsurface drainage 地下排水 03.0474

subsurface excavation method 暗挖法 02.0718

suburban coach 市郊客车 05.0724

suburban passenger car 市郊客车 05.0724

suburban passenger flow 市郊客流 06.0026

suburban passenger train 市郊旅客列车 06.0209

suburban railway 市郊铁路 01.0022

subway 地下铁道，*地铁 01.0023

subway engineering 地铁工程 02.0661

subway motor train unit 地下铁道电动车组
    04.0311

subway station 地铁车站 02.0655

subway tunnel 地铁隧道 02.0648

succesisve route 延续进路 08.0157

successively worked parallel relay network 继电并联
    传递网路 08.0221

suction pressure regulating valve 吸入压力调节阀
    05.0649

sudden rupture of rail 钢轨折断 03.0191

sugar cane car 甘蔗车 05.0823

sulphur cement mortar anchor 硫磺锚固 03.0094

sulphur cement mortar anchorage 硫磺锚固
    03.0094

summary estimate 总概算 02.0022

sum of approximate estimate 总概算 02.0022

sun-shield 遮阳板 04.1068

supercharged diesel engine 增压柴油机 04.0696

supercharging 增压 04.0785

supercharging pressure 增压压力 04.0786

supercharging ratio 增压比 04.0787

superclass bedroom 特等卧室 05.0754

superclass [corridor] compartment [type] sleeping car
    高级包房卧车 05.0731

superclass sleeping compartment 特等卧室
    05.0754

superconducting repulsion force 超导体斥力
    04.1624

superconducting suspension system 超导悬浮系统
    04.1581

superconductor 超导体 04.1602

superelevation 曲线超高 03.0026

super group 超群 07.0309

supergroup distribution frame 超群配线架
    07.0293

superheater chamber 过热蒸汽室 04.1157

superheater header 过热箱 04.1155

superheater tube 过热管 04.1154

super long tunnel 特长隧道 02.0656

super major bridge 特大桥 02.0416

superstructure 桥梁上部结构 02.0443

supervision of construction 工程监理 02.0850

supervision of project 工程监理 02.0850

supervision system 监视系统 07.0505

supplementary reservoir 附加风缸 05.0309

supply transformer 电源变压器 04.0489

supported joint 垫接接头，*承接接头 03.0068

supporter 支持器 04.1398

supporting insulator 支持绝缘子 04.0541

suppression 保持位 04.0146

surface drainage 地表排水 03.0473

surface-hardened switch rail 淬火尖轨 03.0356

surface layer of subgrade 基床表层 02.0242

surface layer of subgrade bed 基床表层 02.0242

surface of grade crossing 道口铺面 03.0274

surge 喘振 04.0795

surplus superelevation 过超高 03.0028

surrounding rock 围岩 02.0704

surveillanced object 监督对象 08.0473

survey and design of bridge crossing 桥渡勘测设计
    02.0084

survey and drawing of geological map 地质图测绘
    02.0049

survey and drawing of investigation 调查测绘
    02.0038

surveying and sketching of geological map 地质图测
    绘 02.0049

survey of bridge axis 桥轴线测量 02.0145

survey of catchment basin characteristics 汇水区流域

特征调查 02.0072

survey of existing railway 既有线测量，＊旧线测量 02.0107

survey of soil and rock composition 土石成分调查 02.0044

survey precision 测量精度 02.0109

survey tunnel 调查坑道 02.0791

suspended ditch 吊沟 02.0298

suspended joint 悬接接头 03.0069

suspended span 悬跨，＊吊孔 02.0442

suspended weight 簧上重量 05.0865

suspender 吊杆 02.0451

suspension bearing 抱轴悬挂装置 04.0461，抱轴轴承 04.1074

suspension bridge 悬索桥，＊吊桥 02.0393

suspension cable 吊缆 02.0494

suspension module 悬浮组件 04.1591

suspension pulley 悬吊滑轮 04.1418

suspension ring 吊环 04.1373

suspension system 悬浮系统 04.1589

swan neck 高鹅头弯管 05.0369

sway bracing 横联 02.0459

swaying [vibration] 侧摆振动 01.0102

swing bolster [复原]摇枕 04.1233

swing bridge 平旋桥 02.0410

swing door [客车]摇门 05.0490

swing hanger 摇枕吊 05.0206

swing hanger cross beam 摇枕吊轴 05.0207

swing rocker 摇鞍 04.1234

swirl combustion chamber 涡流室式燃烧室 04.0798

swirler 旋流器 04.1575

swirl-flow combustion 旋流燃烧 04.1566

switch 转辙器 03.0344

switch-and-lock mechanism 转换锁闭器 08.0302

switch angle 转辙角 03.0341

switch board 配电盘 05.0676

switch circuit controller [自动]开闭器 08.0610

switch cleaner's cabin 道岔清扫房 06.0535

switch closed up 道岔封锁 08.0708

switch control circuit 道岔控制电路 08.0234

switcher 调车机车 04.0016

switches and crossings 道岔 03.0281

switch expansion joint 钢轨伸缩调节器，＊温度调节器 03.0075

switch heater 道岔熔冰器 03.0458

switch indication 道岔表示 08.0161

switch indicator 道岔表示器 08.0094

switch indicator with level 带柄道岔表示器 08.0099

switching 交换 07.0745

switching lead 牵出线 06.0410

switch in transition 道岔转换 08.0182

switch layout 道岔配列 06.0436

switch lever 道岔握柄 08.0291

switch locked indication 道岔锁闭表示 08.0172

switch machine 转辙机 08.0602

switch machine installation 转辙机安装装置 08.0629

switchman's cabin 扳道房 06.0534

switchman's telephone 扳道电话 07.0438

switch normal indication 道岔定位表示 08.0168

switch plate 滑床板 03.0367

switch [point] closure 道岔密贴 08.0132

switch point guard rail 尖轨护轨 03.0361

switch [point] locking 道岔锁闭 08.0131

switch protector 尖轨保护器 03.0362

switch rail 尖轨 03.0313

switch resistance 道岔阻力 06.0476

switch reverse indication 道岔反位表示 08.0169

switch rod 道岔拉杆 03.0363

switch starting 道岔启动 08.0181

switch thrown under moving cars 道岔中途转换 08.0700

switch tie 岔枕 03.0117

switch with follow up movement 带动道岔 08.0195

swiveling drill machine 旋转式钻机 02.0912

symmetrical cable 对称电缆 07.0174

symmetrical cable communication 对称电缆通信 07.0317

symmetrical double curve turnout 单式对称道岔，＊双开道岔 03.0283

symmetrical three throw turnout 三开道岔

03.0286

symmetric half-controlled bridge rectifier 对称半控桥式整流器 04.0334

synchro 自整角机 04.0479

synchronous communication 同步通信 07.0778

synchronous motor 同步电动机 04.0408

synchronous transformer 同步变压器 04.0491

synchronous transmission 同步传输 07.0786

system anchor bolt 系统锚杆 02.0771

system height 结构高度 04.1453

system margin 系统余度 07.0476

system of assigning crew to designated locomotive 机车包乘制 04.0197

T

table plate 水平板，＊水平反射板 04.1148

tablet 路牌 08.0357

tablet exchanger 路签授受器 04.1066

tablet pouch 路牌携带器 08.0360

tachogenerator 测速发电机 04.0480

tachograph 速度记录仪 04.0628

tachometric relay 转速继电器 04.0599

tail lamp 尾灯 05.0680

tail throat of a hump yard 驼峰调车场尾部 06.0511

taking-out and placing-in of cars 取送调车 06.0258

taking wrong train 错乘 06.0056

tamping machine 捣固机械 03.0373

tamping pick 捣镐，＊道镐 03.0467

tandem distributor 音选调度电话汇接分配器，＊调度分配器 07.0427

tandem distributor for dispatching telephone with VF selective calling 音选调度电话汇接分配器，＊调度分配器 07.0427

tandem telephone office 汇接电话所 07.0387

tangent between curves 夹直线 02.0179

tangent retarder 调车线始端减速器 08.0378

tangible wear 有形损耗 06.0699

tank 罐体 05.0468

tank anchor 上鞍 05.0476

tank band 卡带，＊罐带 05.0481

tank car 罐车 05.0799

tank car freight 灌装货物 06.0118

tank container 罐式集装箱 06.0551

tank dome 空气包 05.0471

tank head 罐端板 05.0470

tank saddle 罐体鞍座 05.0469

tank valve 水柜阀 04.1243

tank volume table 罐车容积计表 05.0041

tank washing equipment 洗罐设备 05.0904

tank washing point 洗罐站 05.0945

tank washing shed 洗罐棚，＊洗罐库 05.0902

tank washing siding 洗罐线 05.0939

tap changer 调压开关 04.0555

tapered reverse tilting concrete mixer 锥形反转出料混凝土搅拌机 02.0927

tapered roller bearing 圆锥滚子轴承 05.0140

tapered tilting concrete mixer 锥形倾翻出料混凝土搅拌机 02.0928

tape transmitter 纸带发报机 07.0692

tapped traction transformer 分接牵引变压器 04.0486

tare mass of freight container 集装箱自重 06.0555

tare weight 自重 05.0861

target braking 目的制动 06.0471

target shooting 目标打靶控制 08.0394

tariff kilometerage 货物运价里程 06.0737

TBM 隧道掘进机 02.0981

tea table 茶桌 05.0770

technical design 技术设计 02.0015

technical record book of car 车辆技术履历薄 05.0981

technical record of track 工务设备台帐 03.0262

technical reform of railway 铁路技术改造 02.0002

technical renovation of railway 铁路技术改造 02.0002

technical servicing 技术整备 05.0914

technical speed 技术速度 06.0330

technical station 技术站 06.0395

technical through train 技术直达列车 06.0221

technological regulations for repair and inspection
检修工艺规程 04.0265

tee trap 丁形离水阀, *重力除水阀 05.0594

telecontrol repeat 遥控转发 07.0599

telegraph communication 电报通信 07.0687

telegraph network 电报[通信]网 07.0686

telegraph rate 通报速率 07.0709

telegraph switching equipment 电报交换机
07.0698

telemetry 遥测 08.0456

telephone 电话 07.0063

telephone block system 电话闭塞 08.0321

telephone channel 话路 07.0008

telephone conference of semi-distribution system
半分配制会议电话 07.0300

telephone conference within railway administration
局线会议电话 07.0298

telephone network 电话网 07.0106

telephone set 电话机 07.0090

[telephone] subscriber [电话]用户 07.0113

telephone traffic 话务量 07.0152

telescoping 套车 05.0857

teletype 电传机 07.0689

telex subscriber's telegraph 用户电报 07.0701

telltale 液面高度指示牌 05.0474

temperature regulating valve 温度调节阀
04.1026

temperature regulator 温度调整器 04.0611

temperature relay 温度继电器 04.0595

temperature sensor 温度传感器 04.0634

temperature stress 温度力 03.0231

temperature stress peak 温度力峰 03.0232

temporary bridge 临时性桥, *便桥 02.0430

temporary project 临时工程 02.0856

temporary repair 临修 05.0930

temporary repair siding 临修线 05.0942

temporary support 临时支护 02.0765

ten day car loading plan 旬间装车计划 06.0097

tender 煤水车 04.1239

tender truck 煤水车转向架 04.1242

tendon 预应力筋 02.0479

tension increment 张力增量 04.1437

tension length 锚段 04.1424

tension member 加强构件 07.0482

tension pulley 补偿滑轮 04.1419

terminal effect 终端效应 04.1483

terminal pole 终端杆 07.0167

terminals for power supplies 电源端子 08.0277

terminals of a modular block 组匣端子 08.0280

terminals of a unit block 组合端子 08.0279

terminals of layer 0 of a relay rack 零层端子
08.0278

terminals on distributing board 分线盘端子
08.0276

terminal station 终端站 07.0321

terminal toll office 通信端站 07.0244

termination fitting for catenary 承力索终端锚
固线夹 04.1368

termination fitting for contact wire 接触线终端
锚固线夹 04.1369

test at standstill 定置试验 04.0288

test car 试验车 05.0744

test desk 测试台 07.0131

test for K value of complete car K 值试验 05.0988

test for over-all air-tightness of compressed air
equipments 压缩空气设备全面的气密性试验
04.0658

test for running resistance 运行阻力试验 04.0680

test for sealing of body and external equipment 车体
及外部装备密封试验 04.0659

test loop 测试环线 08.0430

test of vibration caused by carbody bending 车体弯曲
振动试验 05.0872

test on air brake 空气制动试验 04.0666

test on auxiliary machines 辅助机组试验
04.0669

test on coupled operation 重联运行试验
04.0684

test on external overvoltage 外部过电压试验
04.0674

test on filter efficiency 滤尘效果试验
04.0682

test on internal overvoltage 内部过电压试验
04.0673

test on multi unit operation　重联运行试验　04.0684

test on overload protection system of main circuit　主电路过载保护系统试验　04.0672

test on safety equipments　安全设备试验　04.0663

test on sanding gear　撒砂装置试验　04.0683

test on short-circuit protection system of main circuit　主电路短路保护系统试验　04.0671

test on speed regulation　机车调速试验　04.0670

test on ventilation and cooling　通风冷却试验　04.0665

test pile　试桩　02.0611

test pole　试验杆　07.0166

test rack　试验架　07.0352

text　正文　07.0741

theoretical air　理论空气量　04.1573

theoretical lead of turnout　道岔理论导程　03.0328

theoretical length of turnout　道岔理论长度　03.0326

theoretical point of frog　辙叉心轨理论尖端　03.0320

theoretical point of switch rail　尖轨理论尖端　03.0316

thermal blockage　热挂　04.1531

thermal container　保温集装箱　06.0547

thermal crack　热裂纹　05.0256

thermal fatigue　热疲劳　04.1557

thermal insulation layer　隔热层　02.0693

thermal load　热负荷　04.0834

thermal relay　热[力]继电器　08.0589

thermo balance　热平衡　04.0760

thermo-characteristic of steam locomotive　蒸汽机车热工特性　04.1094

thermostatic expansion valve　恒温膨胀阀　05.0634

thickness of ballast bed　道床厚度　03.0132

thin-shelled tubular structure　薄壁筒体结构　05.0383

three-aspect automatic block　三显示自动闭塞　08.0338

three-axle truck　三轴转向架　05.0060

three-hinged arch　三铰拱　02.0399

three phase AC traction motor　三相交流牵引电动机　04.0399

three-phase bridge rectifier　三相桥式整流器　04.0332

three-phase design　三阶段设计　02.0011

three-piece truck　三大件转向架　05.0062

three-pressure equalizing system　三压力机构　05.0333

three states of coupler operation　车钩三态作用　05.0565

three-step design　三阶段设计　02.0011

three-way turnout　三开道岔　03.0286

threshold for channel switching　频道切换阈值　07.0606

threshold level of voice-operated circuit　音控门限电平　07.0151

throat length　咽喉区长度　06.0513

throat of frog　辙叉咽喉　03.0342

throat point　咽喉道岔　06.0512

throat sheet　喉板　04.1122

throttle valve　调整阀，＊汽门　04.1159

through bridge　下承式桥　02.0427

through button circuit　通过按钮电路　08.0251

through express train　旅客直达特别快车　06.0205

through passenger flow　直通客流　06.0024

through passenger train　直通旅客列车　06.0207

throughput　吞吐量　07.0747

through route　通过进路　06.0247

through routing　直通交路　04.0195

through shut-off valve　直通截止阀　05.0648

through signal　通过信号　08.0028

through speed of passenger train　旅客列车直达速度　06.0068

through survey　隧道洞内控制测量，＊贯通测量　02.0143

through traffic　直达运输　06.0103

through train originated from one loading point　始发直达列车　06.0215

through train originated from several adjoining loading points　阶梯直达列车　06.0216

through train with empty cars　空车直达列车　06.0217

through-type freight station　直通式货运站　06.0388

through-type freight yard　通过式货场　06.0084

through-type passenger station　通过式客运站　06.0383

through yard　直通场　06.0442

throwing stones to packing sedimentation　抛石挤淤　02.0269

throw of switch　尖轨动程　03.0337

throw rod　动作杆　08.0613

thrust bearing cap　止推轴承盖　04.0881

thrust bearing shell　止推轴瓦　04.0883

thrusting force　推力　04.1599

thrust ring　止推环　04.0884

thyristor converter electric locomotive　晶闸管变流器电力机车　04.0308

thyristor rectifier device　晶闸管整流装置　04.0338

thyristor rectifier electric locomotive　晶闸管整流器电力机车　04.0307

ticket　车票　06.0034

ticket availability　车票有效期　06.0050

ticket office　售票处　06.0019

tidal river　潮汐河流　02.0636

tie　轨枕　03.0108

tie adzing　削平木枕　03.0216

tied arch　系杆拱，*柔性系杆刚性拱　02.0403

tie plate　缀板　02.0457，垫板　03.0095

tie-plate type journal box positioning device　拉板式轴箱定位装置　05.0153

tie plug　枕木塞　03.0221

tie repairing　修补木枕　03.0217

tie replacing machine　轨枕抽换机　03.0406

tie respacer　方枕器　03.0411

tie respacing　方正轨枕　03.0215

tie-rod type journal box positioning device　拉杆式轴箱定位装置　05.0151

tight joint　接头瞎缝　03.0157

tight-lock coupler　密接式车钩　05.0528

tilting type car body　侧倾车体　05.0377

timber pile　木桩　02.0605

time-area value　时间－面积值　04.0796

time between predetermined repairs　定检时间　04.0259

time division multiplex　时分复用　07.0010

time division multiplex telegraph equipment　时分多路电报设备　07.0696

time interval between trains spaced by automatic block signals　追踪列车间隔时间　06.0302

time interval between two adjacent trains at station　车站间隔时间　06.0298

time interval between two opposing trains arriving at station not at the same time　不同时到达间隔时间　06.0299

time interval for two meeting trains at station　会车间隔时间　06.0300

time interval for two trains despatching in succession in the same direction　同方向列车连发间隔时间　06.0301

time-interval method　时间间隔法　06.0239

time limit for opening to traffic　通车期限　02.0863

time relay　时间继电器　04.0592

time slot　时隙　07.0811

timetable　列车运行时刻表　06.0282

timing　定时　04.0773

timing jitter　定时抖动　07.0380

tipper　翻车机　06.0601

tipping plant　翻车机　06.0601

tire　轮箍　05.0099

tired wheel　有箍车轮　05.0097

tire thickness　轮箍厚度　05.0221

to cancel a block　取消闭塞　08.0350

to cancel a route　取消进路　08.0189

toe end of frog　辙叉趾端　03.0322

toe length of frog　辙叉趾长　03.0333

toe load of fastening　扣件扣压力　03.0045

toe of side slope　坡脚　02.0249

toe spread of frog　辙叉趾宽　03.0335

toggle arm　拉杆　05.0523

toggle linkage　双联杠杆　05.0522

to hold route for shunting　非进路调车　08.0192

toilet　厕所　05.0777

toll automatic dialling　长途自动接续　07.0266

toll automatic switching repeater　长途自动电话中继器　07.0270

toll automatic telephone　长途自动电话　07.0269

toll communication network　长途通信网　07.0238

toll communication within railway administration　局线长途通信　07.0248

toll junction line　长途中继线　07.0272

toll semi-automatic dialling　长途半自动接续　07.0267

toll service desk　长途业务台　07.0254

toll switch board for semi-automatic operation　长途半自动接续台　07.0253

toll telephone office　长途电话所　07.0271

toll telephone switching system　长途电话交换机　07.0251

tone and ringing generator　信号铃流发生器　07.0081

tongue rail　尖轨　03.0313

tongue rail made of special section rail　特种断面尖轨　03.0357

tonnage of freight arrived　货物到达吨数　06.0147

tonnage of freight despatched　货物发送吨数　06.0146

tonnage of freight transported　货物运送吨数　06.0148

tonnage of traction　牵引定数　01.0072

tonnage rating　牵引定数　01.0072

tonne-kilometers charged　计费吨公里　06.0149

tonne-kilometers operated　运营吨公里　06.0150

tool room　工具室　05.0764

top and bottom heading method　上下导坑法　02.0734

top and twin-side bottom heading method　品字形导坑法　02.0741

top chord　[货车]上侧梁　05.0439

top cleaning　找顶　02.0755

top coat　面漆　03.0553

top dead point　上止点　04.0712

top-heading method　上导坑法　02.0735

top of cutting　堑顶　02.0281

top of cutting slope　堑顶　02.0281

topographical survey　地形测量　02.0104

topographic feature survey　地貌调查　02.0040

topographic survey　地形调查　02.0039

top operation anticreep ledge　上防跳台　05.0541

top operation coupler　上作用车钩　05.0530

top picking of tunnel　隧道挑顶　02.0828

toppling　崩塌　03.0480

top rod　上拉杆　05.0359

top water level in front of bridge　桥前壅水高度　02.0083

torch　火炬信号　06.0662

toroidal combustion chamber　ω型燃烧室　04.0801

torpedo　响墩信号　06.0661

torque converter shaft　变扭器轴，＊变矩器轴　04.1015

torque ratio　变扭比，＊变矩比　04.1009

torshear type high strength bolt　扭剪式高强度螺栓　02.0367

torsion bar spring　扭杆弹簧　05.0170

total boiler efficiency　锅炉总效率　04.1111

total budget　总预算　02.0851

total construction time　施工总工期　02.0864

total displacement　总排量　04.0719

total estimate　总概算　02.0022

total internal reflection　全反射　07.0453

total length of turnout　道岔全长　03.0325

total locomotive efficiency　机车效率　04.0075

totally-enclosed motor　全封闭式电动机　04.0411

total output of building industry　建筑业总产值　02.0853

total pressure loss for gas side　燃气侧全压损失　04.1576

total time of construction　施工总工期　02.0864

total track length　线路全长　06.0432

total volume　总容积　05.0039

total wear of rail head　轨头总磨损，＊轨头总磨耗　03.0161

to transfer of lighting indication　灯光转移　08.0351

tourist train　旅游列车　06.0211

towel hanging rod　毛巾杆　05.0769

towel rail　毛巾杆　05.0769

tower crane　塔式起重机　02.0893

track　轨道，＊线路上部建筑　03.0010，线路　03.0015

track alignment　轨道方向　03.0150

track blockade　线路封锁　02.0999

track bolt 接头螺栓，＊鱼尾螺栓 03.0083

track buckling 轨道鼓出，＊跑道 03.0160

track charts 线路平剖面图 03.0264

track circuit 轨道电路 08.0489

track clear 股道空闲 08.0509

track crane 轨道起重机 02.0897

track creeping 线路爬行 03.0159

track cross level 轨道水平，＊左右水平 03.0149

track deformation 轨道变形 03.0146

track depression 轨道明坑 03.0153

track deterioration 轨道几何状态恶化 03.0035

track disorder 轨道变形 03.0146

track distortion 轨道变形 03.0146

track district 工务段 03.0001

track division 工务段 03.0001

track dynamics 轨道动力学 03.0033

track evaluation by recording car 轨检车评分 03.0269

track failure 轨道失效 03.0036

track fastened directly to steel girders 直结轨道 02.0512

track foreman 养路工长 03.0005

track gage 轨距尺，＊道尺 03.0465

track galvanic effect 轨道生电现象 08.0525

track geometry 轨道几何形位 03.0022

track geometry measuring device 轨道检测设备 03.0377

track geometry measuring trolley 轨道检查小车 03.0447

track geometry tolerances 轨道几何尺寸容许公差 03.0145

track group 线束 06.0431

tracking-receiving mode 跟踪接收方式 07.0604

track inspection car 轨道检查车 03.0446

track irregularity 轨道不平顺 03.0142

track jack 起道器 03.0386

track laying machine 铺轨机 03.0439

track laying train 铺轨列车 03.0444

track lead 钢轨引接线 08.0538

trackless transportation 无轨运输 02.0758

track level 轨道水平尺 03.0470

track lifting 起道 03.0199

track lifting and lining machine 起拨道机 03.0387

track lifting and lining tool 起拨道器 03.0388

track lifting machine 起道机 03.0385

track lining 拨道 03.0203

track lining machine 拨道机 03.0383

track lining tool 拨道器 03.0384

track machine 轨道机械 03.0371

track maintenance division 工务段 03.0001

track maintenance section 养路工区 03.0003

track maintenance subdivision 养路领工区 03.0002

track maintenance telephone 养路电话 07.0416

trackman 养路工 03.0006

track master 养路领工员 03.0004

track mechanics 养路工 03.0006，轨道力学 03.0032

track modulus 钢轨基础模量 03.0039

track obstruction indicator 线路遮断表示器 08.0096

track occupancy indication 线路占用表示 08.0170

track occupied 股道占用 08.0510

track panel 轨排 03.0019

track panel laying gantry crane 门式铺轨排机 03.0443

track panel laying machine with cantilever 悬臂式铺轨排机 03.0441

track patrolling man 巡道工 03.0007

track permanent deformation 轨道残余变形 03.0147

track plan 线路平面图 02.0166

track profile 线路纵断面图 02.0167，轨道前后高低，＊前后高低 03.0151

track quality index 轨道质量指数 03.0270

track reactor 轨道电抗器 08.0545

track recording car 轨道检查车 03.0446

track relay transformer 轨道受电变压器 08.0544

track renewal 线路大修 03.0194

track residual deformation 轨道残余变形 03.0147

track rheostat 轨道变阻器 08.0546

track scale test car 检衡车 05.0747

track section 线路区段 08.0204

track shaving machine 开沟平路机 03.0461

track shim for frost heaving roadbed 冻害垫板 03.0098

track shimming 垫冻害垫板 03.0511

track-side telephone 区间电话 07.0417

track-side telephone switching device 区间电话转接机 07.0419

track skeleton 轨排 03.0019

track spike 道钉, * 狗头钉, * 钩头钉 03.0090

track standard 轨道类型 03.0011

track storage effect 轨道电路蓄电现象 08.0524

track strength analysis 轨道强度计算 03.0034

track stresses 轨道应力 03.0041

track structure 轨道结构 03.0012

track subdivision 养路领工区 03.0002

track supervisor 养路领工员 03.0004

track-train interaction 列车与线路相互作用 01.0065

track transformer box 轨道变压器箱 08.0549

track transformer feed end 轨道送电变压器 08.0543

track transportation 有轨运输 02.0757

track treadle 轨道接触器 08.0313

track walker 巡道工 03.0007

track work forms 养路表报 03.0261

track wrenches 螺栓扳手 03.0471

traction and thermodynamic test 牵引热工试验 04.0287

traction autotransformer 自耦牵引变压器 04.0484

traction circuit 牵引电路 04.0372

traction convertor 牵引变流器 04.0326

traction electric network 牵引网 04.1272

traction for train exceed mass norm 机车超重牵引 04.0229

traction frequency convertor 牵引变频器 04.0328

traction invertor 牵引逆变器 04.0327

traction motor 牵引电动机 04.0394

traction motor ammeter 牵引电机电流表 04.0620

traction motor isolating switch 牵引电动机隔离开

关 04.0568

traction motor power supply system 牵引电动机供电制式 04.0364

traction motor voltmeter 牵引电机电压表 04.0619

traction reactor 牵引电抗器 04.0505

traction return current circuits 牵引回流电路 04.1313

traction return current rail 牵引回流轨 04.1337

traction substation 牵引变电所 04.1342

traction test 牵引试验 04.0285

traction transformer 牵引变压器 04.1284

traction transformer of cross connection with three-phase YN, d11 d1 三相 YN, d11 d1 十字交叉接线牵引变压器 04.1291

traction transformer of locomotive 机车牵引变压器 04.0481

traction transformer of Scott connection 斯柯特接线牵引变压器 04.1289

traction transformer of singlephase connection 单相接线牵引变压器 04.1288

traction transformer of singlephase V/V connection 单相 V/V 接线牵引变压器 04.1287

traction transformer of threephase three winding connection 三相三绕组接线牵引变压器 04.1286

traction transformer of threephase YN, d11 connection 三相 YN, d11 接线牵引变压器 04.1285

traction transformer of Wood Bridge connection 伍德布里奇接线牵引变压器 04.1290

tractive effort at coupler 车钩牵引力 04.0091

tractive effort at wheel rim 轮周牵引力 04.0090

tractive force 牵引力 08.0622

tractive force of train 列车牵引力 02.0342

tractor and trailer 牵引车及挂车 06.0596

traffic capacity 输送能力 06.0305

traffic condition 运输条件 06.0137

traffic controller's office 调度所 06.0335

traffic cost 运输成本 06.0721

traffic cost plan 运输成本计划 06.0722

traffic [dispatching] order 调度命令 06.0337

traffic diversion 货物运输变更 06.0141

traffic expenditure 运输支出 06.0723

traffic interruption 线路封锁 02.0999, 行车中断 06.0635

traffic limitation 运输限制 06.0138

traffic program 运输方案 06.0358

traffic restriction 运输限制 06.0138

traffic revenue 运输收入 06.0750

trailable 挤脱 08.0619

trailer 拖车 04.0011

trailing truck 从轮转向架 04.1235

trailing wheel set 从轮对 04.1236

train 列车 06.0199

train accident 列车事故 06.0627

train aerodynamics 列车空气动力学 01.0094

train approaching sensor 列车接近传感器 07.0634

train approach warning device 列车接近报警器 03.0454

train attendant 列车员 06.0064

train block system 行车闭塞法 06.0237

train brake overall test 列车制动全部试验 05.0951

train brake simplified test 列车制动简易试验 05.0952

train brake tester 列车制动试验器 05.0372

train braking 列车制动 04.0111

train class 列车等级 06.0351

train coach supply winding 列车供电绕组 04.0501

train collision 列车正面冲突 01.0084

train conductor 列车长 06.0065

train conductor's station 车长电台 07.0557

train consist list 列车编组顺序表 06.0273

train control section 调度区段 06.0336

train density 列车密度 06.0329

train destination 列车去向 06.0280

train diagram 列车运行图 01.0044

train diagram for automatic block signals 追踪运行图 06.0295

train diagram for doubletrack 双线运行图 06.0292

train diagram for singletrack 单线运行图 06.0291

train diagram in pairs 成对运行图 06.0293

train diagram not in pairs 不成对运行图 06.0294

train diagram plotter 运行图描绘仪 08.0488

train [dispatching] order 调度命令 06.0337

train dispatching section 调度区段 06.0336

train dispatching telephone 列车调度电话 07.0412

train dynamics 列车动力学 01.0093

train examination 列检 05.0934

train guard 运转车长 06.0350

train kilometers 列车公里 06.0693

train list 列车编组顺序表 06.0273

train list information after departure 列车确报 06.0275

train list information in advance 列车预报 06.0274

train movement recording equipment 行车记录设备 08.0487

train number 列车车次 06.0286

train number indication 车次表示 08.0484

train operation accident 行车事故 06.0618

train operation adjustment 列车运行调整 06.0349

train operation control system 列车运行控制系统 08.0431

train out report telegraph 列车确报电报 07.0435

train overturning 列车颠覆 01.0092

train path 列车运行线 06.0283

train pipe 列车管 05.0277

train pipe end valve 暖汽端阀 05.0621

train pipe pressure gradient 列车管压差 04.0155

train police 乘警 06.0066

train position indication 列车位置表示 08.0483

train radio communication 列车无线电通信 07.0580

train radio communication system 列车无线电通信系统 07.0582

train radio dispatching system 列车无线电调度系统 04.0071

train rear end air pressure feedback 列车尾部风压反馈 05.0373

train rear end protection 列车尾部防护 01.0086

transversal type district station 横列式区段站 06.0367

transverse beam 横梁 02.0463

transverse flux linear induction motor 横向磁通直线感应电动机 04.1629

transverse gallery 横洞 02.0788

transverse passage-way 横通道 02.0790

travelling cradle 活动吊篮 02.0542

travelling speed 旅行速度 06.0331

traverse survey 导线测量 02.0102

traversing 导线测量 02.0102

tread brake 踏面制动 04.0123

tread cleaner 踏面清扫器 05.0321

tread conicity 踏面锥度 05.0218

tread contour 踏面外形 05.0215

tread profile 踏面外形 05.0215

tread rolling circle 滚动圆 05.0219

tread slid flat 踏面擦伤 05.0251

tread taping point 踏面基点 05.0217

tread wear 踏面磨耗 05.0249

treated wooden tie 油枕 03.0113

treatment of frost heaving track 整治冻害轨道 03.0510

tree network 树形网[络] 07.0758

trellis arch 格构拱, *钢花拱, *格栅拱 02.0777

trench cutting machine 挖沟机 02.0892

trencher 挖沟机 02.0892

trench excavated 拉槽 02.0745

trenching machine 挖沟机 02.0892

triangle-type junction terminal 三角形枢纽 06.0397

triangulation 三角测量 02.0139

trigonometric leveling 三角高程测量 02.0141

trigonometric survey 三角测量 02.0139

triple valve 三通阀 05.0306

Trofiemov piston valve 分动式汽阀, *特氏阀 04.1203

troop train 军用列车 06.0213

trough for catching falling rocks 落石槽 02.0319

trough girder 槽型梁 02.0387

truck 转向架 05.0057

truck auxiliary transom 构架辅助梁 05.0079

truck bolster 转向架摇枕 05.0090

truck brake rigging 转向架基础制动 05.0184

truck center plate 下心盘 05.0210

truck changing point 转向架换装所 05.0984

truck crane 汽车式起重机 02.0894

truck dead lever 固定杠杆 05.0185

truck end sill 构架端梁 05.0076

truck frame 转向架构架 05.0074

truck frame diagonal 转向架对角线 05.0043

truck live lever 移动杠杆 05.0186

truck longitudinal sill 构架纵梁 05.0078

truck mixer 混凝土搅拌运输车 02.0932

truck-mounted brake assembly 转向架制动组件 05.0325

truck rigidity against distorsion 转向架扭曲刚度 05.0242

truck side frame 转向架侧架 05.0080

truck side sill 构架侧梁 05.0075

truck transom 构架横梁 05.0077

truck with no swing bolster 无摇动台式转向架 05.0064

truncated cone banking 锥体护坡 02.0557

trunk circuit 中继电路, *中继器 07.0261

trunk communication 干线长途通信 07.0247

trunk communication network 干线长途通信网 07.0245

trunk conference telephone 干线会议电话 07.0297

trunk dispatcher switchboard 长途调度台 07.0255

trunk dispatching telephone 干线调度电话 07.0295

trunk line 干线 01.0037

trunk railway 干线铁路 01.0021

truss 桁架 02.0388

truss type steel structure 桁架式钢结构 05.0382

tubular axle 空心车轴 05.0108

tubular column foundation 管柱基础 02.0592

tubular electric heating element 电热管 05.0696

tunnel adit 洞口 02.0662

tunnel announciating device 隧道通知设备 08.0452

tunnel anti-disaster equipment 隧道防灾设施

02.0814

tunnel arch top settlement　隧道拱顶下沉
02.0822

tunnel boring machine　隧道掘进机　02.0981

tunnel boring machine method　掘进机法　02.0724

tunnel construction clearance　隧道建筑限界
01.0058

tunnel efflux ventilation　隧道射流式通风
02.0808

tunnel entrance　隧道进口　02.0663

tunnel exit　隧道出口　02.0664

tunnel fire-fighting system　隧道消防系统
02.0818

tunnel fire hazard　隧道火灾　02.0816

tunnel floor heave　隧道底鼓　02.0823

tunnel for luggage and postbag　行包邮政地道
06.0523

tunnel gas explosion　隧道瓦斯爆炸　02.0819

tunnel ground subsidence　隧道地表沉陷
02.0821

tunnel group　隧道群　02.0660

tunnel holing-through　隧道贯通　02.0762

tunnel illumination　隧道照明　02.0813

tunnel injector type ventilation　隧道射流式通风
02.0808

tunnel leak　隧道漏水　03.0526

tunnel lighting　隧道照明　02.0813

tunnelling shutter jumbo for tunnel lining　隧道衬砌
模板台车　02.0980

tunnel monitoring measurement　隧道监控量测
02.0820

tunnel obstruction signal　隧道遮断信号
08.0451

tunnel opening　洞口　02.0662

tunnel perimeter deflection　隧道周边位移
02.0824

tunnel photographing car　隧道摄影车　05.0749

tunnel portal　洞门　02.0666

tunnel portal frame　洞门框　02.0667

tunnel post　隧道标　03.0253

tunnel reconstruction　隧道改建　02.0826

tunnel repeater　隧道中继器　07.0563

tunnel structure gauge　隧道建筑限界　01.0058

tunnel surrounding mass deflection　隧道地中位移
02.0825

tunnel through error　隧道贯通误差　02.0763

tunnel through plane　隧道贯通面　02.0764

tunnel ventilation　隧道通风　03.0561

tunnel ventilation test　隧道通风试验　02.0809

tunnel warning equipment　隧道报警装置　02.0817

tunnel water handling　隧道防排水　03.0562

tunnel without cover　明洞　02.0694

turbine　涡轮　04.1011

turbine power　涡轮功率　04.1560

turbocharged diesel engine　废气涡轮增压柴油机
04.0697

turbocharger　涡轮增压器　04.0973

turbocharger matching test　增压器配机试验
04.0862

turbo-generator　涡轮发电机　04.1249

turn-around wye　转向线　05.0940

turning facilities　转向设备　04.0273

turnout　道岔　03.0281

turnout branch　道岔侧线　03.0302

turnout from curved track　曲线出岔道岔　03.0319

turnout guard rail　道岔护轨　03.0360

turnout main　道岔主线　03.0301

turnout mast　道岔柱　04.1413

turnout number　道岔号数　03.0310

turnout tie　岔枕　03.0117

turnover of current capital　流动资金周转
06.0749

turnover of freight traffic　货物周转量　06.0683

turnover of passenger traffic　旅客周转量　06.0678

turnround time of passenger train set　旅客列车车底
周转时间　06.0069

turntable　转盘　04.0274

turret　蒸汽塔　04.1180

TV for inspection　检车电视　07.0658

TV for passenger information service　旅客问讯电视
07.0653

TV for railway commerce inspection　商务检查电视
07.0657

TV for record vehicle number　车号抄录电视
07.0659

twinned flat car　双联平车　05.0837

twin-side heading method　双侧导坑法　02.0740

twist　三角坑，＊扭曲　03.0152

two-aspect automatic block　二显示自动闭塞
　　08.0337

two-axle car　二轴车　05.0002

two-axle truck　二轴转向架　05.0059

two coil configuration　双线圈结构　04.1634

two-compartment reservoir　双室风缸　05.0311

two-hinged arch　双铰拱　02.0398

two-phase design　两阶段设计　02.0012

two-pressure equalizing system　二压力机构
　　05.0332

two-stage supercharged diesel engine　两级增压柴油
　　机　04.0699

two stage suspension　两系悬挂　05.0177

two-step design　两阶段设计　02.0012

two stroke diesel engine　二冲程柴油机　04.0691

two-way alternate communication　双向交替通信
　　07.0776

two-way curved arch　双曲拱　02.0402

two way cylinder　双向风缸　05.0519

two way feeding　双边供电　04.1294

two-way repeater　音选双向增音机，＊双向增音机
　　07.0432

two-way repeater for VF selective calling　音选双向
　　增音机，＊双向增音机　07.0432

two-way simultaneous communication　双向同时通信
　　07.0775

two wire system　双线制　05.0714

type of car　车种　05.0015

types of locomotive　机车种类　04.0003

type test　型式试验　04.0280

tyre　轮箍　05.0099

tyred wheel　有箍车轮　05.0097

U

UDE　意外紧急制动　04.0134

ultrasonic inspection for axle　车轴超声探伤
　　05.0946

ultrasonic leak detector　超声波查漏仪　07.0216

un-allocated locomotive　非配属机车　04.0020

unattended repeater　无人增音机　07.0325

unattended repeater with ground temperature com-
　　pensation and powerfeed loop back　地温折返无
　　人增音机　07.0289

unattended repeater with ground temperature compen-
　　sation and powerpassing　地温通过无人增音机
　　07.0290

unbalanced centrifugal acceleration　未被平衡离心加
　　速度　03.0030

unbalanced resistance　不平衡电阻　07.0225

uncoupling device　解钩装置　05.0564

underbreak　欠挖　02.0754

under cut of tunnel　隧道落底　02.0827

under cutting of track　落道　03.0200

underframe　底架　05.0387

underframe mounted traction motor　底架架承式牵
　　引电动机　04.0403

underframe with center sill　有中梁底架
　　05.0388

underframe without center sill　无中梁底架
　　05.0389

underframe with sliding center sill　活动中梁底架
　　05.0390

underground cable terminal box　地中电缆盒
　　08.0285

underground diaphragm wall method　地下连续墙法
　　02.0727

underground path　地道　06.0522

underground railway　地下铁道，＊地铁　01.0023

underground railway tunnel　地铁隧道　02.0648

underground wall method　地下连续墙法
　　02.0727

under lever faucet　下作用水阀　05.0780

underneath clearance　桥下净空　02.0434

under-voltage relay　欠电压继电器　04.0597

underwater concreting　灌筑水下混凝土　02.0616

underwater tunnel　水下隧道，＊水底隧道
　　02.0647

undesirable emergency braking　意外紧急制动
　　04.0134

undesired braking　自然制动　04.0132

undesired release 自然缓解 04.0138

undetected error 未检出差错 07.0871

un-disposal locomotive 非支配机车 04.0022

unequilateral turnout 单式不对称道岔, *不对称双开道岔 03.0284

unfavorable condition for car rolling 溜车不利条件 06.0473

unfavorable geology 不良地质 02.0061

unguarded flange-way 辙叉有害空间 03.0340

unidirectional combined type marshalling station 单向混合式编组站 06.0375

unidirectional longitudinal type marshalling station 单向纵列式编组站 06.0374

unidirectional transmission 单向传输 07.0789

unidirectional transversal type marshalling station 单向横列式编组站 06.0373

uninsulated overlap 非绝缘锚段关节 04.1431

uninsulated transition mast 非绝缘转换柱 04.1410

unintelligible crosstalk 不可懂串音 07.0139

unintended braking 自然制动 04.0132

unintended release 自然缓解 04.0138

union link 结合杆 04.1208

unit block 组合 08.0259

unit block assembly rack 组合架 08.0261

unit-block type relay interlocking 组合式电气集中联锁 08.0117

unit project 单位工程 02.0004

unit resistance 单位阻力 04.0103

unit train 单元列车 06.0219

universal characteristic 万有特性 04.0851

universal rail gage 万能道尺 03.0466

unloading machine with chain buckets 链斗卸车机 06.0598

unloading pipe connection 抽液管座 05.0479

unmechanized hump yard equipment 非机械化驼峰设备 08.0367

unplanned freight traffic 计划外运输 06.0102

unsafe depth foundation of bridge 桥梁浅基 03.0513

unsafe depth foundation protection 桥梁浅基防护 03.0569

unsprung weight 簧下重量 05.0866

unsufficient span of bridge 桥梁孔径不足 03.0530

unsymmetrical double curve turnout 单式不对称道岔, *不对称双开道岔 03.0284

unsymmetrical double curve turnout in the same direction 单式同侧道岔 03.0285

unsymmetrically loading lining eccentrically compressed lining 偏压衬砌 02.0685

unsymmetrical three throw turnout 不对称三开道岔 03.0287

unsymmetrical three-way turnout 不对称三开道岔 03.0287

untreated wooden tie 素枕 03.0114

up direction 上行方向 06.0284

upgrading of bridge and tunnel 桥隧改造 03.0546

upholstered couchette 软卧车 05.0728

upholstered-seat coach 软座车 05.0726

uplifting window 上开窗 05.0504

upper cantilever 拉杆[压管] 04.1384

upper dead center 上止点 04.0712

upward swing door [货车]上开门 05.0493

useful thermal efficiency 有效热效率 04.0828

U-shaped abutment U形桥台 02.0551

utility factor of the position 台位利用系数 05.0965

utilization ratio of construction machinery 施工机械利用率 02.0874

utilization ratio of machine handling 装卸机械利用率 06.0591

V

vacuum brake 真空制动 04.0114

vacuum brake equipment 真空制动装置 05.0267

vacuum chamber 真空缸 05.0368

vacuum pump 真空水泵 02.0917

value insured rail traffic 铁路保价运输 06.0011

valve 气门 04.0925

valve chest 汽室 04.1186

valve cross arm 气门横臂 04.0932

valve diagram 阀动图 04.1261

valve ellipse 阀动椭圆图 04.1262

valve gear 阀装置，＊阀动装置 04.1199

valve guide 气门导管 04.0930

valve push rod 气门推杆 04.0933

valve rocker 气门摇臂 04.0931

valve rotating mechanism 气门旋转机构
04.0936

valve seat 气门座 04.0929

valve setting 阀调整 04.1260

valve spring 气门弹簧 04.0928

valve tappet 气门挺柱 04.0934

valve timing 配气相位，＊配气定时 04.0727

valve travel 阀行程 04.1256

valve type arrester 阀型避雷器 04.0636

valve type track circuit 阀式轨道电路
08.0492

vapor regulater 大气压式暖汽调整阀 05.0598

variable expense 变动支出 06.0727

variable frequency convertor 可调牵引变频器
04.0329

variable frequency speed control 变频调速 04.0346

variable friction type snubbing device 变摩擦式减震
装置 05.0159

variable-geometry gas turbine 变几何燃气轮机
04.1528

variable stator blade 可转静叶 04.1559

variable voltage speed control 变压调速 04.0345

variant train diagram 分号运行图 06.0297

vegetation on slope 边坡植被防护 03.0505

vehicle earth station 车载地球站 07.0673

vehicle gauge 铁路机车车辆限界 01.0059

velocity hump crest of retarder 制动能高
06.0477

velocity triangle 速度三角形 04.1551

vent 通气口 05.0480

ventilated box car 通风车 05.0819

ventilated motor 通风式电动机 04.0412

ventilated transport 通风运输 06.0181

ventilating machine 通风机 02.0977

ventilating set 通风机 02.0977

ventilation by air passage 风道式通风 02.0805

ventilation by ducts 巷道式通风 02.0812

ventilation by pipes 风管式通风 02.0811

ventilation curtain 隧道通风帘幕 02.0806

ventilator 通风机 02.0977，通风器 05.0619

vent valve 放风阀 05.0316

vertical claw rock loader 立爪式装岩机
02.0963

vertical curve 竖曲线 02.0195

vertical dynamic load 垂直动载荷 05.0885

vertical elasticity 铅垂弹性 04.1443

vertical flange 轮缘垂直磨耗 05.0250

vertical horizontal parity 纵横奇偶检验 07.0799

vertical impact 垂向冲击 01.0109

vertical load 垂直载荷 05.0884

vertical loading folded locomotive antenna 加顶垂直
折合机车天线 07.0569

vertical member 竖杆 02.0450

vertical resilience 铅垂弹性 04.1443

vertical split of rail head 轨头垂直劈裂
03.0179

vertical vibration 垂向振动 01.0098

vertical wear of rail head 轨头垂直磨损，＊轨头垂
直磨耗 03.0162

vertical wheel 导线立轮 08.0309

vestibule 通过台 05.0789

vestibule diaphram 风挡 05.0575

vestibule diaphram buffer plate 风挡缓冲板
05.0577

vestibule diaphram face plate 风挡面板 05.0576

vestibule entrance door ［客车］脚蹬门 05.0488

VF selective calling 音频选叫 07.0421

viaduct 高架桥 02.0358

vibrating pile driver 振动打桩机 02.0908

vibration compactor 振动辗压机 02.0887

vibration of rolling stock 机车车辆振动 01.0095

vibration roller 振动辗压机 02.0887

vibratory driver 振动打桩机 02.0908

vibro-driver extractor 振动沉拔桩机 02.0909

video conference 电视会议 07.0065

videophone 可视电话 07.0064

videophone set 可视电话机 07.0096

vigilance device 警惕装置 04.0654

virtual call 虚呼叫 07.0873

virtual circuit 虚电路 07.0780

virtual terminal 虚拟终端 07.0910

visible low spot of track 轨道明坑 03.0153

visible pit of track 轨道明坑 03.0153

visual measurement 目测 02.1045

visual observation 目测 02.1045

visual signal 视觉信号 08.0005

vital circuit 安全电路 08.0214

voice frequency 话频 07.0013

voice-operated anti-singing circuit 音控防鸣电路 07.0150

voltage below level 电压过低 08.0688

voltage regulation mode 调压方式 04.0317

voltage regulator 电压调整器 04.0610

voltage relay 电压继电器 04.0590

voltage sensor 电压传感器 04.0630

voltage stabilizer 稳压器 07.0050

voltage to ground 对地电压 04.1472

voltmeter on side of overhead contact line 网侧电压表 04.0622

volume 音量 07.0069

volume diagram of earth-rock work 土[石]方体积图, ＊土积图 02.0839

volume meter 音量表 07.0147

volume of passenger traffic 旅客运输量 06.0677

volume power 升功率 04.0818

vortex-excited oscillation 涡流激振 02.0524

V-shaped pier V形桥墩 02.0567

V-type diesel engine V型柴油机 04.0703

W

wagon 货车 05.0791

wainscot sheathing 下墙板 05.0447

waist rail 大腰带 05.0435

waist sheet 锅腰托板 04.1220

waiting hall 候车室 06.0020

waiting room 候车室 06.0020

Walschaerts valve gear 华氏阀装置 04.1200

WAN 广域网 07.0755

warehouse 仓库 06.0093

warning portal 限界门 04.1434

warning signal 告警信号 06.0658

warning sign at grade crossing 道口警标 03.0276

warning signs for approaching a station 预告标 08.0101

warning tone 通知音 07.0279

warp 三角坑, ＊扭曲 03.0152

wash basin 洗面器 05.0785

wash bowl 洗手器 05.0779

washing compartment 洗面间 05.0778

washing room 洗面间 05.0778

washing siding for passenger vehicle 客车洗车线 06.0418

washout 水害 03.0497

washout plug 洗炉堵 04.1175

waste bank 弃土堆 02.0283

water collecting area 汇水面积 02.0089

water column 水柱 04.1178

water-cooled diesel engine 水冷柴油机 04.0705

water crane 水鹤 04.0061

water crane indicator 水鹤表示器 08.0097

water gage 水表 04.1177

water hammer 水锤 04.1266

water jacket 水套 04.0874

water level gage 水位表 05.0606

waterproof board 防水板 02.0796

waterproof concrete 防水混凝土 02.0783

waterproofing coating 防水涂层 02.0797

waterproofing of tunnel 隧道防水 02.0793

waterproof layer 防水层 02.0795

waterproof sheet 防水板 02.0796

water separator 水分离器 08.0402

watershed 分水岭 02.0088

water softened and purified 软水与净水 04.0065

water softened in boiler 炉内软水 04.0063

water softened out of boiler 炉外软水 04.0064

water stop tie 止水带 02.0799

water supply 给水 04.0060

water supply air reservoir 给水风缸 05.0660

water supply equipment with lower tank 车下给水装置 05.0658

water supply equipment with roof tank　车上给水装置　05.0655

water supply governer valve　给水调整阀　05.0664

water supply station　给水站，＊给水所　04.0066

water supply system　给水系统　05.0654

water [supply] treatment　给水处理　04.0062

water tank　水柜　04.1240

water tank test cock　水箱验水阀　05.0666

water temperature regulater　水温控制器　05.0614

waveguide line　波导线　07.0593

wavelength devision multiplex　波分复用　07.0506

wave-type deformation of rail head　轨头波形磨损，＊轨头波浪形磨耗　03.0164

way bill　货票　06.0127

wayside equipment　地面设备　08.0418

wayside inductor　地面感应器　08.0422

wayside signaling　区间信号　08.0002

weak electric field area　弱电场区　07.0541

weak electric field area in tunnel　隧道弱电场区　07.0542

weakened field　削弱磁场　04.0433

wearing plate　磨耗板　05.0162

wear resistant rail　耐磨轨　03.0054

weatherproofness　风雨密性　06.0560

web　辐板　05.0121

web hole　辐板孔　05.0126

web member　腹杆　02.0448

web plate　腹板　02.0444

wedge　斜楔　05.0160

wedged rail anchor　穿销防爬器　03.0101

wedging inscribing　楔形内接　03.0139

weed cutter　除草机　03.0463

weed cutting machine　除草机　03.0463

weed killer　喷洒除草机　03.0464

weed killing machine　喷洒除草机　03.0464

weep drain　排水沟　02.0295

weigh bridge test car　检衡车　05.0747

weighing test　称重试验　04.0657

weight bridge track　轨道衡线　06.0087

weight distribution of locomotive　机车重量分配　04.0049

weight/power ratio　比重量　04.0087

weight sensing　测重　08.0388

welded bond　焊接式钢轨接续线　08.0542

welded joint　焊接接头　03.0072

welding frame　焊接机座　04.0453

well-hole car　落下孔车　05.0835

western type toilet　坐式便器　05.0783

wet boring for rock　湿式凿岩　02.1024

wet damage accident　湿损事故　06.0643

wet drilling for rock　湿式凿岩　02.1024

wet shotcreting　湿喷混凝土　02.0780

wet shotcreting machine　湿喷混凝土机　02.0972

wharf　码头　02.0621

wheel　车轮　05.0095

wheelbase　轴距　05.0044

wheelbase of car　车辆全轴距　05.0046

wheelbase of combination truck　转向架组全轴距　05.0047

wheel burn　钢轨擦伤　03.0172

wheel center　轮心　05.0098

wheel detector　车轮检测器　08.0423

wheel diameter　车轮直径　05.0220

wheel flange　轮缘　05.0116

wheel hub　轮毂　05.0118

wheel hub bore　轮毂孔　05.0119

wheel load　轮重　01.0049

wheel loose on axle　轮毂松动　05.0258

wheel-mounted disc brake　轮装盘形制动　05.0328

wheel out of round　车轮不圆　05.0255

wheel pair　轮对　05.0094

wheel-rail interaction　轮轨关系　01.0066

wheel-rail relation　轮轨关系　01.0066

wheel rim　轮辋　05.0117

wheel seat　轮座　05.0113

wheelset　轮对　05.0094

wheelset dynamic balance test　轮对动平衡检验　05.0962

wheelset storing yard　轮对存放场　05.0983

wheel skid　车轮滑行　01.0068，止轮器　06.0668

wheel sliding　车轮滑行　01.0068

wheel slipping　车轮空转　01.0069

wheel tread　踏面　05.0120

whistle 汽笛 04.1181

whistle board 鸣笛标 03.0256

white noise test set 白噪声测试器 07.0355

whole night lamp 终夜灯 05.0685

wide area network 广域网 07.0755

wide-band channel 宽带信道 07.0872

wide joint gap 大轨缝 03.0158

width inside car body 车体内宽 05.0027

width of ballast bed 道床宽度 03.0133

width of door opening 门孔宽度 05.0037

width of the subgrade surface 路基面宽度
    02.0234

width over side sills 底架宽度 05.0026

width over sides of car body 车体宽度
    05.0025

willow fascine 柴排 02.0320

wind break fence 防风栅栏 02.0323

wind-break wall 挡风墙 02.0322

wind fairing 风嘴 02.0528

wind load 风[荷]载 02.0343

window 窗口 07.0868

window blind 窗卷帘 05.0509

window curtain 窗帘 05.0510

window lintel 小腰带 05.0436

window rail 窗台 05.0450

window sash 窗框 05.0511

window sash lock 窗锁 05.0512

window shade 窗卷帘 05.0509

window sill 窗台 05.0450

windscreen wiper 刮雨器 04.1067

windshield wiper 刮雨器 04.1067

wing rail 翼轨 03.0358

wing wall 翼墙 02.0556

wing wall tunnel portal 翼墙式洞门 02.0670

winter season construction 冬季施工 02.0843

wire-adjusting screw 导线反正扣 08.0306

wire carrier 导线导轮 08.0310

wire compensator 导线调整器 08.0307

wire installation 导线装置 08.0305

wire installation curve 导线安装曲线 04.1435

wire lead drop out 线头脱落 08.0692

wireless transmitter 无线话筒 07.0560

withdrawal of train 列车停运 06.0356

wood chip car 刨花车 05.0825

wooden sleeper adzing machine 木枕削平机
    03.0412

wooden strut 木前枕 05.0139

wooden tie 木枕，＊枕木 03.0112

wooden tie drilling machine 木枕钻孔机 03.0407

wood structure 木结构 05.0526

work district for telemechanical system 远动工区
    04.1457

working current 工作电流 08.0623

working cycle 工作循环 04.0722

working-drawing design 施工图设计 02.0017

working jumbo for tunnel lining 隧道衬砌模板台车
    02.0980

working process 工作过程 04.0763

working signal 作业标志 03.0245

working stress rating 容许强度检定 03.0518

working substance 工质 04.0721

working time between trains 列车空隙作业时间
    03.0260

working time of closed section 封闭线路作业时间，
    ＊开天窗作业时间 03.0259

working value 工作值，＊完全吸起值 08.0567

working water level 施工水位 02.0093

work occupation time 封闭线路作业时间，＊开天
    窗作业时间 03.0259

work train with camp cars 工程宿营车 03.0445

worn profile tread 磨耗型踏面 05.0216

wrecking crane 救援起重机，＊救援吊车
    06.0667

writing lamp 记事灯 04.0647

written liaison method 书面联络法 06.0240

wrong clearing of a signal 错误开放信号
    08.0679

wrong handling 错误办理 08.0669

wrong indication 错误显示 08.0047

wrought steel wheel 辗钢车轮 05.0102

Y

Z

汉 英 索 引

A

矮侧板　low side　05.0497

矮端板　low end　05.0498

矮型信号机　dwarf signal　08.0062

爱车点　car caring point　05.0954

安全标志　safety mark, safety symbol, safety indicator　02.1003

安全地点　safety place　02.1008

安全电路　vital circuit, safety circuit　08.0214

安全电压　safety voltage　02.1016

安全阀　safety valve　04.1053

安全防护设备　safety protection equipment　03.0378

安全技术措施　safety technical measures　02.0987

安全教育　safety education　02.0988

安全接点　safety contact, power off contact　08.0628

安全距离　safety distance　02.1005

安全帽　safety cap, safety helmet　02.1009

安全设备试验　test on safety equipments　04.0663

安全绳　safety rope, safety strap　02.1010

安全梯　emergency staircase, fire escape　02.1013

安全网　safety net　02.1011

安全线　catch siding　06.0414

安全性　safety　08.0668

安全用电　safety in utilizing electric energy　02.1015

安全炸药　safety explosion, explosive charge, safety explosive　02.1023

安全装置　safety device, safety equipment　02.1030

安装角　stagger angle　04.1547

按键电话机　key pad telephone set　07.0095

按键开关　button switch, key switch　04.0576

按键式拨号盘　key pad　07.0085

按钮　push-in button　08.0269

按钮表示　button indication　08.0167

暗挖法　subsurface excavation method　02.0718

凹底平车　depressed center flat car　05.0834

B

扒碴机　crib ballast removers　03.0421

拨起高度　height of lifting, lifting height, ascent of elevation　02.0161

白点　flake crack, shatter crack, small nucleus fissure　03.0174

白水表行车　running without water in gage　04.1264

白噪声测试器　white noise test set　07.0355

百叶窗　louver　05.0508

百叶窗油缸　oil cylinder of shutter　04.1054

摆滚振动　rock-roll vibration　01.0099

摆块　centering block　05.0570

摆块吊　centering block hanger　05.0571

摆式车体　pendulum type car body　05.0376

摆式减振器　pendulum damper　04.0897

班长台　chief operator's desk　07.0257

扳道电话　switchman's telephone　07.0438

扳道房　switchman's cabin　06.0534

板梁　plate girder　02.0384

板梁式钢结构　plate girder type steel structure　05.0381

板桥　slab bridge　02.0382

板式车架　plate frame　04.1213

板式轨道　slab-track　03.0111

板式橡胶支座　laminated rubber bearing　02.0521

板桩　sheet pile　02.0604

半补偿链形悬挂　semi-auto-tensioned catenary equipment　04.1359

半穿式桥 half through bridge, midheight deck bridge 02.0426

半堤半堑 part-cut and part-fill section, cut and fill section 02.0240

半叠片机座 semi-laminated frame 04.0454

半段效应 half section effect 04.1504

半分开式扣件 semi-separated rail fastening, mixed holding fastening 03.0088

半分配制会议电话 telephone conference of semi-distribution system 07.0300

半径杆 radius bar, radius rod 04.1206

半路堑式明洞 part cut-type open cut tunnel, part cut-type tunnel without cover, part cut-type gallery 02.0698

*半湿喷混凝土 half wet shotcreting 02.0781

半湿喷混凝土机 half wet shotcreting machine 02.0973

半双工 half-duplex operation 07.0706

半双工传输 half-duplex transmission 07.0782

半双工无线电通信 semi-duplex radio communication 07.0516

半拖车 semi-trailer 05.0828

半循环交路 semi-loop routing 04.0190

半压力式涵洞 inlet submerged culvert 02.0633

半夜灯 evening lamp 05.0686

半自动闭塞 semi-automatic block system 08.0324

半自动闭塞机 semi-automatic block machine 08.0355

半自动闭塞联系电路 liaison circuit with semi-automatic blocks 08.0239

半自动化驼峰 semi-automatic hump 06.0457

半自动化驼峰系统 semi-automatic hump yard system 08.0365

*半自动作业 semi-automatic operation by route 08.0376

办理闭塞 blocking 06.0242

棒式车架 bar frame 04.1214

棒式绝缘子 strut insulator, rod insulator 04.1379

包 packet, package 07.0742

*包板 side sheathing 05.0440

包层直径 cladding diameter 07.0446

包裹 parcel 06.0016

包间式卧车 [corridor] compartment [type] sleeping car 05.0730

包交换 packet switching 07.0850

包交换网 packet switching network 07.0766

包式终端 packet mode terminal 07.0896

包装/拆器 packet assembler/disassembler, PAD 07.0895

包装储运图示标志 pictorial markings for handling of packages 06.0573

包装运输试验 transporting test for package 06.0575

薄壁筒体结构 thin-shelled tubular structure 05.0383

保安器 protector 07.0208

保持位 suppression, maintaining position, holding position 04.0146

保护地线 protective earth wire 02.1033

保护电路 protective circuit 04.0386

保护电容器 protective capacitor 04.0526

保护阀 protective valve 04.0584

保护接地 protective grounding 02.1034

保护区段 overlap protection block section 08.0342

保护套管 protection sleeve 07.0492

保护线 protective wire 04.1316

保护线用连接线 crossbond of protective wire 04.1317

保温集装箱 thermal container 06.0547

保温运输 insulated transport 06.0178

保险带 safety belt 02.1012

保险费 insurance charge 06.0733

保压停车 stopping at maintaining position 04.0156

饱和蒸汽室 saturated chamber 04.1156

抱缸 piston seizure 04.0843

抱轴式牵引电动机 axle hung traction motor, nose suspension traction motor 04.0400

抱轴瓦 axle suspension bush 04.1075

抱轴悬挂装置 suspension bearing 04.0461

抱轴轴承 axle hung bearing, suspension bearing 04.1074

刨花车 wood chip car 05.0825

报警 alarm 08.0175

报警保护系统 alarm and protection system 04.1534

报头 header 07.0740

报头开始信号 start-of-heading signal 07.0861

报文 message 07.0738

爆破排淤 blasting discharging sedimentation, silt arresting by explosion, discharge of sedimentation by bl 02.0268

爆炸事故 explosion accident 06.0636

背负运输车 piggyback car 05.0827

背景噪声场强 background noise field strength 04.1521

倍程衰减 double attenuation 04.1519

倍频衰减 frequency doubling attenuation 04.1520

备电源 stand-by power source 08.0635

备用轨 stock rails per kilometer of track, emergency rail stored along the way 03.0242

备用货车 reserved cars 06.0341

备用机车 locomotive in reserve 04.0025

被串通路 disturbed channel 07.0137

被盗事故 robbery accident, burglary accident 06.0638

被覆层 coating 07.0447

＊被呼 called 07.0818

被叫 called 07.0818

被叫控制复原方式 called subscriber release 07.0408

被叫用户 called party 07.0115

被控点 controlled point 08.0480

本地电池 local battery 07.0089

本地终端 local terminal 07.0907

＊本务机车 leading locomotive 04.0226

本务走行公里 leading locomotive running kilometers 04.0215

本站作业车 local car 06.0235

崩塌 collapse, toppling 03.0480

崩塌地段路基 subgrade in rock fall district, subgrade in collapse zone 02.0224

泵电动机 pump motor 04.0476

泵轮 pump impeller 04.1010

泵支承箱 pump supporting box 04.0877

比功率 power/weight ratio 04.0086

比面积 specific floor area 05.0054

比容[积] specific volume 05.0053

比特 bit 07.0045

比特率 bit rate 07.0046

L/V 比值 L/V ratio 04.0184

比重量 weight/power ratio 04.0087

闭环控制 closed-loop control 04.1616

闭路式轨道电路 close type track circuit 08.0505

闭塞 block system 08.0317

＊闭塞电话 interstation train operation telephone, blocking telephone 07.0415

闭塞分区 block section 08.0340

闭塞机 block instrument 08.0352

闭锁位置 locked position of coupler 05.0566

臂板电锁器联锁 interlocking by electric locks with semaphore 08.0123

臂板接触器 contacts operated by semaphore 08.0315

臂板信号机 semaphore signal 08.0056

臂板转极器 pole changer operated by semaphore 08.0316

避车洞 refuge hole, refuge recess, refuge niche 02.0815

避车台 refuge platform 02.0507

避雷器 lightning arrester 08.0665

避雷线 lightning conductor 07.0211

避难线 refuge siding 06.0415

边沟 side ditch 02.0296

边坡清筛机 ballast shoulder cleaning machine 03.0428

边坡植被防护 vegetation on slope 03.0505

边墙 side wall 02.0678

边走边卸阀 ballast flow control valve 05.0518

编发线 marshalling-departure track 06.0408

编码 coding 07.0016

编组场综合作业自动化 automation of synthetic operations in marshalling yard 01.0112

编组调车 make-up of trains 06.0256

编组能力 make-up capacity 06.0507

编组站 marshalling station, marshalling yard 06.0369

＊扁烟筒 oblong ejector, Giesl ejector 04.1140

＊便桥 temporary bridge 02.0430

便携电台 portable radio set 07.0555

变电所测试车 substation testing car 04.1461

变动支出 variable expense 06.0727

变更径路　route diversion　06.0055

变更设计　altered design　02.0018

变极调速　pole changing speed control　04.0347

变几何燃气轮机　variable-geometry gas turbine　04.1528

* 变矩比　torque ratio　04.1009

* 变矩器轴　torque converter shaft　04.1015

变摩擦式减震装置　variable friction type snubbing device　05.0159

变扭比　torque ratio　04.1009

变扭器轴　torque converter shaft　04.1015

变频器　frequency converter　07.0053

变频调速　variable frequency speed control　04.0346

变坡点　point of gradient change, break in grade　02.0191

变位阀　changeover valve　05.0515

变压器电势　transformer EMF　04.0429

变压器箱　transformer box　08.0283

变压调速　variable voltage speed control　04.0345

变阻调压　rheostatic control　04.0324

标称电压　nominal voltage　04.0428

标称功率　nominal power　04.0083

标称转速　nominal speed　04.0829

标题　heading　07.0739

标志　sign , marker　08.0093

标志灯　marker lamp　04.0643

标志器　marker　07.0400

标准长度钢轨　standard length rail　03.0056

标准大气状况　standard atmospheric condition　04.0845

* 标准分流感度　standard shunting sensitivity　08.0520

标准分路灵敏度　standard shunting sensitivity　08.0520

标准贯入试验　standard penetration test　02.0058

标准煤　standard coal　04.0058

表示　indication　08.0158

表示灯　indication lamp　08.0272

表示灯电源　power source for indication lamp　08.0662

表示电路　indication circuit　08.0235

表示对象　indicated object　08.0467

表示杆　indication rod　08.0614

表示连接杆　connecting rod for indication　08.0627

表示盘　indicating panel　08.0255

表示器　indicator　08.0088

表示周期　indication cycle　08.0482

滨河路堤　embankment on river bank　02.0222

冰凌调查　ice floe survey, frazil ice survey　02.0071

冰压力　ice pressure　02.0346

并励电动机　shunt excited motor　04.0405

并联电容补偿装置　compensator with parallel capacitance　04.1301

并联式轨道电路　multiply connected track circuit　08.0523

并列式枢纽　parallel arrangement type junction　06.0400

并行传输　parallel transmission　07.0784

并置信号点　double signal location　08.0347

播音连接器　public address coupling　05.0673

播音室　public address room　05.0767

播音装置　public address system　05.0692

拨叉　poking fork　05.0524

拨道　track lining　03.0203

拨道机　track lining machine　03.0383

拨道器　track lining tool　03.0384

拨号　dialling　07.0120

拨号脉冲　dial impulse　07.0121

拨号盘　dial　07.0083

拨号音　dialling tone　07.0276

拨号终端　dial-up terminal　07.0909

波导线　waveguide line　07.0593

波导线传输方式　transmission mode with waveguide line　07.0549

波分复用　wavelength devision multiplex　07.0506

波特　baud　07.0047

波纹地板　corrugated floor　05.0403

波纹管　bellows　04.0986

波形辐板　corragated wheel plate　05.0124

补偿电容器　compensation capacitor　04.0528

补偿滑轮　tension pulley　04.1419

补偿线圈　compensating coil　04.0450

补偿装置的电抗比　reactance ratio of compensator　04.1303

补机　banking locomotive　04.0228

不成对运行图　train diagram not in pairs　06.0294

不对称脉冲轨道电路 asymmetrical impulse track circuit 08.0501

不对称三开道岔 unsymmetrical three-way turnout, unsymmetrical three throw turnout 03.0287

*不对称双开道岔 unsymmetrical double curve turnout, unequilateral turnout 03.0284

不分开式扣件 nonseparated rail fastening, direct holding fastening 03.0087

不规则畸变 fortuitous distortion 07.0712

不可懂串音 unintelligible crosstalk 07.0139

不良地质 unfavorable geology 02.0061

不平衡电阻 unbalanced resistance 07.0225

不同时到达间隔时间 time interval between two opposing trains arriving at station not at the same time 06.0299

不完全燃烧热损失 heat loss due to incomplete combustion 04.1107

不限时人工解锁 manual non-time release 08.0148

不摘车修 in-train repair 05.0932

步进制电话交换机 step-by-step telephone switching system 07.0393

步行板 foot plank 02.0506

部分预应力混凝土桥 partially prestressed concrete bridge 02.0374

部件故障检测 inspection of component failure, failure diagnostic 04.0277

C

材料供应计划 material supply plan 06.0719

材料申请计划 material requisition plan 06.0720

材料消耗定额 material consumption norm, material consumption ratings 06.0718

材质不良 bad material 08.0696

财务管理信息系统 financial management information system 06.0762

财务决算审查 financial statements review 06.0761

采暖装置 heating system 05.0591

餐车 dining car 05.0732

餐车洗池 sink 05.0776

餐室 dining room 05.0755

参考当量 reference equivalent 07.0237

残车率 rate of bad order cars 05.0975

残废军人票 disabled armyman ticket 06.0043

残余废气 residual gas 04.0739

残余废气系数 coefficient of residual gas 04.0740

仓库 warehouse 06.0093

操纵台 operating console 08.0254

槽式列车 bunker train 02.0966

槽型梁 trough girder 02.0387

*草测 route reconnaissance 02.0099

厕所 lavatory, toilet 05.0777

侧摆振动 swaying [vibration] 01.0102

侧板 side sheet 04.1123

侧壁导坑法 side heading method 02.0738

侧窗 side window 04.1058

侧灯 side lamp 05.0679

*侧沟 side ditch 02.0296

侧滚振动 rolling [vibration] 01.0101

侧架立柱 side frame column 05.0086

侧架上弦杆 side frame top chord 05.0081

侧架上斜弦杆 side frame top oblique chord 05.0083

侧架弹簧承台 side frame spring seat 05.0085

侧架下弦杆 side frame bottom chord 05.0082

侧架下斜弦杆 side frame bottom oblique chord 05.0084

侧力 [maglev] lateral force 04.1598

侧梁 side sill 05.0394

侧门 side door 05.0496

侧排油阀 side oil outlet valve 05.0484

侧墙 side wall 05.0426

侧墙包板 side sheathing 05.0440

侧倾车体 tilting type car body 05.0377

侧向水平联结系 lateral bracing 02.0458

侧音 side tone 07.0144

侧柱 side post 05.0430

侧柱连铁 side post connecting rail 05.0427

侧柱内补强 inside reinforcement of side post 05.0428

侧撞 conering 08.0371

册页[客]票 coupon ticket 06.0045

测长 distance-to-coupling measurement 08.0387

测段 segment of survey 02.0134

测量放样 staking out in survey 02.0837

测量精度 survey precision, precision of survey 02.0109

测试环线 test loop 08.0430

测试台 test desk 07.0131

测速 speed measurement 08.0386

测速发电机 tachogenerator 04.0480

测重 weight sensing 08.0388

测阻 rollability measurement 08.0389

插板支护 inserting plate support, forepoling 02.0774

插接不良 plug-in trouble 08.0691

插孔 jack 07.0102

插孔排 jack strip 07.0103

插入式混凝土振捣器 immersion type vibrator for concrete 02.0940

插入式继电器 plug-in type relay 08.0586

插塞 plug 07.0101

*插座 jack 07.0102

叉槽 fork pockets 06.0562

叉车 fork-lift truck 06.0600

叉簧 hook switch 07.0078

茶炉 drinking water boiler 05.0670

茶桌 tea table 05.0770

查号台 information desk 07.0258

岔枕 switch tie, turnout tie 03.0117

岔中绝缘 insulated joint within a turnout 08.0536

差错恢复 error recovery 07.0794

差错检测 error detection 07.0843

差错控制 error control 07.0793

差错漏检率 residual error-rate 07.0857

差分调制 differential modulation 07.0832

差分脉码调制 differential pulse-code modulation 07.0028

差分移相键控 differential phase shift keying, DPSK 07.0833

差转电台 radio repeating set 07.0562

拆迁 removing 02.0833

拆装式桁架 demountable truss 02.0389

柴排 firewood raft, mattress, willow fascine 02.0320

柴油打桩机 diesel pile driver 02.0906

柴油机 diesel engine 04.0690

柴油机爆燃 engine detonation 04.0807

*柴油机车 diesel locomotive 04.0006

*柴油机工作粗暴 engine detonation 04.0807

柴油机净重 net weight of diesel engine 04.0838

柴油机起动试验 diesel engine starting test 04.0856

柴油机特性 diesel engine characteristic 04.0848

柴油机支座 engine support 04.0867

柴油机转速表 diesel engine tachometer 04.0965

掺混区 dilution zone 04.1572

铲运机 scoper, scraper, carrying scraper 02.0883

颤振 flutter 02.0526

场间联系电路 liaison circuit between yards 08.0241

场库 storage yard and warehouse 06.0089

场强覆盖区 field strength coverage 07.0540

场强中值 median of field strength 07.0520

常摩擦式减震装置 constant friction type snubbing device 05.0158

常用局减 quick service 04.0152

*常用全制动 full service braking, full service application 04.0130

常用制动 service braking, service application 04.0129

长大货物车 oversize commodity car 05.0833

长大坡道 long steep grade, long heavy grade 02.0186

长定子 long stator 04.1585

长钢轨运输作业列车 long welded rail transporting and working train 03.0438

长轨条 long rail string 03.0021

长轨线路 long welded rail track 03.0018

长交路 long routing 04.0194

长隧道 long tunnel 02.0657

长途半自动接续 toll semi-automatic dialling 07.0267

长途半自动接续台 toll switch board for semi-automatic operation 07.0253

长途电话交换机 toll telephone switching system 07.0251

长途电话所 toll telephone office 07.0271

长途调度台　trunk dispatcher switchboard　07.0255

长途接续台　manual toll switching board　07.0252

长途通信网　toll communication network　07.0238

长途业务台　toll service desk　07.0254

长途中继线　toll junction line　07.0272

长途自动电话　toll automatic telephone　07.0269

长途自动电话中继器　toll automatic switching repeater　07.0270

长途自动接续　toll automatic dialling　07.0266

长线张拉台座　stretching bed for longline production　02.0944

敞车　gondola car, open goods wagon　05.0794

敞顶集装箱　open top container　06.0544

超长超重列车　exceptionally long and heavy train　0i.0083

超长货物　exceptional length freight　06.0161

超导体　superconductor　04.1602

超导体斥力　superconducting repulsion force　04.1624

超导悬浮系统　superconducting suspension system　04.1581

超范围修理　repair beyond the scope of repairing course　04.0261

超负荷试验　over-load test　04.0861

超前导坑　advance heading　02.0743

超前锚杆　advance anchor bolt　02.0773

超前支护　advance support　02.0772

超群　super group　07.0309

超群配线架　supergroup distribution frame　07.0293

超声波查漏仪　ultrasonic leak detector　07.0216

超速停车装置　overspeed trip　04.0968

超挖　overbreak　02.0753

超限货物　out-of-gauge freight　06.0157

超限货物等级　classification of out-of-gauge freight　06.0158

超限货物检查架　examining rack for out-of-gauge freight　06.0159

潮喷混凝土　half wet shotcreting　02.0781

潮汐河流　tidal river　02.0636

车场　yard　06.0438

车长电台　train conductor's station　07.0557

车长阀　caboose valve, conductor's valve, guard's

valve　05.0317

车次表示　train number indication　08.0484

车挡　bumper post　06.0515

车档表示器　buffer stop indicator　08.0098

车底电线管　electric wire conduit underneath the car　05.0702

车底数　number of allocated passenger trains　05.0979

车电分线盒　junction box　05.0704

车顶　roof　05.0454

车顶冰箱冷藏车　overhead brine tank refrigerator car　05.0815

车顶侧梁　roof cant rail　05.0459

车顶端横梁　roof end rail　05.0460

车顶横梁　roof cross beam　05.0461

车顶弯梁　carline　05.0457

车顶纵梁　purline　05.0458

车端冰箱冷藏车　ice-bunker refrigerator car　05.0814

车端缓冲器　end-of-car cushioning device　05.0558

车钩复原装置　coupler centering device　05.0569

车钩缓冲停止器　device for stopping buffer action　06.0165

车钩缓冲装置　coupler and draft gear　05.0527

[车钩缓冲装置]压缩与拉伸　[coupler and draft gear] running-in and running-out　05.0587

车钩间隙　coupler slack　05.0585

车钩连接线　coupling line　05.0574

*车钩连线间长度　length over pulling faces of couplers　05.0019

车钩轮廓　coupler contour　05.0532

车钩牵引力　tractive effort at coupler, drawbar pull　04.0091

车钩三态作用　three states of coupler operation　05.0565

车钩托梁　coupler carrier　05.0572

车钩中心线高度　height of coupler center from top of rail, coupler height　05.0028

车号　number of car　05.0017

车号抄录电视　TV for record vehicle number　07.0659

车号灯　side number plate lamp　04.0645

车号员无线电通信　radio communication for number

taker 07.0624

*车号自动识别 automatic car identification 08.0390

车架 frame 04.1055

车辆报废限度 car condemning limit 05.0950

车辆标记 lettering and marking of car 05.0014

车辆长距比 ratio of car body length to length between truck centers 05.0035

车辆厂修 car repair in works 05.0920

车辆冲击 car impact 05.0588

车辆冲击试验 car impact test 05.0870

车辆存在监测器 presence monitor 08.0408

车辆大修 car heavy repair 05.0925

车辆定距 length between truck centers 05.0034

车辆动力学试验 car dynamics test 05.0871

车辆段 car depot 05.0892

车辆段检修台位利用率 rate of utilization of repair positions in car depot 05.0974

车辆段修 car repair in depot 05.0921

车辆分散供电 separate power supply [system] for car 05.0716

车辆辅修 car auxiliary repair 05.0922

车辆高度 height of car 05.0030

车辆公里 car kilometers 06.0690

车辆构造速度 design speed of car, construction speed of car 05.0842

车辆横向 lateral direction of car 05.0007

车辆互撞 car collision 05.0589

车辆换算长度 converted car length 06.0229

车辆集中供电 centralized power supply [system] for car 05.0715

车辆技术履历薄 technical record book of car 05.0981

车辆计算长度 calculated length of car 05.0967

车辆加速器 car accelerator 08.0398

车辆检修 car inspection and maintenance 05.0890

车辆检修率 rate of cars under repair 05.0971

车辆检修设备 car repair facilities 05.0893

车辆检修停留时间 down time for holding cars for repairing 05.0972

车辆检修限度 car repair limit 05.0948

车辆检修在修时间 down time for car under repair 05.0973

车辆减速器 car retarder 08.0397

车辆交直流供电 AC-DC power supply for car 05.0717

车辆年修 car yearly repair 05.0927

车辆平均长度 average length of car 05.0970

车辆强度试验 car strength test 05.0869

车辆全长 length over pulling faces of couplers 05.0019

车辆全轴距 wheelbase of car 05.0046

车辆设计规范 specifications for design of cars 05.0986

车辆修理厂 car repair works 05.0943

车辆运营 car operation 05.0891

车辆运用维修 car operation and maintenance 05.0915

车辆运用限度 car road service limit 05.0949

车辆制检 car brake examination, car brake inspection 05.0924

车辆制造厂 car manufacturing works 05.0944

车辆中修 car medium repair 05.0926

车辆轴检 car journal and box examination, car journal and box inspection 05.0923

车辆装卸修 car repair before loading or after unloading 05.0928

车辆纵向 longitudinal direction of car 05.0006

车辆最大高度 maximum height of car 05.0029

车辆最大宽度 maximum width of car 05.0024

车辆最大容许速度 maximum permissible speed of car 05.0843

车列 train set 06.0200

车流 car flow 06.0276

车流调整 adjustment of car flow 06.0338

车流径路 car flow routing 06.0279

车流组织 organization of car flow 06.0277

车轮 wheel 05.0095

车轮不圆 wheel out of round 05.0255

车轮厂 [car] wheelset repair factory 05.0982

车轮检测器 wheel detector 08.0423

车轮滑行 wheel sliding, wheel skid 01.0068

车轮静平衡检验 car wheel static balance test 05.0963

车轮空转 wheel slipping 01.0069

车轮扣环 retaining ring [of tire] 05.0100

车轮贴靠　flanging　05.0878

车轮直径　wheel diameter　05.0220

车门自动控制　automatic train door control　08.0435

车票　ticket　06.0034

车票有效期　ticket availability　06.0050

车上给水装置　water supply equipment with roof tank　05.0655

车上水箱　roof water tank　05.0656

车体　car body　05.0374

车体侧倾装置　car body tilting device　05.0378

车体长度　length over ends of body, length of car body　05.0020

车体骨架　body framing　05.0380

车体及外部装备密封试验　test for sealing of body and external equipment　04.0659

车体宽度　width over sides of car body　05.0025

车体内长　length inside car body　05.0023

车体内高　height inside car body　05.0031

车体内宽　width inside car body　05.0027

车体内中心处高度　height inside car body from floor to roof center　05.0032

车体弯曲振动试验　test of vibration caused by car-body bending　05.0872

车下电气插座　car power receptacle　05.0709

车下给水装置　water supply equipment with lower tank　05.0658

车下水箱　lower water tank　05.0659

车型　model of car　05.0016

车载地球站　vehicle earth station　07.0673

车载钢轨涂油器　on-board rail lubricator　03.0404

车站　station　06.0362

车站班计划　station shift operating plan　06.0346

车站办理车数　number of inbound and outbound car handled at station　06.0268

车站等级　class of station　06.0232

车站电台　station radio set　07.0558

车站分布　distribution of stations　02.0153

车站工作组织　organization of station operation　06.0230

车站技术作业表　station technical working diagram　06.0269

车站间隔时间　time interval between two adjacent

trains at station　06.0298

车站阶段计划　station stage operating plan　06.0347

车站控制　station master control　08.0470

车站隧道　station tunnel　02.0654

车站通过能力　carrying capacity of station　06.0505

车站信号　signaling at stations　08.0003

车站行车工作细则　instructions for train operation at station　06.0198

车站咽喉　station throat　06.0506

车站咽喉通过能力　carrying capacity of station throat　06.0502

车站作业计划　station operating plan　06.0345

车种　type of car　05.0015

车轴　axle　05.0107

车轴超声探伤　ultrasonic inspection for axle　05.0946

车轴齿轮箱　axle gear box　04.1002

车轴电磁探伤　magnetic particle inspection for axle, magnaflux inspection for axle　05.0947

车轴发电机　axle generator　05.0711

车轴发电机控制箱　axle generator control box　05.0712

车轴空心轴驱动　quill drive, hollow axle drive　04.0354

车轴模拟试验台　axle analogy test machine　05.0996

车轴驱动方式　mode of axle drive　04.0348

车轴弯曲　bent axle　05.0261

沉管法　immersed tunnelling method　02.0726

沉井挡墙　caisson retaining wall　02.0291

沉井基础　open caisson foundation　02.0583

沉井刃脚　cutting edge of open caisson　02.0585

沉箱基础　pneumatic caisson foundation　02.0591

衬垫　pad　03.0097

衬砌　lining　02.0676

衬砌变形　lining deformation　03.0529

衬砌腐蚀　lining corrosion　03.0528

衬砌裂损　lining cracking　03.0527

称重试验　weighing test　04.0657

成端电缆　formed cable　07.0190

成对运行图　train diagram in pairs　06.0293

成件包装货物　packed freight　06.0116

成组装车　car loading by groups　06.0104

乘警　train police　06.0066

乘座舒适度　riding comfortableness, ride comfort 05.0849

程控电话交换机　stored program controlled telephone switching system　07.0396

*承接接头　supported joint　03.0068

承力索　catenary　04.1361

承力索弛度　catenary sag　04.1449

承力索接头线夹　catenary splice　04.1370

承力索终端锚固线夹　termination fitting for catenary　04.1368

承台　bearing platform　02.0575

承载鞍　adapter　05.0147

承载系数检定　load factor rating　03.0519

持续功率　continuous power　04.0811

持续牵引力　continuous tractive effort　04.0094

持续速度　continuous speed　01.0075

驰振　galloping　02.0525

齿轨[传动]机车　rack locomotive　04.0363

*齿条[传动]机车　rack locomotive　04.0363

充电插头　charging plug　05.0710

充风　charging　04.0140

充风位　charge position　04.0141

充量　charge　04.0720

充量系数　coefficient of charge　04.0741

充量限制阀　filling limiting valve　04.1022

*充气　charging　04.0140

充气避雷器　gas filled arrester　07.0206

充气维护型光缆　gas maintenance type optical fiber cable　07.0487

*充气位　charge position　04.0141

充油型光缆　jelly filled type optical fiber cable 07.0486

冲便阀　flush valve　05.0784

*冲击荷载　impact force of train　02.0337

冲击式钻机　impact-type drill machine, percussion type drill machine　02.0915

冲击系数　coefficient of impact　02.0338

冲击因数　impact factor　03.0535

冲击座　striker, striking casting　05.0414

冲角　angle of attack　05.0873

冲刷　erosion, scouring　03.0481

冲突　collision　06.0629

重传　retransmission　07.0889

重叠区段　overlap section　08.0341

重定向　redirection　07.0880

重复检查　repeated checking　08.0224

重复运输　repeated traffic　06.0108

重联插座　multiple unit socket　04.0608

重联机车走行公里　multi-locomotive running kilometers　04.0217

重联运行试验　test on coupled operation, test on multi unit operation　04.0684

抽验　selective acceptance　02.1049

抽样　sample　07.0372

抽样试验　sampling test　04.0297

抽液管座　unloading pipe connection　05.0479

初步设计　preliminary design　02.0014

初测　preliminary survey　02.0100

初期支护　primary support　02.0687

出碴　mucking and removing　02.0756

出厂试验　predelivery test　04.0283

出发场　departure yard　06.0440

出发线　departure track　06.0407

*出发信号机　starting signal　08.0066

出口角　blade outlet angle　04.1549

出油阀　delivery valve　04.0952

出油阀偶件　delivery valve matching parts 04.0951

出油阀座　delivery valve seat　04.0953

出站信号机　starting signal　08.0066

出中继电路　outgoing trunk circuit　07.0262

*出中继器　outgoing trunk circuit　07.0262

厨房　kitchen　05.0757

厨房车　kitchen car　05.0735

除草机　weed cutting machine, weed cutter 03.0463

除砂机　sand removing machine　03.0459

除雪车　snow plow, snow plough　05.0840

除雪机　snow removing machine, snow remover 03.0457

杆环杆　bar with ball and eye, ball-socket bar 04.1399

杆座鞍子　socket-type saddle　04.1389

储备功率　reserve power　04.0821

储藏室　storage room　05.0763

储风罐　air reservoir　08.0400

储酸室　acid store room　08.0718

触电保安器　electric shock protector　02.1035

穿销防爬器　wedged rail anchor　03.0101

传导干扰　conducted interference　07.0525

传导模　guided modes　07.0454

传动齿轮　transmission gear　04.1076

传动系统　driving system　08.0608

传动轴　transmission shaft　05.0520

传感器　sensor, transducer　04.0629

传输继电器　transmitting relay, transmission relay　08.0584

传输结束信号　end-of-transmission signal　07.0839

传输线　transmission line　07.0058

传输性能　transmission performance　07.0135

传送同步方式　transmission synchronized mode　07.0603

传真发送机　facsimile transmitter　07.0724

传真机　facsimile apparatus, Fax　07.0723

传真接收机　facsimile receiver　07.0725

传真收发机　facsimile transceiver　07.0726

船只或排筏的撞击力　collision force of ship or raft　02.0349

喘振　surge　04.0795

串励电动机　series excited motor　04.0404

串联电容补偿装置　compensator with series capacitance　04.1300

串联式轨道电路　serially connected track circuit　08.0522

串行传输　serial transmission　07.0785

串音　crosstalk　07.0038

串音测试器　crosstalk meter　07.0041

*串音防卫度　signal to crosstalk ratio　07.0140

串音抑制滤波器　crosstalk suppression filter　07.0350

窗间板　pier sheathing　05.0448

窗卷帘　window blind, window shade　05.0509

窗口　window　07.0868

窗框　window sash　05.0511

窗帘　window curtain　05.0510

窗锁　window sash lock　05.0512

窗台　window sill, window rail　05.0450

床头灯　berth lamp　05.0683

炊事室　cooking room　05.0758

垂度　sag　07.0227

垂向冲击　vertical impact　01.0109

垂向振动　vertical vibration　01.0098

垂直板　diaphragm plate, deflecting plate　04.1147

垂直动载荷　vertical dynamic load　05.0885

*垂直反射板　diaphragm plate, deflecting plate　04.1147

垂直载荷　vertical load　05.0884

磁场削弱接触器　field weakening contactor　04.0551

磁场削弱率　field weakening　04.0436

磁场削弱系数　coefficient of field weakening　04.0437

磁浮铁路　magnetic levitation railway, maglev　01.0029

磁感应　magnetic induction　04.1465

磁轨制动　electromagnetic rail brake　04.0122

磁路[系统]　magnetic circuit　08.0574

磁轮　magnet wheel　04.1630

*磁耦合系数　mutual inductance coefficient　04.1473

磁石电话机　magneto telephone set　07.0091

磁石电话交换机　magneto telephone switch board　07.0391

磁石发电机　magneto　07.0080

磁[悬]浮　maglev, magnetic levitation　04.1578

磁[悬]浮车辆　maglev vehicle　04.1584

次应力　secondary stress　03.0538

伺服电机　servomotor　04.0478

*伺服马达　servomotor　04.0478

伺服疲劳试验机　servo fatigue testing machine　05.0992

伺服型加速度计　servo accelerometer　04.1610

从板　follower　05.0573

从动齿轮　driven gear　04.1078

从动轮对　driven wheel set　04.1230

从轮对　trailing wheel set　04.1236

从轮转向架　trailing truck　04.1235

从属信号机　dependent signal　08.0075

从站　slave station　07.0848

粗粒土填料　coarse-grained soil filler, coarse-grained soil fill　02.0255

窜机油　lubricating oil carry-over　04.0840

窜气　blow-by　04.0841

淬火轨　head hardened rail, quenched rail　03.0051

淬火尖轨　surface-hardened switch rail, quenched switch rail　03.0356

存车线　storage siding　06.0411

存储转发　store and forward　07.0864

错乘　taking wrong train　06.0056

错溜　miseroute　08.0368

错误办理　wrong handling　08.0669

错误关闭信号　false stopping of a signal　08.0680

错误解锁　false release　08.0678

错误开放信号　wrong clearing of a signal　08.0679

错误锁闭　false locking　08.0677

错误显示　wrong indication　08.0047

错牙接头　rail ends unevenness in line or surface　03.0170

D

打道钉机　spike driver　03.0408

打磨钢轨　rail grinding　03.0212

打桩机　pile driver　02.0904

大地电阻率　earth resistivity　04.1476

大功率转辙机　heavy duty switch machine　08.0606

大轨缝　excessive joint gap, wide joint gap　03.0158

大横梁　cross bearer　05.0397

大揭盖清筛机　ballast cleaning machine with removed track panels　03.0431

大陆桥　transcontinental railway, inter-continental railway, land-railway　01.0122

大气压式采暖装置　atmospheric pressure steam heating equipment　05.0596

大气压式暖汽调整阀　vapor regulater　05.0598

大桥　major bridge　02.0417

大事故　serious accident　06.0623

大型矿车　large scale ore car　02.0969

大型临时工程　large-scale temporary project　02.0857

大型全断面清筛机　large ballast undercutting cleaners, on-track full section undercutting cleaners　03.0429

大型线路机械　heavy permanent way machine, large permanent way machine　03.0381

大修计划　plan of capital repair　06.0702

大烟管　flue [tube]　04.1135

大腰带　waist rail　05.0435

*大闸　automatic brake valve　04.0167

大站电气集中联锁　relay interlocking for large station　08.0114

大宗货物　mass freight　06.0113

带柄道岔表示器　switch indicator with level　08.0099

带动道岔　switch with follow up movement　08.0195

带回流线的直接供电方式　direct feeding system with return wire　04.1276

带裙鱼尾板　aproned fish plate, fish plate with apron　03.0081

带式输送机　belt conveyer　02.0885

代用票　substituting ticket　06.0046

待避所　refuge place　02.1007

待修机车　locomotive waiting for repair　04.0027

单臂受电弓　single arm pantograph　04.0535

单边带通信　single side band communication　07.0511

单边供电　one way feeding　04.1293

单边型直线感应电动机　single sided linear induction motor　04.1604

单侧导坑法　single side heading method　02.0739

单侧减速齿轮驱动　single reduction gear drive　04.0351

单侧[踏面]制动　single shoe brake　05.0330

单车试验　single car test　05.0953

单车试验器　single car testing device　05.0371

单电动机驱动　monomotor drive　04.0353

单斗挖土机　power shovel　02.0878

单独操纵继电式电气集中联锁　individual level type all-relay interlocking　08.0110

单独操纵作业 manual operation 08.0375

单独制动阀 independent brake valve 04.0168

单断 single break 08.0222

单缸功率 power per cylinder 04.0817

单工 simplex operation 07.0704

单工传输 simplex transmission 07.0783

单工无线电通信 simplex radio communication 07.0512

单管逆变器 individual inverter 05.0699

单管荧光灯 single tube fluorescent lamp 05.0687

单轨条式轨道电路 single rail track circuit 08.0506

单轨铁路 monorail, monorail railway 01.0028

单轨小车 hand cart 03.0472

单呼 individual calling 07.0424

单回路供电 single circuit power supply 04.1279

单机运行 light locomotive running 04.0223

单机走行公里 light locomotive running kilometers 04.0216

单肩回交路 single-arm routing 04.0188

*单节机械冷藏车 mechanical refrigerator car 05.0817

单局制 single-office system 07.0383

单开道岔 simple turnout, lateral turnout 03.0282

单梁式架桥机 single beam girder-erecting machine 02.0948

单流 single current 07.0707

单流液力机械传动 hydromechanical drive with inner ramification 04.0994

单面托盘 single-deck pallet 06.0567

单模光纤 single-mode optical fiber 07.0443

单频感应器 single frequency inductor 08.0428

单曲线 simple curve 02.0173

单式不对称道岔 unsymmetrical double curve turnout, unequilateral turnout 03.0284

单式对称道岔 symmetrical double curve turnout, equilateral turnout 03.0283

单式交分道岔 single slip switches 03.0294

单式同侧道岔 unsymmetrical double curve turnout in the same direction 03.0285

单双工兼容无线电通信 compatible simplex-duplex radio communication 07.0517

单筒壁灯 single cylindrical shade wall lamp 05.0690

单位工程 unit project 02.0004

单位活塞面积功率 piston unit area power 04.0820

单位体积功率 specific volume power 04.0088

单位阻力 unit resistance, specific resistance 04.0103

单线臂板信号机 single wire semaphore signal 08.0059 08.0326

单线桥 single track bridge 02.0420

单线隧道 single track tunnel 02.0651

单线铁路 single track railway 01.0015

单线运行图 train diagram for singletrack 06.0291

单线制 single wire system 05.0713

单相V/V接线牵引变压器 traction transformer of singlephase V/V connection 04.1287

单相低频交流制 single-phase low frequency AC system 04.0315

单相电度表 single-phase wat-hour meter 04.0627

单相工频交流电力机车 single-phase industrial frequency AC electric locomotive 04.0302

单相工频交流电力牵引制 single phase industrial frequency AC electric traction system 04.1268

单相工频交流制 single-phase industrial frequency AC system 04.0314

单相交流电动车组 single-phase industrial frequency AC motor train unit 04.0310

单相交流电力机车 single-phase AC electric locomotive 04.0301

单相交流牵引电动机 single-phase AC traction motor 04.0398

单相接线牵引变压器 traction transformer of single-phase connection 04.1288

单相桥式整流器 single-phase bridge rectifier 04.0331

单项预算 individual budget 02.0852

单向传输 unidirectional transmission 07.0789

单向横列式编组站 unidirectional transversal type marshalling station 06.0373

单向混合式编组站 unidirectional combined type marshalling station 06.0375

单向通信 one-way communication 07.0777

单向自动闭塞 single-directional running automatic block 08.0329

单向纵列式编组站 unidirectional longitudinal type marshalling station 06.0374

单循环液力传动 single-circuit hydraulic transmission 04.0991

单元列车 unit train 06.0219

单元制动 brake unit 05.0327

单胀式蒸汽机车 single expansion steam locomotive 04.1084

单置信号点 single signal location 08.0346

单轴燃气轮机 single-shaft gas turbine 04.1526

单轴转向架 single-axle truck 05.0058

单转子滑片式空压机 single rotary compressor 02.0923

当量跨距 equivalent span length 04.1436

挡车器 stop buffer 06.0468

挡风墙 wind-break wall 02.0322

挡土墙 retaining wall 02.0284

挡烟板 smoke deflector 04.1153

刀把梁 lowered draft sill 05.0423

捣镐 packer, tamping pick, beater 03.0467

捣固道床 ballast tamping 03.0223

捣固机械 tamping machine 03.0373

倒虹吸管 inverted siphon 02.0635

导程 lead 04.1257

导风轮 inducer 04.0975

导管调整器 pipe compensator 08.0294

导管装置 pipe installation 08.0293

导轨与悬浮系统相互作用 guideway suspension interaction 04.1628

导坑 heading 02.0733

导框式转向架 pedestal truck 05.0065

导流堤 diversion dike 02.0317

导流建筑物 regulating structure 02.0617

导轮 guide wheel 04.1012

导轮对 leading truck wheel set 04.1232

导轮转向架 leading truck 04.1231

导纳电桥 admittance bridge 07.0222

导频 pilot freqency 07.0305

导频放大器 pilot amplifier 07.0339

导频无人增音机 pilot unattended repeater 07.0288

导曲线 lead curve 03.0303

导曲线半径 radius of lead curve 03.0304

导曲线支距 offset of lead curve 03.0305

导热系数 coefficient of thermal conductivity 05.0652

导线安装曲线 wire installation curve 04.1435

导线测量 traversing, traverse survey 02.0102

导线导轮 wire carrier 08.0310

导线反正扣 wire-adjusting screw 08.0306

导线立轮 vertical wheel 08.0309

导线平轮 horizontal wheel 08.0308

导线平轮组 horizontal wheel assembly 08.0311

导线调整器 wire compensator 08.0307

导线装置 wire installation 08.0305

导向安全 failure to the safe side 08.0670

导向力 guidance force 04.1625

导向系统 guidance system 04.1588

导向线 leading line, alignment guiding line 02.0160

导音频信号 pilot audio fequency signal 07.0612

到达场 receiving yard, arriving yard 06.0439

到达路 destination railway 06.0186

到达线 receiving track, arriving track 06.0406

到发场 receiving-departure yard 06.0441

到发线 arrival and departure track 06.0405

到发线出岔电路 protection circuit with switch lying in receiving-departure track 08.0247

到发线通过能力 carrying capacity of receiving-departure track 06.0504

道碴 ballast 03.0124

道碴槽 ballast tub 03.0486

道碴层 ballast layer 03.0123

道碴巢 ballast nest 03.0489

*道碴床 ballast layer 03.0123

道碴袋 ballast pocket 03.0488

道碴电阻 ballast resistance 08.0517

道碴机械 ballast machine 03.0372

道碴级配 ballast grading 03.0125

道碴犁 ballast plow 03.0420

*道碴漏泄电阻 ballast resistance 08.0517

道碴桥面 ballasted deck, ballasted floor 02.0501

道碴清筛机械 ballast cleaning machine 03.0375

道碴箱 ballast box 03.0487

道岔　turnout, switches and crossings　03.0281

道岔表示　switch indication　08.0161

道岔表示电源　power source for switch indication　08.0660

道岔表示器　switch indicator　08.0094

道岔侧线　branch line of turnout, branch track of turnout, turnout branch　03.0302

道岔错误表示　false indication of a switch　08.0682

道岔定位表示　switch normal indication　08.0168

道岔动作电源　power source for switch operation　08.0659

道岔反位表示　switch reverse indication　08.0169

道岔封锁　switch closed up　08.0708

道岔号数　turnout number　03.0310

道岔后理论长度　rear part theoretical length of turnout　03.0330

道岔后部实际长度　rear part actual length of turnout　03.0331

道岔护轨　turnout guard rail　03.0360

道岔基线　reference line of turnout　03.0324

道岔绝缘段　insulated switch section　06.0435

道岔控制电路　switch control circuit　08.0234

道岔控制电源　power source for switch control　08.0658

道岔拉杆　switch rod, stretcher bar　03.0363

道岔理论长度　theoretical length of turnout　03.0326

道岔理论导程　theoretical lead of turnout　03.0328

道岔连接杆　connecting bar, following stretcher bar　03.0364

道岔密贴　switch [point] closure　08.0132

道岔配列　switch layout　06.0436

道岔启动　switch starting　08.0181

道岔前部理论长度　front part theoretical length of turnout　03.0329

道岔清扫房　switch cleaner's cabin　06.0535

道岔区段　section with a switch or switches　08.0202

道岔区坡　gradient within the switching area　06.0497

道岔全长　total length of turnout　03.0325

道岔人工解锁　manual release of a locked switch　08.0146

道岔熔冰器　switch heater　03.0458

道岔失去表示　loss of indication of a switch　08.0681

道岔实际长度　actual length of turnout　03.0327

道岔始端　beginning of turnout　03.0311

道岔顺序启动　sequential starting of switches　08.0183

道岔顺序转换　sequential transiting of switches　08.0184

道岔锁闭　switch [point] locking　08.0131

道岔锁闭表示　switch locked indication　08.0172

道岔握柄　switch lever　08.0291

道岔中途转换　switch thrown under moving cars　08.0700

道岔中心　center of turnout　03.0306

道岔终端　end of turnout　03.0312

道岔主线　main line of turnout, main track of turnout, turnout main　03.0301

道岔柱　turnout mast　04.1413

道岔转换　switch in transition　08.0182

道岔阻力　switch resistance　06.0476

*道尺　track gage　03.0465

道床　ballast bed　03.0120

道床边坡夯实机　ballast shoulder consolidating machine　03.0424

道床碴肩　shoulder of ballast bed　03.0134

道床底碴夯实机　subballast consolidating machine　03.0426

道床厚度　thickness of ballast bed, depth of ballast　03.0132

道床宽度　width of ballast bed　03.0133

道床系数　ballast coefficient, ballast modulus　03.0038

道床阻力　ballast resistance　03.0044

道钉·track spike, rail spike, dog spike　03.0090

道钉锤　spike hammer　03.0468

*道镐　packer, tamping pick, beater　03.0467

道口　grade crossing, level crossing　03.0273

道口防护无线电通信　radio communication for highway crossing protection　07.0632

道口护桩　protective stake at grade crossing　03.0279

道口监视电视　monitor TV for highway crossing

07.0660

道口接近区段 approach section of a highway level crossing 08.0442

道口警标 warning sign at grade crossing 03.0276

道口看守工 grade crossing watchman, level crossing watchman 03.0008

道口栏木 cross barrier at grade crossing 03.0277

道口平台 level stretch of grade crossing 03.0275

道口铺面 grade crossing pavement, surface of grade crossing 03.0274

道口闪光信号 highway level crossing flashing signal 08.0447

道口室外音响器 highway level crossing out door audible device 08.0441

道口通知设备 highway level crossing announcing device 08.0438

道口信号机 highway level crossing signal 08.0444

道口信号控制盘 highway level crossing signal control panel 08.0445

道口遥信遥测设备 remote surveillance and telemetering for highway level crossing 08.0448

道口栅栏 side barrier at grade crossing 03.0278

道口遮断信号 highway level crossing obstruction signal 08.0443

道口自动信号 automatic level crossing signal 08.0439

灯光转移 to transfer of lighting indication 08.0351

灯丝断丝 filament burn-out 08.0689

灯丝断丝报警 alarm for burnout of filaments 08.0177

等效干扰电流 equivalent disturbing current 04.1493

低高度梁 shallow girder 02.0376

低摩合成闸瓦 low friction composite brake shoe 05.0199

低频干扰防卫度 signal to low frequency interference rate 07.0283

低速柴油机 low speed diesel engine 04.0693

低压保安阀 low pressure safety valve 04.0172

低压侧调压 low voltage regulation 04.0319

低压电路 low voltage circuit, low tension circuit 04.0370

低压调压开关 low voltage tap changer, low tension tap changer 04.0556

低压绕组 low voltage winding, low tension winding 04.0497

低音风喇叭 low tone air horn 04.0652

低值易耗品 low value and easily wornout articles 06.0698

敌对进路 conflicting route 06.0251

敌对信号 conflicting signal 08.0045

底板 floor 02.0680

底碴 subballast 03.0131

底架 underframe 05.0387

底架长度 length over end sills, length of underframe 05.0021

底架架承式牵引电动机 underframe mounted traction motor 04.0403

底架宽度 width over side sills 05.0026

底门 bottom door 05.0495

底漆 prime coat 03.0551

底圈 mud ring 04.1124

地板面高度 height of floor from rail top, floor height 05.0033

地板面积 floor area 05.0038

地板排水装置 floor draining device 05.0669

地表排水 surface drainage 03.0473

地层压力 ground pressure, stratum pressure 02.0706

地层柱状图 column diagram of stratum, graphic logs of strata, drill log of stratum 02.0068

地道 underground path 06.0522

地电流影响 influence of ground current 04.1467

地方铁路 local railway, regional railway 01.0008

地方性编组站 local marshalling station 06.0372

地基 foundation, foundation soil, subgrade 02.0577

地基承载力 bearing capacity of foundation, bearing capacity of ground, bearing capacity of subgrade 02.0047

地貌调查 topographic feature survey, geomorphologic survey 02.0040

地面感应器 wayside inductor 08.0422

地面钢轨涂油器 on-track rail lubricator 03.0403

地面设备 wayside equipment 08.0418

地面信号　ground signal　08.0013

地球站　earth station　07.0672

地区电话交换机　local telephone switching system　07.0390

地区电话网　local telephone network　07.0381

*地铁　subway, metro, underground railway　01.0023

地铁车站　subway station, metro station　02.0655

地铁工程　subway engineering, metro engineering　02.0661

地铁隧道　subway tunnel, underground railway tunnel, metro tunnel　02.0648

地温梯度　geothermal gradient　02.0715

地温通过无人增音机　unattended repeater with ground temperature compensation and powerpassing　07.0290

地温折返无人增音机　unattended repeater with ground temperature compensation and powerfeed loop back　07.0289

地下电缆　ground cable　07.0179

地下连续墙法　underground diaphragm wall method, underground wall method, diaphragm wall method　02.0727

地下排水　subsurface drainage　03.0474

地下铁道　subway, metro, underground railway　01.0023

地下铁道电动车组　subway motor train unit　04.0311

地形测量　topographical survey　02.0104

地形调查　topographic survey　02.0039

地应力　crustal stress　02.0716

地震基本烈度　basic intensity of earthquake, seismic basic intensity　02.0066

地震力　seismic force　02.0350

地震烈度　earthquake intensity　02.0351

地震系数法　seismic coefficient method　02.0209

地震震级　earthquake magnitude　02.0352

地址　address　07.0900

地质调查　geologic survey　02.0041

地质图测绘　survey and drawing of geological map, surveying and sketching of geological map　02.0049

地中电缆盒　underground cable terminal box　08.0285

第二接近区段　second approach section　08.0207

第二离去区段　second departure section　08.0210

第二系悬挂　secondary suspension　05.0179

第二限度　2nd limit　04.0270

第三轨受电器　conductor rail collector　04.0542

第一第二系弹簧挠度比　ratio of spring deflections of primary and secondary suspension　05.0240

第一接近区段　first approach section　08.0206

第一离去区段　first departure section　08.0209

第一系悬挂　primary suspension　05.0178

第一限度　1st limit　04.0269

递热器　combining chamber　04.1170

点对点长途自动接续　point to point toll automatic dialling　07.0268

点对点连接　point to point connection　07.0801

点火　ignition　04.0775

点火器　spark lighter　05.0610

点连式调速系统　point-continued type speed control system　06.0488

点式机车信号　intermittent type cab signaling　08.0410

点式调速系统　point type speed control system　06.0486

点头振动　pitching [vibration], nodding [vibration]　01.0103

垫板　tie plate　03.0095

垫冻害垫板　track shimming　03.0511

垫接接头　supported joint　03.0068

垫砂起道　measured shovel packing　03.0201

电报交换机　telegraph switching equipment　07.0698

电报通信　telegraph communication　07.0687

电报[通信]网　telegraph network　07.0686

电测仪表　electrical measuring instrument　04.0618

电传机　teletype　07.0689

电磁阀　electromagnetic valve　04.0581

电磁环境　electromagnetic environment　04.1516

*电磁计数器　message register　07.0105

电磁继电器　electromagnetic relay　08.0578

电磁兼容[性]　electromagnetic compatibility　07.0575

电磁接触器　electromagnetic contactor　04.0545

电磁屏蔽暗室　electromagnetic shielding darkroom 04.1515

电磁铁驱动器　magnet driver, chopper　04.1593

电磁吸引式系统　electromagnetic attraction system 04.1583

电磁悬浮系统　electromagnetic suspension system, EMS　04.1579

电动臂板电锁器联锁　interlocking by electric locks with electric semaphore　08.0124

电动臂板信号机　electric semaphore signal 08.0058

电动车组　electric multiple unit, motor coach set, electric motor train unit　04.0309

电动传送设备　electric motor operated conveyer 08.0407

电动吊车　electric hoist　02.0902

* 电动葫芦　electric hoist　02.0902

电动机超速　runaway speed　04.0419

电动机独立供电　motor individual power supply 04.0367

* 电动机扭矩　motor torque　04.0421

电动机特性　motor characteristic　04.0420

电动机转矩　motor torque　04.0421

电动机转速　motor speed　04.0418

电动卷扬机　electric winch　02.0901

电动排斥式系统　electrodynamic repulsion system, EDS　04.1580

电动水泵　electric water pump　05.0602

电动转辙机　electric switch machine　08.0603

电分段　sectioning　04.1336

电分段装置　section point　04.1309

电杆　pole　07.0164

电感式加速度计　inductive accelerometer　04.1608

电感应　electric induction　04.1466

* 电镐　portable electric tamper, hand electric tamper 03.0413

电化学加固土壤　electro-chemical treatment of soil 03.0512

电话　telephone　07.0063

电话闭塞　telephone block system　08.0321

电话机　telephone set　07.0090

电话集中器　concentrated telephone unit　07.0439

电话网　telephone network　07.0106

[电话]用户　[telephone] subscriber　'07.0113

电机集中联锁　electro-mechanical interlocking 08.0107

电机空心轴驱动　hollow shaft motor drive 04.0355

电加热器　electric heater　05.0695

电加湿器　electric moistening device　05.0697

电键　key　07.0691

电空传送设备　electropneumatic conveyer 08.0406

电空阀　electropneumatic valve　04.0582

电空接触器　electropneumatic contactor　04.0548

电空制动　electropneumatic brake　04.0119

电空制动电路　electropneumatic brake circuit, E-P brake circuit　04.0388

电空制动控制器　E-P brake controller 04.0575

电空制动装置　electropneumatic brake equipment 05.0269

电空转辙机　electropneumatic switch machine 08.0604

电缆　cable　07.0171

电缆标石　cable marking stake　07.0203

电缆充气维护设备　cable gas-feeding equipment 07.0204

电缆管道　cable duct　07.0200

电缆屏蔽系数　shielding factor of cable　04.1481

电缆套管　cable sleeve　07.0201

电缆障碍探测器　cable fault detector　07.0217

电力传动方式　mode of electric drive　04.0340

电力传动内燃机车　diesel-electric locomotive 04.0688

电力电缆及连接器　power supply cable and coupling 05.0675

电力调度电话　power dispatching telephone 07.0414

电力机车　electric locomotive　04.0007

电力牵引干扰　electric traction interference 07.0533

电力牵引供电系统　power supply system of electric traction　04.1270

电力牵引远动系统　electric traction telemechanical system, electric traction remote control system

04.1271

*电力铁道 electrified railway, electric railway 01.0020

电连接 electrical connector 04.1374

电连接器 electric coupler 04.0391

电连接线夹 electrically connecting clamp 04.1375

电流传感器 current sensor 04.0631

电流继电器 current relay 04.0591

电流－力特性 current-force characteristic 04.1626

电流制 current system 04.0312

电流制转换开关 current system changeover switch 04.0563

电路 circuit 07.0110

电路交换 circuit switching 07.0721

电码 code 07.0688

电码轨道电路 coded track circuit 08.0495

电码继电器 code relay 08.0579

电码孔 code holes 07.0720

电码自动闭塞 automatic block with coded track circuit 08.0331

电耦合系数 electric coupling coefficient 04.1523

电平 level 07.0030

电平表 level meter 07.0356

电瓶车 storage battery car 02.0967

电器室 electric apparatus room 04.1062

电气化干扰 electrification interference 04.1491

电气化铁路 electrified railway, electric railway 01.0020

电气集中[联锁] electric interlocking 08.0108

电气路牌闭塞 electric tablet block system 08.0323

电气路牌机 electric tablet instrument 08.0354

电气路签闭塞 electric staff system 08.0322

电气路签机 electric staff instrument 08.0353

电气锁闭 electric locking 08.0135

电气系统 electric system 05.0671

电气制动试验 electric braking test 04.0681

电热玻璃 electric heating glass 04.0617

电热采暖装置 electric heating equipment 05.0615

电热管 tubular electric heating element 05.0696

电容式加速度计 capacitive accelerometer 04.1609

电扇 electric fan 05.0706

电视会议 video conference 07.0065

电枢线圈 armature coil 04.0441

电刷 brush 04.0459

电刷装置 brush gear 04.0456

电锁器 electric lock 08.0314

电锁器联锁 interlocking by electric locks 08.0121

电文交换 message switching 07.0722

电务维修无线电通信 radio communication for maintenance of signal and communication equipment 07.0638

电压传感器 voltage sensor 04.0630

电压过低 voltage below level 08.0688

电压继电器 voltage relay 04.0590

电压调整器 voltage regulator 04.0610

电压自动补偿装置 autoregulation voltage compenator 04.1305

电压自动调整器 automatic voltage regulator 08.0655

电液阀 electro-hydraulic valve 04.0583

电液转辙机 electrohydraulic switch machine 08.0605

电源变压器 supply transformer 04.0489

电源端子 terminals for power supplies 08.0277

电源屏 power supply panel 08.0650

电源室 power supply room 08.0716

电源转换屏 power switching over panel 08.0654

电站列车[车组] power plant car train-set 05.0839

电子电话交换机 electronic telephone switching system 07.0395

电子控制电路 electronic control circuit 04.0387

电阻焊 flash butt welding 03.0239

电阻制动 rheostatic brake 04.0117

吊杆 suspender, hanger 02.0451

吊沟 suspended ditch 02.0298

吊环 suspension ring 04.1373

*吊孔 suspended span 02.0442

吊缆 suspension cable 02.0494

*吊桥 suspension bridge 02.0393

吊索 dropper 04.1403

吊弦 hanger 04.1401

吊弦线夹 hanger ear 04.1402

*吊装口 hatch 05.0462

调查测绘 survey and drawing of investigation, investigation survey, investigation surveying and sketching 02.0038

调查坑道 investigation gallery, survey tunnel 02.0791

调车 shunting, resorting, car classification 06.0254

调车表示器 shunting indicator 08.0091

调车表示器电路 shunting indicator circuit 08.0250

调车场 marshalling yard, shunting yard, classification yard 06.0443

调车呼叫信号音 shunting calling tone 07.0628

调车机车 shunting locomotive, switcher 04.0016

调车进路 shunting route 06.0246

调车控制器 shunting controller 04.0574

调车区电气集中联锁 interlocking for shunting area 08.0116

调车设备 marshalling facilities, classification facilities 06.0446

调车事故 accident in shunting operation 06.0628

调车无线电通信 radio communication for shunting 07.0622

调车线 shunting track, classification track 06.0409

调车线始端减速器 tangent retarder 08.0378

调车信号 shunting signal 08.0019

调车信号机 shunting signal 08.0078

调车信号音 shunting tone 07.0626

调车作业计划 shunting operation plan 06.0348

*调度分配器 tandem distributor for dispatching telephone with VF selective calling, tandem distributor 07.0427

调度集中 centralized traffic control, CTC 08.0460

调度集中分机 field equipment of CTC 08.0462

调度集中总机 control office equipment of CTC 08.0461

调度监督 dispatcher's supervision system 08.0471

调度控制 dispatcher's control 08.0468

调度命令 traffic [dispatching] order, train [dispatching] order 06.0337

调度区段 train dispatching section, train control section 06.0336

调度日班计划 daily and shift traffic plans 06.0343

调度所 traffic controller's office, dispatcher's office 06.0335

调度所选叫通话箱 selective calling and talking box for dispatching office 07.0428

跌水 hydraulic drop 02.0299

叠板弹簧 laminated spring 05.0168

叠加轨道电路 overlap track circuit 08.0496

丁坝 spur dike 02.0618

丁形离水阀 tee trap 05.0594

顶车 lift one end of car 05.0936

顶车座 jacking pad 05.0425

顶灯 ceiling lamp 05.0678

顶盖座 hatch cover seat 05.0464

顶进法 jack-in method 02.0543

顶推式架设法 erection by incremental launching method 02.0532

定测 location survey, alignment, final location survey 02.0101

定点停车 stopping a train at a target point 08.0434

定反位锁闭 normal and reverse locking 08.0139

定滑轮装置 fixed pulley 04.1420

定检公里 running kilometers between predetermined repairs 04.0258

定检时间 time between predetermined repairs 04.0259

定界符 delimiter 07.0898

定期检修 repair based on time or running kilometers 04.0251

定期票 periodical ticket 06.0047

定期维修 periodical maintenance 08.0724

定期修 periodic repair 05.0918

定时 timing 04.0773

定时抖动 timing jitter 07.0380

定位管 registration arm 04.1391

定位环 steady ring 04.1394

定位接点 normal contact 08.0554

定位器 steady arm 04.1392

定位索 registration wire 04.1364

定位锁闭 normal locking 08.0137

定位线夹 steady ear 04.1395

定位柱　registration mast　04.1414

定位装置　registration device　04.1390

定向爆破　directional blasting　02.0271

定向天线　directional antenna　07.0566

定修　periodical repair, light repair　04.0243

定压风缸　constant pressure reservoir　05.0313

定压增压　constant pressure charging　04.0790

定员　seating capacity　05.0018

定置试验　stationary test, test at standstill　04.0288

定子　stator　04.0444

丢失事故　loss accident　06.0639

冬季施工　cold weather construction, winter season construction　02.0843

动板　movable plate　05.0562

动车　motor car　04.0010

动车组　motor train unit, motor train set　04.0009

动程　stroke　08.0618

＊动荷系数　coefficient of dynamic force　05.0883

动接点　contact heel, movable contact　08.0551

动力触探试验　dynamic penetration test　02.0057

动力回动机　power reverse gear　04.1210

动力内接　dynamic inscribing　03.0141

动力坡度　momentum grade　02.0187

动力室　power room　04.1060

动力试验车　dynamometer car　05.0745

动力稳定机　dynamic track stabilizer　03.0427

动力涡轮　power turbine　04.1539

动力系数　coefficient of dynamic force　05.0883

动力学试验　dynamics test　04.0289

动力制动　dynamic brake　04.0115

＊动轮轴箱　driving box　04.1223

动态不平顺　dynamic track irregularity　03.0144

动态长度　distance-to-go　08.0372

动态耦合　dynamic coupling　04.1620

动叶损失　moving blade loss　04.1554

动载试验　dynamic test　03.0534

动轴箱　driving box　04.1223

＊动轴箱槽铁　driving box shoe　04.1224

动轴箱平铁　driving box shoe　04.1224

动轴箱楔铁　driving box wedge　04.1225

＊动轴箱斜铁　driving box wedge　04.1225

动作杆　throw rod　08.0613

动作连接杆　operating rod for driving a switch 08.0626

冻害　frost heaving　03.0484

冻害垫板　frost heave board, track shim for frost heaving roadbed, frost shim　03.0098

冻结货物　frozen freight　06.0174

冻结接头　frozen joint　03.0073

冻结深度　freezing depth　02.0065

冻胀力　frost heaving force　02.0347

洞口　tunnel adit, tunnel opening　02.0662

洞口段衬砌　lining of tunnel portal section 02.0684

洞门　tunnel portal　02.0666

洞门框　tunnel portal frame　02.0667

洞室药包爆破　chamber explosive package blasting, chamher blasting　02.0274

洞穴地段路基　subgrade in cavity zone, subgrade in cavern zone　02.0228

抖动　jitter　07.0869

斗车　bucket loader　02.0968

斗式提升机　bucket elevator　02.0986

毒品车　poison car, poisonous goods wagon 05.0831

＊独轨铁路　monorail, monorail railway　01.0028

独立传动机车　individual drive locomotive 04.0362

独立驱动　individual drive　04.0358

独立同步方式　individual synchronized mode 07.0602

渡板　gangway foot plate　05.0453

渡轮　ferry boat　02.0642

渡线　crossover　03.0296

端部塞门　end cock　05.0283

端盖　end shield　04.0460

端横梁　end floor beam　02.0465

端接管　pipe nipple　05.0276

端梁　end sill　05.0395

端门　end door　05.0486

端墙包板　end sheathing　05.0445

端墙式洞门　end wall tunnel portal　02.0668

端柱　end post　05.0429

短波单边带无线电台　single side-band short wave station　07.0649

短波通信车　short wave radio communication vehicle 07.0650

短定子　short stator　04.1586

短轨　short rail　03.0057

短交路　short routing　04.0193

短路器　short-circuiting device　04.0566

短隧道　short tunnel　02.0659

短枕　short tie, block tie　03.0118

段管线　depot siding　06.0423

段修循环系数　circulating factor of repair in depot　05.0968

断高　broken height　02.0129

断轨保障　broken rail protection　08.0521

断链　broken chain　02.0127

＊断汽　cut-off　04.1092

＊堆焊钢轨　resurfacing of rail　03.0211

堆货场　storage yard　06.0090

堆码能力　stacking capability　06.0557

堆码作业　stacking operation　06.0594

堆装货物　stack-loading freight　06.0115

对称半控桥式整流器　symmetric half-controlled bridge rectifier　04.0334

对称电缆　symmetrical cable　07.0174

对称电缆通信　symmetrical cable communication　07.0317

对地电压　voltage to ground　04.1472

对活塞式柴油机　opposed-piston type diesel engine　04.0702

对角撑　diagonal brace　05.0401

对流运输　cross-haul traffic　06.0106

对向重叠进路　route with overlapped section in the opposite direction　08.0155

墩帽　pier coping　02.0573

墩身　pier body, pier shaft　02.0572

墩台防撞　collision prevention around pier, pier protection against collision　03.0573

墩台基础挖验　excavating foundation for checking purpose, foundation examination by excavation　03.0523

墩台基础钻探　drilling foundation for checking purpose, foundation examination by drilling　03.0524

墩周冲淤　scouring and depositing around pier　03.0570

蹲式便器　eastern type toilet, squat-across type water closet　05.0782

钝角辙叉　obtuse frog　03.0351

盾构　shield　02.0982

盾构法　shield method　02.0725

多磁铁系统　multimagnet system　04.1600

多次磨耗车轮　multiple wear wheel　05.0104

＊多灯信号机　multi-lenses signal　08.0054

多点连接　multipoint connection　07.0800

多电流制电力机车　multiple system electric locomotive　04.0305

多电平编码　multilevel encoding　07.0737

多段桥［联结］　bridges in cascade, multi rectifier bridge　04.0336

多滑面滑板　multiple bearing type guide　04.1195

多机牵引　multi-locomotive traction　04.0225

多机牵引无线电通信　radio communication for multiple-operated locomotive units　07.0587

多径传播　multipath propagation　07.0537

多局制　multi-office system　07.0384

多链路　multilink　07.0807

多路复用　multiplex　07.0009

多模渐变型光纤　graded index multimode optical fiber　07.0444

多年冻土路基　subgrade in permafrost soil zone　02.0219

多线桥　multi-track bridge　02.0422

多线隧道　multiple track tunnel　02.0653

多线铁路　multiple track railway　01.0017

多循环液力传动　multi-circuit hydraulic transmission　04.0992

多轴车　multi-axle car　05.0004

多轴转向架　multi-axle truck　05.0061

惰行泵　idle running pump　04.1028

惰行阻力　idle running resistance, coasting resistance　04.0097

E

鹅颈槽　goose neck tunnel　06.0564

*额定功率　nominal power　04.0083

额定值　rated value　08.0568

扼流变压器　impedance transformer　08.0548

扼流圈　choke　07.0051

耳机　earphone　07.0072

耳墙式洞门　ear wall tunnel portal　02.0671

耳墙式桥台　abutment with cantilevered retaining wall　02.0552

二冲程柴油机　two stroke diesel engine　04.0691

二次参数　secondary parameter　08.0534

二次衬砌　secondary lining　02.0688

二次空气　secondary air　04.1569

二次流损失　secondary flow loss　04.1552

二次群　secondary group　07.0377

二进制编码　binary encoding　07.0735

二三压力混合机构　composite two and three-pressure equalizing system　05.0334

二位侧　right side of car　05.0011

二位端　"A" end of car　05.0009

二显示自动闭塞　two-aspect automatic block　08.0337

二压力机构　two-pressure equalizing system　05.0332

二轴车　two-axle car　05.0002

二轴转向架　two-axle truck　05.0059

F

发车表示器　departure indicator　08.0090

发车表示器电路　departure indicator circuit　08.0249

发车进路　departure route　06.0249

发车进路信号机　route signal for departure　08.0071

发车线路表示器　departure track indicator　08.0092

发车信号　departure signal　08.0027

发电车　generator car　05.0741

发电轨道车　power generating rail car　03.0435

发电室　power plant compartment　05.0765

发电走行两用车　self-propelled power generating car　03.0436

发光二极管　light-emitting diode, LED　07.0471

发火次序　firing order　04.0772

发码器　code sender　07.0398

*发热量　heat value　04.0052

发送路　originating railway　06.0185

发送信道　transmit channel　07.0865

发信放大器　transmitting amplifier　07.0341

乏汽喷口　exhaust nozzle　04.1141

乏汽喷口座　exhaust nozzle seat　04.1142

阀动图　valve diagram, Zeuner valve diagram　04.1261

阀动椭圆图　valve ellipse　04.1262

*阀动装置　valve gear　04.1199

阀式轨道电路　valve type track circuit　08.0492

阀调整　valve setting　04.1260

阀型避雷器　valve type arrester　04.0636

阀行程　valve travel　04.1256

阀装置　valve gear　04.1199

翻板　platform trap door　05.0466

翻车机　tipper, tipping plant, dumper　06.0601

翻浆冒泥　mud-pumping　03.0155

反铲挖土机　backhoe　02.0880

反超高　reverse superelevation, counter superelevation, negative superelevation　03.0029

反定位　push-off mode　04.1397

反动度　degree of reaction　04.1543

反馈重发纠错　error correction by feed-back repetition　07.0842

反滤层　reverse filtration layer, inverted filter, protective filter　02.0308

反台阶法 negative benching tunnelling method 02.0732

反位接点 reverse contact 08.0555

反位锁闭 reverse locking 08.0138

反向传播 reverse propagation 07.0536

反向反射信号 backscattered signal 07.0498

反向器 reverser 04.0562

反向曲线 reverse curve, curve of opposite sense 02.0178

反向信道 backward channel 07.0803

反向行车 train running in reverse direction 06.0352

反压护道 berm with superloading, berm for back pressure, counter swelling berm 02.0265

反应板 reaction plate 04.1592

反应轨 reaction rail 04.1603

反装 left-handed machine 08.0617

返还系数 release factor 08.0571

方案比选 scheme comparison, route alternative 02.0154

方向电源 directional traffic power source 08.0644

方向滤波器 directional filter 07.0348

方向转接器 directional switch 08.0427

方枕器 tie respacer 03.0411

方正轨枕 tie respacing, squaring of ties 03.0215

房建大修 major repair of buildings and structures 03.0590

房建检修 inspection and repair of buildings and structures 03.0591

房建维修 regular maintenance of buildings and structures 03.0589

防爆设施 blasting protection facilities 02.1036

防尘 dust prevention, dust control 02.1025

防尘板 dust guard 05.0136

防尘板座 dust guard seat 05.0112

防串装置 anticreeping device 04.1427

防风栅栏 wind break fence 02.0323

防洪 flood control 03.0493

防洪预抢工程 precautionary work against flood 03.0576

防护 protection 04.1503

防护变压器 protective transformer 08.0547

防护道岔 protective turnout 08.0194

防护间距 protection distance 04.1522

防护栏杆 protection railing 02.1014

防护率 protection ratio 04.1510

防护滤波器 protection filter 04.1500

防护区段 protected section 08.0343

防护信号 protection signal 06.0656

防护音响信号 protecting acoustic signal 02.1000

防滑器 anti-skid device 05.0320

防滑鞋 antiskid shoe 02.1026

防火净距 fire protection distance 02.1021

防火门 fire protection gate 02.1018

防空转防滑行保护电路 anti-slip/slide protection circuit 04.0389

防空转防滑行保护装置 anti-slip/slide protection device 04.0587

防空转撒砂电空阀 anti-slip sanding valve 04.0586

防雷装置 lightning protection device 02.1032

防爬器 anti-creeper, rail anchor 03.0100

防爬支撑 anti-creep strut 03.0103

防砂 sand-drift control, sand protection 03.0494

防水板 waterproof board, waterproof sheet 02.0796

防水层 waterproof layer 02.0795

防水等级 classification of waterproof 02.0794

防水混凝土 waterproof concrete 02.0783

防水林 forestation against flood 03.0587

防水涂层 waterproofing coating 02.0797

防雪 snow-drift control, snow protection 03.0495

防雪树篱 snow protection hedge 03.0502

防雪栅 snow fence, snow guard 03.0500

防雪障 snow protection bank 03.0501

防止重复 prevention for repetitive clear of a signal 08.0134

防撞破凌 breaking up ice run, breaking up ice floe prevent collision 03.0588

防坠器 falling protector 02.1028

妨害故障 hindrance fault 08.0740

仿真线 artificial line 07.0346

访问 access 07.0749

放大器 amplifier 07.0052

放电器 discharger 04.0637

放风阀 vent valve 05.0316

放热规律　law of heat release　04.0759

放热率　rate of heat release　04.1101

放散温度力　destressing, stress liberation　03.0230

放水阀　blow off valve　04.1174

放线　setting-out of route, lay out of route　02.0119

非涅尔反射　Fresnel reflection　07.0497

非安全电路　non-vital circuit　08.0215

非对称半控桥式整流器　asymmetric half-controlled bridge rectifier　04.0335

非机械化驼峰　non-mechanized hump　06.0455

非机械化驼峰设备　unmechanized hump yard equipment　08.0367

非集中道岔　locally operated switch　08.0199

非集中联锁　non-centralized interlocking　08.0119

非接触式传感器　noncontacting proximity sensor　04.1601

非紧坡地段　section of unsufficient grade, section of insufficient grade　02.0159

非进路调车　to hold route for shunting　08.0192

非进路调车电路　circuit to hold a route for shunting　08.0244

非绝缘锚段关节　uninsulated overlap　04.1431

非绝缘转换柱　uninsulated transition mast　04.1410

非联锁道岔　non-interlocked switch　08.0197

非联锁区　non-interlocking area　08.0201

非粘着制动　non-adhesion braking　05.0324

非配属机车　un-allocated locomotive　04.0020

非平行运行图　non-parallel train diagram　06.0290

非渗水土路基　non-permeable soil subgrade, impervious embankment　02.0213

非踏面制动　off tread braking　05.0326

非吸上式注水器　non-attraction injector　04.1167

非线性电阻器　non-linear resistor　04.0523

非营业站　non-operating station　06.0081

非运用车　non-serviceable car, car not for traffic use　06.0272

非运用车系数　coefficient of cars not in service　05.0976

非责任事故　nonresponsible accident　06.0617

非支配机车　un-disposal locomotive　04.0022

非自复式按键开关　non-self-reset push-key switch　04.0578

非自复式按钮　stick button　08.0271

飞车　run away　04.0844

飞轮　flywheel　04.0893

废气净化　exhaust purification　04.0809

废气涡轮　exhaust turbine　04.0977

废气涡轮增压　exhaust turbocharging　04.0788

废气涡轮增压柴油机　turbocharged diesel engine　04.0697

废弃工程　abandoned project, abandoned construction work　02.0841

分贝　decibel　07.0044

分布[式]网[络]　distributed network　07.0763

分部工程　part project　02.0005

分部开挖法　partial excavation method　02.0729

*分层网[络]　hierarchical network　07.0757

分电话所　branch telephone office　07.0385

分动式汽阀　adjustable piston valve, Trofiemov piston valve　04.1203

分段减压　split reduction　04.0150

分段绝缘器　section insulator　04.1415

分割区段　cut section　08.0531

分隔　sever　07.0125

分号运行图　variant train diagram　06.0297

分机　extension　07.0134

分集接收　diversity reception　07.0577

分级保护　cascade protection　07.0236

分级调压　stepped voltage regulation　04.0320

分级网[络]　hierarchical network　07.0757

分检室　mail sorting room　05.0761

分接牵引变压器　tapped traction transformer　04.0486

分界点　train spacing point　06.0359

分开式扣件　separated rail fastening, indirect holding fastening　03.0086

分类折旧率　classified depreciation rate　06.0707

分流电抗器　divert shunt reactor, inductive shunt　04.0510

分流电阻器　divert shunt resistor, shunting resistor　04.0518

*分流感度　shunting sensitivity　08.0519

分路　shunt　08.0526

分路道岔　branching turnout　08.0373

分路灵敏度　shunting sensitivity　08.0519

分路效应 shunting effect 08.0511

分配阀 distributing valve 05.0291

分配阀试验台 distributing valve test rack 05.0370

分品复接 grading 07.0156

分坡平段 level stretch between opposite sign gradient 02.0196

*分歧道岔 branching turnout 08.0373

分区所 section post, SP 04.1349

分散供电方式 diversified power supply system 04.1297

分水岭 watershed, dividing ridge 02.0088

分通信枢纽 sectional communication center of railway branch administration 07.0242

分线盒 distribution box without protectors 07.0196

分线盘 distributing terminal board 08.0265

分线盘端子 terminals on distributing board 08.0276

分线箱 distribution box with protectors 07.0195

分相回流线 split return wire 04.1315

分相绝缘器 neutral section insulator 04.1416

分相装置 neutral section 04.1308

分项工程 item project 02.0006

分向电缆盒 cable branching terminal box 08.0287

分支接线 branch connection 04.1281

分轴燃气轮机 split-shaft gas turbine 04.1527

分转向角 auxiliary deflection angle 02.0123

*分组 packet, package 07.0742

*分组交换 packet switching 07.0850

*分组交换网 packet switching network 07.0766

*分组式终端 packet mode terminal 07.0896

*分组装拆器 packet assembler/disassembler, PAD 07.0895

粉末货物车 powdered goods car 05.0801

封闭式通风集装箱 closed ventilated container 06.0543

封闭线路作业时间 work occupation time, working time of closed section 03.0259

封存机车 locomotive stored up 04.0026

封锁 close up 08.0706

封锁区间 closing the section 06.0670

峰顶 hump crest 06.0448

峰顶调车员室 shunter's cabin at hump crest 06.0531

峰顶平台 platform of hump crest 06.0452

*峰高 hump height 06.0493

峰高计算点 calculate point of hump height 06.0449

峰下减速器 master retarder 08.0379

峰值电流 peak current 04.0427

峰值话音功率 peak speech power 07.0068

峰值检波器 peak detector 07.0570

峰值转矩 peak torque 04.0423

*风泵 air compressor 04.0161

风挡 vestibule diaphram 05.0575

风挡缓冲板 vestibule diaphram buffer plate 05.0577

风挡面板 vestibule diaphram face plate 05.0576

风道式通风 ventilation by air passage 02.0805

风动石碴[漏斗]车 pneumatic ballast hopper car 05.0830

风动摇炉装置 pneumatic grate shaking rigging 04.1250

风动凿岩机 pneumatic rock drill 02.0975

风镐 air pick, pneumatic pick 02.0961

风管路调压设备 air pipeline pressure governor 08.0404

风管式通风 ventilation by pipes 02.0811

风[荷]载 wind load 02.0343

风口 air port 05.0642

风喇叭 air horn 04.0650

风轮 fan 05.0613

风沙地段路基 subgrade in windy and sandy zone, subgrade in desert 02.0229

风压继电器 air pressure relay 04.0602

风压调整器 manometer regulator 08.0403

风雨密性 weatherproofness 06.0560

风嘴 wind fairing 02.0528

否认 negative acknowledge, NAK 07.0809

扶手 hand hold, grab iron 05.0418

辐板 plate, web 05.0121

辐板径向裂纹 radial crack in plate 05.0259

辐板孔 plate hole, web hole 05.0126

辐板圆周裂纹 circumferential crack in plate 05.0260

辐射干扰 radiated interference 07.0526

辐射模　radiation modes　07.0455

浮沉振动　bouncing [vibration]　01.0100

浮充供电　floating charge power supply　08.0633

＊浮吊　floating crane　02.0898

浮力　buoyancy　02.0334

浮桥　pontoon bridge, floating bridge, bateau bridge　02.0395

浮式沉井基础　floating caisson foundation　02.0584

浮式起重机　floating crane　02.0898

浮箱　floating box, pontoon　02.0954

浮运架桥法　bridge erection by floating　02.0530

辅机　assisting locomotive　04.0227

辅助编组站　auxiliary marshalling station　06.0380

辅助车场　auxiliary yard　06.0444

辅助承力索　auxiliary catenary　04.1362

辅助电动机　auxiliary motor　04.0471

辅助电路　auxiliary circuit　04.0375

辅助电路电压表　auxiliary circuit voltmeter　04.0623

辅助电路库用插座　auxiliary circuit socket for shed supply　04.0606

辅助电路库用转换开关　auxiliary circuit transfer switch for shed supply　04.0565

辅助发电机　auxiliary generator　04.0465

辅助机油泵　auxiliary lubricating oil pump　04.1039

辅助机组试验　test on auxiliary machines　04.0669

辅助接触器　auxiliary contactor　04.0554

辅助坑道　service gallery　02.0785

辅助梁　floor stringer　05.0399

辅助绕组　auxiliary winding　04.0500

辅助设施　auxiliary facilities　02.0862

辅助所　auxiliary block post　06.0361

辅助走行公里　auxiliary running kilometers　04.0214

腐坏事故　decay accident　06.0641

副按键开关组　secondary push-key switch group, secondary key switch set　04.0580

副风缸　auxiliary reservoir　05.0308

副交点　auxiliary intersection point　02.0121

副连杆　auxiliary connecting rod　04.0918

副前照灯　subhead lamp, dim head light　04.0642

副司机　assistant driver　04.0202

覆盖防护　covered protection　02.1001

复轨器　re-railer, rerailing device　06.0669

＊复合衬砌　composite lining　02.0686

复合式柴油机　compound-supercharged diesel engine　04.0700

复励电动机　compound excited motor　04.0406

复曲线　compound curve　02.0176

复式交分道岔　double slip switches　03.0295

复示信号　repeating signal　08.0031

复示信号机　repeating signal　08.0083

[复原]摇枕　swing bolster　04.1233

复原装置　centering device　04.1237

复胀式蒸汽机车　compound expansion steam locomotive　04.1085

腹板　web plate　02.0444

腹杆　web member　02.0448

负荷特性　load characteristic　04.0849

负载试验　loaded test, load test　04.0286

＊负载特性　load characteristic　04.0849

附加导线　additive wire　04.1366

附加风缸　supplementary reservoir　05.0309

附加力　subsidiary load, secondary load　02.0340

附加阻力　additional resistance　04.0099

附着式混凝土振捣器　attached type vibrator for concrete, attached type vibrator　02.0941

G

改道　gage correction, gaging of track　03.0202

改建铁路　reconstructed railway　01.0034

概率极限状态设计法　probabilistic limit state design method　02.0208

概算定额　rating of approximate estimate, rating form for estimate　02.0027

盖板　cover plate　05.0444

盖板涵　slab culvert　02.0630

盖板式棚洞　slab shed tunnel, slad shed gallery　02.0700

干捣水泥砂浆　dry tamped cement mortar　03.0556

干摩擦式轴箱定位装置装置　dry friction type journal box positioning device 05.0149

干喷混凝土 dry shotcreting 02.0779

干喷混凝土机 dry shotcreting machine 02.0971

干扰 disturbance, interference 07.0042

干扰场强 interference field strength 07.0522

干扰电压 interference voltage 07.0521

干扰影响 disturbing influence 04.1492

干扰源 interference source 04.1508

干散货集装箱 dry bulk container 06.0552

干涉场 interference field 07.0539

干燥管 dry pipe 04.1163

干燥器 dryer 05.0635

甘蔗车 sugar cane car 05.0823

杆件 member 05.0386

杆上电缆盒 cable terminal box on a post 08.0286

杆上工作台 pole balcony 07.0213

感应传输方式 inductive transmission mode 07.0548

*感应分路 divert shunt reactor, inductive shunt 04.0510

感应式机车信号 inductive cab signaling 08.0415

感应式列车无线电通信 inductive train radio communication 07.0581

感应线圈 induction coil 07.0075

干线 trunk line, main line 01.0037

干线长途通信 trunk communication 07.0247

干线长途通信网 trunk communication network 07.0245

干线调度电话 trunk dispatching telephone 07.0295

干线供电 main linely connected power supply 08.0637

干线会议电话 trunk conference telephone 07.0297

干线铁路 main line railway, trunk railway 01.0021

*刚构桥 rigid frame bridge 02.0390

刚架桥 rigid frame bridge 02.0390

刚架式棚洞 framed shed tunnel, framed shed gallery 02.0701

刚性齿轮驱动 solid gear drive 04.0350

刚性定位轮对 rigidly positioned wheelset 05.0881

*刚性系杆柔性拱桥 Langer bridge, flexible arch bridge with rigid tie 02.0404

钢板网式空气滤清器 steel sheet mesh type air filter 04.1047

钢板桩 steel sheet pile 02.0603

钢拱支撑 steel arch support 02.0767

钢管桩 steel pipe pile 02.0602

钢轨 rail 03.0047

钢轨擦伤 engine burn, wheel burn 03.0172

钢轨打标记 branding and stamping of rails 03.0182

钢轨低接头 depressed joint, battered joint of rail 03.0156

钢轨对地电位 rail potential to ground 04.1338

钢轨工作边 gage line 03.0062

钢轨基础模量 rail supporting modulus, track modulus 03.0039

钢轨接头 rail joint 03.0065

钢轨接续线 rail bond 08.0540

钢轨绝缘 rail insulation 08.0535

钢轨绝缘不良 bad rail insulation 08.0698

钢轨拉伸器 rail tensor 03.0396

钢轨连接器 rail joint bond 04.1339

钢轨裂纹 rail cracks 03.0188

钢轨落锤试验 drop test of rail 03.0192

钢轨磨损检查车 rail profile measuring car 03.0451

钢轨磨损检查仪 rail profile gauge 03.0452

钢轨刨边机 rail-head edges planing machine 03.0399

钢轨屏蔽系数 shielding factor of track 04.1479

钢轨伤损 rail defects and failures 03.0171

钢轨伸缩调节器 expansion rail joint, rail expansion device, switch expansion joint 03.0075

钢轨探伤车 rail flaw detection car 03.0449

钢轨探伤仪 rail flaw detector 03.0450

钢轨推凸机 rail weld seam shearing machine 03.0400

钢轨推凸器 rail shearing device 03.0401

钢轨位移观测桩 rail creep indication posts 03.0255

钢轨锈蚀 rail corrosion 03.0173

钢轨引接线　track lead　08.0538

钢轨折断　brittle fractures of rail, suddenr upture of rail　03.0191

钢轨支点弹性模量　modulus of elasticity of rail support　03.0040

钢轨阻抗　rail impedance　08.0518

钢轨组合辙叉　bolted rigid frog, assembled frog　03.0348

钢轨钻孔机　rail drilling machine　03.0389

钢轨钻孔器　rail drilling tool　03.0390

*钢花拱　trellis arch, lattice arch　02.0777

钢绞线　steel strand　02.0472

钢结构　steel structure　05.0379

钢筋　reinforcement, steel bar　02.0473

钢筋调直机　bar straightener　02.0936

钢筋混凝土电杆　reinforced concrete pole　07.0165

钢筋混凝土桥　reinforced concrete bridge　02.0370

钢筋混凝土桩　reinforced concrete pile　02.0606

钢筋冷拉机　bar cold-drawing machine　02.0939

钢筋切断机　bar cutter　02.0937

钢筋弯曲机　bar bender　02.0938

钢梁腐蚀裂纹　corrosion cracking of steel bridge　03.0558

钢梁加固　strengthening of steel bridge　03.0574

钢梁疲劳损伤　steel bridge fatigue damage　03.0539

钢梁应力腐蚀裂纹　stress corrosion cracking of steel bridge　03.0559

钢梁油漆　protective coating of steel bridge　03.0547

钢桥　steel bridge　02.0362

钢丝　steel wire　02.0470

钢丝束　bundled steel wires　02.0471

钢丝刷除锈　brush cleaning　03.0550

钢线　steel wire　07.0159

钢芯铝绞线　steel-cored aluminum stranded wire　07.0161

钢枕　steel tie　03.0115

钢桩　steel pile　02.0601

缸盖出口废气温度　exhaust temperature at cylinder head outlet　04.0733

港湾站　harbour station　06.0391

杠杆上拉杆　lever connecting rod　05.0188

高度调整阀　leveling valve　05.0181

高鹅头弯管　high level bend, swan neck　05.0369

高级包房卧车　superclass [corridor] compartment [type] sleeping car　05.0731

高架候车厅　overhead waiting hall　06.0021

高架桥　viaduct　02.0358

高架铁路　elevated railway　01.0027

高磷闸瓦　high phosphor cast iron brake shoe　05.0196

高锰钢整铸辙叉　solid manganese steel frog, cast manganese steel frog　03.0347

高摩合成闸瓦　high friction composite brake shoe　05.0197

高频分线盒　high frequency terminal box　07.0197

高频转接段　high frequency section　07.0281

高速柴油机　high speed diesel engine　04.0695

高速列车　high speed train　01.0082

高速铁路　high speed railway　01.0019

高压保安阀　high pressure safety valve　04.0171

高压侧调压　high voltage regulation　04.0318

高压电路　high voltage circuit, high tension circuit　04.0369

高压调压开关　high voltage tap changer, high tension tap changer　04.0557

高压罐车　high pressure tank car　05.0800

高压机油泵　high pressure lubricating oil pump　04.1040

高压绕组　high voltage winding, high tension winding　04.0496

高压油管　high pressure fuel pipe　04.0959

高音风喇叭　high tone air horn　04.0651

高柱信号机　high signal　08.0061

告警系统　alarm system　07.0331

告警信号　warning signal　06.0658

告警信号电路　alarm circuit　07.0082

格构拱　trellis arch, lattice arch　02.0777

*格栅拱　trellis arch, lattice arch　02.0777

蛤壳式抓斗　clamshell bucket　02.0881

隔断层　insulating course, insulating layer　02.0311

隔火带　fire protection strip　02.1020

隔离变压器　isolating transformer　04.0490

隔离层　isolation layer　02.0689

隔热层 thermal insulation layer 02.0693

隔热棚车 insulated box car 05.0811

个别概算 individual approximate estimate 02.0020

个人通信 personal communication 07.0677

各缸均匀性试验 cylinder power equalizing test 04.0857

*各站电话 interstation telephone 07.0411

给风阀 feed valve 04.0163

给气调整阀 pressure regulating valve 05.0517

跟踪接收方式 tracking-receiving mode 07.0604

更新改造计划 plan of renewal and upgrading 06.0704

工厂漆 shop coat 03.0555

工程报价 project quoted price 02.0847

工程承包 contracting of project 02.0849

工程船舶 engineering ship, engineering vessel 02.0952

工程地质 engineering geology 02.0060

工程地质条件 engineering geologic requirement, engineering geologic condition 02.0063

工程地质图 engineering geological map 02.0067

工程地质选线 engineering geologic location of line 02.0151

工程发包 contracting out of project 02.0848

工程监理 supervision of construction, supervision of project 02.0850

工程间接费 indirect expense of project, indirect cost of project 02.0032

工程施工无线电通信 radio communication for engineering construction 07.0639

工程宿营车 work train with camp cars 03.0445

工程投标 bidding for project 02.0846

工程预备费 reserve fund of project 02.0033

工程运输 engineering transportation 02.0842

工程招标 calling for tenders of project, calling for tending of project 02.0845

工程直接费 direct expense of project, direct cost of project 02.0031

工程质量检验 inspection of engineering quality 02.1042

工程质量评定 evaluation of engineering quality 02.1050

工程质量验收 acceptance of engineering quality 02.1051

工地漆 field coat 03.0554

工法制度 construction method system 02.0871

工具室 tool room 05.0764

工矿机车 industrial and mining locomotive 04.0013

工伤事故 accident on duty 06.0649

工务段 track division, track district, track maintenance division 03.0001

工务设备台帐 technical record of track, bridge and other equipments 03.0262

工务维修无线电通信 radio communication for track maintenance 07.0636

工形梁 I-beam 02.0385

工业干扰 industrial interference 07.0534

工业企业铁路 industry railway 01.0024

工业站 industrial station 06.0390

工质 working substance, medium 04.0721

工作波长 operating wavelength 07.0467

工作电流 working current 08.0623

工作过程 working process 04.0763

工作轮 blade wheel 04.1013

工作循环 working cycle 04.0722

工作值 working value 08.0567

功率储备系数 coefficient of reserve power 04.0822

功率调节系统 power regulating system 04.0963

功率因数补偿装置 power factor compensation device 04.0589

供电臂 feeding section 04.1292

供电臂短路电流 short-circuit current of feeding section 04.1327

供电臂干扰计算电流 disturbing calculation current of feeding section 04.1326

供电臂平均电流 average current of feeding section 04.1324

供电臂瞬时最大电流 instantaneous maximum current of feeding section 04.1328

供电臂有效电流 r.m.s current of feeding section, effective current of feeding section 04.1325

供电臂最大负荷电流 maximum load current of feeding section 04.1329

供电段 section for power supply 04.1454

*AT供电方式 autotransformer feeding system 04.1278

*BT供电方式 booster transformer feeding system 04.1277

供电领工区 fore work district for power supply 04.1455

供电线 feeder 04.1310

供汽量 evaporation capacity for engine 04.1096

供汽率 rate of evaporation for engine 04.1098

供油泵 oil feed pump 04.1018

供油提前角 advance angle of fuel supply 04.0780

公里标 kilometer post 03.0249

公铁两用检修车 road-railway repairing vehicle 04.1458

公铁两用桥 combined bridge, combined highway and railway bridge, combined rail-cum-road bridge 02.0356

公铁两用线路机械 rail/road permanent way machine 03.0380

公务车 officer's car, service car 05.0742

公用乘车证 service pass 06.0048

公用电话交换网 public switched telephone network 07.0109

公用电话网 public telephone network 07.0107

公用数据网 public data network 07.0753

弓角 pantograph horn 04.0539

弓头 pantograph bow 04.0537

弓网关系 pantograph-contact line relation 04.0185

拱背线 extrados 02.0489

拱顶 arch crown 02.0484

拱度 camber 02.0438

拱腹 soffit 02.0487

拱腹线 intrados 02.0488

拱涵 arch culvert 02.0629

拱肋 arch rib 02.0483

拱桥 arch bridge 02.0396

拱圈封顶 closing the top of lining 02.0759

拱矢 rise of arch 02.0485

拱形明洞 arch open cut tunnel, arch tunnel without cover, arch gallery 02.0695

拱砖 arch brick 04.1128

拱砖管 arch tube 04.1129

共电电话机 common battery telephone set 07.0092

共电电话交换机 common battery telephone switch board 07.0392

共线信令系统 common channel signaling system 07.0836

共线自动电话 party-line automatic telephone 07.0420

共用箱 cab signal box 08.0424

钩车 cars per cut 06.0500

钩耳 coupler pivot lug 05.0540

钩肩 coupler horn 05.0537

钩颈 coupler neck 05.0536

钩螺栓 claw bolt, hook bolt, anchor bolt 03.0084

钩舌 coupler knuckle 05.0543

钩舌销 knuckle pivot pin 05.0546

钩身 coupler shank 05.0534

钩锁 coupler lock 05.0544

钩锁销 coupler lock lift 05.0545

钩体 coupler body 05.0533

钩头鞍子 hook-type saddle 04.1388

*钩头钉 track spike, rail spike, dog spike 03.0090

钩头正面 coupler front face 05.0539

钩腕外臂 coupler guard arm 05.0538

钩尾 coupler tail 05.0535

钩尾销 draft key 05.0547

*勾贝 piston 04.0899

*勾贝杆 piston rod 04.1190

*勾贝行程 piston travel 05.0342

*狗头钉 track spike, rail spike, dog spike 03.0090

构架侧梁 truck side sill 05.0075

构架端梁 truck end sill 05.0076

构架辅助梁 truck auxiliary transom 05.0079

构架横梁 truck transom 05.0077

构架纵梁 truck longitudinal sill 05.0078

构件 structural member 05.0385

构件支撑 member support 02.0766

构造钢筋 constructional reinforcement 02.0478

构造轨缝 structural joint gap, maximum joint gap structurally obtainable 03.0064

构造速度 construction speed, design speed 01.0078

箍筋 stirrup 02.0474

估算指标 index of estimate 02.0029

鼓筒式混凝土搅拌机 drum concrete mixer 02.0926

鼓形位置转换开关 drum position changeover switch 04.0559

股道空闲 track clear 08.0509

股道占用 track occupied 08.0510

故障－安全 fail-safe 08.0666

故障办理 emergency treatment after failure 08.0702

*故障电流 frictional working current 08.0612

故障复原 restoration after a failure 08.0703

故障积累 failure accumulation 08.0704

故障升级 progression of failure 08.0705

故障修 corrective maintenance 08.0727

*故障诊断 inspection of component failure, failure diagnostic 04.0277

固定备用方式 fixed reservation system 04.1299

固定电台 base station 07.0553

固定杠杆 truck dead lever 05.0185

固定区 nonbreathing zone, fixed zone, deformation-free zone 03.0236

固定式汽阀 rigid piston valve 04.1202

固定信号 fixed signal 08.0010

固定支座 fixed bearing 02.0515

固定轴距 rigid wheelbase 05.0045

固定资产残值 scrap value of fixed assets 06.0745

固定资产大修 capital repair of fixed assets 06.0701

固定资产更新改造 renewal and reconstruction of fixed assets, renewal and upgrading of fixed assets 06.0703

固定资产更新率 rate of fixed assets renewal 06.0747

固定资产投资 fixed asset investment 06.0710

固定资产退废率 rate of fixed assets retirement 06.0746

固定资产原价 original value of fixed assets 06.0744

固定资金 fixed capital 06.0743

固定座椅 fixed seat 05.0771

固端拱 fixed-end arch 02.0397

固沙造林 stabilization for sands by afforestation 02.0321

固有畸变 inherrent distortion 07.0844

刮雨器 windshield wiper, windscreen wiper 04.1067

挂断 hanging up 07.0117

挂机状态 on-hook 07.0119

挂瓦室 journal bearing babbit metal lining room 05.0896

关闭信号 closing signal 06.0253

关键工程 key project 02.0861

关节货车 articulated freight car 05.0838

关节客车 articulated passenger car, articulated coach 05.0723

管道电缆 duct cable 07.0183

管涵 pipe culvert 02.0627

管界标 section sign 03.0254

管内工作车 local cars to be unloaded 06.0320

管内客流 local passenger flow 06.0025

管内旅客列车 local passenger train 06.0208

管内装卸率 local loading and unloading rate 06.0325

管棚支护 pipe-shed support, pipe roofing support 02.0776

管形燃烧室 can-type combustor 04.1563

管柱挡墙 cylindrical shaft retaining wall 02.0290

管柱基础 tubular column foundation 02.0592

罐车 tank car 05.0799

罐车容积计表 tank volume table 05.0041

*罐带 tank band 05.0481

罐端板 tank head 05.0470

罐式集装箱 tank container 06.0551

罐体 tank 05.0468

罐体鞍座 tank saddle 05.0469

罐体长度 length of tank 05.0022

罐装货物 tank car freight 06.0118

灌浆压力 grouting pressure 02.0712

灌木切割机 brush cutting machine, brush cutter 03.0462

灌筑水下混凝土 underwater concreting, concreting with tremie method 02.0616

*贯通测量 intunnel control survey, through survey 02.0143

光传输模式 optical transmission mode 07.0468

光带 light strip 08.0274

光带式表示 stript indication light 08.0174

光点式表示 spotted indication light 08.0173

光电导线 photoelectric traverse 02.0103

光功率计 optical power meter 07.0493

光接口 optical interface 07.0478

光缆 optical fiber cable 07.0480

光缆接头 optical fiber cable joint closure 07.0491

光连接器 optical connector 07.0473

光面爆破 smooth blasting 02.0748

光时域反射仪 optical time domain reflectometer 07.0495

光适配器 optical adapter 07.0474

光数字段 optical digital section 07.0503

光衰减器 optical attenuator 07.0475

光探测器 optical detector 07.0472

光纤 optical fiber 07.0442

光纤剥除器 optical fiber stripper 07.0490

光纤带宽 bandwidth of an optical fiber 07.0462

光纤分配架 optical fiber distribution frame 07.0479

光纤接续损耗 optical fiber splice loss 07.0499

光纤切断器 optical fiber cutter 07.0489

光纤融接机 optical fiber fusion splicing machine 07.0488

光纤数字线路系统 optical fiber digital line system 07.0501

光纤通信系统 optical fiber communication system 07.0500

光线路保护切换设备 optical line protection switching equipment 07.0504

光线路终端设备 optical line terminal equipment 07.0463

光源 optical source 07.0469

光再生[中继]段 optical regenerator section 07.0502

光[再生]中继器 optical [regenerative] repeater 07.0477

广播呼叫 broadcast call 07.0816

广厅 public hall, concourse 06.0017

广域网 wide area network, WAN 07.0755

规程 procedure 07.0846

硅油减振器 silicon oil damper 04.0895

硅油弹簧减振器 silicon oil spring damper 04.0896

硅整流器电力机车 silicon rectifier electric locomotive 04.0306

硅整流装置 silicon rectifier device 04.0337

轨撑 rail brace 03.0104

轨道 track 03.0010

轨道暗坑 loose tie 03.0154

轨道变形 track deformation, track disorder, track distortion 03.0146

轨道变压器箱 track transformer box 08.0549

轨道变阻器 track rheostat 08.0546

轨道不平顺 track irregularity 03.0142

轨道残余变形 track residual deformation, track permanent deformation 03.0147

轨道电抗器 track reactor 08.0545

轨道电路 track circuit 08.0489

轨道电路电码化 coding of continuous track circuit 08.0508

轨道电路调整状态 regulated state of a track circuit 08.0513

轨道电路分割 cut-section of a track circuit 08.0530

轨道电路分路状态 shunted state of a track circuit 08.0512

轨道电路蓄电现象 track storage effect 08.0524

轨道动力学 track dynamics 03.0033

轨道方向 track alignment 03.0150

轨道鼓出 track buckling 03.0160

轨道鼓出临界温度 critical temperature of track buckling 03.0233

轨道衡线 weight bridge track 06.0087

轨道机械 track machine 03.0371

轨道几何尺寸容许公差 track geometry tolerances 03.0145

轨道几何形位 track geometry 03.0022

轨道几何状态恶化 track deterioration 03.0035

轨道检测设备 track geometry measuring device 03.0377

轨道检查车 track recording car, track inspection car

03.0446

轨道检查小车　track geometry measuring trolley 03.0447

轨道接触器　track treadle　08.0313

轨道结构　track structure　03.0012

轨道框架刚度　rigidity of track panel　03.0037

轨道类型　classification of track, track standard 03.0011

轨道力学　track mechanics　03.0032

轨道明坑　visible pit of track, visible low spot of track, track depression　03.0153

轨道平车　rail flat car　03.0437

轨道起重机　track crane, rail crane　02.0897

轨道前后高低　longitudinal level of rail, track profile 03.0151

轨道强度计算　track strength analysis　03.0034

轨道生电现象　track galvanic effect　08.0525

轨道失效　track failure　03.0036

轨道受电变压器　track relay transformer　08.0544

轨道水平　track cross level　03.0149

轨道水平尺　track level　03.0470

轨道送电变压器　track transformer feed end 08.0543

轨道稳定性　stability of track　03.0042

轨道养护标准　standards of track maintenance 03.0267

轨道应力　track stresses　03.0041

轨道质量指数　track quality index　03.0270

轨底　rail base, rail bottom　03.0050

轨底崩裂　burst of rail base, burst of rail bottom, broken rail base　03.0190

轨底坡　rail cant　03.0061

轨端崩裂　rail end breakage　03.0181

＊轨端倒棱　rail end chamfering　03.0183

轨端马鞍形磨损　rail end batter, saddle wear of rail end　03.0168

轨端削角　rail end chamfering　03.0183

轨缝　rail gap, joint gap　03.0063

轨缝调整器　rail gap adjuster, rail puller　03.0395

轨腹上圆弧接触夹板　head free flat joint bar 03.0080

轨检车评分　track evaluation by recording car, evaluation by track inspection car　03.0269

轨节　rail link　03.0020

轨距　rail gage, rail gauge　01.0048

轨距尺　track gage　03.0465

轨距杆　gage tie bar, gage rod, gage tie　03.0105

轨距加宽　gauge widening　03.0023

轨排　track panel, track skeleton　03.0019

轨头　rail head　03.0048

轨头剥离　gage line shelly cracks　03.0176

＊轨头波浪形磨耗　wave-type deformation of rail head　03.0164

轨头波纹磨损　corrugation of rail head, rail corrugation　03.0165

轨头波形磨损　wave-type deformation of rail head 03.0164

＊轨头侧面磨耗　side wear of rail head　03.0163

轨头侧面磨损　side wear of rail head　03.0163

轨头长波浪磨损　long wave undulation of rail head 03.0167

＊轨头垂直磨耗　vertical wear of rail head 03.0162

轨头垂直磨损　vertical wear of rail head　03.0162

轨头垂直劈裂　vertical split of rail head　03.0179

轨头底面接触夹板　head contact flat joint bar 03.0079

轨头掉块　spalling of rail head　03.0177

轨头短波浪磨损　short wave undulation of rail head 03.0166

轨头发裂　head checks, hair crack of rail head 03.0189

轨头非对称断面打磨　asymmetrical rail head profile grinding　03.0214

轨头肥边　flow of rail head, lipping of rail head 03.0169

轨头水平劈裂　horizontal split of rail head 03.0186

轨头微细裂纹　detail fracture of rail head 03.0180

轨头压溃　crushing of rail head　03.0178

轨头整形　rail head reprofiling　03.0213

＊轨头总磨耗　total wear of rail head　03.0161

轨头总磨损　total wear of rail head　03.0161

轨下基础　sub-rail foundation, sub-rail track bed 03.0107

轨行式起重机　crane on-track, rail-mounted crane　03.0440

轨行式装运机械　rail-mounted handling and transportation machine　03.0376

轨腰　rail web　03.0049

轨腰劈裂　piped rail, split of rail web　03.0187

轨枕　tie, cross tie, sleeper　03.0108

* 轨枕板　broad concrete tie　03.0109

轨枕抽换机　tie replacing machine　03.0406

滚动试验台　rolling rig　05.0990

滚动圆　tread rolling circle　05.0219

滚动轴承故障自动检测　automatic roller bearing defect detection　05.0961

滚动轴承间　roller bearing shop　05.0894

滚轮摇臂　roller rocker　04.0935

滚圈　rolling ring　05.0367

滚轴支座　roller bearing　02.0519

滚柱轴承试验台　roller bearing test stand　05.0994

锅胴　boiler barrel course　04.1134

锅炉安全装置　boiler safety device　04.1176

锅炉过热面积　boiler super heating surface　04.1252

锅炉借水　running at dropping water level　04.1263

锅炉净效率　net boiler efficiency　04.1112

锅炉牵引力　boiler tractive effort　04.1088

锅炉热平衡　boiler heat balance　04.1106

锅炉散热面积　boiler heat dissipating surface　04.1253

锅炉水位　boiler water level　04.1255

锅炉蒸发面积　boiler evaporative heating surface　04.1251

锅炉总效率　total boiler efficiency　04.1111

锅腰托板　waist sheet　04.1220

国际标准集装箱　ISO freight container　06.0540

国际货物联运　international through freight traffic　06.0183

国际联运车辆过轨　transferring of car from one railway to another for international through traffic　06.0194

国际联运车辆交接单　acceptance and delivery list of car for international through traffic　06.0192

国际联运货物换装　transhipment of international through goods　06.0193

国际联运货物交接单　acceptance and delivery list of freight for international through traffic　06.0191

国际联运货物票据　international through freight shipping documents　06.0190

国际联运旅客车票　passenger ticket for international through traffic　06.0044

国际联运旅客特别快车　international express train　06.0206

国际联运协定　agreement of international through traffic　01.0123

国际联运议定书　protocol of international through traffic　01.0124

国际联运站　international through traffic station　06.0393

国际铁路货物联运协定　agreement of international railway through freight traffic　06.0189

国际铁路联运　international railway through traffic　01.0121

国际铁路联运公约　convention of international railway through traffic　01.0125

国际铁路协定　agreement of frontier railway　06.0188

国家标准集装箱　GB freight container　06.0539

国境站　frontier station　06.0392

国有铁路　national railway, state railway　01.0007

过超高　surplus superelevation, excess elevation　03.0028

过充风　overcharging　04.0142

* 过充气　overcharging　04.0142

过电流继电器　over-current relay　04.0598

过渡电抗器　transition reactor　04.0507

过渡电阻器　transition resistor　04.0517

过渡工程　transition project　02.0859

过境路　transit railway　06.0187

过量充风　overcharge　04.0144

* 过量充气　overcharge　04.0144

过量减压　over reduction　04.0148

过量空气系数　excess air factor　04.0743

过滤层　filter　03.0475

过滤器　filter　05.0636

过热管　superheater tube　04.1154

过热箱　superheater header　04.1155
过热蒸汽室　superheater chamber　04.1157

过远运输　excessively long-distance traffic　06.0107

H

海底电缆　submarine cable　07.0180
涵洞孔径　aperture of culvert　02.0626
旱桥　dry bridge　02.0359
焊轨机　rail welding machine　03.0405
焊接机座　**welding frame**　04.0453
焊接接头　welded joint　03.0072
焊接式钢轨接续线　welded bond　08.0542
焊修钢轨　resurfacing of rail　03.0211
夯实道床　ballast ramming, ballast consolidating　03.0224
夯实机械　ballast consolidating machine　03.0374
航测外控点　field control point of aerophotogrammetry　02.0136
航测选线　aerial surveying alignment　02.0135
航带设计　flight strip design, design of flight strip　02.0132
巷道式通风　ventilation by ducts　02.0812
4毫米锁闭　check 4mm opening of a switch point　08.0133
荷载组合　loading combination　02.0354
核伤　nucleus flaw, oval flaw　03.0175
核心车次　scheduled train number　06.0287
核子密度湿度测定　determination of nuclear density-moisture　02.0262
合并杆　combination lever　04.1207
合格工程　qualified project　02.1055
合金淬火轨　head hardened alloy steel rail, quenched alloy steel rail　03.0053
合金轨　alloy steel rail　03.0052
合理运输　rational traffic　06.0105
合造客车　composite passenger car, composite coach　05.0722
合资铁路　joint investment railway, jointly owned railway　01.0010
河床冲刷　channel erosion　03.0585
河床铺砌　river bed paving　02.0620
河道调查　river course survey　02.0070
河调－防护失修　disrepair of flow regulating and shoreprotectingstructure, river bank protection out of repair　03.0515
河流比降　slope of river, comparable horizon of river　02.0077·
河滩路堤　embankment on plain river beach　02.0221
横承力索　headspan wire　04.1363
横承力索线夹　headspan wire clamp　04.1372
横带　rail　05.0437
横担　cross arm　07.0163
横电磁波小室　cross electromagnetic wave small room　04.1514
横洞　transverse gallery　02.0788
横断面测量　cross leveling, cross-section survey, cross-section leveling　02.0105
横断面选线　cross-section method of railway location, location with cross-section method, cross-section method for location of line　02.0162
横联　sway bracing　02.0459
横梁　floor beam, transverse beam　02.0463
横列式区段站　transversal type district station　06.0367
横通道　transverse passage-way　02.0790
横向冲击　lateral impact　01.0108
横向传播特性　cross propagation characteristic　04.1518
横向磁通直线感应电动机　transverse flux linear induction motor　04.1629
横向拉杆　lateral connecting rod　05.0213
横向力　lateral force　05.0886
横向弹性　lateral elasticity, lateral resilience　04.1444
横向稳定性　lateral stability　05.0850
横向振动　lateral vibration　01.0097
衡重式挡土墙　balance weight retaining wall, gravity retaining wall with relieving platform, balanced type retaining wall　02.0286
恒功调速比　speed ratio on constant power

04.0438

恒速　constant speed　04.0108

*恒温控制器　room thermostat　05.0645

恒温膨胀阀　thermostatic expansion valve　05.0634

恒载　dead load　02.0331

桁架　truss　02.0388

桁架式钢结构　truss type steel structure　05.0382

洪水标记　flood mark　03.0584

洪水调查　flood survey　02.0069

洪水频率　flood frequency　02.0090

洪水淹没　flood inundation on tracks　03.0586

红外线轴温检测所　infrared journal temperature detection point　05.0958

红外线轴温探测系统　infrared journal temperature detection system　05.0959

喉板　throat sheet　04.1122

候车室　waiting room, waiting hall　06.0020

后板　back sheet　04.1121

后车架片　rear frame　04.1216

后从板座　rear draft lug, rear draft stop　05.0412

后挡板　back shield　05.0137

后缓冲铁　rear bumper　04.1219

后接点　back contact　08.0553

后膨胀板　rear expansion sheet　04.1222

后圈　back coil　08.0577

后燃期　after burning period　04.0784

后退信号　backing signal　08.0039

后张法预应力梁　post-tentioned prestressed concrete girder　02.0373

呼叫检查　calling check　07.0597

呼叫信号　calling signal　07.0275

呼叫装置　call button　05.0694

呼救信号　calling help signal　06.0659

呼损率　percent of call loss　07.0153

*呼吸区　breathing zone　03.0234

护岸　revetment, shore protection　02.0316

护背距离　guard rail face gage, back gage　03.0339

护道　berm　02.0250

护轨　guard rail, check rail　03.0060

护轨与心轨的查照间隔　[guard rail] check gage　03.0338

护坡　slope protection, revetment, pitching　02.0294

护墙　guard wall　02.0293

护套　sheath　07.0483

护筒　pile casing　02.0613

互不控制复原方式　called and calling subscriber release　07.0409

互感系数　mutual inductance coefficient　04.1473

互感阻抗　mutual inductive impedance　04.1474

互换　interchange　07.0744

互调干扰　intermodulation interference　07.0527

华氏阀装置　Walschaerts valve gear　04.1200

滑板　guide, slide bar, slip sheet　04.1194

滑床板　slide plate, switch plate　03.0367

滑动模板　slip form　02.0958

滑动式盘车机构　sliding type barring mechanism　04.0971

滑动轴承　plain bearing　05.0128

滑阀　slide valve　05.0297

*滑门　sliding door　05.0492

滑坡　landslip, landslide　03.0477

滑坡地段路基　subgrade in slide　02.0226

滑台　expansion shoe　04.1221

话路　telephone channel　07.0008

话频　voice frequency　07.0013

话务量　telephone traffic　07.0152

话音侧音　speech side tone　07.0145

环境温度　ambient temperature　04.0846

环境噪声侧音　ambient noise side tone　07.0146

环路电阻　loop resistance　07.0223

环线　loop　06.0421

环形交路　circular routing　04.0192

环形枢纽　loop-type junction terminal　06.0401

环形网[络]　ring network　07.0761

环状供电　looply connected power supply　08.0638

缓冲层　buffer layer　07.0484

缓冲杆　buffer rod　05.0578

缓冲梁　buffer beam　05.0400

缓冲器反弹　draft gear recoil　05.0584

缓冲器能量吸收率　rate of energy absorbed by draft gear　05.0583

缓冲器容量　draft gear capacity　05.0579

缓冲器箱体　draft gear housing　05.0559

缓冲器行程　draft gear travel　05.0580

缓冲器预紧力　draft gear initial compression

05.0581

缓冲器阻抗力 draft gear reaction force at rating travel 05.0582

缓冲区 buffer zone, transition zone 03.0235

缓动继电器 slow-acting relay 08.0596

缓放继电器 slow release relay 08.0598

缓放时间 slow release time 08.0565

缓和坡度 slight grade, flat grade, easy grade 02.0197

缓和曲线 transition curve, easement curve, spiral transition curve 02.0174

缓和曲线半径变更率 rate of easement curvature, rate of transition curve 02.0175

缓解 release 04.0135

缓解波速 release propagation rate 04.0158

缓解部 release portion 05.0294

缓解阀 release valve 05.0315

缓解停车 stopping at release 04.0157

缓坡地段 section of easy grade, section of gentle slope 02.0157

缓吸继电器 slow pick-up relay 08.0597

缓吸时间 slow pick-up time 08.0564

*换边 change side of double line 02.0204

换侧 change side of double line 02.0204

换挡 changeover governor 04.1019

换件大修 component exchange repair 04.0253

换气过程 scavenging period 04.0728

换算吨公里成本 cost of converted ton-kilometer 06.0726

换算均布活载 equivalent uniform live load 02.0328

换算线路长度 equivalent track kilometerage 03.0268

换算周转量 converted turnover 06.0687

换算走行公里 converted running kilometers 04.0212

换算阻力 converted resistance 04.0104

换土 change soil, soil replacement 02.0267

换相电抗器 commutation reactor 04.0514

换相电容器 commutating capacitor 04.0525

换相联接 exchange phase connection, phase alternating connection 04.1283

换向极铁心 interpole core 04.0448

换向极线圈 interpole coil 04.0449

换向片 commutator segment 04.0440

换向器 commutator 04.0439

换向限止阀 standstill detector valve 04.1025

换向轴 reversing shaft 04.1017

换行 line feed 07.0715

换装站 transhipment station 06.0389

幻通谐振变压器 phantom resonant transformer 04.1496

簧上重量 suspended weight, sprung weight 05.0865

簧下重量 non-suspended weight, unsprung weight 05.0866

灰土换填夯实法 method of lime-soil replacement and tamping 02.0615

灰箱 ashpan 04.1133

恢复规程 recovery procedure 07.0855

恢复通车 restoring traffic 06.0671

回波效应 echo effect 07.0834

回车 carriage return 07.0714

回动拉杆 reverse pull rod 04.1212

回动手把 reverse lever 04.1211

回风道 air return duct 05.0640

回铃音 ring back tone 07.0278

回流区 recirculating zone 04.1571

回流线 return wire 04.1314

回热度 regenerator effectiveness 04.1574

回热器 regenerator 04.1533

回声 echo 07.0143

回转式盘车机构 rotary type barring mechanism 04.0972

会车间隔时间 time interval for two meeting trains at station 06.0300

会让站 passing station 06.0363

会议电话分机 conference telephone subset 07.0303

会议电话汇接机 conference telephone tandem board 07.0302

会议电话总机 conference telephone central board 07.0301

汇接电话所 tandem telephone office 07.0387

汇流条 bus-bar 08.0275

汇水面积 catchment area, water collecting area,

drainage area 02.0089

汇水区流域特征调查 survey of catchment basin characteristics 02.0072

混合电源 AC-battery power source 08.0642

混合供电制 AC-battery power supply system 08.0632

混合列车 mixed train 06.0210

混合式给水预热装置 mixed feed water heater 04.1169

混合式货场 mixed-type freight yard 06.0085

*混合室 combining chamber 04.1170

混合线圈 hybrid coil 07.0347

混合形枢纽 combined type junction terminal 06.0402

混合液力机械传动 hydromechanical drive with direct step 04.0996

混煤 mixed coal 04.1102

混凝土泵车 concrete pump truck 02.0934

混凝土吊斗 concrete lifting bucket 02.0935

混凝土搅拌机 concrete mixer 02.0925

混凝土搅拌楼 concrete mixing plant 02.0930

混凝土搅拌运输车 concrete mixing and transporting car, truck mixer, transit mixer 02.0932

混凝土裂纹 concrete cracks 03.0557

混凝土桥 concrete bridge 02.0369

混凝土输送泵 concrete pump 02.0933

混凝土枕 concrete tie 03.0116

混凝土枕螺栓钻取机 concrete tie dowel drilling and pulling machine 03.0410

混凝土振动台 concrete vibrating stand 02.0943

混频器 mixer 07.0054

活顶盖 removable hatch cover 05.0463

活顶棚车 sliding roof box car, sliding roof goods van 05.0796

活动吊篮 travelling cradle 02.0542

活动模架逐跨施工法 segmental span-by-span construction using form traveller 02.0539

活动桥 movable bridge 02.0408

活动支座 expansion bearing, movable bearing 02.0516

活动中梁底架 underframe with sliding center sill, cushioning underframe 05.0390

活动座椅 self-folding seat 05.0775

活墙棚车 sliding side box car 05.0797

活塞 piston 04.0899

活塞衬套 piston bush 04.0905

活塞杆 piston rod 04.1190

活塞环 piston ring 04.0906

活塞冷却喷嘴 piston cooling nozzle 04.0871

活塞面积 piston area 04.0819

活塞平均速度 mean piston speed 04.0833

活塞裙 piston skirt 04.0904

活塞体 piston body 04.0902

活塞头 piston head 04.0903

活塞销 piston pin 04.0909

活塞行程 piston stroke 04.0708

活鱼车 live fish car 05.0822

火花间隙 spark gap 04.1340

火警信号 fire alarm signal 06.0660

火炬信号 torch 06.0662

火箱 firebox 04.1116

火箱管板 firebox tube sheet 04.1132

火星网 spark arrester netting 04.1150

火焰除锈 flame cleaning 03.0549

火焰管 flame tube 04.1556

火灾事故 fire accident 06.0637

霍尔片 Hall plate 04.1622

货差率 mistake rate of goods 06.0647

货场 freight yard, goods yard 06.0082

货场监视电视 monitor TV for freight yard 07.0656

货车 freight car, wagon 05.0791

货车保有量 number of freight cars on hand 05.0980

货车标记载重量 marked loading capacity of car 06.0152

货车动载重 dynamic load of car 06.0154

货车技术交接所 freight car technical condition handing-over post 05.0917

货车检修率 ratio of freight cars under repair 06.0694

[货车]脚蹬 sill step 05.0406

货车静载重 static load of car 06.0153

货车溜放风阻力 rolling car resistance due to wind effects 06.0475

货车溜放基本阻力 basic rolling car resistance

06.0474

[货车]门锁　door latch　05.0503

[货车]内墙板　lining　05.0452

货车日产量　serviceable work-done per car day
06.0156

货车日车公里　car kilometers per car per day
06.0328

[货车]上侧梁　top chord　05.0439

[货车]上开门　upward swing door　05.0493

货车施封　car seal　06.0139

货车洗刷所　freight car washing point　05.0905

[货车]下开门　downward swing door　05.0494

[货车]摇枕挡　column guide　05.0092

货车载重量利用率　coefficient of utilization for car
loading capacity　06.0155

货车站修所　freight car repairing point　05.0906

货车中转距离　average car-kilometers per transit
operation　06.0324

货车周转距离　average car-kilometers in one turnround　06.0323

货车周转时间　car turnround time　06.0326

货车装载清单　car loading list　06.0128

货垛　stack of freight　06.0595

货流　freight flow　06.0143

货流量　freight flow volume　06.0144

货流图　freight flow diagram　06.0145

货棚　freight shed, goods shed　06.0092

货票　way bill, freight invoice　06.0127

货区　freight area, goods area　06.0088

货损率　damage rate of goods　06.0646

货位　freight section, goods section　06.0094

货物标记　freight label　06.0136

货物承运　acceptance of freight　06.0130

货物到达吨数　tonnage of freight arrived　06.0147

货物到达作业　freight operation at destination station
06.0134

货物发送吨数　tonnage of freight despatched
06.0146

货物发送作业　freight operation at originated station
06.0133

货物换装整理　transhipment and rearrangement of
goods　06.0140

货物积累损伤指数　rate of accumulated freight dam-

age　05.0852

[货物]计费重量　charged weight　06.0738

货物交付　dilivery of freight　06.0131

货物交接所　freight transfer point　06.0195

货物列车　freight train, goods train　06.0214

货物列车编组计划　freight train formation plan
06.0278

货物列车检修所　freight train inspection and service
point　05.0907

货物列车区段检修所　transit freight train inspection
and service point　05.0909

货物列车一般检修所　freight train ordinary inspection and service point　05.0910

货物列车主要检修所　freight train main inspection
and service point　05.0908

货物品类　goods category　06.0100

货物平均运程　average haul of freight traffic
06.0684

货物途中作业　freight operation en route　06.0135

货物托运　consigning of freight　06.0129

货物运单　consignment note　06.0126

货物运到期限　freight transit period　06.0125

货物运价　freight rate　06.0734

货物运价号　freight tariff No.　06.0739

货物运价里程　tariff kilometerage　06.0737

货物运价率　freight rate　06.0740

货物运输变更　traffic diversion　06.0141

货物运输计划　freight traffic plan　06.0681

货物运输量　freight traffic volume　06.0682

货物运输系数　coefficient of freight traffic
06.0685

货物运送吨数　tonnage of freight transported
06.0148

货物站台　freight platform, goods platform
06.0091

货物重心的横向位移　lateral shift for center of gravity of goods　06.0167

货物重心的纵向位移　longitudinal shift for center of
gravity of goods　06.0168

货物周转量　turnover of freight traffic　06.0683

货物转向架　freight turning rack　06.0162

货物转向架支距　distance between centers of freight
turning rack　06.0163

货源　freight traffic source　06.0142

货运波动系数　fluctuating coefficient of freight traffic　06.0306

货运调度电话　freight dispatching telephone　07.0413

货运机车　freight locomotive, goods locomotive　04.0015

货运经济调查　economic investigation of freight traffic　06.0686

货运密度　density of freight traffic　06.0151

货运事故　freight traffic accident　06.0619

货运杂费　miscellaneous fees of goods traffic　06.0741

货运站　freight station　06.0386

货运站综合作业自动化　automation of synthetic operations at freight station　01.0110

货运专线　railway line for freight traffic, freight special line, freight traffic only line　01.0040

货主　owner of freight, consignor, consignee　06.0132

J

基本轨　stock rail　03.0354

基本建设计划　plan of capital construction　06.0708

基本建设投资　capital construction investment　06.0709

基本建筑界限　fundamental construction clearance, fundamental structure gauge　01.0056

基本交叉间隔　fundamental transposition interval　07.0230

基本进路　basic route　08.0153

基本联锁电路　fundamental interlocking circuit　08.0228

基本票价　basic fare　06.0732

基本型链路控制规程　basic link control procedure　07.0887

基本运行图　primary train diagram　06.0296

基本折旧率　basic depreciation rate　06.0705

基本阻力　basic resistance　04.0095

基础制动装置　foundation brake rigging　05.0272

基床　subgrade bed, formation　02.0241

基床表层　surface layer of subgrade bed, formation top layer, surface layer of subgrade　02.0242

基床底层　bottom layer of subgrade, formation base layer, bottom layer of subgrade bed　02.0243

基带　baseband　07.0791

基带传输　baseband transmission　07.0792

基底　foundation base, base　02.0247

＊基地台　base station　07.0553

＊基平　benchmark leveling　02.0116

基群　basic group　07.0308

基群配线架　basic group distribution frame　07.0292

奇偶检验　parity check　07.0798

奇偶校验码　odd-even check code　07.0015

机车　locomotive　04.0002

机车包乘制　system of assigning crew to designated locomotive　04.0197

机车保养　locomotive maintenance　04.0238

机车报废　locomotive retirement　04.0029

机车比率　locomotive ratio　04.0004

＊机车比值　locomotive ratio　04.0004

机车长度　locomotive overall length　04.0040

机车厂修　locomotive repair in works　04.0245

机车超重牵引　traction for train exceed mass norm　04.0229

机车车辆冲击　impact of rolling stock　01.0106

机车车辆共振　resonance of rolling stock　01.0105

机车车辆溜逸　runaway of locomotive or car　06.0634

机车车辆破损　rolling stock damage　06.0632

机车车辆上部限界　clearance limit for upper part of rolling stock　01.0060

机车车辆下部限界　clearance limit for lower part of rolling stock　01.0061

机车车辆运用计划　rolling stock utilization plan　06.0689

机车车辆振动　vibration of rolling stock　01.0095

机车乘务制度　locomotive crew working system　04.0196

机车乘务组　locomotive crew　04.0200

机车出入段作业　preparation of locomotive for leaving and arriving at depot 04.0206

机车出租 leased locomotive 04.0028

机车储备 locomotive reservation, locomotive storage 04.0030

机车传动效率 transmission efficiency of locomotive 04.0077

机车大修 locomotive overhaul [repair], locomotive general overhaul 04.0241

机车电台 locomotive station 07.0556

机车调度命令 locomotive dispatching order 04.0230

机车定期修 locomotive periodical repair 04.0247

机车段修 locomotive repair in depot 04.0246

机车分配阀 locomotive distributing valve 04.0165

机车感应器 locomotive inductor 08.0421

机车高度 locomotive height 04.0042

机车功率 locomotive power 04.0082

机车功率试验 locomotive [traction] power test 04.0675

机车功率因数测定 measurement of power factor 04.0676

机车公里 locomotive kilometers 06.0692

机车故障 locomotive failure 04.0073

机车固定轴距 locomotive rigid wheel base 04.0039

机车锅炉 locomotive boiler 04.1115

机车集中供电 locomotive centralized power supply 04.0365

机车技术规范 locomotive technical specification 04.0033

机车计算重量 calculated weight of locomotive 04.0043

机车监控记录装置 locomotive supervise and record apparatus 04.0072

机车检修 locomotive inspection and repair 04.0239

机车检修段 locomotive repair depot 04.0069

机车检修率 ratio of locomotives under repair 06.0696

机车检修修程 classification of locomotive repair 04.0240

机车交路 locomotive routing 04.0187

机车接近通知 approaching announcing in cab 08.0437

机车紧急放风阀 locomotive emergency vent valve 04.0170

机车库 locomotive shed 06.0537

机车宽度 locomotive width 04.0041

机车临修 locomotive temporary repair 04.0250

机车履历簿 locomotive logbook 04.0299

机车轮乘制 locomotive crew pooling system 04.0198

机车轮周功率曲线 locomotive power curve at wheel rim 04.0079

机车轮周效率 efficiency of locomotive at wheel rim 04.0076

机车每轴闸瓦作用力 brake shoe force per axle of locomotive 04.0127

机车能耗 locomotive energy consumption 06.0728

机车平均牵引总重 average gross weight hauled by locomotive 04.0219

机车牵引变压器 traction transformer of locomotive 04.0481

机车牵引力 locomotive tractive effort 04.0089

机车牵引力曲线 locomotive tractive effort curve 04.0080

[机车]牵引梁 draw beam 04.1056

机车牵引区段 locomotive tractive district 04.0186

机车牵引特性 locomotive tractive characteristic 04.0074

机车牵引特性曲线 locomotive tractive characteristic curve 04.0078

机车全周转 complete turnround of locomotive 04.0207

机车全周转距离 distance of one complete turnround of locomotive 04.0209

机车全周转时间 period of one complete turnround of locomotive 04.0210

机车全轴距 locomotive total wheel base 04.0037

机车日产量 average daily output of locomotive 04.0220

机车日车公里 average daily locomotive running kilometers 04.0218

机车设备 locomotive equipment 08.0419

机车试运转 locomotive trial run 04.0279

机车速度表 locomotive speedmeter 04.0626

机车随乘制 locomotive caboose crew system 04.0199

机车调速试验 test on speed regulation 04.0670

机车万吨公里能耗 energy consumption per 10 000 t·km of locomotive 04.0053

* 机车无线电操纵 radio telecontrol for locomotive, locomotive radio-control 07.0679

机车无线电遥控 radio telecontrol for locomotive, locomotive radio-control 07.0679

机车效率 total locomotive efficiency 04.0075

机车效率测定 measurement of efficiency of locomotive 04.0677

机车信号 cab signal 08.0012

机车信号测试区段 cab signaling testing section 08.0416

机车信号设备 cab signaling equipment 08.0409

机车信号作用点 cab signaling inductor location 08.0417

机车需要系数 coefficient of locomotive requirment 04.0222

机车验收 acceptance of locomotive 04.0278

机车用柴油 diesel oil for locomotive 04.0055

机车用电 electricity for locomotive 04.0056

机车用换算煤 converted coal for locomotive 04.0057

机车用煤 coal for locomotive 04.0054

机车用润滑剂 lubricant for locomotive 04.0059

机车预期牵引特性曲线 predetermined tractive characteristic curve of locomotive 04.0081

机车运用段 locomotive running depot 04.0068

机车运用指标 index of locomotive operation 04.0205

机车在段停留时间 detention time of locomotive at depot 04.0208

机车噪声 locomotive noise 04.0051

机车粘着重量 locomotive adhesive weight 04.0046

机车振动参数测试 measurements of vibration parameters 04.0668

机车整备 locomotive servicing, locomotive running preparation 04.0031

机车整备能力 locomotive service capacity 04.0032

机车整备重量 locomotive service weight 04.0044

机车制动机 locomotive brake gear 04.0160

机车制动距离 locomotive braking distance 04.0126

机车制动周期 locomotive braking period 04.0125

机车种类 types of locomotive 04.0003

机车重联电连接器 multi-locomotive electric coupler 04.0393

机车重量 locomotive weight 04.0045

机车重量分配 weight distribution of locomotive 04.0049

机车周转图 locomotive working diagram 06.0288

机车专用设备 special equipment for locomotive operation 04.0276

机车转向架轴距 locomotive wheel base of bogie 04.0038

机车自动操纵 automatic locomotive operation 04.0237

* 机车自动信号 cab signal 08.0012

机车走行公里 locomotive running kilometers 04.0211

机车走行线 locomotive running track 06.0412

机车组装后的检查与试验 inspection and test of locomotive after completion of construction 04.0655

机待线 locomotive waiting track 06.0413

机动螺钉-螺栓搬手 rail screw-bolt power wrench 03.0402

机体 engine block, main frame 04.0863

机务段 locomotive depot 04.0067

机务段联系电路 liaison circuit with a locodepot 08.0242

机务段运行揭示 running service-bulletin of depot 04.0231

机务设备通过能力 carrying capacity of locomotive facilities 04.0275

机务维修无线电通信 radio communication for maintenance of locomotive 07.0637

机务折返段 locomotive turnaround depot 04.0070

机械保温车辆段 mechanical refrigerator car depot 05.0898

机械臂板信号机 mechanically operated semaphore signal 08.0057

机械不完全燃烧热损失 heat loss due to combustibles in refuse 04.1108

机械传动内燃机车 diesel-mechanical locomotive 04.0686

机械负荷 mechanical load 04.0835

机械化驼峰 mechanized hump 06.0456

机械化驼峰设备 mechanized hump yard equipment 08.0366

机械集中联锁 mechanical interlocking 08.0106

机械冷藏车 mechanical refrigerator car 05.0817

机械冷藏车组 mechanical refrigerator car group 05.0818

机械损失功率 mechanical loss power 04.0815

机械锁闭 mechanical locking 08.0136

机械台班定额 rating per machine per team, rating per machine-team 02.0030

机械通风 mechanical ventilation 02.0804

机械通风装置 mechanical ventilation equipment 05.0618

机械效率 mechanical efficiency 04.0826

机械增压 engine-driven supercharging 04.0789

机械增压柴油机 engine-driven supercharged diesel engine 04.0698

机油泵 lubricating oil pump 04.1037

机油粗滤器 lubricating prefilter 04.1042

机油滤清器 lubricating oil filter 04.1041

机油热交换器 lubricating oil heat exchanger 04.1045

机油消耗量 oil consumption 04.0769

机油消耗率 specific oil consumption 04.0768

机座 frame 04.0451

畸变 distortion 07.0814

积炭 carbon deposit 04.0805

激光器 laser [diode], LD 07.0470

极频轨道电路 polar-frequency pulse track circuit 08.0497

极频自动闭塞 automatic block with polar frequency impulse track circuit 08.0334

*极限误差 maximum error, limiting error 02.0111

极限状态设计法 limit state design method 02.0207

极性检查电路 polarity checking circuit 08.0225

极性交叉 polar transposition 08.0532

集尘器 dirt collector 05.0287

集电靴 collector shoe 04.0544

集结时间 car detention time under accumulation 06.0263

集中道岔 centrally operated switch 08.0198

集中电源 centrally connected power source 08.0645

集中供电 centrally connected power supply 08.0636

集中供电方式 centralized power supply system 04.1296

集中化修理 centralization of repair 04.0263

集中联锁 centralized interlocking 08.0105

集中器 concentrator 07.0830

集中式逆变器 centralized inverter 05.0700

集中[式]网[络] centralized network 07.0762

*集中修 centralization of repair 04.0263

集重货物 concentrated weight goods 06.0169

集装袋 flexible freight container 06.0572

集装化运输 containerized traffic 06.0576

集装箱 freight container 06.0538

集装箱车 container car 05.0826

集装箱额定质量 rating of freight container, gross mass of freight container 06.0554

集装箱运输 freight container traffic 06.0578

集装箱载重 payload of freight container 06.0556

集装箱自重 tare mass of freight container 06.0555

*集装箱总重 rating of freight container, gross mass of freight container 06.0554

集总加感 lumped loading 07.0234

给水 water supply 04.0060

给水处理 water [supply] treatment 04.0062

给水调整阀 water supply governer valve 05.0664

给水风缸 water supply air reservoir 05.0660

[给水]减压阀 [pressure] reducing valve 05.0663

*给水所 water supply station 04.0066

给水系统 water supply system 05.0654

给水预热装置 feed water heater 04.1168

给水站 water supply station 04.0066

给水装置 feed water rigging 04.1165

急流槽 chute 02.0301

挤岔　forcing open of the point　06.0630

挤岔报警　alarm for a trailed switch　08.0180

挤轨　gage widening　05.0875

挤密砂桩　sand compaction pile　02.0608

挤切　dissectible　08.0620

挤切销　dissectible pin　08.0621

挤脱　trailable　08.0619

挤压混凝土衬砌　extruding concrete tunnel lining　02.0692

挤压涡流　extruding swirl　04.0803

技术设计　technical design　02.0015

技术速度　technical speed　06.0330

技术站　technical station　06.0395

技术整备　technical servicing　05.0914

技术直达列车　technical through train　06.0221

计费吨公里　tonne-kilometers charged　06.0149

计费吨公里成本　cost of charged ton-kilometer　06.0725

计划内运输　planned freight traffic　06.0101

计划外运输　out-of-plan freight traffic, unplanned freight traffic　06.0102

计划修　planned maintenance　08.0728

计算供汽率　calculated rate of evaporation for engine　04.1099

计算机网[络]　computer network　07.0764

计算遮断比　calculated cut-off　04.1093

计轴自动闭塞　automatic block with axle counter　08.0336

记发器　register　07.0397

记录台　recording desk　07.0256

记录制　record [demand] working　07.0273

记事灯　writing lamp　04.0647

既有铁路　existing railway　01.0032

既有线测量　survey of existing railway　02.0107

继电半自动闭塞　all-relay semi-automatic block system　08.0325

继电并联传递网路　successively worked parallel relay network　08.0221

继电并联网路　parallel relay network　08.0219

继电串联网路　series relay network　08.0220

继电器　relay　08.0550

继电器防震架　shock absorber base for relays　08.0282

继电器控制电源　power source for relay control　08.0661

*继电器励磁　relay energized　08.0558

继电器灵敏度　relay sensitivity　08.0572

*继电器失磁　relay released　08.0559

继电器释放　relay released　08.0559

继电器室　relay room　08.0715

继电器吸起　relay energized　08.0558

继电器箱　relay case　08.0281

继电式电气集中联锁　all-relay interlocking　08.0109

夹直线　intermediate straight line, tangent between curves　02.0179

家禽车　poultry car　05.0821

家畜车　stock car　05.0820

加冰所　re-icing point　06.0176

加顶垂直折合机车天线　vertical loading folded locomotive antenna　07.0569

加顶圆盘机车天线　disc-loading locomotive antenna　07.0568

加封　sealing　08.0711

加感电缆　loaded cable　07.0192

加感节距　loading coil spacing　07.0235

加感箱　loading coil box　07.0194

加固地基　improved foundation, improved ground　02.0578

加减速顶　dowty accelerator-retarder　06.0463

加筋土挡土墙　reinforced earth retaining wall, reinforced soil retaining wall　02.0288

加劲杆　stiffener　02.0452

加快票　fast extra ticket　06.0036

加力牵引坡度　pusher grade, assisting grade　02.0183

加煤机　stoker　04.1246

加强构件　tension member　07.0482

加强线　line feeder　04.1367

加热集装箱　heated container　06.0550

加速　acceleration　04.0106

加速顶　dowty accelerator　06.0462

加速度反馈　acceleration feedback　04.1606

加速缓坡　easy gradient for acceleration, accelerating grade　02.0199

加速力　acceleration force　04.0109

加速坡　accelerating grade　06.0495

加速推送信号　humping fast signal　08.0036

加速制动阀部　accelerated application valve portion
　05.0295

＊加算阻力　converted resistance　04.0104

加温车　heater car　05.0812

加温套　steam jacket　05.0482

加温运输　heating transport　06.0180

加桩　additional stake, plus stake　02.0114

假线　building-out network　07.0345

架车　jack up car body　05.0935

架承式牵引电动机　frame mounted traction motor
　04.0401

架空地线　aerial earth wire　07.0212

架空地线屏蔽系数　shielding factor of aerial earth
　wire　04.1482

架空电缆　aerial cable　07.0178

架空明线　open wire　07.0157

架立钢筋　erection bar　02.0477

架桥机架设法　erection by bridge girder erecting
　equipment　02.0531

架修　intermediate repair　04.0242

监测信号　monitor signal　07.0609

监督对象　surveillanced object　08.0473

监视系统　supervision system　07.0505

监听　monitoring　07.0126

尖端杆　front rod of a point　08.0303

尖轨　switch rail, tongue rail, blade　03.0313

尖轨保护器　switch protector　03.0362

尖轨补强板　reinforcing bar　03.0366

尖轨长度　length of switch rail　03.0332

尖轨动程　throw of switch　03.0337

尖轨跟端　heel of switch rail　03.0318

尖轨护轨　switch point guard rail　03.0361

尖轨尖端　actual point of switch rail　03.0317

尖轨理论尖端　theoretical point of switch rail
　03.0316

间壁　partition　05.0790

间隔　space　07.0717

间隔铁　filler, spacer block　03.0365

间隔铁式尖轨转辙器　loose heel switch　03.0346

间隔制动　spacing braking　06.0470

间隙效应　slack action　04.0183

检波　detection　07.0020

检波器　detecter　07.0056

检测　check-out, inspection and measurement
　08.0735

检查井　inspection well, manhole　02.0309

检车电视　TV for inspection　07.0658

检错码　error detecting code　07.0795

检衡车　weigh bridge test car, track scale test car
　05.0747

检票监视电视　monitor TV for ticket check
　07.0654

检修不良　not well inspected and repaired
　08.0694

检修范围　scope of repairing course, scope of repair
　04.0260

检修工艺规程　technological regulations for repair
　and inspection　04.0265

检修基本技术条件　fundamental technical require-
　ments for repair and inspection　04.0264

检修机车　locomotive under repairing　04.0024

检修停时　standing time under repair　04.0262

检修限度　locomotive repair limit　04.0268

检修周期　period of inspection and repair　04.0256

检修作业程序　repair procedure, shop program
　04.0266

简单悬挂　tramway type suspension equipment
　04.1353

简易混凝土搅拌站　simple concrete mixing plant
　02.0931

简易驼峰　simplified hump　06.0454

简支梁桥　simply supported beam bridge　02.0379

减价票　reduced-fare ticket　06.0040

减速　deceleration　04.0107

减速顶　dowty retarder　06.0461

减速力　deceleration force　04.0110

减速器　retarder　06.0460

减速器出口速度　release speed at retarder　06.0484

减速器工作状态　retarder in working state
　08.0382

减速器缓解状态　retarder released　08.0383

减速器接近限界　clearance of a retarder　08.0384

减速器入口速度　entrance speed at retarder
　06.0483

减速器制动状态 retarder in closed state 08.0381

减速推送信号 humping slow signal 08.0037

减速信号 restriction signal 08.0022

减压阀 reducing valve 04.0164

减压中继阀 reduction relay valve 05.0318

减振器 shock absorber, damper 05.0156

减振器试验台 damper test stand 05.0997

减振指数 damping index 05.0246

鉴定试验 homologation test 04.0282

键盘 keyboard 07.0694

箭翎线 herringbone track 06.0445

建筑业总产值 total output of building industry 02.0853

降压电阻 step down resistance 05.0609

降压气室 pressure reducing reservoir 05.0310

胶结绝缘接头 glued insulated joint 03.0071

交叉 crossing 03.0290

交叉渡线 scissors crossing, double crossover 03.0297

交叉跨越 crossing 04.1490

交叉偏差 deviation from transposition interval 07.0231

交叉区 transposition section 07.0233

交叉疏解 crossing untwining 06.0518

交叉指数 transposition index 07.0232

交叉制式 transposition system 07.0229

交出空车数 number of empty cars delivered 06.0318

交出重车数 number of loaded cars delivered 06.0316

交错排列的磁铁布置 staggered magnet configuration 04.1633

交点 intersection point 02.0120

交分道岔 slip switch 03.0293

交互式协议 interactive protocol 07.0882

交换 exchange, switching 07.0745

交接箱 cross-connecting box 07.0193

交接桩 delivery-receiving stake 02.0835

交流电源屏 AC power supply panel 08.0651

交流二元二位继电器 AC two element two position relay 08.0595

交流感应子发电机 inductor type alternator 05.0698

交流供电制 AC power supply system 08.0631

交流轨道电路 AC track circuit 08.0491

交流换向器电动机 alternating current commutator motor 04.0410

交流计数电码轨道电路 AC counting coded track circuit 08.0502

交流计数电码自动闭塞 automatic block with AC counting code track circuit 08.0335

交流继电器 AC relay 08.0581

交流接触器 AC contactor 04.0547

交流牵引变电所 AC traction substation 04.1344

交流牵引电动机 AC traction motor 04.0397

交流主发电机 main alternator 04.0464

交调干扰 cross modulation interference 07.0528

交－直－交流传动 AC-DC-AC drive 04.0343

交－直流传动 AC-DC drive 04.0342

交直流继电器 AC-DC relay 08.0582

*铰型接头夹板 head free flat joint bar 03.0080

矫直钢轨 straightening of kinked rail 03.0210

脚手架 scaffold 02.0623

角灯 corner lamp 05.0682

角杆 angular pole 07.0169

角件 corner fittings 06.0561

角深 pull 07.0228

角柱 corner post 05.0431

教育车 education car 05.0743

较大限制信号 more restrictive signal 08.0042

较大允许信号 more favorable signal 08.0044

接车进路 receiving route 06.0248

接车进路信号机 route signal for receiving 08.0070

接车信号 receiving signal 08.0026

接触不良 bad contact 08.0699

接触网 overhead contact line equipment 04.1350

接触网标称电压 nominal voltage of overhead contact line 04.1332

接触网测试车 measuring car for overhead contact line equipment 04.1459

接触网工区 maintenance gang for catenary 04.1456

接触网故障探测装置 fault locator for overhead contact line equipment 04.1417

接触网架空地线 overhead earth wire 04.1377

接触网架线车 installation vehicle for contact wire 02.0983

接触网检修车 repairing car for overhead contact line equipment 04.1460

*接触网锚段衬砌 anchor-section lining 02.0691

接触网限界 clearance limit for overhead contact wire, clearance limit for overhead catenary system, overhead catenary system gauge 01.0064

接触网悬挂 overhead contact line, catenary 04.1351

接触网最低电压 minimum voltage of overhead contact line 04.1333

接触网最高电压 maximum voltage of overhead contact line 04.1331

接触网作业车 operation vehicle for contact wire 02.0984

接触线 contact wire 04.1365

接触线弛度 contact wire sag 04.1450

接触线电连接线夹 electrically connecting clamp for contact wire 04.1376

接触线接头线夹 contact wire splice 04.1371

接触线预留弛度 contact wire pre-sag 04.1451

接触线终端锚固线夹 termination fitting for contact wire 04.1369

接触线最大水平偏移值 maximum horizontal displacement of contact wire 04.1452

接地安全棒 earthing pole 04.0639

接地保护放电装置 earth protection discharger 04.1341

接地报警 grounding alarm 08.0179

接地电抗器 earthing reactor, grounding reactor 04.0508

接地回流电刷 earth return brush 04.0640

接地继电器 earth fault relay 04.0601

接地开关 earthing switch 04.0567

接地线夹 earth clamp 04.1378

接点闭合 contact closed 08.0556

接点断开 contact open 08.0557

接点系统 contact system 08.0575

接点压力 contact pressure 08.0573

接发车进路信号机 route signal for receiving-departure 08.0072

接发车无线电通信 radio communication for train reception and starting 07.0584

接发列车 train reception and departure 06.0236

接杆 pipe jaw 08.0295

接近表示 approach indication 08.0159

接近长度 approach length 04.1484

接近发码 coding during train approaching 08.0344

接近距离 separation distance 04.1485

接近连续式机车信号 approach continuous cab signaling 08.0412

接近区段 approach section 08.0205

接近锁闭 approach locking 08.0128

*接入 access 07.0749

接入空车数 number of empty cars received 06.0317

接收机动态范围 receiver dynamic range 07.0466

接收机灵敏度 receiver sensitivity 07.0465

接收线圈 receiving coil 08.0420

接收信道 receive channel 07.0866

接受者 receptor 07.0906

接通率 percent of call completed 07.0155

*接头缝 rail gap, joint gap 03.0063

接头夹板 joint bar, splice bar, fish plate 03.0077

接头联结零件 rail joint accessories, rail joint fastenings 03.0076

接头螺栓 track bolt, fish bolt 03.0083

接头瞎缝 closed joint, tight joint 03.0157

接头阻力 joint resistance 03.0043

接线端子 connection terminal 05.0705

接线盒 connection box 05.0703

接运重车数 number of loaded cars received 06.0315

阶段缓解 graduated release 04.0137

阶段提升 graduated increasing 04.0145

阶段制动 graduated application 04.0131

阶梯直达列车 through train originated from several adjoining loading points 06.0216

截断塞门 cut-out cock 05.0288

截水沟 intercepting ditch, catch-drain 02.0300

截止波长 cut-off wavelength 07.0461

节点 panel point 02.0453

节点板 gusset plate 02.0454

节间 panel 02.0440

节间长度　panel length　02.0436

结点　node　07.0767

结构高度　system height, encumbrance　04.1453

结合杆　union link, connecting link　04.1208

结合梁桥　composite beam bridge　02.0375

结胶　caking　04.0806

解除闭塞　block cleared　08.0349

解钩装置　uncoupling device　05.0564

解码　decode　07.0017

解锁　release　08.0140

解锁按钮盘　manual release button panel　08.0257

解锁电路　release circuit　08.0231

解锁进路　released route　08.0190

解锁力　releasing force　08.0625

解体调车　break-up of trains　06.0255

解调　demodulation　07.0019

介电强度试验　dielectric test　04.0661

紧急部　emergency portion　05.0293

紧急风缸　emergency reservoir　05.0312

紧急呼叫　emergency call　07.0615

紧急局减　quick action　04.0153

紧急撒砂　emergency sanding　04.0181

紧急停车装置　emergency stop mechanism 04.0969

紧急信号　emergency signal　07.0614

紧急制动　emergency braking, emergency application 04.0133

紧急制动阀　emergency brake valve　05.0290

紧急制动信号音　emergency braking tone　07.0627

紧坡地段　section of sufficient grade　02.0158

进风道　air inlet duct　05.0638

进局电缆　entrance cable　07.0186

进局设备　incoming equipment　07.0294

进口角　blade inlet angle　04.1548

进路　route　06.0243

进路表　route sheet　08.0213

进路表示器　route indicator　08.0089

进路表示器电路　route indicator circuit　08.0248

进路操纵作业　semi-automatic operation by route 08.0376

进路储存器　route storaging devices　08.0405

进路电路　route selecting circuit　08.0229

进路分段解锁　sectional release of a locked route 08.0143

进路继电式电气集中联锁　route type all-relay interlocking　08.0111

进路交叉　crossing of routes　06.0517

进路解锁　route release　08.0141

进路人工解锁　manual route release　08.0145

进路锁闭　route locking　08.0127

进路锁闭表示　route locking indication　08.0165

进路信号机　route signal　08.0069

进路一次解锁　route release at once　08.0142

进气道　intake duct　04.1536

进气阀　air inlet valve　05.0478

进气管　air inlet pipe　04.0981

进气门　inlet valve　04.0926

进气凸轮　inlet cam　04.0938

进气温度　intake temperature　04.0731

进气稳压箱　air inlet pressure stabilizing chamber 04.0982

进气涡流　intake swirl　04.0804

进气行程　intake stroke　04.0723

进气压力　intake pressure　04.0732

进汽余面　steam lap　04.1258

进行信号　proceed signal　08.0020

进站信号机　home signal　08.0065

禁溜车停留线　no-humping car storage　06.0437

近场　near field　07.0459

近场区　near field area　07.0545

近程监督分区　directly surveillanced subsection 08.0475

近程网路　directly surveillanced network　08.0477

* 近程遥信分区　directly surveillanced subsection 08.0475

近端串音　near-end crosstalk　07.0039

近端串音衰减　near-end crosstalk attenuation 07.0141

近体防护　nearby protection　02.1002

浸水路堤　immerseable embankment　02.0313

尽端式枢纽　stub-end type junction terminal 06.0403

尽头式货场　stub-end type freight yard　06.0083

尽头式货运站　stub-end freight station　06.0387

尽头式客运站　stub-end passenger station 06.0384

尽头线　stub-end siding　06.0416

尽头信号机　signal for stub-end track　08.0086

*劲性骨架混凝土梁　girder with rolled steel section encased in concrete, skeleton reinforced concrete girder　02.0378

晶闸管变流器电力机车　thyristor converter electric locomotive　04.0308

晶闸管整流器电力机车　thyristor rectifier electric locomotive　04.0307

晶闸管整流装置　thyristor rectifier device　04.0338

精密导线测量　precise traverse survey, accurate traverse survey　02.0140

经济调查　economic investigation, economic survey　02.0042

经济核算　economic accounting　06.0756

经济效果　economic effects　06.0757

经纬距　plane rectangular coordinate　02.0126

警冲标　fouling post　06.0514

警笛　siren　04.0653

警惕按钮　acknowledgment button　08.0426

警惕手柄　acknowledgment lever　08.0425

警惕装置　vigilance device　04.0654

静电[感应]电流　electrostatic induced current　04.1469

静力触探　static sounding, static probing, cone penetration test　02.0056

静力内接　static inscribing　03.0140

静强度试验台　static strength test rack　05.0989

静水压力　hydrostatic pressure　02.0333

静态不平顺　static track irregularity, irregularity without load　03.0143

静态长度　car space　08.0369

静压力拔桩机　static pressure pile extractor, static pressure pile drawing machine　02.0910

静叶损失　stationary blade loss　04.1553

静液压泵　hydrostatic pump　04.1051

静液压马达　hydrostatic motor　04.1052

静载试验　static test　03.0533

静止图象可视电话　still picture videophone　07.0066

净跨　clear span　02.0432

净制动率　net braking ratio　05.0341

纠错码　error correcting code　07.0796

酒吧车　buffet car　05.0733

救生设施　lifesaving appliance　02.1031

*救援吊车　wrecking crane　06.0667

救援队　breakdown gang　06.0665

救援机车　breakdown locomotive　06.0666

救援列车　breakdown train, rescue train　06.0664

救援列车无线电通信　radio communication for train relieving　07.0645

救援起重机　wrecking crane　06.0667

*旧线测量　survey of existing railway　02.0107

就地灌筑法　cast-in-place method, cast-in-situ method　02.0541

就地灌注桩　cast-in-place concrete pile, cast-in-situ concrete pile　02.0595

局部冲刷　local scour, partial scour　02.0097

局部电源　locally supplied power source　08.0646

局部减压　local reduction　04.0149

局部控制　local control　08.0185

局部控制电路　local control circuit　08.0245

局部控制盘　local control panel　08.0284

局间通信枢纽　communication center between several railway administration　07.0241

局间直通中继方式　inter office through trunk　07.0406

局通信枢纽　communication center of railway administration　07.0240

局线长途通信　toll communication within railway administration　07.0248

局线长途通信网　railway administration toll communication network　07.0246

局线调度电话　dispatching telephone within railway administration　07.0296

局线会议电话　telephone conference within railway administration　07.0298

局用电缆　central office cable　07.0175

局域网　local area network, LAN　07.0756

矩形桥墩　rectangular pier　02.0570

矩形通风装置　oblong ejector, Giesl ejector　04.1140

聚四氟乙烯支座　poly-tetrafluoroethylene bearing, PTFE bearing　02.0523

聚液窝　liquid trap　05.0477

锯轨机 rail cutting machine, rail sawing machine 03.0394

俱乐部车 club car 05.0734

掘进机法 tunnel boring machine method 02.0724

绝对加速度反馈 absolute acceleration feedback 04.1607

绝对信号 absolute signal 08.0024

绝热集装箱 insulated container 06.0548

*绝缘变压器 isolating transformer 04.0490

绝缘不良 bad insulation 08.0697

绝缘电阻 insulation resistance 07.0224

绝缘轨距杆 insulated gage rod 03.0106

绝缘间隙 insulation gap 04.1438

绝缘接头 insulated joint 03.0070

绝缘锚段关节 insulated overlap 04.1430

绝缘梯车 insulated ladder trolley 04.1463

绝缘鞋 insulant shoe 02.1027

绝缘转换柱 insulated transition mast 04.1409

绝缘子 insulator 07.0162

绝缘子清洗车 insulator cleaning car 04.1462

均方[误]差 mean square error 02.0110

均方根值检波器 root mean square detector 07.0573

均衡风缸 equalizing reservoir 05.0304

均衡杠杆 equalizing lever 05.0362

均衡拉杆 equalizing pull rod 05.0361

均衡梁 equalizer 05.0214

均衡坡度 balanced grade 02.0188

均衡速度 balancing speed 01.0077

均衡系统 equalizing system 07.0328

均流电抗器 sharing reactor 04.0512

均压线 equalizer, cable bond 04.0442

*均匀轨缝 adjusting of rail gaps, evenly distributing joint gaps 03.0208

军用列车 military train, troop train 06.0213

竣工报告 completion report of construction work 02.0867

竣工决算 final accounts of completed project 02.0035

K

喀斯特地段路基 subgrade in karst zone 02.0227

喀音 click 07.0280

卡带 tank band 05.0481

开闭所 sub-section post, SSP 04.1347

开敞式卧车 open type sleeping car 05.0729

开放系统互连 open system interconnection, OSI 07.0770

开放信号 clearing signal 06.0252

开工报告 report on starting of construction work, commencement report of construction work, construction starting report 02.0866

开沟机 ditcher 03.0460

开沟平路机 track shaving machine, ditching and grading machine 03.0461

开路式轨道电路 open type track circuit 08.0504

开式回热循环燃气轮机 open regenerative cycle gas turbine 04.1525

开式简单循环燃气轮机 open simple cycle gas turbine 04.1524

开锁位置 lockset position of coupler 05.0567

*开天窗作业时间 work occupation time, working time of closed section 03.0259

开通 put into operation 08.0710

开挖工作面 excavated surface, excavated work face 02.0752

铠装电缆 armoured cable 07.0176

勘测设计无线电通信 radio communication for survey and design 07.0640

勘探 exploration, prospecting 02.0050

抗滑明洞 anti-skid-type open cut tunnel, anti-skid-type tunnel without cover, anti-skid-type gallery 02.0699

抗滑桩 anti-slide pile, counter-sliding pile 02.0292

抗挤出稳定性 stability against forcing out during train buckling 05.0856

*抗剪结合件 shear connector 02.0513

抗剪连接件 shear connector 02.0513

抗倾覆稳定性 stability against overturning 05.0854

抗扰度 immunity to interference 04.1509

抗脱轨稳定性 stability against derailment 05.0855

壳式牵引变压器 shell-type traction transformer 04.0483

可钉地板 nailable floor 05.0404

可懂串音 intelligible crosstalk 07.0138

可懂度 intelligibility 07.0148

可动心轨辙叉 movable-point frog 03.0349

可动翼轨辙叉 movable-wing frog 03.0350

*可靠度设计法 probabilitic limit state design method 02.0208

可靠性 reliability 08.0667

可靠性试验 reliability test 04.0293

可控电磁铁 controlled electromagnet 04.1590

可控减速顶 dowty controllable retarder 06.0464

可控桥式整流器 controlled bridge rectifier 04.0333

可视电话 videophone 07.0064

可视电话机 videophone set 07.0096

可躺座椅 reclining seat 05.0773

可调拐肘 adjustable crank 08.0299

可调牵引变频器 variable frequency convertor 04.0329

可弯式尖轨转辙器 flexible switch 03.0345

可行性研究 feasibility study 02.0009

可转静叶 variable stator blade 04.1559

*克服高度 height of lifting, lifting height, ascent of elevation 02.0161

客车 passenger car, carriage, coach 05.0718

[客车]摆门 spring butt rocking door 05.0491

客车保有量 number of passenger cars on hand 05.0977

[客车]隔门 partition door 05.0489

客车技术整备所 passenger train technical servicing point 05.0916

[客车]脚蹬 entrance door step 05.0405

[客车]脚蹬门 vestibule entrance door 05.0488

客车客座利用率 percentage of passenger seats utilization per car 06.0074

[客车]门锁 door lock 05.0502

[客车]内墙板 panel 05.0451

客车配属辆数 number of allocated passenger cars 05.0978

客车平均日车公里 average car-kilometers per car-day 06.0071

[客车]上侧梁 cant rail 05.0438

客车洗车线 washing siding for passenger vehicle 06.0418

[客车]摇门 swing door 05.0490

客车整备库 coach servicing shed 05.0900

*客车整备棚 coach servicing shed 05.0900

客车整备所 passenger car servicing depot 06.0525

客车轴温报警 passenger car journal temperature warning 05.0960

客货车厂段修规程 regulations for passenger/freight car repair in factory/depot 05.0985

客货运混合线路 railway line for mixed passenger and freight traffic 01.0042

客货运站 mixed passenger and freight station 06.0385

客流 passenger flow 06.0023

客流调查 passenger flow investigation 06.0028

客流量 passenger flow volume 06.0027

客流图 passenger flow diagram 06.0029

客票 passenger ticket 06.0035

客室 passenger room, passenger compartment 05.0752

[客室]通道 aisle, gangway 05.0788

客厅车 saloon car 05.0738

客运机车 passenger locomotive 04.0014

客运监视电视 monitor TV for passenger service 07.0652

客运密度 passenger traffic density 06.0067

客运事故 passenger traffic accident 06.0620

客运业务无线电通信 radio communication for passenger service 07.0621

客运站 passenger station 06.0382

客运专线 railway line for passenger traffic, passenger special line, passenger traffic only line 01.0041

客站货场无线电通信 station and freight yard radio communication 07.0642

坑道自稳时间 self-stabilization time of tunnel 02.0710

空白 blank 07.0718

空车　empty car　05.0012

空车调整　adjustment of empty cars　06.0340

空车直达列车　through train with empty cars　06.0217

空车走行率　percentage of empty to loaded car kilometers　06.0322

空档　stop short　06.0485

*空吊板　loose tie　03.0154

空腹拱　open-spandrel arch　02.0401

空间[波]传播方式　space-wave propagation mode　07.0547

空间分集　space diversity　07.0578

空间间隔法　space-interval method　06.0238

空陆水联运集装箱　air/surface container, air/intermodal container　06.0553

空气包　tank dome　05.0471

空气干燥器　air dryer　04.0176

空气控制装置　pneumatic control device　05.0514

空气幕沉井法　air curtain method for sinking caisson　02.0590

空气调节装置　air conditioning equipment　05.0626

空气消耗量　air consumption　04.0770

空气消耗率　specific air consumption　04.0771

空气压缩机　air compressor　04.0161

空气压缩机室　air compressor room　08.0720

空气预冷装置　air precooler　05.0643

空气预热装置　air preheater　04.1164

空气真空两用制动装置　air and vacuum dual brake equipment　05.0268

空气制动　air brake　04.0113

空气制动试验　test on air brake　04.0666

空气制动装置　air brake equipment　05.0266

空气主断路器　line air-blast circuit-breaker　04.0531

空气自动控制装置　automatic air control device　05.0627

空气阻力　air resistance　04.0102

空燃比　air/fuel ratio　04.0744

空调客车　air conditioned passenger car, air conditioned coach　05.0720

空心板桥　hollow slab bridge　02.0383

空心车轴　tubular axle, hollow axle　05.0108

空心桥墩　hollow pier　02.0560

空压机电动机　air compressor motor　04.0474

空载特性　no-load characteristic　04.0853

空中索道　aerial ropeway　02.0900

空重车制动装置　empty/load brake equipment　05.0301

空重车转换塞门　empty and load changeover cock　05.0289

空走距离　idling stopping distance　05.0348

空走时间　idling braking time　05.0344

控温运输　transport under controlled temperature　06.0177

控制爆破　controlled blasting　02.0747

控制泵　control pump　04.1027

控制变压器　control transformer　04.0493

控制点　controlling point　08.0479

控制电机　control electric machine　04.0477

控制电缆连接器　control cable coupling　05.0674

控制电路　control circuit　04.0377

控制电路库用插座　control circuit socket for shed supply　04.0607

控制电源　control source　04.0378

控制电源电流表　control supply ammeter　04.0625

控制电源电压表　control supply voltmeter　04.0624

控制电源隔离开关　isolating switch for control supply　04.0572

控制对象　controlled object　08.0466

控制工程　dominant project　02.0860

控制机构　control mechanism　04.0967

控制盘　control panel　08.0256

控制区间　control section, controlling section　02.0170

控制室　control cabin　05.0646

控制台　control desk　08.0253

控制台单元　control desk element　08.0266

控制台室　control room　08.0719

控制箱　control box　05.0608

控制信号　control signal　07.0608

控制信号检波器　control signal detector　07.0592

控制站　control station　07.0824

控制周期　control cycle　08.0481

控制字符　control character　07.0823

扣车条件　specified conditions for detaining cars　05.0964

扣件 rail fastening 03.0085

扣件扣压力 toe load of fastening 03.0045

库检 examination in depot 05.0933

*跨度 span 02.0431

跨径 span 02.0431

跨线桥 overpass bridge, grade separation bridge, flyover 02.0357

跨越杆 cross-over pole 07.0168

跨装 straddle 06.0164

快动继电器 quick-acting relay 08.0599

快速接头 quick coupling 05.0485

快速转辙机 quick-acting switch machine 08.0607

快吸继电器 quick pick-up relay 08.0600

宽带干扰 broadband interference 04.1506

宽带信道 wide-band channel 07.0872

宽轨铁路 broad-gage railway 01.0014

宽混凝土轨枕 broad concrete tie 03.0109

矿山法 mining method, mine tunnelling method 02.0720

矿山铁路 mine railway 01.0025

矿石车 ore car 05.0807

矿渣道碴 slag ballast 03.0128

馈电电流 feeding current 07.0087

扩大 enlargement 02.0744

扩大初步设计 enlarged preliminary design, expanded preliminary design 02.0016

扩大基础 spread foundation 02.0581

扩频通信 spread spectrum communication 07.0674

扩压器 diffuser 04.0979

扩音柱 speaking post in yard 07.0441

扩音转接机 control set for sound amplifying in yard 07.0440

阔大货物 exceptional dimension freight 06.0160

阔大货物限界 clearance limit for freight with exceptional dimension, clearance limit for oversize commodities 01.0063

L

拉板式轴箱定位装置 tie-plate type journal box positioning device 05.0153

拉槽 pull trough, trench excavated 02.0745

拉撑 brace 04.1131

拉出值 stagger 04.1400

拉杆 toggle arm 05.0523

拉杆底座 bracket base 04.1385

拉杆式轴箱定位装置 tie-rod type journal box positioning device 05.0151

拉杆[压管] upper cantilever 04.1384

拉缸 piston scraping 04.0842

拉钩检查距离 car spacing for uncoupled inspection 05.0969

拉沟 slotting 03.0503

拉门 sliding door 05.0492

拉钮 pull-out button 08.0268

拉线 stay 07.0170

栏杆 railing, handrail, handrailing 02.0510

拦石墙 stone cut off wall, stone falling wall, buttress wall for intercepting falling rocks 02.0318

缆索 cable 02.0492

缆索起重机 cable crane 02.0899

朗格尔式桥 Langer bridge, flexible arch bridge with rigid tie 02.0404

劳动定额 labor norm, labor ratings 06.0717

劳动工资计划 plan of labor and wages 06.0714

雷达测速器 radar speedometer 08.0396

雷电干扰 lightning interference 08.0676

冷藏集装箱 refrigerated container 06.0549

冷藏加温车 refrigerator and heater car 05.0813

冷藏运输 refrigerated transport 06.0179

冷冻板冷藏车 freezing-plate refrigerator car 05.0816

冷滑 cold-running 04.1440

冷凝器 condenser 05.0631

冷凝器风道 condenser air concentrator 05.0650

冷却风扇 cooling fan 04.1032

冷却货物 cooled freight 06.0175

冷却室 cooling room 04.1061

冷却水泵 cooling water pump 04.1029

冷水泵 cold water pump 04.1171

冷铸铁轮 chilled cast iron wheel 05.0105

离合器操纵装置 clutch operating device 04.1050

离去表示 departure indication 08.0160

离去区段　departure section　08.0208

离水格子　floor rack　05.0525

离线　contact loss　04.1447

离线率　contact loss rate　04.1448

离心式机油滤清器　centrifugal oil filter　04.1043

离心水泵　centrifugal pump　02.0916

理论空气量　theoretical air　04.1573

励磁变压器　excitation transformer　04.0495

励磁电流表　excitation ammeter　04.0621

励磁电路　energizing circuit　08.0216

励磁接触器　excitation contactor　04.0553

励磁绕组　excitation winding　04.0499

励磁整流装置　excitation rectifier device　04.0339

历史洪水位　historic flood level　02.0078

例行试验　routine test　04.0294

*立交桥　overpass bridge, grade separation bridge, flyover　02.0357

立接制　no-delay [demand] working　07.0274

立体交叉　grade separation　03.0272

立柱式托盘　post pallet　06.0570

立爪式装岩机　vertical claw rock loader　02.0963

沥青道床　asphalt cemented ballast bed　03.0122

联轨站　junction station　06.0394

联合集尘截断塞门　combined dirt collector and cutout cock　05.0286

联络线　connecting line　06.0419

*联票　coupon ticket　06.0045

联锁　interlocking　08.0103

联锁表　interlocking table　08.0212

联锁道岔　interlocked switch　08.0196

联锁电路　interlocking circuit　04.0383

联锁区　interlocking area　08.0200

联锁设备　interlocking equipment　08.0104

联锁图表　interlocking chart and table　08.0211

联锁箱　point detector　08.0290

联锁箱联锁　interlocking by point detector　08.0120

联系电路　liaison circuit　08.0236

连杆　connecting rod　04.0910

连杆比　connecting rod length/crank radius ratio　04.0711

连杆长度　length of connecting rod　04.0710

连杆衬套　connecting rod bush　04.0912

连杆盖　connecting rod cap　04.0914

连杆螺母　connecting rod nut　04.0916

连杆螺栓　connecting rod bolt　04.0915

连杆驱动　rod drive　04.0360

连杆体　connecting rod body　04.0911

连杆轴瓦　connecting rod bearing shell　04.0913

连挂速度　coupling speed　06.0482

连接　connection　07.0116

连接方式　connection mode　07.0903

连接杆　pipe link　08.0296

连接拉杆　cylinder lever connecting rod　05.0358

连接箱　connecting box　04.0876

连续梁桥　continuous beam bridge　02.0380

连续式调速系统　continued type speed control system　06.0487

连续式轨道电路　continuous track circuit　08.0493

连续式机车信号　continuous type cab signaling　08.0411

链斗卸车机　unloading machine with chain buckets　06.0598

链斗装车机　loading machine with chain buckets　06.0597

链接电缆　link cable　07.0189

链路管理　link management　07.0806

链路控制规程　link control procedure　07.0886

链路协议　link protocol　07.0845

链形悬挂　overhead contact line with catenary, longitudinal suspension　04.1352

链子钩　screw coupling　05.0556

梁端缓冲梁　auxiliary girder for controlling angle change　02.0467

梁高　depth of girder　02.0437

梁腋　haunch　02.0481

两点检查　released by checking two sections　08.0150

两级增压柴油机　two-stage supercharged diesel engine　04.0699

两阶段设计　two-step design, two-phase design　02.0012

两系悬挂　two stage suspension　05.0177

量化　quantizing　07.0373

了望车　observation car　05.0739

了望窗　observation window　05.0507

列车　train　06.0199
列车保留　train stock reserved　06.0355
列车闭路电视　cable TV on train　07.0661
列车编成辆数　number of cars in a train　06.0281
列车编组顺序表　train consist list, train list　06.0273
列车车次　train number　06.0286
列车车底需要数　number of passenger train set required　06.0070
列车冲击力　impact force of train　02.0337
列车出发正点率　percentage of punctuality of trains despatched to total trains　06.0332
列车带电平均电流　average current of charging train　04.1323
列车带电运行时分　train running time on load　04.1274
列车等级　train class　06.0351
列车等线　train waiting for a receiving track　06.0354
列车颠覆　train overturning　01.0092
列车调度电话　train dispatching telephone　07.0412
列车动力学　train dynamics　01.0093
列车防护无线电通信　radio communication for train protection　07.0633
列车分离　train separation　01.0091
列车供电电路　power supply circuit for train　04.0376
列车供电绕组　train coach supply winding　04.0501
列车公里　train kilometers　06.0693
列车管　train pipe　05.0277
列车管压差　train pipe pressure gradient　04.0155
列车广播　broadcasting for train　07.0586
列车横向摇摆力　lateral swaying force of train　02.0344
列车活塞作用　piston action of train　02.0807
列车活载　live load of train　02.0335
列车加开　running of extra train　06.0357
列车接近报警器　train approach warning device　03.0454
列车接近传感器　train approaching sensor　07.0634
列车接近告警无线电通信　radio communication for warning of train approaching　07.0635

列车进路　train route　06.0245
列车客座利用率　percentage of passenger seats utilization per train　06.0073
列车空气动力学　train aerodynamics　01.0094
列车空隙作业时间　working time between trains　03.0260
列车扣除系数　coefficient of train removal　06.0311
列车拉伸　train running out　01.0090
列车离心力　centrifugal force of train　02.0336
列车旅客无线电话　passenger radiotelephone on train　07.0588
列车密度　train density　06.0329
列车平均电流　average current of train　04.1321
列车平均载客人数　average number of passengers carried per train　06.0072
列车平均总重　average train gross weight　06.0695
列车牵引力　tractive force of train　02.0342
列车去向　train destination　06.0280
列车确报　train list information after departure　06.0275
列车确报电报　train out report telegraph　07.0435
列车事故　train accident　06.0627
列车速度检测仪　train speed monitoring device　03.0453
列车停运　withdrawal of train　06.0356
列车尾部防护　train rear end protection　01.0086
列车尾部风压反馈　train rear end air pressure feedback　05.0373
列车尾追　train tail collision　01.0085
列车位置表示　train position indication　08.0483
列车无线电调度通信　radio dispatching communication for train　07.0583
列车无线电调度系统　train radio dispatching system　04.0071
列车无线电调度转接分机　transfer branch set for radio train-dispatching　07.0590
列车无线电调度总机　office equipment for radio train-dispatching　07.0589
列车无线电通信　train radio communication　07.0580
列车无线电通信系统　train radio communication system　07.0582
列车压缩　train running in　01.0089

列车业务无线电通信 radio communication for train service 07.0585

列车有效电流 effective current of train, r.m.s current of train 04.1322

列车与线路相互作用 track-train interaction 01.0065

列车预报 train list information in advance 06.0274

列车员 train attendant 06.0064

列车员室 attendant's room 05.0766

列车运缓 train running delay 06.0353

列车运行调整 train operation adjustment 06.0349

列车运行控制系统 train operation control system 08.0431

列车运行时刻表 timetable 06.0282

列车运行图 train diagram 01.0044

列车运行线 train path 06.0283

列车运行正点率 percentage of punctuality of trains running to total trains 06.0333

列车长 train conductor 06.0065

列车正面冲突 train collision 01.0084

列车制动 train braking 04.0111

列车制动简易试验 train brake simplified test 05.0952

列车制动力 braking force of train 02.0341

[列车制动]溜放试验 coasting braking test 05.0879

列车制动全部试验 train brake overall test 05.0951

列车制动试验器 train brake tester 05.0372

列车重量标准 railway train load norm 06.0228

列车自动调速 automatic train speed regulation 08.0436

列车自动限速 automatic train speed restriction 08.0432

列车自动运行 automatic train operation 08.0433

列车纵向动力 longitudinal dynamic force of train 05.0586

列检 train examination 05.0934

*列检所 freight train inspection and service point 05.0907

列检无线电通信 radio communication for train inspection 07.0623

*裂土地区路基 subgrade in swelling soil zone, sub-grade in expansive soil region 02.0217

*裂相回流线 split return wire 04.1315

临界高度 critical height 02.0246

临界坡度 critical grade 02.0185

临界牵引力 critical tractive effort 04.1090

临界速度 critical speed 01.0080

临界阻尼 critical damping 05.0245

临界阻尼值 critical damping value 05.0247

临时便线 shoofly 03.0575

临时工程 temporary project 02.0856

临时旅客列车 extra passenger train, additional passenger train 06.0212

临时性桥 temporary bridge 02.0430

临时支护 temporary support 02.0765

临险抢护 emergency rush engineering, emergency repairs 03.0580

临修 temporary repair 05.0930

临修线 temporary repair siding 05.0942

邻道干扰 adjacent channel interference 07.0530

邻线干扰 interference from neighboring line 08.0672

菱形交叉 diamond crossing 03.0291

零层端子 terminals of layer 0 of a relay rack 08.0278

零担办公车 office car for peddler train 05.0832

零担仓库 scattered freight storehouse 06.0528

零担货物 less-than-carload freight 06.0112

零担货物中转站 less-than-carload freight transhipment station, part-load transhipment station 06.0079

零应力轨温 stress free rail temperature 03.0228

铃流 ring current 07.0124

灵敏度 sensitivity 07.0034

溜车不利条件 unfavorable condition for car rolling 06.0473

溜车有利条件 favorable condition for car rolling 06.0472

溜放调车 fly-shunting, coasting, jerking 06.0260

*溜放进路程序控制 automatic switching control of humping yard by routes 08.0392

溜放进路自动控制 automatic switching control of humping yard by routes 08.0392

溜放速度 rolling speed 06.0498

溜放速度自动控制 automatic rolling down speed control 08.0385

硫磺锚固 sulphur cement mortar anchor, sulphur cement mortar anchorage 03.0094

流冰 ice drift 02.0638

流动资金 current capital, liquid fund 06.0748

流动资金周转 turnover of current capital 06.0749

流量控制 flow control 07.0867

流砂 quick sand, drift sand 02.0609

流水压力 pressure of water flow 02.0345

流速继电器 flow relay 04.0600

流线型车体 streamlined car body 05.0375

漏乘 missing a train 06.0057

漏斗 hopper 05.0467

漏斗车 hopper car 05.0808

漏斗门风手动传动装置 pneumatic/manual hopper door operating device 05.0513

漏斗棚架法 hopper-shed support tunnelling method 02.0736

漏解锁 missing release 08.0685

漏锁闭 missing locking 08.0684

漏泄 leakage 04.0139

漏泄电流 leakage current 04.1470

漏泄模 leaky modes 07.0456

漏泄同轴电缆 leaky coaxial cable 07.0173

漏泄同轴电缆传输方式 transmission mode with leaky coaxible cable 07.0550

炉撑 stay 04.1130

炉床 grate 04.1125

炉床面积 grate area 04.1254

炉口 fire hole 04.1126

炉门 fire door 04.1127

炉内软水 water softened in boiler 04.0063

炉外软水 water softened out of boiler 04.0064

炉灶 range 05.0756

卤素查漏仪 halogen leak detector 07.0215

路堤 embankment, fill 02.0238

路堤边坡 side slope of embankment, fill slope 02.0248

路堤填料 embankment fill material, embankment filler, filling material of embankment 02.0252

路拱 road crown, [subgrade] crown 02.0235

路基 subgrade, road bed, formation subgrade 02.0210

路基承载板测定 determination of bearing slab of subgrade, bearing slab method for subgrade testing, bearing plate test on subgrade 02.0263

路基横断面 subgrade cross-section 02.0232

路基机械 subgrade machine, machine for roadway work 03.0370

路基挤起 subgrade bulge, subgrade squeeze-out 03.0492

路基面 subgrade surface, formation 02.0233

路基面宽度 width of the subgrade surface, formation width 02.0234

路基松软 soft spots of road bed 03.0476

路基下沉 subgrade settlement 03.0490

路际串音 inter-channel crosstalk 07.0362

路肩 [road] shoulder, subgrade shoulder 02.0236

路肩高程 formation level, shoulder level 02.0237

路内人员伤亡 casualty of railway man, on-duty casualty 06.0650

路牌 tablet 08.0357

路牌携带器 tablet pouch 08.0360

路牌自动授收机 automatic tablet exchanger 08.0361

路签 train staff 08.0356

路签灯 train staff lamp 04.0646

路签授受器 staff exchanger, tablet exchanger 04.1066

路签携带器 staff pouch 08.0358

路签自动授收机 automatic staff exchanger 08.0359

路堑 cut, road cutting 02.0239

路堑边坡 cutting slope, side slope of cut 02.0280

路堑平台 platform of cutting, berm in cutting 02.0282

路堑石方爆破 rock cutting blasting, rock blasting in cut 02.0270

路堑式明洞 cut-type open cut tunnel, cut-type tunnel without cover, cut-type gallery 02.0697

路外人员伤亡 casualty of non-railway man, not on-duty casualty 06.0651

路网性编组站 network marshalling station 06.0370

路用机车 locomotive of service train, service locomotives 04.0018

路用列车 railway service train 06.0227

录音电话机 recording phone set 07.0098

铝热焊 [alumino-]thermit welding 03.0240

旅客乘车系数 coefficient of passengers travelling by trains 06.0680

旅客乘降所 passenger stop point 06.0526

旅客到达人数 number of passengers arrived 06.0031

旅客发送人数 number of passengers despatched, number of passengers originated 06.0030

旅客换乘 passenger transference 06.0054

旅客快车 fast passenger train 06.0203

旅客列车 passenger train 06.0202

旅客列车包车制 responsibility crew system of passenger train 06.0063

旅客列车包乘制 assigning crew system of passenger train 06.0062

旅客列车编组 passenger train formation 06.0201

旅客列车车底周转时间 turnround time of passenger train set 06.0069

旅客列车乘务制度 crew working system of passenger train 06.0060

旅客列车乘务组 passenger train crew 06.0059

旅客列车检修所 passenger train inspection and service point 05.0913

旅客列车轮乘制 crew pooling system of passenger train 06.0061

旅客列车直达速度 through speed of passenger train 06.0068

旅客疲劳时间 passenger fatigue time 05.0853

旅客票价 passenger fare 06.0731

旅客平均运程 average journey per passenger 06.0679

旅客人公里成本 cost per passenger kilometer 06.0724

旅客伤亡 passenger casualty 06.0652

旅客特别快车 express train 06.0204

旅客问讯电视 TV for passenger information service 07.0653

旅客运输计划 passenger traffic plan 06.0676

旅客运输量 volume of passenger traffic 06.0677

旅客运送人数 number of passengers transported 06.0032

旅客站舍 passenger building 06.0519

旅客直达特别快车 through express train 06.0205

旅客周转量 turnover of passenger traffic 06.0678

旅客最高聚集人数 maximum number of passengers in peak hours 06.0033

旅行速度 travelling speed, commercial speed 06.0331

旅游列车 tourist train 06.0211

履带式起重机 crawler crane, caterpiller crane 02.0896

履带式桩架 crawler pile frame 02.0911

滤波电抗器 filter reactor 04.0511

滤波电路 filter circuit 04.0374

滤波电容器 filter capacitor 04.0524

滤波器 filter 07.0057

滤波装置 filter 04.1304

滤尘效果试验 test on filter efficiency 04.0682

滤尘止回阀 strainer check valve 04.0169

卵石道碴 gravel ballast 03.0129

乱显示 false indication 08.0048

轮/轨接触应力 rail/wheel contact stress 03.0046

*轮渡斜引道 ferry slip 02.0643

轮渡引线 ferry slip 02.0643

轮渡栈桥 ferry trestle bridge 02.0641

轮渡站 ferry station 02.0640

轮对 wheelset, wheel pair 05.0094

轮对存放场 wheelset storing yard 05.0983

轮对动平衡检验 wheelset dynamic balance test 05.0962

轮对横动量 lateral play of wheel set 04.0050

轮对内侧距 distance between backs of wheel flanges 05.0048

轮箍 tire, tyre 05.0099

轮箍厚度 tire thickness 05.0221

轮毂 wheel hub 05.0118

轮毂长度 hub length 05.0227

轮毂厚度 hub thickness 05.0225

轮毂孔 wheel hub bore 05.0119

轮毂孔直径 hub bore diameter 05.0228

轮毂破裂 burst hub 05.0257

轮毂松动 wheel loose on axle 05.0258

轮毂直径 hub diameter 05.0226

轮轨关系 wheel-rail relation, wheel-rail interaction 01.0066

轮轨游间 clearance between wheel flange and gage line 03.0135

轮胎式起重机 rubber tired crane, rough-terrain crane 02.0895

轮辋 wheel rim 05.0117

轮辋宽度 rim width 05.0222

轮辋烧伤 burnt rim 05.0253

轮辋辗出 spread rim 05.0254

轮心 wheel center 05.0098

轮修 alternative maintenance 08.0731

轮询 polling 07.0852

轮缘 wheel flange 05.0116

轮缘槽 flange-way, flange clearance 03.0343

轮缘垂直磨耗 vertical flange 05.0250

轮缘高度 flange height 05.0223

轮缘厚度 flange thickness 05.0224

轮缘喷油器 flange lubricator 04.0638

轮重 wheel load 01.0049

轮重减载率 rate of wheel load reduction 05.0858

轮周功率 output power at wheel rim 04.0085

轮周牵引力 tractive effort at wheel rim 04.0090

轮装盘形制动 wheel-mounted disc brake 05.0328

轮座 wheel seat 05.0113

螺杆式空气压缩机 screw compressor 02.0922

螺杆式制冷压缩机 screw type refrigeration compressor 05.0629

螺孔裂纹 bolt hole crack 03.0193

螺栓扳手 track wrenches 03.0471

* 螺栓孔倒棱 bolt hole chamfering 03.0184

螺栓孔加强 bolt hole cold-working strenthening 03.0185

螺栓孔削角 bolt hole chamfering 03.0184

螺栓螺纹钉 bolt-screw spike 03.0093

螺栓示功扳手 bolt wrench with indicator 02.0368

螺纹道钉 screw spike 03.0092

螺旋喷射桩 auger injected pile 02.0596

螺旋曲线 spiral curve, clothoid curve 03.0024

螺旋弹簧 coil spring, helical spring 05.0167

螺旋卸车机 spiral unloading machine 06.0599

逻辑信道 logical channel 07.0879

裸露药包爆破 adobe blasting, contact blasting 02.0279

落锤试验 drop test 05.0590

落道 under cutting of track, lowering of track 03.0200

落石 rock fall 03.0482

落石槽 stone falling channel, trough for catching falling rocks 02.0319

落石冲击力 impact force of falling stone 02.0713

落下孔车 well-hole car 05.0835

* 落下时间 drop away time 08.0561

* 落下值 release value 08.0569

洛杉矶磨损试验 Los Angeles abrasion test 03.0126

洛泽式桥 Lohse bridge, rigid arch bridge with rigid tie and vertical suspenders 02.0405

M

码速调整 code rate justification 07.0378

码头 wharf 02.0621

码型发生器 pattern generator 07.0371

码元 element 07.0810

码组 block 07.0743

码组结束信号 end-of-block signal 07.0837

码组起始信号 start-of-block signal 07.0860

码组校验 block check 07.0815

马口 excavation of side wall at intervals 02.0746

* 埋深 buried depth of tunnel 02.0714

埋置式桥台 buried abutment 02.0549

脉冲 impulse, pulse 07.0025

脉冲编码调制 pulse code modulation 07.0027

脉冲变压器 pulse transformer 04.0492

脉冲继电器 impulse relay 08.0583

脉冲式轨道电路 pulse track circuit 08.0494

脉冲增压 pulse charging 04.0791

脉冲自动闭塞 automatic block with impulse track circuit 08.0332

脉流牵引电动机 pulsating current traction motor 04.0396

脉码调制终端机 PCM terminal 07.0367

满磁场 full field 04.0432

曼彻斯特编码 Manchester encoding 07.0734

慢行牌 yellow board, speed indicator 03.0247

盲沟 blind ditch 02.0800

忙时串杂音 busy hour crosstalk and noise 07.0359

忙音 busy tone 07.0277

锚定板挡土墙 anchored retaining wall by tie rods, anchored bulkhead retaining wall, anchored plate retaining wall 02.0287

锚定板式桥台 anchor slab abutment 02.0550

锚碇 anchorage 02.0498

锚段 tension length 04.1424

锚段关节 overlap 04.1429

锚杆挡墙 anchored bolt retaining wall, anchored retaining wall by tie rods 02.0289

锚杆支护 anchor bolt support 02.0769

锚固装置 anchor fitting 04.1428

*锚孔 anchor span 02.0441

锚跨 anchor span 02.0441

*锚喷支护 shotcrete and rock bolt support 02.0770

锚柱 anchor mast 04.1412

锚座 socket 02.0497

毛巾杆 towel hanging rod, towel rail 05.0769

铆接钢桥 riveted steel bridge 02.0363

冒顶 roof fall 02.0993

冒进信号 overrunning of signal 06.0631

煤槽 fuel space, coal bin 04.1241

煤车 coal car 05.0803

煤的技术当量 equivalence of coal 04.1105

煤粉机车 pulverized coal locomotive, fine coal locomotive 04.1082

煤矿沉陷 mining subsidence 03.0491

煤气机车 gas fired locomotive 04.1083

煤水车 tender 04.1239

煤水车转向架 tender truck 04.1242

每延米重量 load per meter of track 05.0860

门滑轮 door guide roller 05.0501

门孔宽度 width of door opening 05.0037

门式铺轨排机 track panel laying gantry crane 03.0443

门柱 door post 05.0432

迷宫式密封 labyrinth seal 05.0144

*迷流 leakage current 04.1470

米轨铁路 meter-gage railway 01.0013

密封圈 seal ring 05.0146

密接式车钩 tight-lock coupler 05.0528

密贴尖轨 closed switch rail, close contact between switch point and stock rail 03.0355

密贴调整杆 adjustable switch operating rod 08.0304

面漆 top coat 03.0553

面向比特协议 bit-oriented protocol 07.0884

面向字符协议 character-oriented protocol 07.0883

灭火花电路 spark extinguishing circuit 08.0227

敏感系数 sensitivity coefficient 04.1494

明洞 open-cut tunnel, tunnel without cover, gallery 02.0694

明洞门 open-cut-tunnel portal, gallery portal 02.0675

明桥面 open deck, ballastless deck, open floor 02.0499

明渠 open channel, open ditch, open drain 02.0634

明挖法 open-cut method 02.0719

明挖基础 open-cut foundation, open excavation foundation 02.0582

明线通信 open-wire communication 07.0316

鸣笛标 whistle board 03.0256

蘑菇形开挖法 mushroom-type tunnelling method 02.0737

模板 form 02.0956

模场直径 mode field diameter 07.0450

模拟试验 analogue test 04.0298

模拟微波中继通信 analog microwave relay communication 07.0664

模数牵引力 modulus of tractive effort 04.1086

模筑衬砌 moulded lining 02.0683

膜板 diaphragm 05.0299

膜式空气弹簧 diaphragm type air spring 05.0171

磨轨车 rail grinding car 03.0392

磨轨机 rail grinding machine 03.0391

磨轨列车 rail grinding train 03.0393

磨耗板 wearing plate 05.0162

磨耗限度 limit of wear 04.0267

磨耗型踏面 worn profile tread 05.0216

磨合 running-in 04.0855
摩擦板 friction plate 05.0161
摩擦电流 frictional working current 08.0612
摩擦结合式高强度螺栓 high strength friction grip bolt 02.0366
摩擦离合器 friction clutch 04.1049
摩擦联结器 frictional clutch 08.0611
摩擦式减振器 snubber 05.0157
摩擦制动 friction braking 05.0322
摩擦桩 friction pile 02.0597
莫尔斯电码 Morse code 07.0702
母车 car with axle generator 05.0750
木结构 wood structure 05.0526
木前枕 wooden strut 05.0139

木枕 wooden tie 03.0112
木枕防腐 preservation of wooden tie 03.0218
木枕防裂装置 anti splitting device 03.0222
木枕刻痕 incising of wooden tie 03.0220
木枕削平机 wooden sleeper adzing machine 03.0412
木枕预钻孔 preboring of wooden tie, preboring of spike holes 03.0219
木枕钻孔机 wooden tie drilling machine 03.0407
木桩 timber pile 02.0605
目标打靶控制 target shooting 08.0394
目测 visual measurement, visual observation 02.1045
目的制动 target braking 06.0471

N

耐腐蚀轨 corrosion resistant rail 03.0055
耐久性试验 durability test 04.0292
耐磨轨 wear resistant rail 03.0054
难行车 hard rolling car 06.0489
难行线 hard running track 06.0429
囊式空气弹簧 bellows type air spring 05.0172
挠度 deflection 02.0439
挠度裕量系数 coefficient of spring deflection reservation 05.0241
内部过电压试验 test on internal overvoltage 04.0673
内顶板 ceiling 05.0456
内端墙 inside end wall 05.0434
内构架转向架 inside-frame truck 05.0073
内火箱 inside firebox, inner firebox 04.1118
内火箱顶板 crown sheet 04.1120
内燃动车组 diesel coach set 04.0689
内燃机车 diesel locomotive 04.0006
内燃牵引干扰 diesel traction interference 08.0675
内梯 manhole ladder 05.0475
内烟筒 stack extension 04.1144
内止点 inner dead point, inner dead center 04.0714
内走廊 inside corridor, gangway 04.1063
能力储备系数 coefficient of reserved capacity 06.0307

泥浆套沉井法 slurry jacket method for sinking caisson 02.0589
泥石流 earth flow, debris flow, mud and rock flow 03.0483
泥石流地段路基 subgrade in debris flow zone 02.0231
泥石流流域 catchment basin of debris flow 02.0087
泥沼地区路基 subgrade in bog [soil] zone, subgrade in morass region, subgrade in swampland 02.0216
尼尔森式洛泽梁桥 Nielsen type Lohse bridge, rigid arch bridge with rigid tie and inclined suspenders 02.0407
尼尔森体系桥 Nielsen system bridge 02.0406
逆变器 inverter 07.0049
逆流燃烧 counter-flow combustion 04.1565
逆向牵引 backward haulage 04.1267
粘着牵引力 adhesive tractive effort 04.0093
粘着系数 adhesion coefficient 01.0067
粘着制动 adhesion braking 05.0323
粘着重量利用系数 adhesive weight utility factor 04.0048
粘着重量增加器 booster 04.1238
扭杆弹簧 torsion bar spring 05.0170
扭剪式高强度螺栓 torshear type high strength bolt

02.0367

*扭曲 twist, warp 03.0152

暖汽端阀 train pipe end valve 05.0621

暖汽软管 heater hose 05.0624

暖汽软管连接器 heater hose coupler 05.0625

暖汽支管 heater branch pipe 05.0623

暖汽主管 heater train pipe 05.0622

P

爬轨 climb on rail 05.0874

爬升模板 climbing shuttering 02.0959

排斥力 repulsion force 04.1596

排风道 air exhausting duct 05.0641

排架 bent 02.0622

排架式桥墩 pile bent pier 02.0571

排列进路 route setting 08.0187

排流线圈 drainage coil 07.0210

排气背压 exhaust back pressure 04.0738

排气道 exhaust duct 04.1537

排[气]风扇 exhaust fan 05.0620

排气门 exhaust valve 04.0927

排气速度 exhaust velocity [for a turbine] 04.1542

排气凸轮 exhaust cam 04.0939

排气温度 exhaust temperature 04.0737

排气箱 exhaust box 04.0985

排气行程 exhaust stroke 04.0726

排气烟度 exhaust smoke density 04.0810

排气支管 exhaust branch pipe 04.0984

排气总管 exhaust manifold 04.0983

排汽余面 exhaust lap 04.1259

排水槽 drainage channel 02.0302

*排水阀 blow off valve 04.1174

排水沟 weep drain, drainage ditch, drain ditch 02.0295

排水砂垫层 sand filled drainage layer, drainage sand blanket 02.0314

*排水砂井 sand drain 02.0310

排烟热损失 heat loss due to exhaust gas 04.1109

排障器 pilot, life guard 04.1065

盘车 barring 04.1530

盘车机构 barring mechanism, barring gear 04.0970

盘型制动 disc brake 04.0124

盘形缓冲器 side buffer 05.0554

判决值 decision value 07.0374

旁承 side bearing 05.0182

旁承承载 side bearing loading 05.0055

旁承间隙 side bearing clearance 05.0036

旁承载荷 load on side bearing 05.0234

旁承支重转向架 side bearing truck 05.0072

旁弯 sidewise bending 05.0443

抛石防护 riprap protection 03.0579

抛石挤淤 throwing stones to packing sedimentation, packing sedimentation by throwing stones, packing up sedimentation by dumping 02.0269

抛掷爆破 pin-point blasting 02.0276

*跑道 track buckling 03.0160

配碴整形机 ballast distributing and regulating machine 03.0419

配电柜 electric power distribution cabinet 05.0677

配电盘 switch board 05.0676

配件互换修 repair with interchangeable component 04.0254

*配气定时 valve timing 04.0727

配气相位 valve timing 04.0727

配属机车 allocated locomotive 04.0019

配线电缆 distribution cable 07.0188

喷锚衬砌 shotcrete bolt lining 02.0690

喷锚构筑法 shotcrete-bolt construction method 02.0723

喷锚支护 shotcrete and rock bolt support 02.0770

喷洒除草机 weed killing machine, weed killer 03.0464

喷砂除锈 sand blasting 03.0548

喷射钢纤维混凝土 steel fiber shotcrete, steel fiber reinforced shotcrete 02.0782

喷射混凝土 shotcrete, spray concrete 02.0778

喷射混凝土回弹 rebound of shotcrete 02.0761

喷射混凝土机械手 shotcrete manipulator 02.0974

喷射混凝土修理 shotcrete repair 03.0563

喷射混凝土支护 shotcrete support 02.0768

喷油泵　injection pump　04.0947

* 喷油泵传动齿轮　injection pump drive gear　04.0944

喷油泵传动装置　injection pump transmission mechanism　04.0945

喷油泵驱动齿轮　injection pump drive gear　04.0944

喷油泵凸轮　injection pump cam　04.0940

喷油持续角　continuous injection angle　04.0779

喷油规律　law of injection　04.0778

喷油器　fuel injector　04.0954

喷油器滤芯　injector filter core　04.0958

喷油速率　injection rate　04.0777

喷油提前角　injection advance angle　04.0781

喷油嘴偶件　injector nozzle matching parts　04.0955

喷油嘴针阀　nozzle needle valve　04.0956

喷油嘴针阀体　nozzle needle valve body　04.0957

喷嘴环　nozzle ring　04.0978

盆式橡胶支座　pot rubber bearing　02.0522

盆形燃烧室　bowl [shaped] combustion chamber　04.0802

棚车　box car, covered goods wagon　05.0795

棚洞　shed tunnel, shed gallery　02.0696

膨胀比　expansion ratio　04.1544

膨胀多变指数　polytropic index of expansion　04.0758

膨胀水箱　expansion tank, expansion drum　04.1030

膨胀土地区路基　subgrade in swelling soil zone, subgrade in expansive soil region　02.0217

膨胀行程　expansion stroke　04.0725

膨胀终点温度　expansion terminal temperature　04.0756

膨胀终点压力　expansion terminal pressure　04.0757

劈相机　Arno converter, phase splitter　04.0470

劈相机故障隔离开关　fault isolating switch for phase splitter　04.0571

偏畸变　bias distortion　07.0710

偏极继电器　polar biased relay　08.0593

偏心杆　eccentric rod　04.1204

偏心曲拐　eccentric crank　04.1229

偏压衬砌　unsymmetrically loading lining eccentrically compressed lining　02.0685

片间平均电压　mean voltage between segments　04.0430

片间最高电压　maximum voltage between segments　04.0431

票货分离　separation of waybill from shipment　06.0644

票据传送设备　classification list conveyer system　08.0395

拼接板　splice plate　02.0455

拼装式桥墩　assembly pier, pier constructed with precast units　02.0564

频带　frequency band　07.0006

频道　frequency channel　07.0005

频道空闲信号　path free signal　07.0610

频道切换阈值　threshold for channel switching　07.0606

频道选择器　channel selector　07.0564

频分复用　frequency division multiplex　07.0011

频率　frequency　07.0007

频率定点切换　fixed-point frequency switching　07.0607

频率分集　frequency diversity　07.0579

频率跟踪切换方式　frequency tracking switching mode　07.0605

频[率]偏[差]　frequency offset　07.0284

频谱特性　spectrum character　04.1517

频移键控　frequency shift keying, FSK　07.0732

频组方式　frequency group mode　07.0600

频组选择器　group frequency selector　07.0565

品字形导坑法　top and twin-side bottom heading method　02.0741

平板地板　flat floor　05.0402

平板式混凝土振捣器　plate vibrator　02.0942

平板支座　plate bearing　02.0517

平波电抗器　smoothing reactor　04.0506

平车　flat car　05.0798

平地机　grader　02.0886

平调　flat regulation　07.0311

平轨器　rail bending tool, rail bender　03.0398

平均工作气隙　mean operating air gap　04.1631

平均机械损失压力　mean mechanical loss pressure

04.0825

平均减速度　average retardation rate　05.0350

平均输出光功率　average optical output power　07.0464

平均有效压力　mean effective pressure　04.0824

平均值检波器　average detector　07.0572

平均指示压力　mean indicated pressure　04.0823

平孔排水　horizontal hole drainage　02.0307

平面调车场　flat marshalling yard　06.0508

平面交叉　level crossing　03.0271

平面无线电调车信号　radio operated signal for level shunting　07.0685

平台集装箱　platform container　06.0546

平托盘　flat pallet　06.0566

平稳性指标　riding index　05.0851

平型双头夹板　flat joint bar　03.0078

平行导坑　parallel heading　02.0789

平行渡线　parallel crossover　03.0298

平行接近　parallelism approach　04.1486

平行进路　parallel route　06.0250

平行运行图　parallel train diagram　06.0289

平旋桥　swing bridge　02.0410

平原地区选线　location in plain region, plain location　02.0147

屏蔽　shielding　04.1513

屏蔽变压器　reduction transformer　04.1497

屏蔽电缆　shielded cable　07.0181

屏蔽体　shield　04.1512

屏蔽系数　shielding factor　04.1477

坡道阻力　gradient resistance　04.0100

坡度　grade, gradient, slope　02.0180

坡度标　grade post　03.0251

坡度[代数]差　algebraic difference between adjacent gradients　02.0194

坡度牵出线　draw-out track at grade　06.0434

坡度折减　compensation of gradient, gradient compensation, grade compensation　02.0200

坡段　grade section　02.0192

坡段长度　length of grade section　02.0193

坡脚　toe of side slope　02.0249

坡面防护　slope protection　02.0315

破冰体　ice apron, ice-breaking cutwater, ice guard　02.0576

破封　break a seal　08.0712

破损阶段设计法　plastic stage design method　02.0206

迫导向径向转向架　forced-steering radial truck　05.0070

铺草皮　sodding　03.0509

铺轨机　track laying machine　03.0439

铺轨列车　track laying train　03.0444

普通运价　general rate　06.0735

Q

起拨道机　track lifting and lining machine　03.0387

起拨道器　track lifting and lining tool　03.0388

起道　raising of track, track lifting　03.0199

起道钉机　spike puller　03.0409

起道机　track lifting machine　03.0385

起道器　rail jack, track jack　03.0386

* 起动变矩器　starting torque converter　04.1003

起动变扭器　starting torque converter　04.1003

起动电流　starting current　04.0426

起动电路　starting circuit　04.0385

起动电容器　starting capacitor　04.0527

起动电阻器　starting resistor　04.0515

起动缓坡　flat gradient for starting　02.0198

起动机油泵　starting lubricating oil pump　04.1038

起动继电器　starting relay　04.0604

起动加速器　starting accelerator　04.0966

起动加速试验　starting and acceleration test　04.0679

起动牵引力　starting tractive effort　04.0092

起动转矩　starting torque　04.0422

起动阻力　starting resistance　04.0098

起拱点　springing　02.0486

起止式　start-stop type　07.0713

起止信号　start-stop signal　07.0863

起止信号发生器　start-stop signal generator　07.0699

起止信号畸变测试器　start-stop signal distortion

tester 07.0700

起重横梁 jacking floor beam 02.0466

企业自备车 private car 06.0095

砌拱支架 soffit scaffolding 02.0960

气闭头 gas-tight block 07.0199

气垫 air-cushion 04.1612

气缸 cylinder 04.0872

气缸盖 cylinder head 04.0919

气缸盖螺母 cylinder head nut 04.0921

气缸盖螺栓 cylinder head stud 04.0920

气缸工作容积 stroke volume 04.0718

气缸套 cylinder liner 04.0873

气缸体 cylinder block 04.0864

气缸直径 cylinder bore diameter 04.0706

气缸最大容积 maximum stroke volume 04.0716

气缸最小容积 minimum stroke volume 04.0717

气环 gas ring 04.0907

气门 valve 04.0925

气门导管 valve guide 04.0930

气门横臂 valve cross arm 04.0932

气门弹簧 valve spring 04.0928

气门挺柱 valve tappet 04.0934

气门推杆 valve push rod 04.0933

气门旋转机构 valve rotating mechanism 04.0936

气门摇臂 valve rocker 04.0931

气门座 valve seat 04.0929

气密性 air-tightness 05.0653

气敏查漏仪 gas-sensitive leak detector 07.0218

气隙磁通密度 air gap flux density 04.1613

气隙－力特性 distance-force characteristic 04.1627

气象资料 meteorological data 02.0064

气压焊 oxyacetylene pressure welding 03.0238

弃土堆 waste bank, bankette, spoil bank 02.0283

汽车式起重机 automobile crane, autocrane, truck crane 02.0894

汽笛 whistle 04.1181

汽阀 piston valve 04.1201

汽缸鞍 cylinder saddle 04.1185

汽缸后盖 back cylinder head 04.1184

汽缸排水阀 cylinder drain valve 04.1189

汽缸牵引力 cylinder tractive effort 04.1089

汽缸前盖 front cylinder head 04.1183

汽缸套 cylinder bushing 04.1182

*汽门 regulator valve, throttle valve 04.1159

汽室 steam chest, valve chest 04.1186

汽室套 steam chest bushing 04.1187

汽水共腾 priming 04.1265

牵出线 switching lead, shunting neck, lead track 06.0410

牵出线改编能力 resorting capacity of lead track 06.0503

牵引变电所 traction substation, SS 04.1342

牵引变电所标称电压 nominal voltage of traction substation 04.1330

牵引变流器 traction convertor 04.0326

牵引变频器 traction frequency convertor 04.0328

牵引变压器 traction transformer 04.1284

牵引车及挂车 tractor and trailer 06.0596

牵引电动机 traction motor 04.0394

牵引电动机隔离开关 traction motor isolating switch 04.0568

牵引电动机供电制式 traction motor power supply system 04.0364

牵引电机电流表 traction motor ammeter 04.0620

牵引电机电压表 traction motor voltmeter 04.0619

牵引电抗器 traction reactor 04.0505

牵引电路 traction circuit 04.0372

牵引定数 tonnage rating, tonnage of traction 01.0072

牵引方式 mode of traction 01.0071

牵引杆 draw bar 04.1080

牵引钩 hauling hook 05.0420

牵引回流电路 traction return current circuits 04.1313

牵引回流轨 traction return current rail 04.1337

牵引力 tractive force 08.0622

牵引梁 draft sill 05.0421

牵引逆变器 traction invertor 04.0327

牵引热工试验 traction and thermodynamic test 04.0287

牵引试验 traction test 04.0285

牵引网 traction electric network 04.1272

牵引网阻抗 impedance of traction electric network 04.1335

牵引–制动位转换开关 power/brake changeover switch 04.0561

牵引种类 kinds of traction, category of traction 01.0070

牵引装置 draw gear 04.1079

牵纵拐肘 escapement 08.0300

铅垂弹性 vertical elasticity, vertical resilience 04.1443

*铅堵 fusible plug 04.1179

千斤顶 jack，02.0903

钳夹车 schnabel car 05.0836

钳形梁 schnabel 05.0424

前窗 front window 04.1057

前垂板 draft plate 04.1149

前从板座 front draft lug, front draft stop 05.0411

前后风压差 false gradient 04.0154

*前后高低 longitudinal level of rail, track profile 03.0151

前缓冲铁 front bumper 04.1218

前接点 front contact 08.0552

前墙 front wall 02.0554

前圈 front coil 08.0576

前群 pregroup 07.0307

前照灯 head lamp, head light 04.0641

前置机 front-end processor 07.0891

潜孔钻机 diving drill 02.0889

潜水泵 diving pump 02.0920

潜水钻机 diving drill machine 02.0913

浅孔爆破 shallow hole blasting 02.0272

浅埋隧道 shallow tunnel, shallow-depth tunnel, shallow burying tunnel 02.0649

堑顶 top of cutting slope, top of cutting 02.0281

欠超高 deficient superelevation 03.0027

欠电压继电器 under-voltage relay 04.0597

欠挖 underbreak 02.0754

墙板压筋 flute or rib on sheathing 05.0449

墙顶封口 seal, seal at the top of wall 02.0760

强拆 forced releasing 07.0127

强电干扰 high voltage interference 08.0673

强度试验 strength test 04.0290

强化系数 coefficient of intensification 04.0836

强迫导向循环 forced guided circulation 04.0504

强迫通风式电动机 force ventilated motor 04.0414

强迫循环 forced circulation 04.0503

强制内接 compulsory inscribing 03.0138

强制式混凝土搅拌机 forced concrete mixer 02.0929

敲缸 knock 04.0808

桥渡勘测设计 survey and design of bridge crossing 02.0084

桥墩 pier 02.0559

[桥涵]拱圈 arch ring 02.0482

桥涵扩孔 opening enlargement of bridge and culvert 03.0568

桥涵水文 hydrology of bridge and culvert 02.0081

桥梁标 bridge post 03.0252

桥梁标准活载 standard live load for bridge 02.0326

桥梁道碴槽 ballast trough 02.0500

桥梁荷载谱 bridge load spectrum 02.0327

桥梁合龙 closure 02.0540

桥梁横向刚度 lateral rigidity of bridge 03.0542

桥梁护轨 guard rail of bridge 02.0502

桥梁护木 guard timber of bridge 02.0503

桥梁基础 bridge foundation 02.0580

桥梁检定承载系数 rated load-bearing coefficient for bridge, rated load-bearing coefficient for bridge as compared with standard live loading 03.0521

桥梁检定试验 bridge rating test 03.0517

桥梁建筑限界 bridge construction clearance, bridge structure gauge 01.0057

桥梁孔径不足 unsufficient span of bridge 03.0530

桥梁挠度 deflection of bridge span 03.0532

桥梁疲劳剩余寿命 fatigue residual life of bridge 03.0540

桥梁疲劳剩余寿命评估 evaluating fatigue residual life of bridge 03.0541

桥梁浅基 shallow foundation of bridge，unsafe depth foundation of bridge 03.0513

桥梁浅基防护 unsafe depth foundation protection, bridge shallow foundation protection 03.0569

桥梁全长 overall length of bridge 02.0433

桥梁上部结构 superstructure 02.0443

桥梁上拱度 camber of bridge span 03.0531

桥梁实验车 bridge test car 03.0448

桥梁通知设备 bridge announciating device 08.0450

桥梁下部结构 substructure 02.0546

桥梁遮断信号 bridge obstruction signal 08.0449

桥梁自振频率 self-excited vibrational frequency of bridge span, natural frequency of bridge span 03.0544

桥梁自振周期 natural vibration period of bridge 02.0529

桥梁最大横向振幅 maximum lateral amplitude of bridge 03.0543

桥门架 portal frame 02.0461

桥面系 floor system 02.0464

桥前壅水高度 backwater height in front of bridge, top water level in front of bridge 02.0083

桥上人行道 sidewalk on bridge 02.0505

桥隧病害整治 damage repair for bridge and tunnel, repair bridge and tunnel fault 03.0567

桥隧大修 major repair of bridge and tunnel, capital repair of bridge and tunnel 03.0545

桥隧改造 upgrading of bridge and tunnel 03.0546

桥隧经常保养 regular maintenance of bridge and tunnel 03.0565

桥隧屏蔽系数 shielding factor of bridge and tunnel 04.1480

桥隧守护电话 bridge and tunnel guarder telephone 07.0418

桥隧巡守 bridge and tunnel patrolling 03.0571

桥隧巡守工 bridge and tunnel watchman, bridge and tunnel patrolling man 03.0009

桥隧养护 maintenance of bridge and tunnel 03.0564

桥隧综合维修 comprehensive maintenance of bridge and tunnel structure 03.0566

桥塔 bridge tower, pylon 02.0490

桥台 abutment 02.0547

桥下净空 underneath clearance 02.0434

桥形接线 bridge connection 04.1282

桥枕 bridge tie, bridge sleeper 02.0504

桥址水文观测 hydrologic observation of bridge site 03.0516

桥轴线测量 survey of bridge axis 02.0145

撬棍 lining bar, claw bar 03.0469

切断音响按钮 button for cut-off an audible signal 08.0446

切换阀 transfer valve 04.0166

侵入限界绝缘 insulated joint located within the clearance limit 08.0537

轻浮货物 light and bulk freight 06.0121

轻轨铁路 light railway, light rail 01.0026

轻型轨道车 light rail motor car, light motor trolley 03.0433

＊轻型线路机械 light permanent way machine, small permanent way machine 03.0382

清筛道床 ballast cleaning 03.0225

清算收入 clearing revenue 06.0752

清晰度 articulation 07.0149

请求数据传送 request data transfer 07.0856

丘陵地段选线 hilly land location, location of line on hilly land 02.0150

球面支座 spherical bearing 02.0520

球芯折角塞门 ball type angle cock 05.0284

球型燃烧室 spherical combustion chamber 04.0800

球形滚子轴承 spherical roller bearing 05.0142

球形心盘 spherical center plate 05.0408

球形轴箱定位装置 ball type journal box positioning device 05.0154

区段 district 01.0047

区段列车 district train 06.0223

区段锁闭 section locking 08.0130

区段通信 division communication 07.0410

区段小运转列车 district transfer train 06.0225

区段遥控 remote control for a section 08.0458

区段占用表示 section occupancy indication 08.0171

区段站 district station 06.0366

区间 section 01.0046

区间闭塞 section blocked 08.0318

区间电话 track-side telephone 07.0417

区间电话转接机 track-side telephone switching device 07.0419

区间封锁 section closed up 08.0707

区间空闲 section cleared 08.0319

区间联系电路 liaison circuit with block signaling 08.0237

区间通过能力　carrying capacity of the block section　06.0308

区间信号　wayside signaling　08.0002

区间占用　section occupied　08.0320

区域地质　regional geology　02.0059

区域性编组站　regional marshalling station　06.0371

曲柄　crank　04.0891

曲柄半径　crank radius　04.0707

曲柄臂　crank arm　04.0890

曲柄销　crank pin　04.0889

曲梁　curved beam　02.0415

曲梁式转向架　curved-beam truck　05.0068

曲线标　curve post　03.0250

曲线超高　superelevation, cant, elevation of curve　03.0026

曲线出岔道岔　turnout from curved track　03.0319

曲线尖轨　curved switch　03.0315

曲线控制点　curve control point　02.0118

曲线内接　inscribed to curves　03.0136

曲线桥　curved bridge　02.0414

曲线通过　curve negotiating　05.0845

曲线通过试验　curve negotiation test　04.0667

曲线折减　compensation of curve, curve compensation　02.0201

曲线辙叉　curved frog　03.0353

曲线正矢　curve versine　03.0031

曲线阻力　curve resistance　04.0101

曲轴　crankshaft　04.0887

曲轴减振器　crankshaft vibration damper　04.0894

曲轴平衡块　crankshaft counter balance　04.0892

曲轴箱　crankcase　04.0865

曲轴箱防爆门　explosion-proof door of crankcase　04.0869

曲轴箱呼吸器　crankcase breather　04.0866

曲轴正时齿轮　crankshaft timing gear　04.0941

取送调车　taking-out and placing-in of cars　06.0258

取土坑　borrow pit　02.0251

取消闭塞　to cancel a block　08.0350

取消进路　to cancel a route　08.0189

取样试验检查　sampling inspection　02.1044

去禁溜线信号　shunting signal to prohibitive humping line　08.0040

去滞曲线　demagnetisation curve　04.1619

全补偿链形悬挂　auto-tensioned catenary equipment　04.1358

全车振动试验台　full car vibration test rig　05.0993

* 全磁场　full field　04.0432

全叠片机座　full-laminated frame　04.0455

全断面道床夯实机　full section ballast consolidating machine　03.0425

全断面开挖法　full-face tunnelling method　02.0728

全反射　total internal reflection　07.0453

全分配制会议电话　conference telephone of full-distribution system　07.0299

全封闭式电动机　totally-enclosed motor　04.0411

全焊钢桥　all welded steel bridge　02.0365

全呼　general calling　07.0422

全回转式架梁起重机　full circle girder erecting crane　02.0946

全金属客车　all metal passenger car, all metal coach　05.0719

全开位置　full open position of coupler　05.0568

全面起道捣固　out-of-face surfacing　03.0198

全球定位系统　global positioning system, GPS　02.0137

全向天线　omnidirectional antenna　07.0567

全验　overall acceptance　02.1048

缺省值　default value　07.0751

缺陷工程　defect project, drawback project　02.1054

确认　acknowledge, ACK　07.0808

裙板　apron　05.0391

裙筒　apron shell　04.1145

群放大器　group amplifier　07.0336

群解调器　group demodulator　07.0335

群调制器　group modulator　07.0334

群转接站　group through-connection station　07.0320

R

燃煤独立温水采暖装置 coal-burning [heater type] hot water heating equipment 05.0600

燃煤机车 coal fired locomotive 04.1081

燃煤温水锅炉 coal-burning heater 05.0604

燃气侧全压损失 total pressure loss for gas side 04.1576

燃气发生器 gas generator 04.1532

燃气轮机车 gasturbine locomotive 04.0008

燃烧过程 combustion process 04.0774

燃烧率 rate of combustion, rate of firing 04.1100

燃烧区 combustion zone 04.1570

燃烧稳定性 combustion stability 04.1577

燃烧效率 combustion efficiency 04.1113

燃油粗滤器 fuel prefilter 04.1035

燃油独立温水采暖装置 oil-burning [heater type] hot water heating equipment 05.0601

燃油精滤器 fuel precision filter 04.1036

燃油输送泵 fuel feed pump 04.1034

燃油温水锅炉 oil burning heater 05.0605

燃油箱 fuel tank 04.1069

燃油消耗量 fuel consumption 04.0767

燃轴 severe hot box 05.0955

扰流板 spoiler 02.0527

绕行地段 detouring section, round section 02.0203

热泵 heat pump 05.0630

热风采暖装置 hot air heating equipment 05.0616

热腐蚀 hot corrosion 04.1558

热负荷 thermal load 04.0834

热工试验室 heat engineering laboratory 05.0991

热挂 thermal blockage 04.1531

热滑 hot-running 04.1441

热[力]继电器 thermal relay 08.0589

热裂纹 thermal crack 05.0256

热疲劳 thermal fatigue 04.1557

热平衡 heat balance, thermo balance 04.0760

热平衡试验 heat balance test 04.0860

热水泵 hot water pump 04.1172

热值 heat value 04.0052

热轴 hot box 05.0956

人工电报机 manual telegraph set 07.0690

人工电话所 manual telephone office 07.0389

人工分路 manual shunt 08.0527

人工解锁 manual release 08.0144

人工解锁表示 manual release indication 08.0166

人孔 manhole, manway 05.0472

人孔盖 manhole cover 05.0473

人为故障 human failure 08.0701

人行道荷载 sidewalk loading 02.0339

人行桥 foot bridge, pedestrian bridge 02.0360

人员掩蔽所 hiding-place for personnel 02.1006

人字坡 double spur grade 02.0181

日常检查 routine inspection 04.0248

日常维修 current maintenance 08.0725

日降雨量 daily precipitation 03.0582

日要车计划 daily car requisition plan 06.0099

熔断器 fuse 08.0664

熔断器断丝 fuse burn-out 08.0690

熔断器断丝报警 fuse break alarm 08.0178

容许冲刷 allowable scour 02.0095

容许强度检定 working stress rating 03.0518

容许信号 permissive signal 08.0025

容许应力设计法 allowable stress design method 02.0205

容许运输期限 permissive period of transport 06.0182

柔性[桥]墩 flexible pier 02.0563

*柔性系杆刚性拱 tied arch 02.0403

蠕变 creep, crawl 03.0479

蠕滑 creep 05.0846

入中继电路 incoming trunk circuit 07.0263

*入中继器 incoming trunk circuit 07.0263

软定位器 pull-off arm 04.1393

软横跨 headspan suspension 04.1406

软水与净水 water softened and purified 04.0065

软土地区路基 subgrade in soft soil zone, subgrade in

soft clay region 02.0215

软卧车 cushioned berth sleeping car, upholstered couchette 05.0728

软座车 cushioned-seat coach, upholstered-seat coach 05.0726

瑞利散射 Rayleigh scattering 07.0496

锐角辙叉 end frog, acute frog 03.0352

弱电场区 weak electric field area 07.0541

S

撒砂阀 sanding valve 04.0179

撒砂器 sanding sprayer 04.0180

撒砂装置 sanding device 04.0177

撒砂装置试验 test on sanding gear 04.0683

塞钉式钢轨接续线 plug bond 08.0541

塞入门 plug door 05.0499

[塞入门]导轮 door guide wheel 05.0500

三次抛物线曲线 cubic parabola curve 03.0025

三大件转向架 three-piece truck 05.0062

三点检查 released by checking three sections 08.0151

三铰拱 three-hinged arch 02.0399

三角测量 trigonometric survey, triangulation 02.0139

三角高程测量 trigonometric leveling 02.0141

三角坑 twist, warp 03.0152

三角形枢纽 triangle-type junction terminal 06.0397

三阶段设计 three-step design, three-phase design 02.0011

三开道岔 symmetrical three throw turnout, three-way turnout 03.0286

三通阀 triple valve 05.0306

三显示自动闭塞 three-aspect automatic block 08.0338

三相 YN, d11 d1 十字交叉接线牵引变压器 traction transformer of cross connection with threephase YN, d11 d1 04.1291

三相 YN, d11 接线牵引变压器 traction transformer of threephase YN, d11 connection 04.1285

三相交流牵引电动机 three phase AC traction motor 04.0399

三相桥式整流器 three-phase bridge rectifier 04.0332

三相三绕组接线牵引变压器 traction transformer of threephase three winding connection 04.1286

三压力机构 three-pressure equalizing system 05.0333

三轴转向架 three-axle truck 05.0060

散热管 radiator 05.0607

散热器 radiator 04.1031

散热损失 heat dissipation 04.1110

散装货物 bulk freight 06.0114

散装水泥车 bulk cement car 05.0810

扫气泵 scavenging pump 04.0988

扫气泵扫气 blower scavenging 04.0729

扫气系数 coefficient of scavenging 04.0742

扫气箱 scavenging box 04.0875

扫气效率 scavenging efficiency 04.0745

扫气压力 scavenging pressure 04.0730

色灯电锁器联锁 interlocking by electric locks with color light-signals 08.0122

色灯信号机 color-light signal 08.0053

色散常数 chromatic dispersion constant 07.0458

森林铁路 forest railway 01.0030

砂道碴 sand ballast 03.0130

砂害 sand blockade, sand drift 03.0498

砂井 sand drain 02.0310

砂石车 gravel car 05.0806

砂土液化 sand liquefaction 02.0324

砂箱 sand box 04.0178

砂桩 sand pile 02.0607

山岭隧道 mountain tunnel 02.0645

山区河谷选线 mountain and valley region location, location of line in mountain and valley region 02.0149

山区铁路 mountain railway 01.0031

闪光电源 flashing power source 08.0663

闪光信号 flashing signal 08.0015

扇形齿板 quadrant 04.1162

伤亡率 casualty rate 06.0654

商务检查电视 TV for railway commerce inspection

07.0657

上鞍 tank anchor 05.0476

上承式桥 deck bridge 02.0425

上导坑法 top-heading method 02.0735

上防跳台 top operation anticreep ledge 05.0541

上开窗 uplifting window 05.0504

上拉杆 top rod 05.0359

上挠度 camber 05.0441

上旁承 body side bearing 05.0410

上墙板 cornice sheathing 05.0446

上下导坑法 top and bottom heading method 02.0734

上心盘 body center plate 05.0407

上行方向 up direction 06.0284

上止点 top dead point, upper dead center 04.0712

上作用车钩 top operation coupler 05.0530

蛇行运动 hunting, nosing 01.0088

射频 radio frequency 07.0014

设备故障 breakdown of equipment, equipment failure 06.0633

设备停用 equipment out-of use 08.0709

*设计标高 design elevation 02.0076

设计概算 approximate estimate of design, budgetary estimate of design 02.0019

设计高程 design elevation 02.0076

设计荷载 design load 02.0329

设计洪水过程线 designed flood hydrograph 02.0094

设计鉴定 certification of design, appraisal of design 02.0034

设计阶段 design phase, design stage 02.0010

设计流量 design discharge 02.0091

设计流速 design current velocity 02.0075

设计水位 design water level 02.0092

伸缩缝 expansion joint 02.0508

伸缩区 breathing zone 03.0234

伸缩运动 fore and aft motion 01.0087

深井水泵 deep well pump 02.0918

深孔爆破 deep hole blasting 02.0273

深埋隧道 deep tunnel, deep-depth tunnel, deep burying tunnel 02.0650

深盆辐板 deep dish wheel plate 05.0123

渗管 leaky pipe 02.0306

渗井 leaching well, seepage well 02.0305

渗水暗沟 blind drain 02.0303

渗水隧洞 leak tunnel, permeable tunnel, drainage tunnel 02.0304

渗水土路基 permeable soil subgrade, pervious embankment 02.0212

升功率 volume power 04.0818

升降桥 lift bridge 02.0411

绳度整正曲线计算器 string lining computer, stringline calculator 03.0207

绳路 cord circuit 07.0259

绳栓 rope lug 05.0416

绳正法整正曲线 string lining of curve 03.0206

失去联锁 loss of interlocking 08.0686

失速 stall 04.1562

施工产值 construction output value 02.0872

施工调查 construction investigation 02.0830

施工防护 construction protection 02.0998

施工防护无线电通信 radio communication forprotection of construction 07.0631

施工妨碍 construction interference 08.0695

施工封闭线路 line occupation for works 03.0258

施工复测 construction repetition, repetive survey, construction repetition survey 02.0836

施工工艺流程 construction technology process, construction process 02.0870

施工荷载 constructional loading 02.0353

施工机械利用率 utilization ratio of construction machinery 02.0874

施工机械完好率 ratio of construction machinery in good condition 02.0875

施工计划管理 planned management of construction 02.0868

施工利润 construction profit 02.0873

施工水位 construction level, construction water level, working water level 02.0093

施工图设计 construction detail design, working-drawing design 02.0017

施工形象进度 construction figure progress, figurative progress of construction work 02.0876

施工验收 delivery-receiving acceptance 02.1058

施工准备 construction preparation 02.0831

施工总工期 total time of construction, total construction time 02.0864

施工组织方案 construction scheme 02.0865

湿喷混凝土 wet shotcreting 02.0780

湿喷混凝土机 wet shotcreting machine 02.0972

湿式凿岩 wet boring for rock, wet drilling for rock 02.1024

湿损事故 wet damage accident 06.0643

十字头 crosshead 04.1191

十字头扁销 crosshead key 04.1192

十字头圆销 crosshead pin 04.1193

十字形枢纽 cross-type junction terminal 06.0398

石灰砂桩 lime sand pile 02.0266

石笼 gabion 03.0506

时分多路电报设备 time division multiplex telegraph equipment 07.0696

时分复用 time division multiplex 07.0010

时间继电器 time relay 04.0592

时间间隔法 time-interval method 06.0239

时间－面积值 time-area value 04.0796

时隙 time slot 07.0811

实腹拱 spandrel-filled arch, solid-spandrel arch 02.0400

实施性施工组织设计 design for practical construction organization, practical design for construction scheme, operative construction organization design 02.0855

实体桥墩 solid pier 02.0561

实制动距离 actual stopping distance 05.0347

实制动时间 actual braking time, instantaneous application time, IAT 05.0345

使用限度 operation limit 04.0271

始发者 originator 07.0905

始发直达列车 through train originated from one loading point 06.0215

示功阀 indicator valve 04.0924

示功图 indicator diagram 04.0761

示功图丰满系数 fullness coefficient of indicator diagram 04.0762

事故报告 accident report 02.0989

事故处理 settlement of accident, accident disposal 06.0609

事故调查 accident investigation 06.0611

事故分析 accident analysis 06.0615

事故记录 accident record 06.0610

事故救援 accident rescue 06.0663

事故率 accident rate 06.0653

事故赔偿 accident indemnity 06.0613

事故信号 accident signal 06.0657

事故信息管理 accident information management 06.0614

事故隐患 accident threat 06.0626

事故预测 accident forecast 06.0612

事故预防 prevention of accident, accident averting 06.0655

事故预防措施的检查 checks for prevention of accidents 04.0662

事故照明 emergency lighting 08.0648

事故状态的接触网最低电压 minimum voltage of overhead contact line at accident condition 04.1334

释放时间 drop away time 08.0561

释放值 release value 08.0569

市郊客车 suburban passenger car, suburban coach 05.0724

市郊客流 suburban passenger flow 06.0026

市郊旅客列车 suburban passenger train 06.0209

市郊铁路 suburban railway 01.0022

室内测试 indoor test, laboratory test 02.0054

室温控制器 room thermostat 05.0645

视觉信号 visual signal 08.0005

视在大地导电率 apparent earth conductivity 04.1475

视在单位能耗 specific apparent energy consumption 04.1273

试验车 test car 05.0744

试验杆 test pole 07.0166

试验架 test rack 07.0352

试桩 test pile 02.0611

收碴机 ballast recollecting machine 03.0422

收信放大器 receiving amplifier 07.0342

手柄 handle, knob 08.0267

手持式凿岩机 portable rock drill, jack hammer 02.0976

手动水泵 hand water pump 05.0603

手动调压 manual voltage regulation 08.0657

*手动作业 manual operation 08.0375

手机　handset　07.0074

手孔　hand hole　07.0202

手提电动捣固机　portable electric tamper, hand electric tamper　03.0413

手提风动捣固机　portable pneumatic tamper　03.0414

手提内燃捣固机　portable gasoline-powered tamper　03.0415

手信号　hand signal　08.0014

手制动机　hand brake　05.0265

手制动链轮　hand brake sheave wheel　05.0364

手制动曲拐　hand brake bell crank　05.0363

守车　caboose, brake van, guard's van　05.0802

售票处　booking office, ticket office　06.0019

受电端　receiving end　08.0515

受电弓　pantograph　04.0534

受电弓标称电压　nominal voltage at pantograph　04.0316

受电弓电空阀　pantograph valve　04.0585

受电弓隔离开关　pantograph isolating switch　04.0569

受电弓滑板　pantograph pan　04.0538

受电弓空气动力效应　aerodynamic effects of pantograph　04.1446

受电弓气缸　pantograph cylinder　04.0540

受电弓上抬力　pantograph upthrust, pantograph static contact force　04.1445

受电弓试验　pantograph test　04.0660

受电器　current collector　04.0533

受电靴装置　shoegear　04.0543

受话器　receiver　07.0071

受控站　secondary station　07.0858

枢纽小运转列车　junction terminal transfer train　06.0226

枢纽遥控　remote control of a junction terminal　08.0459

枢纽直径线　diametrical line of junction terminal　06.0422

梳妆架　comb rack　05.0786

输出轴　output shaft　04.1016

输入轴　input shaft　04.1014

输送能力　traffic capacity　06.0305

输纸孔　feed holes　07.0719

书面联络法　written liaison method　06.0240

树形网[络]　tree network　07.0758

竖杆　vertical member　02.0450

竖井　shaft　02.0786

竖井联系测量　shaft connection survey　02.0792

竖曲线　vertical curve　02.0195

竖旋桥　bascule bridge　02.0409

数据　data　07.0826

数据传送率　data transfer rate　07.0731

数据电路端接设备　data circuit terminating equipment, DCE　07.0828

数据链路　data link　07.0805

数据通信　data communication　07.0729

数据通信网　data communication network　07.0061

数据网[络]　data network　07.0752

数据信道　data channel　07.0827

数据信号速率　data signaling rate　07.0730

数据站　data station　07.0831

数据终端设备　data terminal equipment　07.0829

数值孔径　numerical aperture　07.0449

数字编码选呼　selective call with digital pulse coding　07.0618

数字电话网　digital telephone network　07.0060

数字复接设备　digital multiplex equipment　07.0368

数字微波中继通信　digital microwave relay communication　07.0665

数字显示器　digital display　04.0613

刷盒　brush box　04.0458

刷坡　slope cutting　03.0504

刷握　brush-holder　04.0457

衰减　attenuation　07.0032

衰减常数　attenuation constant　07.0457

衰落　fading　07.0538

栓固能力　restraint capability　06.0558

栓焊钢桥　bolted and welded steel bridge　02.0364

双壁钢围堰钻孔基础　double wall steel cofferdam bored foundation　02.0587

双臂受电弓　double arm pantograph　04.0536

双边供电　two way feeding　04.1294

双边型直线感应电动机　double sided linear induction motor　04.1605

双侧导坑法　twin-side heading method　02.0740

双侧减速齿轮驱动 double reduction gear drive 04.0352

双侧[踏面]制动 clasp brake 05.0331

双层轿车平车 double deck sedan car 05.0829

双层客车 double deck passenger car, double deck coach, bi-level coach 05.0721

双层桥 double-deck bridge 02.0428

双重绝缘棒式绝缘子 double strut insulator, double rod insulator 04.1380

双重控制 dual control 08.0186

双重作业 double freight operations 06.0266

双电压制电力机车 dual voltage electric locomotive 04.0303

双断 double break 08.0223

双二进制编码 doubinary encoding 07.0736

双工 duplex operation 07.0705

双工传输 duplex transmission 07.0781

双工无线电通信 duplex radio communication 07.0515

双管荧光灯 double tube fluorescent lamp 05.0688

双轨条式轨道电路 double rail track circuit 08.0507

双回路供电 double circuit power supply 04.1280

双机牵引 double locomotive traction 04.0224

*双机重联牵引 double locomotive traction 04.0224

双肩回交路 double-arm routing 04.0189

双铰拱 two-hinged arch 02.0398

*双开道岔 symmetrical double curve turnout, equilateral turnout 03.0283

双联杠杆 toggle linkage 05.0522

双联平车 twinned flat car 05.0837

双链形悬挂 compound catenary equipment 04.1360

双梁式架桥机 double beam girder-erecting machine 02.0949

双流 double current 07.0708

双流液力机械传动 hydromechanical drive with outer ramification 04.0995

双面调车信号机 signal for shunting forward and backward 08.0084

双面托盘 double-deck pallet 06.0568

双频感应器 double frequency inductor 08.0429

双频率制电力机车 dual frequency electric locomotive 04.0304

双曲拱 two-way curved arch, cross-curved arch 02.0402

双室风缸 two-compartment reservoir 05.0311

双筒壁灯 double cylindrical shade wall lamp 05.0691

双推单溜 single rolling on double pushing track 06.0478

双推双溜 double rolling on double pushing track 06.0479

双线臂板信号机 double wire semaphore signal 08.0060

双线继电半自动闭塞 double track all-relay semiautomatic block system 08.0327

双线桥 double track bridge 02.0421

双线圈结构 two coil configuration 04.1634

双线隧道 double track tunnel 02.0652

双线铁路 double track railway 01.0016

双线运行图 train diagram for doubletrack 06.0292

双线制 two wire system 05.0714

双向传输 bidirectional transmission 07.0788

双向风缸 two way cylinder 05.0519

双向横列式编组站 bidirectional transversal type marshalling station 06.0376

双向混合式编组站 bidirectional combined type marshalling station 06.0378

双向交替通信 two-way alternate communication 07.0776

双向同时通信 two-way simultaneous communication 07.0775

*双向增音机 two-way repeater for VF selective calling, two-way repeater 07.0432

双向自动闭塞 double-direction running automatic block 08.0330

双向纵列式编组站 bidirectional longitudinal type marshalling station 06.0377

双悬臂式架桥机 double cantilever girder-erecting machine 02.0947

水表 water gage 04.1177

水锤 water hammer 04.1266

水底电缆 subaqueous cable 07.0185

*水底隧道 subaqueous tunnel, underwater tunnel

02.0647

水分离器 water separator 08.0402

水柜 water tank 04.1240

水柜阀 tank valve 04.1243

水害 flood damage, washout 03.0497

水害断道 railroad break down due to flood, line blockade due to flood 03.0577

水害复旧 restoration work for flood damage, restoration of flood damaged structures 03.0581

水害抢修 rush repair of flood damage to open for traffic 03.0578

水鹤 water crane 04.0061

水鹤表示器 water crane indicator 08.0097

水库路基 subgrade in reservoir, embankment crossing reservoir 02.0223

水冷柴油机 water-cooled diesel engine 04.0705

水力半径 hydraulic radius 02.0082

水煤浆 coal water slurry, coal water mixture 04.1104

水面坡度 slope of water surface 02.0085

水泥输送泵 cement pump 02.0924

水盘 drip pan 05.0657

水平板 horizontal plate, table plate 04.1148

*水平反射板 horizontal plate, table plate 04.1148

水套 water jacket 04.0874

水位表 water level gage 05.0606

水温控制器 water temperature regulater 05.0614

水文测量 hydrological survey 02.0086

水文地质调查 hydrogeologic survey 02.0043

水文断面 hydrologic sectional drawing, hydrologic section, hydrologic cross-section 02.0073

水下监视电视 monitor TV under water 07.0662

水下隧道 subaqueous tunnel, underwater tunnel 02.0647

水箱验水阀 water tank test cock 05.0666

水柱 water column 04.1178

水准点高程测量 benchmark leveling 02.0116

瞬时调速率 instantaneous speed change rate 04.0832

瞬时分路 instantaneous shunt 08.0528

瞬时分路不良 instantaneous loss of shunting 08.0671

顺坝 longitudinal dam 02.0619

顺列式枢纽 longitudinal arrangement type junction terminal 06.0399

顺坡 run-off elevation 03.0244

顺向重叠进路 route with overlapped section in the same direction 08.0156

斯柯特接线牵引变压器 traction transformer of Scott connection 04.1289

私有铁路 private railway 01.0009

司机 driver 04.0201

司机操纵台 driver's desk 04.0234

司机控制器 driver controller 04.0573

司机模拟操纵装置 simulator for driver train-handling 04.0235

司机室 driver's cab 04.0233

司机室工作条件检查 check on working conditions in the driver's cab 04.0685

司机室空调装置 driver's cab air conditioner 04.0616

司机室取暖电炉 driver's cab electric heater 04.0614

司机室热风装置 driver's cab air heater 04.0615

司机运转报单 driver's service-report, driver's log 04.0232

司机座椅 driver's chair 04.1059

司炉 fireman 04.0204

死区段 dead section 08.0529

四冲程柴油机 four stroke diesel engine 04.0692

四点检查 released by checking four sections 08.0152

四频组方式 four-frequency group mode 07.0601

四显示自动闭塞 four-aspect automatic block 08.0339

四用广播机 four-function public address equipment 05.0693

四轴车 four-axle car 05.0003

松弛压力 relaxation pressure 02.0707

松动爆破 blasting for loosening rock 02.0277

送电端 feed end 08.0514

送风道 air delivery duct 05.0639

送风器 blower 04.1146

送话器 transmitter 07.0070

送受分开电路 sending and receiving separated circuit 08.0226

送桩 pile follower 02.0610

素枕 untreated wooden tie 03.0114

速差制信号 speed signaling 08.0016

速度 speed 01.0074

速度传感器 speed sensor 04.0633

速度记录仪 tachograph 04.0628

速度继电器 speed relay 04.0594

速度控制系统 speed control system 04.0105

速度三角形 velocity triangle 04.1551

速度特性 speed characteristic 04.0850

塑料轴瓦头 plastic wearing end for plain bearing 05.0135

宿营车 dormitory car 05.0740

随工验收 acceptance following construction, follow-up acceptance 02.1052

碎石道碴 stone ballast 03.0127

碎石机 crusher 02.0891

隧道报警装置 tunnel warning equipment 02.0817

隧道标 tunnel post 03.0253

隧道冰害 frost damage in tunnel 03.0499

隧道衬砌模板台车 working jumbo for tunnel lining, tunnelling shutter jumbo for tunnel lining 02.0980

隧道出口 tunnel exit 02.0664

隧道底鼓 tunnel floor heave 02.0823

隧道地表沉陷 tunnel ground subsidence 02.0821

隧道地中位移 tunnel surrounding mass deflection 02.0825

隧道洞口投点 horizontal point of tunnel portal, geodetic control point of portal location of adit 02.0144

隧道洞内控制测量 intunnel control survey, through survey 02.0143

隧道洞外控制测量 outside tunnel control survey 02.0142

隧道防火措施 measures against tunnel fire 03.0560

隧道防排水 tunnel water handling 03.0562

隧道防水 waterproofing of tunnel 02.0793

隧道防灾设施 tunnel anti-disaster equipment 02.0814

隧道改建 tunnel reconstruction 02.0826

隧道功率分配器 power divider in tunnel 07.0591

隧道拱顶下沉 tunnel arch top settlement 02.0822

[隧道]拱圈 arch 02.0677

隧道贯通 tunnel holing-through 02.0762

隧道贯通面 tunnel through plane 02.0764

隧道贯通误差 tunnel through error 02.0763

隧道火灾 tunnel fire hazard 02.0816

隧道监控量测 tunnel monitoring measurement 02.0820

隧道建筑限界 tunnel construction clearance, tunnel structure gauge 01.0058

隧道进口 tunnel entrance 02.0663

隧道掘进机 tunnel boring machine, TBM 02.0981

隧道漏水 tunnel leak 03.0526

隧道落底 under cut of tunnel 02.0827

隧道埋置深度 buried depth of tunnel 02.0714

隧道坡度折减 compensation of gradient in tunnel, compensation grade in tunnel 02.0202

隧道群 tunnel group 02.0660

隧道弱电场区 weak electric field area in tunnel 07.0542

隧道摄影车 tunnel photographing car 05.0749

隧道射流式通风 tunnel efflux ventilation, tunnel injector type ventilation 02.0808

隧道施工通风 construction ventilation of tunnel 02.0810

隧道套拱 cover arch of tunnel 02.0829

隧道挑顶 top picking of tunnel 02.0828

隧道通风 tunnel ventilation 03.0561

隧道通风帘幕 ventilation curtain 02.0806

隧道通风试验 tunnel ventilation test 02.0809

隧道通知设备 tunnel announciating device 08.0452

隧道瓦斯爆炸 tunnel gas explosion 02.0819

隧道围岩分级 classification of tunnel surrounding rock 02.0048

隧道消防系统 tunnel fire-fighting system 02.0818

隧道压浆 pressure grouting of tunnel 02.0798

隧道运营通风 permanent ventilation of tunnel 02.0802

隧道照明 tunnel illumination, tunnel lighting 02.0813

隧道遮断信号 tunnel obstruction signal 08.0451
隧道中继器 tunnel repeater 07.0563
隧道周边位移 tunnel perimeter deflection 02.0824
隧道专家系统 expert system of tunnel 02.0703
损耗 loss 07.0033
损失事故 damage accident 06.0640
梭式矿车 shuttle car 02.0965
缩短轨 standard shortened rail, fabricated short rail used on curves, standard curtailed rail 03.0058
缩短装置 shortening device 05.0563
缩位地址 abbreviated address 07.0874
索鞍 cable saddle 02.0495

索铲挖土机 dragline 02.0879
索夹 cable band, cable clamp 02.0496
索平面 cable plane 02.0491
锁闭 locking 08.0126
锁闭电路 locking circuit 08.0230
锁闭杆 locking rod 08.0615
锁闭力 locking force 08.0624
锁闭系统 locking system 08.0609
锁定轨温 fastening-down temperature of rail 03.0227
锁紧中心销 locking center pin 05.0212

T

塌方落石报警器 land slide warning device 03.0456
*他动轮对 driven wheel set 04.1230
他励电动机 separately excited motor 04.0407
塔式起重机 tower crane, column crane 02.0893
踏面 wheel tread 05.0120
踏面剥离 shelled tread 05.0252
踏面擦伤 flat sliding, tread slid flat 05.0251
踏面基点 tread taping point 05.0217
踏面磨耗 tread wear 05.0249
踏面清扫器 tread cleaner 05.0321
踏面外形 tread contour, tread profile 05.0215
踏面制动 tread brake 04.0123
踏面锥度 tread conicity 05.0218
台后填方 filling behind abutment 02.0558
台架式集装箱 platform based container 06.0545
台架试验 stand test, bench test 04.0854
台间联络线 interposition trunk 07.0405
台阶法 benching tunnelling method 02.0730
台阶式洞门 bench tunnel portal 02.0672
台帽 abutment coping 02.0555
台身 abutment body 02.0553
台位利用系数 utility factor of the position 05.0965
坍方 slide , slip 03.0478
弹簧补偿器 spring tensioner 04.1423
弹簧道钉 elastic rail spike 03.0091
弹簧垫圈 spring washer 03.0099

弹簧动挠度 dynamic spring deflection 05.0239
弹簧防爬器 spring rail anchor 03.0102
弹簧刚度 spring stiffness 05.0237
弹簧静挠度 static spring deflection 05.0238
弹簧摩擦式缓冲器 spring friction draft gear 05.0551
弹簧柔度 spring flexibility 05.0236
弹簧式减振器 spring damper 04.0898
弹簧托板 spring plank 05.0209
弹簧托梁 spring plank carrier 05.0208
弹簧悬挂装置 spring suspension 05.0164
弹力继电器 spring-type relay 08.0588
弹性车轮 elastic wheel 05.0106
弹性齿轮驱动 resilient gear drive 04.0349
*弹性垫板 rubber tie plate 03.0096
弹性定位轮对 elastically positioned wheelset 05.0880
弹性挤开 gage elastically widened, elastic squeeze-out 03.0148
弹性简单悬挂 stitched tramway type suspension equipment 04.1354
弹性抗力 elastic resistance 02.0711
弹性扣件 elastic rail fastening 03.0089
弹性链形悬挂 stitched catenary equipment 04.1357
弹性旁承 elastic side bearing 05.0183
弹性止挡 elastic bolster guide 05.0093
碳当量 carbon equivalent 02.0469

* 探询　polling　07.0852

探照式色灯信号机　searchlight signal　08.0055

炭精避雷器　carbon arrester　07.0205

掏底开挖　cut the vertical earthwork bottom
02.0997

陶瓷避雷器　ceramic arrester　07.0207

套车　telescoping　05.0857

套管　sheath　02.0480

套管铰环　sleeve with clevis and ring　04.1387

套管钻机　drill machine with casing　02.0914

套线　overlapping line　02.0165

套线道岔　mixed gage turnout　03.0300

特长隧道　super long tunnel　02.0656

特大桥　super major bridge　02.0416

特等卧室　superclass bedroom, superclass sleeping
compartment　05.0754

特定运价　special rate　06.0736

特快加快票　express extra ticket　06.0037

* 特氏阀　adjustable piston valve, Trofiemov piston
valve　04.1203

特殊地质　special geology　02.0062

特殊荷载　particular load　02.0348

特殊试验　special test　04.0295

特殊条件下的路基　subgrade under special condition
02.0220

特殊土路基　subgrade of special soil　02.0214

特性畸变　characteristic distortion　07.0711

特性阻抗　characteristic impedance　07.0226

特种车辆活载　special live load, live load of special
type wagon　03.0522

特种断面尖轨　special heavy section switch rail,
tongue rail made of special section rail, full-web
section switch　03.0357

特种用户　special subscriber　07.0403

梯　ladder　05.0419

梯线　ladder track　03.0299

提升阀　poppet valve　05.0300

天窗　skylight　05.0506

天沟　gutter, overhead ditch, intercepting ditch
02.0297

* 天花板　ceiling　05.0456

天桥　over-bridge, passenger foot-bridge　06.0521

天然地基　natural foundation, natural ground
02.0579

填充线　interstitial wire　07.0485

填料　packing　04.1188

填料分类　classification of filling material　02.0253

* 挑水坝　spur dike　02.0618

条件电源　conditional power source　08.0643

条件电源屏　conditional power supply panel
08.0653

调幅　amplitude modulation　07.0022

调节板　regulating plate　04.1386

调节电阻器　regulating resistor　04.0520

调节放大器　regulating amplifier　07.0340

调节轨　buffer rail　03.0237

调梁设备　beam straightening equipment　05.0901

调频　frequency modulation　07.0023

调频轨道电路　frequency modulated track circuit
08.0499

调速伺服机构　speed-governing servomechanism
04.0962

调速方式　mode of speed control　04.0344

调速器　governor　04.0960

* 调速器传动齿轮　governor drive gear　04.0943

调速器驱动齿轮　governor drive gear　04.0943

调速设备　speed control device　06.0459

调速特性　speed regulation characteristics　04.0852

调速系统　speed-governing system　04.0961

调相　phase modulation　07.0024

调压方式　voltage regulation mode　04.0317

调压开关　tap changer　04.0555

调压器　pressure regulator　04.0162

调压牵引变压器　regulating traction transformer
04.0485

调压绕组　regulating winding　04.0498

调整阀　regulator valve, throttle valve　04.1159

调整轨缝　adjusting of rail gaps, evenly distributing
joint gaps　03.0208

调整总概算　adjusted sum of approximate estimate
02.0024

调制　modulation　07.0018

调制解调器　modem　07.0849

跳轨　jump on rail　05.0877

跳频通信　frequency hopping communication
07.0675

跳线　jumper　08.0539

铁磁吸力　ferromagnetic attraction force　04.1623

铁道　railway, railroad　01.0001

铁道车辆　railway vehicle, railway car　05.0001

铁道法规　Railway Act　01.0117

铁道科学　railway science　01.0004

铁道牵引动力　railway traction power, railway motive power　04.0001

铁道通信　railway communication, railroad communication　07.0001

铁道信号　railway signaling, railroad signaling　08.0001

* 铁路　railway, railroad　01.0001

铁路保价运输　value insured rail traffic　06.0011

铁路保险运输　insured rail traffic　06.0010

铁路财务　railway finance　06.0742

铁路财务成果　railway financial result　06.0758

铁路财务状况　railway financial condition　06.0759

铁路测量　railway survey　02.0098

铁路长期计划　long-term railway plan　06.0674

铁路长途电话网　railway long distance telephone network　07.0249

铁路长途字冠　prefix number for railway toll call　07.0250

铁路的连带责任　joint responsibility of railway　06.0184

铁路等级　railway classification　01.0006

铁路地区电话　railway local telephone　07.0382

铁路电视　railway TV　07.0651

铁路短波通信　short wave communication for railway　07.0647

铁路法　Railway Law　01.0116

铁路防护无线电通信　radio communication for railway protection　07.0630

铁路干线机车　railway trunk line locomotive　04.0012

铁路感应无线电通信　inductive radio communication for railway　07.0510

铁路高速运输　railway high speed traffic　06.0009

铁路告警无线电通信　radio communication for railway warning　07.0629

铁路工程地质遥感　remote sensing of railway engineering geology　02.0133

铁路公安无线电通信　radio communication for railway public security　07.0643

铁路固定资产　railway fixed assets　06.0697

铁路涵洞　railway culvert　02.0625

* 铁路航测　railway aerial photogrammetry　02.0130

铁路航空勘测　railway aerial surveying　02.0131

铁路航空摄影测量　railway aerial photogrammetry　02.0130

铁路货物运输　railway freight traffic　06.0075

铁路货物运输规程　regulations for railway freight traffic　06.0007

铁路货运组织　railway freight traffic organization　06.0076

铁路机车车辆限界　rolling stock clearance [for railway], vehicle gauge　01.0059

铁路技术　railway technology　01.0005

铁路技术改造　technical reform of railway, technical renovation of railway, betterment and improvement of railway　02.0002

铁路技术管理规程　regulations of railway technical operation　01.0119

铁路计划　railway plan　06.0673

铁路建设基金　railway construction fund　06.0754

铁路建筑长度　construction length of railway　01.0045

铁路建筑限界　railway construction clearance, structure clearance for railway, railway structure gauge　01.0055

铁路军事运输　railway military service　06.0012

铁路勘测　railway reconnaissance　02.0037

铁路客运组织　railway passenger traffic organization　06.0014

铁路旅客运输　railway passenger traffic　06.0013

铁路旅客运输规程　regulations for railway passenger traffic　06.0006

铁路轮渡　railway car ferries　02.0639

铁路年度计划　annual railway plan　06.0675

铁路桥　railway bridge　02.0355

铁路枢纽　railway junction terminal　06.0396

铁路数据交换系统　railway data exchange system　01.0114

铁路隧道　railway tunnel　02.0644

铁路条例　railway code　01.0118

＊铁路通信　railway communication, railroad communication　07.0001

铁路网　railway network, railroad network　01.0003

铁路微波中继通信　microwave relay communication for railway　07.0663

铁路无线电通信　railway radio communication　07.0507

铁路无线电遥控　radio telecontrol for railway　07.0678

铁路无线电中继通信　relay radio communication for railway　07.0509

铁路线　railway line, railroad line　01.0002

铁路新线建设　newly-built railway construction　02.0001

＊铁路信号　railway signaling, railroad signaling　08.0001

铁路行车组织　organization of train operation　06.0196

铁路行车组织规则　rules for organization of train operation　06.0197

铁路选线　railway location, approximate railway location, location of railway route selection　02.0146

铁路移动无线电通信　railway mobile communication　07.0508

铁路应急短波通信　short wave communication for railway emergency　07.0648

铁路用地　right-of-way　02.0036

铁路运价　railway tariff, railway rate　06.0730

铁路运输　railway transportation, railway traffic　06.0001

铁路运输安全　safety of railway traffic　06.0602

铁路运输调度　railway traffic control, railway traffic dispatching　06.0334

铁路运输管理　railway transport administration　06.0002

铁路运输经济　railway transport economy　06.0672

铁路运输利润　railway traffic profit　06.0760

铁路运输全员劳动生产率　labor productivity of railway transport　06.0716

铁路运输质量管理　quality control of railway trans-portation　06.0005

铁路运输周转基金　railway traffic turnover fund　06.0755

铁路运输组织　railway traffic organization　06.0004

铁路运营　railway operation　06.0003

铁路运营长度　operating length of railway, operating distance, revenue length　01.0043

铁路运营信息系统　railway operation information system　01.0113

铁路站内电话　railway station telephone　07.0437

铁路职工数　number of railway staff and workers　06.0715

铁路重载运输　railway heavy haul traffic　06.0008

铁路主要技术条件　main technical standard of railway, main technical requirement of railway　02.0003

铁路专用线　railway special line　01.0039

＊铁线　steel wire　07.0159

听觉信号　audible signal　08.0006

停车防护　stopping train protection, standing train protection　02.1004

停车牌　red board, stop indicator　03.0246

停车器　stopping device　06.0467

停车信号　stop signal　08.0023

停电　power failure　08.0687

停驻制动　parking braking　05.0365

通报速率　telegraph rate　07.0709

通车期限　time limit for opening to traffic　02.0863

＊通道　inside corridor, gangway　04.1063

通风车　ventilated box car　05.0819

通风机　ventilating set, ventilating machine, ventilator　02.0977

通风机电动机　blower motor　04.0475

通风冷却试验　test on ventilation and cooling　04.0665

通风器　ventilator　05.0619

通风式电动机　ventilated motor　04.0412

通风效率　drafting efficiency　04.1114

通风运输　ventilated transport　06.0181

通风装置　drafting apparatus　04.1139

通过按钮电路　through button circuit　08.0251

通过进路　through route　06.0247

通过能力　carrying capacity　06.0304

通过能力限制区间 restriction section of carrying capacity 06.0310

通过式货场 through-type freight yard 06.0084

通过式客运站 through-type passenger station 06.0383

通过台 vestibule 05.0789

通过信号 through signal 08.0028

通过信号机 block signal 08.0073

通过最小曲线半径 minimum radius of curvature negotiable 05.0844

通航水位 navigation water level, NWL 02.0080

通话计数器 message register 07.0105

通话命令 calling order 07.0595

通话请求 calling request 07.0594

通话锁闭 calling block 07.0596

通流部分 flow passage 04.1545

通路 path 07.0004

通路串杂音防卫度 channel signal to crosstalk and noise ratio 07.0365

通路固有杂音 channel basic noise 07.0358

通路净衰耗 channel net loss 07.0314

通路频率特性 channel frequency characteristic 07.0363

通路线性 channel linearity 07.0285

通路振幅特性 channel amplitude characteristic 07.0364

通路振鸣边际 channel singing margin 07.0366

通气口 vent 05.0480

通信 communication 07.0002

通信处理机 communication processor 07.0892

通信端站 terminal toll office 07.0244

通信接口 communication interface 07.0894

通信控制器 communication controller 07.0893

通信控制字符 communication control character 07.0888

通信盲区 communication blind district 07.0543

通信枢纽 communication center 07.0239

通信网 communication network 07.0059

通信卫星 communication satellite 07.0671

通信协议 communication protocol 07.0774

通信子网 communication subnet 07.0765

通信总枢纽 master communication center of railway whole administration 07.0243

通用货车 general-purpose freight car 05.0792

通用集装箱 general purpose container 06.0541

通知音 warning tone 07.0279

同步变压器 synchronous transformer 04.0491

同步传输 synchronous transmission 07.0786

同步电动机 synchronous motor 04.0408

同步通信 synchronous communication 07.0778

同侧下锚 same-side anchor 04.1432

同方向列车连发间隔时间 time interval for two trains despatching in succession in the same direction 06.0301

同频单工无线电通信 same-frequency simplex radio communication 07.0513

同频干扰 same frequency interference 07.0529

同频干扰区 same frequency interference area 07.0544

同向曲线 curves of same sense, adjacent curves in one direction 02.0177

同意按钮盘 agreement button panel 08.0258

同轴电缆 coaxial cable 07.0172

同轴电缆通信 coaxial cable communication 07.0318

铜包钢线 copper-clad steel wire 07.0160

铜线 copper wire 07.0158

投影断链 projection of broken chain 02.0128

投资估算 investment estimate 02.0028

投资回收期 repayment period of capital cost 02.0155

投资检算 checking of investment 02.0025

投资效果 effect of investment 06.0711

头戴送受话器 head set 07.0100

透镜式色灯信号机 multi-lenses signal 08.0054

透明传送 transparent transfer 07.0746

透水路堤 pervious embankment, permeable embankment 02.0312

凸轮位置转换开关 cam position changeover switch 04.0560

凸轮轴 camshaft 04.0937

凸轮轴正时齿轮 camshaft timing gear 04.0942

突发差错 burst error 07.0840

突发传输 burst transmission 07.0790

突泥 projecting mud soil 02.0991

突水 gushing water 02.0992

W

挖沟机　trench cutting machine, trenching machine, trencher　02.0892

挖掘机　excavating machine, excavator　02.0877

挖孔桩　dug pile　02.0600

*挖神仙土　cut the vertical earthwork bottom　02.0997

挖探　excavation prospecting　02.0051

瓦斯浓度　gas density, gas consistency　02.0996

瓦斯突出　gas projection　02.0995

瓦斯治理　gas control　02.1022

外部过电压试验　test on external overvoltage　04.0674

外顶板　roof sheet　05.0455

外端门　gangway door　05.0487

外端墙　outside end wall　05.0433

外火箱　outside firebox, outer firebox　04.1117

外火箱顶板　roof sheet　04.1119

*外特性　speed regulation characteristics　04.0852

外烟筒　smokestack, chimney　04.1143

外移桩　shift out stake, stake outward, offset stake　02.0115

外止点　outer dead point, outer dead center　04.0715

外走廊　outside corridor, running board　04.1064

弯起钢筋　bent-up bar　02.0476

弯形止阀　bend stop valve　05.0668

*完全吸起值　working value　08.0567

*碗扣式脚手架　soffit scaffolding　02.0960

晚点表示　delaying time indication　08.0486

万能道尺　universal rail gage　03.0466

万能杆件　fabricated universal steel members　02.0955

万向轴驱动　cardan shaft drive　04.0356

万有特性　universal characteristic　04.0851

腕臂　cantilever　04.1381

腕臂底座　cantilever base　04.1383

网侧电路　circuit on side of overhead contact line　04.0371

网侧电压表　voltmeter on side of overhead contact line, overhead side voltmeter　04.0622

网关　gateway　07.0771

网络　network　07.0029

网络管理　network management　07.0769

网络互连　inter operation　07.0768

网络计划技术　network planning technique　02.0869

网形网[络]　mesh network　07.0759

往复式空气压缩机　reciprocating compressor　02.0921

往复式制冷压缩机　reciprocating type refrigeration compressor　05.0628

微波通信　microwave communication　07.0666

微波通信车　microwave communication vehicle　07.0668

微机－继电式电气集中联锁　microcomputer-relay interlocking　08.0112

微机联锁　microcomputer interlocking　08.0113

危险货物　dangerous freight, dangerous goods　06.0172

危险货物包装标志　labels for packages of dangerous goods　06.0574

危险影响　dangerous influence　04.1464

围栏　railing around coping of pier or abutment　02.0574

围岩　surrounding rock　02.0704

围岩压力　pressure of surrounding rock　02.0705

围岩自承能力　self-supporting capacity of surrounding rock　02.0709

围堰　cofferdam　02.0586

维修不良　not well maintained　08.0693

维修车　maintenance car　05.0746

尾巴电缆　stub cable　07.0191

尾灯　tail lamp　05.0680

尾框　coupler yoke　05.0548

未被平衡离心加速度　unbalanced centrifugal acceleration　03.0030

未检出差错　undetected error　07.0871

位同步　bit synchronizing　07.0835

位置指示器　notch indicator　04.0612

位置转换开关　position changeover switch　04.0558

卫星通信　satellite communication　07.0669

温度传感器　temperature sensor　04.0634

温度继电器　temperature relay　04.0595

温度力　temperature stress　03.0231

温度力峰　temperature stress peak　03.0232

温度调节阀　temperature regulating valve　04.1026

＊温度调节器　expansion rail joint, rail expansion device, switch expansion joint　03.0075

温度调整器　temperature regulator　04.0611

温水采暖装置　hot water heating equipment　05.0599

温水箱　hot water tank　05.0661

文件传真机　document facsimile apparatus　07.0728

稳定电阻器　stabilizing resistor　04.0519

稳定光源　stabilized light source　07.0494

稳定性安全系数　safety factor of stability　05.0889

稳定运行工况　steady running condition　05.0051

稳压器　voltage stabilizer　07.0050

问讯处　information office, inquiry office　06.0022

涡流激振　vortex-excited oscillation　02.0524

涡流室式燃烧室　swirl combustion chamber　04.0798

涡流制动　eddy current brake　04.0121

涡轮　turbine　04.1011

涡轮发电机　turbo-generator　04.1249

涡轮功率　turbine power　04.1560

涡轮膨胀比　expansion ratio of turbine　04.0736

涡轮入口废气温度　exhaust temperature at turbine inlet　04.0734

涡轮入口废气压力　exhaust pressure at turbine inlet　04.0735

涡轮增压器　turbocharger　04.0973

卧铺票　berth ticket　06.0038

卧室　bedroom, sleeping compartment　05.0753

圬工梁裂损　cracking of concrete and masonry beam　03.0525

圬工桥　masonry bridge　02.0361

污染事故　contamination accident　06.0642

无碴轨道　ballastless track　03.0014

无碴无枕梁　girder without ballast and sleeper　02.0377

无岔区段　section without a switch　08.0203

无导框式转向架　non-pedestal truck　05.0067

无调中转车　transit car without resorting　06.0233

无调中转车停留时间　detention time of car in transit without resorting　06.0264

无缝线路　continuously welded rail track, jointless track　03.0017

无轨运输　trackless transportation　02.0758

无害地段　harmless district　02.0190

无极继电器　neutral relay　08.0591

无级调压　stepless voltage regulation　04.0321

无间隙牵引杆　slackless drawbar　05.0557

＊无铰拱　fixed-end arch　02.0397

无绝缘轨道电路　jointless track circuit　08.0498

无连接方式　connectionless mode　07.0904

无人增音机　unattended repeater　07.0325

无线电测距　radio distance-measurement　07.0681

无线电调车信号　radio shunting signal　07.0683

无线电干扰　radio interference, radio disturbance　07.0524

[无线电]干扰测量仪　[radio] interference meter　04.1511

无线电干扰测量仪　radio interference meter　07.0574

无线电机车信号　radio locomotive signal　07.0682

无线电台　radio station　07.0552

无线电台守候状态　radio set in stand-by state　07.0576

无线电噪声　radio noise　04.1505

无线话筒　wireless transmitter　07.0560

无线寻呼机　radio paging set　07.0561

无效运输　ineffective traffic　06.0110

无形损耗　intangible wear　06.0700

无压力涵洞　inlet unsubmerged culvert　02.0631

无摇动台式转向架　truck with no swing bolster　05.0064

无摇枕转向架　bolsterless truck　05.0066

无载起动电空阀　no-load starting electro pneumatic valve　04.0173

无中梁底架　underframe without center sill　05.0389

五单位数字保护电码 protected 5-unit numerical code 07.0703

五通塞门 five-way cock 05.0667

伍德布里奇接线牵引变压器 traction transformer of Wood Bridge connection 04.1290

雾化 atomization 04.0776

雾化杯 atomizing cup 05.0612

雾化轮 atomizing wheel 05.0611

物探 geophysical prospecting 02.0053

误比特率 bit-error rate 07.0875

误交付 delivery mistake 06.0645

误码测试仪 code error tester 07.0369

*误码率 bit-error rate 07.0875

误用故障 misuse fault 08.0739

误字率 character-error rate 07.0876

误组率 block-error rate 07.0877

X

吸浮力 attraction [lift] force 04.1595

吸流变压器 booster transformer, BT 04.1318

吸流变压器供电方式 booster transformer feeding system 04.1277

吸起时间 pick-up time 08.0560

吸起值 pick-up value 08.0566

吸入压力调节阀 suction pressure regulating valve 05.0649

吸上式注水器 attraction injector 04.1166

吸上线 boosting cable 04.1320

席别灯 car class indicating lamp 05.0681

*洗罐库 tank washing shed 05.0902

洗罐棚 tank washing shed 05.0902

洗罐设备 tank washing equipment 05.0904

洗罐线 tank washing siding 05.0939

洗罐站 tank washing point 05.0945

洗炉堵 washout plug 04.1175

洗面间 washing room, washing compartment 05.0778

洗面器 wash basin 05.0785

洗手器 wash bowl 05.0779

系杆拱 tied arch 02.0403

系统锚杆 system anchor bolt 02.0771

系统余度 system margin 07.0476

细粒土填料 fine-grained soil filler, fine-grained soil fill 02.0256

下承式桥 through bridge 02.0427

下垂 drooping 05.0442

下防跳台 rotary operation anticreep ledge 05.0542

下峰信号 down hump trimming signal 08.0038

下开窗 dropping window 05.0505

下锚段衬砌 anchor-section lining 02.0691

下坡道防护电路 protection circuit for approaching heavy down grade 08.0246

下墙板 wainscot sheathing 05.0447

下心盘 truck center plate 05.0210

下行方向 down direction 06.0285

下止点 bottom dead point, lower dead center 04.0713

下作用车钩 bottom operation coupler 05.0531

下作用水阀 under lever faucet 05.0780

先拱后墙法 arch first lining method, flying arch method 02.0751

先开阀 pilot valve 04.1160

先墙后拱法 side wall first lining method 02.0750

先张法预应力梁 pretensioned prestressed concrete girder 02.0372

鲜活货物 fresh and live freight 06.0117

弦杆 chord member 02.0447

显燃期 sensible combustion period 04.0783

显示距离 range of a signal 08.0049

险性事故 bad accident, dangerous accident 06.0624

现在车 cars on hand 06.0270

限界 clearance, gauge 01.0053

限界改善 clearance improvement 03.0572

限界架 clearance limit frame 03.0280

限界检测车 clearance car, clearance inspection car 03.0455

限界检查 checking of clearance, clearance check measurement 03.0514

限界检查器 clearance treadle 08.0401

限界门 warning portal 04.1434

限界图 clearance diagram 01.0054

限流电抗器 inductive reactor, current limiting reactor 04.0513

限时人工解锁 manual time release 08.0147

限压阀 limiting valve 05.0305

限制坡度 ruling grade, limiting grade 02.0182

限制速度 limited speed, speed restriction 01.0076

限制用户 limited subscriber 07.0402

限制轴重 axle load limited 01.0052

线岔 overhead crossing 04.1404

线间距[离] distance between centers of tracks, midway between tracks 02.0152

线路 track, permanent way 03.0015

线路标志 road way signs, permanent way signs 03.0248

线路测量 route survey, profile survey, longitudinal survey 02.0106

线路大修 major repair of track, overhaul of track, track renewal 03.0194

线路点 field location 08.0464

线路放大器 line amplifier 07.0337

线路封锁 track blockade, closure of track, traffic interruption 02.0999

线路复测 repetition survey of existing railway, resurvey of existing railway 02.0108

线路机具 permanent way tool 03.0369

线路机械 permanent way machine 03.0368

线路接触器 line contactor 04.0550

线路利用率 line efficiency 07.0154

线路滤波器 line filter 07.0349

线路爬行 track creeping 03.0159

线路平面图 track plan, line plan 02.0166

线路平剖面图 track charts 03.0264

线路清理机械 permanent way clearing machine 03.0379

线路区段 track section 08.0204

线路全长 total track length 06.0432

*线路上部建筑 track 03.0010

线路试验 road test, running test 05.0867

线路所 block post 06.0360

线路踏勘 route reconnaissance 02.0099

线路维修 maintenance of track 03.0196

线路维修规则 rules of maintenance of way 03.0266

线路有效长 effective track length 06.0433

线路杂音 line noise 07.0360

线路占用表示 track occupancy indication 08.0170

线路遮断表示器 track obstruction indicator 08.0096

线路中断 line interruption 03.0257

线路中心线 central lines of track 06.0425

线路中修 intermediate repair of track 03.0195

线路纵断面图 track profile, line profile 02.0167

线圈高度 coil height 04.1617

*线群出发信号机 group starting signal 08.0068

线群出站信号机 group starting signal 08.0068

线束 track group 06.0431

线束减速器 group retarder 08.0380

线头脱落 wire lead drop out 08.0692

*线性电动机 linear motor 04.0415

线障脉冲测试器 pulse echo fault locator 07.0214

相错式接头 alternate joint, staggered joint, broken joint 03.0067

相对密度 relative density 02.0258

相对摩擦系数 relative friction coefficient 05.0243

相对湿度 relative humidity 04.0847

相对式接头 opposite joint, square joint 03.0066

相对阻尼系数 damping factor 05.0248

相干调制 coherent modulation 07.0821

*相互式接头 alternate joint, staggered joint, broken joint 03.0067

相控调压 phase control 04.0322

相敏轨道电路 phase detecting track circuit 08.0503

相片传真机 photographic facsimile apparatus 07.0727

相位抖动测量仪 phase jitter tester 07.0370

相移键控 phase shift keying, PSK 07.0733

镶辑复照图 index of photography 02.0138

箱底承载能力 floor loading capability 06.0559

箱涵 box culvert 02.0628

箱门封条 door seal gasket 06.0563

箱式托盘 box pallet 06.0569

箱形梁 box girder 02.0386

响度 loudness 07.0067

响墩信号 torpedo 06.0661

项目管理　project management　06.0712

项目建议书　proposed task of project　02.0008

项目评估　project appraisal　06.0713

橡胶垫　rubber pad　05.0175

橡胶垫板　rubber tie plate　03.0096

橡胶堆　rubber-metal pad　05.0174

橡胶缓冲器　rubber draft gear　05.0549

橡胶摩擦式缓冲器　rubber friction draft gear
05.0550

橡胶弹簧　rubber spring　05.0173

橡胶弹性导柱式轴箱定位装置　rubber elastic guide
post type journal box positioning device　05.0150

*像片索引图　index of photography　02.0138

象限角　quadrantal angle　02.0125

削平木枕　tie adzing　03.0216

削弱磁场　weakened field　04.0433

消侧音器　anti-sidetone device　07.0076

消防器材　fire-fighting apparatus and materials
02.1019

消防设施　fire-fighting equipment　02.1017

消声器　noise silencer, muffler　04.0987

小导管预注浆　pre-grouting with small duct
02.0775

小孩票　child ticket　06.0042

小横梁　crosstie　05.0398

小卖部　snack counter, buffet　05.0759

小桥　minor bridge　02.0419

小型临时工程　small-scale temporary project
02.0858

小型线路机械　light permanent way machine, small
permanent way machine　03.0382

小型液压捣固机　light hydraulic tamping machine
03.0416

小型枕底清筛机　small ballast undercutting cleaners
03.0432

小烟管　[smoke] tube　04.1136

小腰带　window lintel　05.0436

小运转机车　locomotive for district transfer, transfer
locomotive train　04.0017

*小闸　independent brake valve　04.0168

小站电气集中联锁　relay interlocking for small station　08.0115

*楔型接头夹板　head contact flat joint bar

03.0079

楔形内接　wedging inscribing　03.0139

协议　protocol　07.0881

协议规范　protocol specification　07.0885

携带电话机　portable telephone set　07.0097

*斜撑　brace　04.1131, diagonal brace　05.0401

斜撑桅杆式架梁起重机　cross stay derrick girder
erecting machine　02.0945

斜调　slope regulation　07.0312

斜洞门　skew tunnel portal　02.0674

斜对称载荷　diagonally symmetrical loading force
05.0235

斜杆　diagonal member　02.0449

斜交桥　skew bridge　02.0413

斜接近　oblique exposure　04.1487

斜接近段长度　oblique exposure length　04.1488

斜接近段的等效距离　equivalent distance of the
oblique exposure　04.1489

斜井　inclined shaft　02.0787

斜拉桥　cable-stayed bridge　02.0394

斜缆　stay cable, inclined cable　02.0493

斜链形悬挂　inclined catenary, srew catenary
04.1355

*斜腿刚构桥　strutted beam bridge, slant-legged
rigid frame bridge　02.0391

斜腿刚架桥　strutted beam bridge, slant-legged rigid
frame bridge　02.0391

斜楔　wedge　05.0160

*斜悬杆式刚性拱刚性梁桥　Nielsen type Lohse
bridge, rigid arch bridge with rigid tie and inclined
suspenders　02.0407

*斜张桥　cable-stayed bridge　02.0394

斜桩　batter pile, raking pile, spur pile　02.0612

谐波电流百分比测定　measurement of percentage of
harmonic current　04.0678

谐波干扰　harmonic interference　08.0674

卸车数　number of car unloadings　06.0314

蟹爪式装岩机　crab rock loader　02.0964

泄水洞　drain cavern, drain tunnel　02.0801

泄水孔　drainage opening　02.0511

芯[包层表面]不圆度　non-circularity of core [clad-
ding]　07.0452

芯[包层表面]同心度　core [cladding] concentricity

07.0451

芯径 core diameter 07.0445

新奥法 New Austrian Tunnelling Method, NATM
02.0722

新建铁路 newly-built railway 01.0033

心轨 point rail, nose rail 03.0359

心盘面自由高 free height of center plate wearing
surface from rail top 05.0042

心盘载荷 load on center plate 05.0233

心盘座 center filler 05.0409

心式牵引变压器 core-type traction transformer
04.0482

信串比 signal to crosstalk ratio 07.0140

信道空闲信号 channel free signal 07.0611

信道容量 channel capacity 07.0804

信号 signal 08.0004

信号变压器 signal transformer 04.0494

信号表示 signal indication 08.0162

信号大修 signal overhaul repair, signal major repair
08.0737

信号灯 signal lamp 07.0104

信号点 signal location 08.0345

信号电路 signal circuit 04.0384

信号发生器 signal generator 07.0079

信号复示器 signal repeater 08.0273

信号故障 signal fault 08.0738

信号关闭 signal at stop 08.0051

信号关闭表示 stop signal indication 08.0164

信号机 signal 08.0052

信号机点灯电路 signal lighting circuit 08.0233

信号机点灯电源 signal lighting power source
08.0647

信号机后方 in rear of a signal 08.0714

* 信号机内方 in rear of a signal 08.0714

信号机前方 in advance of a signal 08.0713

* 信号机外方 in advance of a signal 08.0713

信号集中修 signal centralized maintenance
08.0732

信号检修 signal inspection, signal check-out maintenance 08.0733

信号开放 signal at clear 08.0050

信号开放表示 cleared signal indication 08.0163

信号控制电路 signal control circuit 08.0232

信号铃流发生器 tone and ringing generator
07.0081

信号楼 signal tower 06.0536

信号桥 signal bridge 08.0064

信号托架 signal bracket 08.0063

信号维修 signal maintenance 08.0723

信号握柄 signal lever 08.0292

信号无效标 signal out of order sign 08.0102

信号显示 signal aspect and indication 08.0046

信号选别器 signal slot 08.0312

信号整治 signal renovation 08.0734

信号质量检测 signal quality detection 07.0859

信号中修 signal intermediate repair 08.0736

信令电流 signaling current 07.0086

信令电路 signaling circuit 07.0077

信息 information 07.0003

信息反馈重发纠错 error correction by information
feed-back repetition 07.0841

信噪比 signal to noise ratio 07.0043

星形网[络] star network 07.0760

H 型柴油机 H-type diesel engine 04.0704

V 型柴油机 V-type diesel engine 04.0703

型钢混凝土梁 girder with rolled steel section encased in concrete, skeleton reinforced concrete girder
02.0378

型煤 moulded coal 04.1103

ω 型燃烧室 toroidal combustion chamber 04.0801

型式试验 type test 04.0280

形变压力 deformation pressure 02.0708

s 形辐板 s-type wheel plate, s-plate 05.0122

V 形桥墩 V-shaped pier 02.0567

U 形桥台 U-shaped abutment 02.0551

行包事故 luggage and parcel traffic accident
06.0621

行包邮政地道 tunnel for luggage and postbag
06.0523

行车闭塞法 train block system 06.0237

行车记录设备 train movement recording equipment
08.0487

行车凭证 running token 06.0241

行车事故 train operation accident 06.0618

行车信号 train signal 08.0018

行车信号机 train signal 08.0077

行车指挥自动化 automation of traffic control 01.0111

行车中断 traffic interruption 06.0635

行程缸径比 stroke/bore ratio 04.0709

行李 luggage, baggage 06.0015

行李包裹承运 acceptance of luggages and parcels 06.0052

行李包裹交付 dilivery of luggages and parcels 06.0053

行李包裹托运 consigning of luggages and parcels 06.0051

行李车 baggage car, luggage van 05.0736

行李房 luggage office, baggage office 06.0018

行李架 baggage rack, luggage rack 05.0768

行李票 luggage ticket, baggage ticket 06.0049

行李室 baggage room, luggage compartment 05.0762

*行修 running inspection 04.0249

性能试验 performance test 04.0281

修补木枕 tie repairing 03.0217

修车库 freight car temporary repairing shed 05.0899

*修车棚 freight car temporary repairing shed 05.0899

修车台位长度 length of repair position 05.0903

修车线 repair siding 05.0941

修正功率 corrected power 04.0812

修正总概算 amended sum of approximate estimate, revised general estimate 02.0023

虚电路 virtual circuit 07.0780

虚呼叫 virtual call 07.0873

虚拟终端 virtual terminal 07.0910

蓄电池 storage battery, accumulator 05.0707

蓄电池充电系统试验 checks of battery charging-arrangement 04.0664

蓄电池电路 battery circuit 04.0379

蓄电池供电 storage battery power supply 08.0640

蓄电池室 battery room 08.0717

蓄电池箱 storage battery box, accumulator box 05.0708

蓄能制动 energy-storing brake 04.0120

悬板桥 stressed ribbon bridge 02.0392

悬臂灌注法 cast-in-place cantilever construction, free cantilever segmental concreting with suspended formwork 02.0536

悬臂架设法 cantilever erection, erection by protrusion 02.0535

悬臂梁桥 cantilever beam bridge 02.0381

悬臂拼装法 cantilevered assembling construction, free cantilever erection with segments of precast concrete 02.0537

悬臂式棚洞 cantilever shed tunnel, cantilever shed gallery 02.0702

悬臂式铺轨机 rail laying machine with cantilever 03.0442

悬臂式铺轨排机 track panel laying machine with cantilever 03.0441

*悬带桥 stressed ribbon bridge 02.0392

悬吊滑轮 suspension pulley 04.1418

悬浮导向分别控制 separate control lift/guidance 04.1632

悬浮导向兼用 combined lift and guidance 04.1618

悬浮力 lift force 04.1597

悬浮系统 suspension system 04.1589

悬浮系统动力学 dynamics of suspension system 04.1621

悬浮组件 suspension module 04.1591

悬接接头 suspended joint 03.0069

悬距 clearance, air-gap 04.1615

悬空脚手架 hanging stage, hanging scaffold 02.0624

悬跨 suspended span 02.0442

悬球 ball suspension 04.1614

悬索桥 suspension bridge 02.0393

旋风燃烧 cyclone-combustion 04.1567

旋风筒式空气滤清器 cyclone type air filter 04.1048

旋流器 swirler 04.1575

旋流燃烧 swirl-flow combustion 04.1566

旋转法施工 erection by swing method 02.0544

旋转号盘电话机 rotary dial telephone set 07.0094

旋转式拨号盘 rotary dial 07.0084

旋转式钻机 swiveling drill machine, rotary drill machine 02.0912

旋转腕臂 hinged cantilever 04.1382

选呼 selective call 07.0616

选件 option 07.0750

选控信号 selectivity signal 07.0613

选路 route selection 08.0191

选路制信号 route signaling 08.0017

选频电平表 selective level meter 07.0357

* 选项 option 07.0750

选择 selecting 07.0853

选择性 selectivity 07.0035

学生票 student ticket 06.0041

穴蚀 cavitation 04.0839

雪崩 snow slip, snow slide, snow avalanche 03.0485

雪害 snow blockade, snow drift 03.0496

雪害地段路基 subgrade in snow damage zone, sub-grade in snow disaster zone 02.0230

循环挡板 circulating baffle plate 05.0644

循环检查制 cyclic scanning system 08.0474

循环交路 loop routing 04.0191

循环码 cyclic code 07.0825

循环冗余检验 cyclic redundancy check, CRC 07.0797

循环直达列车 shuttled block train 06.0218

旬间装车计划 ten day car loading plan 06.0097

巡道工 track walker, track patrolling man 03.0007

巡道工无线电通信 radio communication for track walker 07.0641

Y

压电式加速度计 piezoelectric accelerometer 04.1611

压钩坡 coupler compression grade 06.0480

压紧装置 pinch device 05.0521

压块 pressing block 05.0560

压扩律 companding law 07.0375

压力传感器 pressure sensor 04.0632

压力风缸 pressure reservoir 05.0314

压力继电器 pressure relay 04.0596

压力升高比 pressure step-up ratio, rate of pressure rise 04.0755

压力式涵洞 outlet submerged culvert 02.0632

压力主管 pressure pipe, second brake pipe 05.0278

压路机 roller 02.0888

压气机耗功 power input to compressor 04.1561

压气机叶轮 blower impeller 04.0976

压气涡轮 compressor turbine 04.1540

压实标准 compacting criteria 02.0257

压实系数 compacting factor, compacting coefficient 02.0259

压式水阀 compression faucet 05.0781

压缩比 compression ratio 04.0751

压缩多变指数 polytropic index of compression 04.0750

压缩空气设备全面的气密性试验 test for over-all air-tightness of compressed air equipments 04.0658

压缩始点温度 compression beginning temperature 04.0748

压缩始点压力 compression beginning pressure 04.0746

压缩行程 compression stroke 04.0724

压缩终点温度 compression terminal temperature 04.0749

压缩终点压力 compression terminal pressure 04.0747

压油机 mechanical lubricator 04.1248

咽喉道岔 throat point 06.0512

咽喉区长度 throat length 06.0513

咽喉信号机 signal in throat section 08.0085

烟度 limit of smoke 04.0837

烟箱 smokebox 04.1137

烟箱大门 smokebox front 04.1151

烟箱管板 smokebox tube sheet 04.1138

烟箱小门 smokebox door 04.1152

盐渍土地区路基 subgrade in salty soil zone, sub-grade in saline soil region 02.0218

研究性试验 investigation test 04.0296

岩爆 rock burst 02.0994

岩堆地段路基 subgrade in rock deposit zone, sub-grade in talus zone, subgrade in scree zone

02.0225

岩块填料 rock block filler, rock filler, rock fill 02.0254

*岩溶地段路基 subgrade in karst zone 02.0227

岩石路基 rock subgrade 02.0211

延长杆 extension rod 05.0662

延续进路 successive route 08.0157

沿线走行公里 running kilometers on the road 04.0213

眼镜式开挖法 spectacles type tunnelling method 02.0742

扬弃爆破 abandoned blasting, abandonment blasting 02.0275

扬声电话机 loud speaking telephone set 07.0099

扬声器 loud-speaker 07.0073

阳[电]极 anode 04.1502

仰拱 invert, inverted arch 02.0679

仰坡 front slope 02.0665

养路表报 maintenance-of-way report and forms, track work forms 03.0261

养路电话 track maintenance telephone 07.0416

养路费用 maintenance of way expenditures 03.0265

养路工 trackman, machine operator, track mechanics 03.0006

养路工长 track foreman 03.0005

养路工区 track maintenance section, permanent way gang 03.0003

养路管理电脑系统 computer aided track maintenance and management system 03.0263

养路领工区 track subdivision, track maintenance subdivision 03.0002

养路领工员 track master, track supervisor 03.0004

样板工程 sample project 02.1056

摇鞍 swing rocker 04.1234

摇杆 main rod 04.1196

摇头振动 yawing [vibration], hunting [vibration] 01.0104

摇枕挡 bolster guide 05.0091

摇枕挡间隙 bogie bolster play 05.0230

摇枕吊 bolster hanger, swing hanger 05.0206

摇枕吊轴 swing hanger cross beam 05.0207

摇枕弹簧 bolster spring 05.0166

摇轴支座 rocker bearing 02.0518

遥测 telemetry 08.0456

遥调 remote regulation 08.0457

遥控 remote control 08.0454

遥控区段 remotely controlled section 08.0465

遥控转发 telecontrol repeat 07.0599

遥信 remote surveillance 08.0455

遥信区段 remotely surveillanced section 08.0472

药壶爆破 pot hole blasting 02.0278

要车计划表 car planned requisition list 06.0098

业务通信系统 service communication system 07.0330

叶型 blade profile 04.1546

叶型折转角 camber 04.1550

夜间信号 night signal 08.0008

*液力变矩器 hydraulic torque converter 04.1008

液力变扭器 hydraulic torque converter 04.1008

液力传动 hydraulic transmission 04.0990

液力传动内燃机车 diesel-hydraulic locomotive 04.0687

液力传动系统 hydraulic transmission system 04.0997

液力传动箱 hydraulic transmission gear box 04.0998

液力换向传动箱 hydrodynamic reverser 04.0999

液力机械传动 hydromechanical drive 04.0993

液力偶合器 hydraulic coupler 04.1006

液力气动式缓冲器 hydropneumatic draft gear 05.0553

液力循环元件 hydraulic unit 04.1005

液力制动 hydraulic brake 04.0116

液力制动操纵阀 hydrodynamic brake operating valve 04.1021

液力制动阀 hydrodynamic brake valve 04.1020

液力制动器 hydraulic brake 04.1007

液面高度指示牌 liquid level indicating plate, telltale 05.0474

液压传动 hydrostatic drive 04.0989

液压打桩机 hydraulic pile driver 02.0907

*液压捣固车 hydraulic tamping machine 03.0417

液压捣固机 hydraulic tamping machine 03.0417

液压减振器 hydraulic damper, oil damper 05.0163

液压式缓冲器 hydraulic draft gear 05.0552

液压式张拉千斤顶 hydraulic tensioning jack 02.0545

一般冲刷 general scour 02.0096

一般路基 general subgrade, ordinary subgrade 02.0244

一般事故 ordinary accident 06.0625

一般性检查 general inspection 04.0656

一次参数 primary parameter 08.0533

一次电池供电 primary cell power supply 08.0639

* 一次缓解 direct release 04.0136

一次货物作业平均停留时间 average detention time of local car for loading or unloading 06.0267

一次空气 primary air 04.1568

一次磨耗车轮 one-wear wheel 05.0103

一次群 primary group 07.0376

一阶段设计 one-step design, one-phase design 02.0013

一批货物 consignment 06.0124

一送多受 single feeding and multiple receiving track circuit 08.0516

一体从板座 rear draft check casting 05.0413

一体构架转向架 rigid frame truck 05.0063

一位侧 left side of car 05.0010

一位端 "B" end of car 05.0008

一系悬挂 single stage suspension 05.0176

一致性测试 conformance testing 07.0890

医疗车 hospital car 05.0748

移动备用方式 movable reservation system 04.1298

移动电台 mobile station 07.0554

移动杠杆 truck live lever 05.0186

移动牵引变电所 movable traction substation 04.1346

移动牵引变压器 movable traction transformer 04.1345

移动通信 mobile communication 07.0676

移动微波通信 mobile microwave communication 07.0667

移动信号 movable signal 08.0011

移交车 loaded cars to be delivered at junction sta-tions 06.0321

移频轨道电路 frequency-shift modulated track cir-cuit 08.0500

移频自动闭塞 automatic block with audio frequency shift modulated track circuit 08.0333

仪表电路 instrument circuit 04.0382

仪器检查 inspect by instrument 02.1047

以太网 ethernet 07.0773

易冻货物 freezable freight 06.0120

易腐货物 perishable freight 06.0173

易燃货物 inflammable freight 06.0119

易熔塞 fusible plug 04.1179

易行车 easy rolling car 06.0491

易行线 easy running track 06.0430

意外紧急制动 undesirable emergency braking, UDE 04.0134

溢呼 overflow call 07.0264

译码器 decoder 07.0399

异步传输 asynchronous transmission 07.0787

异步电动机 asynchronous motor 04.0409

异步辅助电动机 asynchronous auxiliary motor 04.0473

异步通信 asynchronous communication 07.0779

异侧下锚 different-side anchor 04.1433

异频单工无线电通信 different-frequency simplex radio communication 07.0514

异型接头 compromise joint 03.0074

异型接头夹板 compromise joint bar 03.0082

异形轨 compromise rail 03.0059

翼轨 wing rail 03.0358

翼墙 wing wall 02.0556

翼墙式洞门 wing wall tunnel portal 02.0670

翼缘 flange 02.0445

翼缘板 flange plate 02.0446

音控防鸣电路 voice-operated anti-singing circuit 07.0150

音控门限电平 threshold level of voice-operated cir-cuit 07.0151

音量 volume 07.0069

音量表 volume meter 07.0147

音频 audio frequency 07.0012

音频通讯电路 audio communication circuit 04.0390

音频选叫　VF selective calling　07.0421

音频终端装置　audio frequency terminating set
07.0332

音频转接段　audio frequency section　07.0282

音频组合选呼　selective call with audio frequency
coding　07.0617

音选调度电话分机　dispatching telephone subset
with VF selective calling　07.0426

音选调度电话汇接分配器　tandem distributor for
dispatching telephone with VF selective calling,
tandem distributor　07.0427

音选调度电话滤波器　bridging filter for dispatching
telephone with VF selective calling　07.0433

音选调度电话总机　dispatching telephone control
board with VF selective calling　07.0425

音选双向增音机　two-way repeater for VF selective
calling, two-way repeater　07.0432

音选同线电话分机　party-line telephone subset with
VF selective calling　07.0431

音选同线电话分配器　party-line telephone distributor for VF selective calling　07.0430

音选同线电话总机　party-line telephone control
board with VF selective calling　07.0429

引导信号　calling-on signal　08.0029

引前相供电臂　leading phase feeding section
04.1307

引桥　approach spans　02.0424

引入架　lead-in rack　07.0351

引入试验架　lead-in test rack　07.0353

隐蔽工程　hidden project, hidden construction work
02.0840

隐蔽工程检查　hidden project inspection　02.1043

印制电路板插座　printed circuit board socket
04.0609

鹰架式架设法　erection with scaffolding　02.0534

*应变计　strain gage, strainometer　03.0536

应变时效　strain ageing　02.0468

应变仪　strain gage, strainometer　03.0536

应急电缆　emergency cable　07.0182

应急救灾无线电通信　radio communication for
emergency purpose　07.0644

应急卫星通信　emergency satellite communication
07.0670

营业外支出　non-operating outlay　06.0729

营业站　operating station　06.0080

影响电流　influencing current　04.1468

硬点　hard spot　04.1442

硬横跨　portal structure　04.1405

硬卧车　semi-cushioned berth sleeping car, semi-cushioned couchette　05.0727

硬座车　semi-cushioned seat coach　05.0725

永磁悬浮系统　permanent magnet suspension system, PMM　04.1582

永久性桥　permanent bridge　02.0429

永久虚[拟]电路　permanent virtual circuit
07.0851

用户电报　telex subscriber's telegraph　07.0701

用户引入线　subscriber's lead-in　07.0132

用户主机　subscriber's main station　07.0133

优化操纵　optimum handling, optimum operation
04.0236

*优良工程　high grade project, high quality project
02.1057

优质工程　high grade project, high quality project
02.1057

邮政车　postal car, mail van　05.0737

邮政间　post office compartment　05.0760

油尺　fuel dipstick　04.1071

油导筒式轴箱定位装置　oil guide cylinder type journal box positioning device　05.0148

油底壳　oil pan, oil sump　04.0868

油分离器　oil extractor　05.0637

油环　oil ring　04.0908

油浸式牵引变压器　oil-immersed type traction transformer　04.0487

油卷　lubricating roll　05.0138

油气分离器　oil separator　04.0870

油水分离器　oil-water separator　04.0174

油位表　fuel level gage　04.1070

油线室　journal box packing room　05.0895

油压动力室　hydraulic pressure engine room
08.0722

油压继电器　oil-pressure relay　04.0603

*油压减振器　hydraulic damper, oil damper
05.0163

油浴式空气滤清器　oil bath air filter　04.1046

油枕　treated wooden tie　03.0113

游车　idle car　06.0166

有碴轨道　ballasted track　03.0013

有调中转车　transit car with resorting　06.0234

有调中转车停留时间　detention time of car in transit with resorting　06.0262

有缝线路　jointed track　03.0016

有盖漏斗车　covered hopper car　05.0809

有箍车轮　tired wheel, tyred wheel　05.0097

有轨运输　track transportation　02.0757

有害地段　harmful district　02.0189

有极继电器　polarized relay　08.0592

有人增音机　attended repeater　07.0324

有效功率　effective power　04.0814

有效燃油消耗率　effective specific fuel consumption　04.0766

有效热效率　useful thermal efficiency　04.0828

有效容积　effective volume　05.0040

有效压缩比　effective compression ratio　04.0752

有形损耗　tangible wear　06.0699

有源降压装置　active degrade voltage apparatus　04.1501

有载自动调节　autoregulation on load　04.1302

有中梁底架　underframe with center sill　05.0388

右开道岔　right hand turnout　03.0289

迂回进路　detour route, alternative route　08.0154

迂回线　round about line　06.0420

迂回运输　round about traffic, circuitous traffic　06.0109

迂回中继　alternative trunking　07.0265

淤积　silting, siltation　02.0637

余速损失　leaving velocity loss　04.1555

鱼腹梁　fish-belly sill　05.0422

*鱼尾板　joint bar, splice bar, fish plate　03.0077

*鱼尾螺栓　track bolt, fish bolt　03.0083

雨季施工　raining season construction, rainy season construction　02.0844

雨量计　rain gage　03.0583

雨檐　eaves　05.0465

预办闭塞　preworking a block　08.0348

预防维修　preventive maintenance　08.0726

预防维修制　preventive maintenance system　04.0255

预告标　warning signs for approaching a station　08.0101

预告信号　approaching signal　08.0030

预告信号机　distant signal　08.0076

预可行性研究　pre-feasibility study　02.0007

预裂爆破　presplit blasting　02.0749

预留变形量　deformation allowance　02.0717

预留沉落量　reserve settlement, settlement allowance　02.0264

预排进路　presetting of a route　08.0188

预切槽　precutting trough　02.0784

预燃室　precombustion chamber　04.0922

预燃室喷嘴　precombustion chamber nozzle　04.0923

预燃室式燃烧室　precombustion chamber, pre-chamber　04.0799

预热锅炉　preheating boiler　04.1033

预算定额　rating of budget, rating form for budget　02.0026

预应力混凝土桥　prestressed concrete bridge　02.0371

预应力筋　tendon　02.0479

预制钢壳钻孔基础　prefabricated steel shell bored foundation　02.0588

预制混凝土构件　precast concrete units, precast concrete members　02.0538

预制桩　precast pile　02.0594

原位测试　*in situ* test　02.0055

圆端形桥墩　round-ended pier　02.0568

圆曲线　circular curve　02.0172

圆形桥墩　circular pier　02.0569

圆柱滚子轴承　cylindrical roller bearing　05.0141

圆锥滚子轴承　tapered roller bearing　05.0140

源点地址　source address　07.0901

远场　far field　07.0460

远场区　far field area　07.0546

远程监督分区　relayed surveillanced subsection　08.0476

远程网路　relayed surveillanced network　08.0478

*远程遥信分区　relayed surveillanced subsection　08.0476

远程终端　remote terminal　07.0908

远动工区　work district for telemechanical system

Z

杂音测试器　psophometer　07.0354

杂音抑制器　noise suppressor　04.1499

杂音抑制线圈　noise suppression coil　04.1498

灾害性地质　disaster geology　02.0990

载波　carrier [wave]　07.0021

载波电报终端机　carrier telegraph terminal　07.0695

载波电话增音机　carrier telephone repeater　07.0323

载波电话终端机　carrier telephone terminal　07.0322

载波调度电话中继器　carrier adaptor for dispatching telephone　07.0287

载波通信　carrier communication　07.0315

载波遥接话路　carrier channel connected to telephone line　07.0286

载供系统　carrier supply system　07.0326

载荷比例阀　load proportional valve　05.0303

载荷传感阀　load sensor valve　05.0302

载流承力索　current-carrying catenary　04.1312

载漏　carrier leak　07.0306

载频　carrier frequency　07.0304

载频放大器　carrier amplifier　07.0338

载频同步　carrier frequency synchronization　07.0313

载噪比　carrier-to-noise ratio　07.0523

载重　loading capacity　05.0862

再充风　recharging　04.0143

*再充气　recharging　04.0143

再生制动　regenerative brake　04.0118

再生中继　regeneration and repetition　07.0379

再用轨　second hand rail, relaying rail　03.0243

再振铃　re-ringing　07.0123

在线修　on line maintenance　08.0729

凿岩机　air hammer drill, rock drill　02.0890

凿岩台车　rock drilling jumbo　02.0962

噪声　noise　07.0037

造桥机　bridge fabrication machine　02.0951

责任事故　responsible accident　06.0616

增量调制　delta modulation　07.0026

增压　supercharging　04.0785

增压比　supercharging ratio　04.0787

增压柴油机　supercharged diesel engine　04.0696

增压度　degree of turbocharging, degree of supercharging　04.0794

增压器出口温度　discharge temperature of supercharger　04.0793

增压器机油滤清器　oil precision filter for turbocharger　04.1044

增压器配机试验　turbocharger matching test　04.0862

[增压器]压气机　blower　04.0974

增压压力　supercharging pressure　04.0786

增压中冷　charge inter-cooling　04.0792

增益　gain　07.0031

增音站　repeater station　07.0319

闸板阀　gate valve　04.1023

闸板转换阀　gate change-over valve　04.1024

闸片　brake lining, brake pad　05.0202

闸瓦　brake shoe　05.0193

闸瓦背　brake shoe back　05.0198

闸瓦插销　brake shoe key　05.0200

闸瓦间隙自动调节器　automatic slack adjuster　05.0360

闸瓦试验台　[brake shoe] inertia dynamometer　05.0995

闸瓦托　brake head　05.0192

闸瓦压力　brake shoe pressure　05.0335

摘车修　car detached repair　05.0931

摘挂调车　detaching and attaching of cars　06.0257

摘挂列车　pick-up and drop train　06.0224

摘机状态　off-hook　07.0118

窄带干扰　narrowband interference　04.1507

窄轨铁路　narrow-gage railway　01.0012

斩波调压 chopper control 04.0323

*斩波器 magnet driver, chopper 04.1593

辗钢车轮 wrought steel wheel, rolled steel wheel 05.0102

展线 extension of line, development of line, line development 02.0163

展线系数 coefficient of extension line, coefficient of developed line 02.0164

占线表示 occupancy indication 08.0485

站场感应通信 inductive communication in yard and station 07.0620

站场监视电视 monitor TV for yard and station 07.0655

站场排水 yard drainage 06.0527

站场通信 station-yard communication 07.0436

站场无线电通信 station-yard radio communication 07.0619

站场无线电中继转发台 radio relay set in yard and station 07.0625

站场型网路 geographical circuitry 08.0218

站调楼 yard controller's tower 06.0530

站间联系电路 liaison circuit between stations 08.0240

站间行车电话 interstation train operation telephone, blocking telephone 07.0415

站界 station limit 06.0231

站界标 station limit sign 08.0100

站内道口联系电路 liaison circuit with highway crossings within the station 08.0243

站坪 station site 06.0516

站坪长度 length of station site 02.0168

站坪坡度 grade of station site 02.0169

站前广场 station square 06.0524

站台 platform 06.0520

站台票 platform ticket 06.0039

站线 siding, station track, yard track 06.0404

站修 repair track maintenance 05.0929

*站修所 freight car repairing point 05.0906

站修线 repair track in station, repair siding 05.0938

张力增量 tension increment 04.1437

涨圈式密封 piston ring type seal 05.0145

丈量 measure 02.1046

找顶 top cleaning 02.0755

找小坑 spot surfacing 03.0197

照查锁闭 check locking 08.0129

照明灯 illuminating lamp 04.0648

照明电路 lighting circuit 04.0381

照明稳压器 illumination voltage stabilizer 05.0701

照明装置 illumination equipment 05.0672

兆欧表 megger 07.0219

遮断比 cut-off 04.1092

遮断信号 obstruction signal 08.0032

遮断信号按钮 obstruction signal button 08.0453

遮断信号机 obstruction signal 08.0081

遮断预告信号机 approach obstruction signal 08.0082

遮阳板 sun-shield 04.1068

折叠座椅 folding seat 05.0774

折角塞门 angle cock 05.0282

*折棚门 gangway door 05.0487

折射率分布 refractive-index profile 07.0448

辙叉 frog, crossing 03.0307

辙叉跟长 heel length of frog 03.0334

辙叉跟端 heel end of frog, frog heel 03.0323

辙叉跟宽 heel spread of frog 03.0336

辙叉号数 frog number 03.0309

辙叉角 frog angle 03.0308

辙叉心轨尖端 actual point of frog 03.0321

辙叉心轨理论尖端 theoretical point of frog 03.0320

辙叉咽喉 throat of frog 03.0342

辙叉有害空间 gap in the frog, open throat, unguarded flange-way 03.0340

辙叉趾长 toe length of frog 03.0333

辙叉趾端 toe end of frog, frog toe 03.0322

辙叉趾宽 toe spread of frog 03.0335

真空缸 vacuum chamber 05.0368

真空水泵 vacuum pump 02.0917

真空制动 vacuum brake 04.0114

真空制动装置 vacuum brake equipment 05.0267

真空主断路器 line vacuum circuit-breaker 04.0530

枕盒 crib 03.0119

枕间夯实机 crib consolidating machine 03.0423

枕梁　body bolster　05.0396

＊枕木　wooden tie　03.0112

＊枕木盒　crib　03.0119

枕木塞　tie plug　03.0221

枕形壁灯　pillow shaped wall lamp　05.0689

振荡器　oscillator　07.0055

振动沉拔桩机　vibro-driver extractor　02.0909

振动打桩机　vibrating pile driver, vibratory driver　02.0908

振动辗压机　vibration compactor, vibration roller　02.0887

振铃　ringing　07.0122

振铃器　signaling equipment　07.0333

振铃信号振荡器　ringing signal oscillator　07.0344

蒸发量　evaporation capacity　04.1095

蒸发率　rate of evaporation　04.1097

蒸发器　evaporator　05.0632

蒸汽采暖装置　steam heating equipment　05.0592

蒸汽打桩机　steam pile driver, steam pile hammer　02.0905

蒸汽机车　steam locomotive　04.0005

蒸汽机车连杆　steam locomotive side rod　04.1197

蒸汽机车热工特性　thermo-characteristic of steam locomotive　04.1094

蒸汽机车洗修　steam locomotive boiler washout repair　04.0244

蒸汽塔　turret　04.1180

征地　expropriation　02.0832

整备品重量　servicing weight　05.0864

整备线　servicing siding　06.0424

整备线配置系数　allocation factor of service track　05.0966

整车分卸　car load freight unloaded at two or more stations　06.0123

整车货物　car load freight　06.0111

整理道床　ballast trimming　03.0226

整流变压器　rectifier transformer　04.0488

整流方式　mode of rectification　04.0325

整流继电器　rectifier relay　08.0590

整流器　rectifier　07.0048

整流器供电　rectifier power supply　08.0641

整体车轮　solid wheel, mono-bloc wheel　05.0096

整体承载结构　monocoque structure, integral loadcarrying structure　05.0384

整体道床　solid bed, integrated ballast bed, monolithic concrete bed　03.0121

整体活塞　one-piece piston　04.0900

整体列车　integral train　05.0841

整体式衬砌　integral lining　02.0681

整正轨缝　dispersal of rail gaps, adjusting joint gaps up to standard　03.0209

整正曲线　curve adjusting, curve lining　03.0205

整正水平　adjusting of cross level　03.0204

整治冻害轨道　treatment of frost heaving track　03.0510

正常动作继电器　normal acting relay　08.0601

正定位　pull-off mode　04.1396

正洞门　orthonormal tunnel portal, straight tunnel portal　02.0673

正交桥　right bridge　02.0412

正交异性板　orthotropic plate　02.0509

＊正面坡　front slope　02.0665

正桥　main bridge　02.0423

正台阶法　positive benching tunnelling method　02.0731

正文　text　07.0741

正文结束信号　end-of-text signal　07.0838

正文开始信号　start-of-text signal　07.0862

正向传播　forward propagation　07.0535

正向信道　forward channel　07.0802

正装　right-handed machine　08.0616

帧格式　frame format　07.0897

帧首定界符　frame start delimiter　07.0899

支承桩　bearing pile　02.0598

支持绝缘子　supporting insulator　04.0541

支持器　supporter　04.1398

支电话所　minor telephone office　07.0388

支配机车　disposal locomotive　04.0021

支线　branch line　01.0038

支柱侧面限界　mast gauge　04.1439

支座　bearing　02.0514

直插用户　direct plug-in subscriber　07.0401

直达运输　through traffic　06.0103

直辐板　straight wheel plate　05.0125

直轨器　rail straightening tool, rail straightener

03.0397

直角拐肘　right angle crank　08.0298

直角交叉　rectangular crossing, square crossing
03.0292

直角截止阀　angle shut-off valve　05.0651

直接供电方式　direct feeding system　04.1275

直接缓解　direct release　04.0136

直接进入阀　direct admission valve　05.0366

直接喷射燃烧室　direct injection combustion chamber　04.0797

直接驱动　gearless drive, direct drive　04.0357

直结轨道　track fastened directly to steel girders
02.0512

直链形悬挂　polygonal catenary　04.1356

直列式柴油机　straight type diesel engine, in line
type diesel engine　04.0701

直流传动　DC drive　04.0341

直流电力机车　DC electric locomotive　04.0300

直流电力牵引制　DC electric traction system
04.1269

直流电流互感器　DC current transformer　04.0635

直流电桥　direct current bridge　07.0220

直流电源屏　DC power supply panel　08.0652

直流辅助电动机　DC auxiliary motor　04.0472

直流辅助发电机　DC auxiliary generator　04.0466

直流高速断路器　DC high speed circuit-breaker
04.0532

直流供电制　DC power supply system　08.0630

直流轨道电路　DC track circuit　08.0490

直流继电器　DC relay　08.0580

直流接触器　DC contactor　04.0546

直流控制发电机　DC control generator　04.0468

直流励磁机　DC exciter　04.0469

直流起动发电机　DC starting generator, dynastarter　04.0467

直流牵引变电所　DC traction substation　04.1343

直流牵引电动机　DC traction motor　04.0395

直流燃烧　straight-flow combustion　04.1564

直流斩波器　DC chopper　04.0330

直流制　DC system　04.0313

直流主发电机　DC main generator　04.0463

直埋电缆　buried cable　07.0184

直通场　through yard　06.0442

直通交路　through routing　04.0195

直通截止阀　through shut-off valve　05.0648

直通客流　through passenger flow　06.0024

直通空气制动装置　straight air brake equipment
05.0271

直通列车　transit train　06.0222

直通旅客列车　through passenger train　06.0207

直通式货运站　through-type freight station
06.0388

直通制动管　direct air brake pipe　05.0279

直线电动机　linear motor　04.0415

*直线感应电动机　linear asynchronous motor,
linear induction motor　04.0417

直线尖轨　straight switch　03.0314

直线同步电动机　linear synchronous motor
04.0416

直线异步电动机　linear asynchronous motor, linear
induction motor　04.0417

*直悬杆式刚性拱刚性梁桥　Lohse bridge, rigid
arch bridge with rigid tie and vertical suspenders
02.0405

直压式采暖装置　direct pressure steam heating equipment　05.0593

直压式暖汽调整阀　car pressure regulater　05.0595

K值试验　test for K value of complete car
05.0988

指导司机　driver instructor　04.0203

指导性施工组织设计　design for guiding construction organization, guiding design for construction
scheme　02.0854

指挥电话总机　command telephone control board
07.0434

指令电路　command circuit　04.0380

指示灯　indicator lamp　04.0649

指示功　indicated work　04.0764

指示功率　indicated power　04.0813

指示牵引力　indicated tractive effort　04.1087

指示燃油消耗率　indicated specific fuel consumption
04.0765

指示热效率　indicated thermal efficiency　04.0827

指示压力系数　coefficient of indicated pressure
04.1091

止回阀　boiler check valve　04.1173

止轮器　wheel skid　06.0668

止水带　water stop tie　02.0799

止推环　thrust ring　04.0884

止推轴承盖　thrust bearing cap　04.0881

止推轴瓦　thrust bearing shell　04.0883

纸带发报机　tape transmitter　07.0692

纸浆车　pulp car　05.0824

纸上定线　paper location of line　02.0156

制动　braking　04.0128

制动安定性　service stability　05.0353

制动倍率　leverage ratio　05.0338 ·

制动波速　braking propagation rate　04.0159

制动不衰竭性　inexhaustibility　05.0355

制动撑架　braking bracing　02.0460

制动初速　initial speed at brake application
　　05.0349

制动电抗器　braking reactor　04.0509

制动电路　braking circuit　04.0373

制动电阻柜　braking resistor cubicle　04.0521

制动电阻器　braking resistor　04.0516

制动电阻元件　braking resistor grid　04.0522

制动墩　braking pier　02.0565

制动方式　brake mode　04.0112

制动缸　brake cylinder　05.0307

制动缸后杠杆　auxiliary lever　05.0357

制动缸活塞行程　piston travel　05.0342

制动缸前杠杆　cylinder lever　05.0356

[制动缸压力]保持阀　retaining valve　05.0319

制动杠杆传动效率　brake rigging efficiency
　　05.0340

制动管减压量　brake pipe pressure reduction
　　04.0147

制动管路　brake piping　05.0273

制动机　brake　制动检修所　brake inspection point
　　05.0912

制动接触器　braking contactor　04.0552

制动距离　stopping distance　05.0346

制动拉杆　lever connection　05.0187

制动力　braking force　05.0336

制动梁　brake beam　05.0189

制动梁槽钢　brake beam compression channel
　　05.0203

制动梁弓形杆　brake beam tension rod　05.0204

制动梁拉杆　brake beam pull rod　05.0190

制动梁下拉杆　brake beam bottom rod　05.0191

制动梁支柱　brake beam strut　05.0205

制动灵敏度　sensitivity　05.0351

制动率　braking ratio　05.0337

制动能高　velocity hump crest of retarder　06.0477

制动盘　brake disc　05.0201

制动软管　hose　05.0281

制动软管连接器　[brake] hose coupling　05.0280

制动时间　braking time　05.0343

制动室　brake repair room　05.0897

制动试验　brake test　04.0284

制动衰竭性　exhaustibility　05.0354

制动铁鞋　brake shoe, skate　06.0465

制动位　retarder location　06.0469

制动稳定性　insensitivity　05.0352

制动系统　brake system　05.0262

制动效率　brake efficiency　05.0339

制动员室　brakeman's cabin　06.0533

制动支管　brake branch pipe　05.0275

制动支管三通　branch pipe tee　05.0285

制动主管　brake pipe　05.0274

[制动]柱塞　spool　05.0298

制动转矩　braking torque　04.0424

制动装置　brake equipment, brake gear　05.0263

制际串音　inter-system crosstalk　07.0361

制冷加温装置　refrigeration and heating equipment
　　05.0647

质量保证　quality assurance　02.1038

质量管理　quality management　02.1037

质量监督　quality superintendence, quality surveil-
　　lance　02.1041

质量控制　quality control　02.1039

质量体系　quality system　02.1040

滞后相供电臂　lagging phase feeding section
　　04.1306

滞燃期　combustion lagging period, combusting delay
　　period　04.0782

中长隧道　medium tunnel　02.0658

*中承式桥　half through bridge, midheight deck
　　bridge　02.0426

中和变压器　neutralizing transformer　04.1495

中和轨温　neutral temperature　03.0229

中－活载　CR-live loading, China railway standard live loading　02.0325
中继传输方式　relay transmission mode　07.0551
中继电路　trunk circuit　07.0261
*中继器　trunk circuit　07.0261
中继站　repeater station　08.0463
中间齿轮箱　intermediate gear box　04.1001
中间电缆盒　intermediate cable terminal box　08.0288
中间工艺检验　intermediate inspection at the technological process　04.0272
中间缓冲装置　intermediate buffer　04.1245
中间继电器　intermediate relay　04.0593
*中间联结零件　rail fastening　03.0085
中间配线架　intermadiate distributing frame　07.0129
中间坡　intermediate grade　06.0496
中间漆　intermediate coat　03.0552
中间牵引装置　intermediate draw gear　04.1244
中间设备　intermediate equipment　07.0878
中间试验　intermediate test　05.0987
中间体　pipe bracket　05.0296
中间线路滤波器　intermediate line filter　07.0291
中间验收　intermediate acceptance　02.1053
中间站　intermediate station　06.0365
中间站公务电话　interstation telephone　07.0411
中间柱　single suspension mast　04.1407
中冷器　intercooler　04.0980
中梁　center sill　05.0392
中梁悬臂部　overhang, cantilever portion of center sill　05.0393
中磷闸瓦　medium phosphor cast iron brake shoe　05.0195
*中平　center stake leveling　02.0117
中桥　medium bridge　02.0418
中速柴油机　medium speed diesel engine　04.0694
中途返回　midway return operation　08.0193
*中途折返　midway return operation　08.0193
中文译码机　Chinese character code translation equipment　07.0693
*中误差　mean square error　02.0110
中线测量　center line survey　02.0112
中线桩　center line stake　02.0113

中心电话所　central telephone office　07.0386
中心电台　center station　07.0559
中心控制　centralized control　08.0469
中心锚结　mid-point anchor　04.1425
中心锚结线夹　mid-point anchor clamp　04.1426
中心排油阀　central oil outlet valve　05.0483
中心销　center pin, king pin, pivot pin　05.0211
中心柱　center mast　04.1411
中型清筛机　medium ballast undercutting cleaners　03.0430
中行车　middle rolling car　06.0490
中央电池　central battery　07.0088
中央缓冲器　central draft gear　05.0555
中央楔块　center wedge　05.0561
中转车平均停留时间　average detention time of car in transit　06.0265
中桩高程测量　center stake leveling　02.0117
终点地址　destination address　07.0902
终端电缆盒　cable terminal box　08.0289
终端杆　terminal pole　07.0167
终端效应　terminal effect　04.1483
终端站　terminal station　07.0321
终夜灯　whole night lamp　05.0685
重车　loaded car　05.0013
重车重心　center of gravity for car loaded　06.0170
重车重心高　center height of gravity for car loaded　06.0171
重锤夯实法　heavy tamping method　02.0614
重大事故　grave accident　06.0622
*重力除水阀　tee trap　05.0594
重力继电器　gravitation type relay　08.0587
重力式挡土墙　gravity retaining wall　02.0285
重力式桥墩　gravity pier　02.0562
重力式桥台　gravity abutment　02.0548
重型轨道车　heavy rail motor car, heavy motor trolley　03.0434
*重型线路机械　heavy permanent way machine, large permanent way machine　03.0381
重载列车　heavy haul train　01.0081
重载铁路　heavy haul railway　01.0018
重质货物　heavy freight　06.0122
舟桥　bateau bridge　02.0953

周转轨 inventory stock rails 03.0241

*洲际铁路 transcontinental railway, inter-continental railway, land-railway 01.0122

轴承端盖 roller bearing end cap 05.0143

轴端轴承 axle end bearing 04.1073

轴颈 journal 05.0110

轴颈后肩 journal back fillet 05.0111

轴颈中心距 distance between journal centers 05.0049

轴距 wheelbase 05.0044

轴列式 axle arrangement 04.0034

轴领 end collar, axle collar 05.0109

轴流式涡轮 axial flow turbine 04.1538

轴流式压气机 axial flow compressor 04.1541

轴流水泵 axial flow pump 02.0919

轴身 axle body 05.0114

*轴式 axle arrangement 04.0034

轴瓦 plain journal bearing 05.0133

轴瓦垫 journal bearing wedge 05.0134

轴温 journal temperature 05.0957

轴箱 journal box, axle box 05.0127

轴箱承台 [side frame] pedestal bearing boss 05.0089

轴箱挡 axle box guide 05.0088

轴箱导框 side frame pedestal 05.0087

轴箱导框间隙 axle box play 05.0229

轴箱盖 journal box lid 05.0129

轴箱后盖 journal box rear cover 05.0131

轴箱前盖 journal box front cover 05.0130

轴箱弹簧 journal box spring 05.0165

轴箱弹簧支柱 journal box spring guide post 05.0155

轴箱体 journal box body 05.0132

轴箱托板 pedestal brace 04.1217

轴箱轴承 axle box bearing 04.1072

轴向推力 axial thrust 04.1529

轴向位移保护装置 axial displacement limiting device 04.1535

轴载荷 load on axle journals 05.0231

*轴载荷转移 axle load transfer 04.0047

轴中央部 axle center 05.0115

轴重 axle load 01.0050

轴重转移 axle load transfer 04.0047

轴重转移补偿装置 axle load transfer compensation device 04.0588

轴装盘形制动 axle-mounted disc brake 05.0329

肘销 knuckle pin 04.1198

昼间信号 day signal 08.0007

昼夜通用信号 signal for day and night 08.0009

主按键开关组 main button switch group, main key switch set 04.0579

主车架片 main frame 04.1215

主串通路 disturbing channel 07.0136

主灯丝断丝报警 alarm for burnout of a main filament 08.0176

主电路 power circuit, main circuit 04.0368

主电路短路保护系统试验 test on shortcircuit protection system of main circuit 04.0671

主电路过载保护系统试验 test on overload protection system of main circuit 04.0672

主电路库用插座 main circuit socket for shed supply 04.0605

主电路库用转换开关 main circuit transfer switch for shed supply 04.0564

主电源 main power source 08.0634

主动齿轮 driving gear 04.1077

主动轮对 main driving wheel set 04.1227

主断路器 line circuit-breaker 04.0529

主发电机 main generator 04.0462

主阀 main valve 04.1161

主干电缆 main cable 07.0187

主河槽 main river channel 02.0074

*主呼 calling 07.0817

主机 leading locomotive 04.0226

主极铁心 mainpole core 04.0446

主极线圈 mainpole coil 04.0447

主叫 calling 07.0817

主叫控制复原方式 calling subscriber release 07.0407

主叫用户 calling party 07.0114

主控站 primary station 07.0854

主力 principal load 02.0330

主连杆 main connecting rod 04.0917

主梁中心距 center to center distance between main girder 02.0435

*主桥 main bridge 02.0423

主曲拐销 main crank pin 04.1228

主群 master group 07.0310

主弹性联轴节 main elastic coupling 04.1000

主体信号机 main signal 08.0074

主要编组站 main marshalling station 06.0379

主站 master station 07.0847

主振器 master oscillator 07.0343

主蒸汽管 steam pipe 04.1158

主整流柜隔离开关 isolating switch for main silicon rectifier cubicle 04.0570

主轴承 main bearing 04.0878

主轴承盖 main bearing cap 04.0880

主轴承螺母 main bearing nut 04.0886

主轴承螺栓 main bearing stud 04.0885

主轴承座 main bearing seat, main bearing housing 04.0879

主轴颈 crank journal, main journal 04.0888

主轴瓦 main bearing shell 04.0882

柱插 stake pocket 05.0415

柱塞 plunger 04.0949

柱塞偶件 plunger matching parts 04.0948

柱塞套 plunger sleeve 04.0950

柱式洞门 post tunnel portal 02.0669

柱式桥墩 column pier 02.0566

贮液桶 liquid receiver 05.0633

铸钢车轮 cast steel wheel 05.0101

铸铁闸瓦 cast iron brake shoe 05.0194

铸造机座 cast frame 04.0452

注浆泵 grouting pump, injection pump 02.0979

注浆机 grouting machine, grouter 02.0978

注水口 filling pipe end 05.0665

注意信号 caution signal 08.0021

专业性货运站 specialized freight station 06.0078

专用电话网 private telephone network 07.0108

专用电路 private circuit 07.0112

专用货车 special-purpose freight car 05.0793

专用基金 special fund 06.0753

专用集装箱 specific purpose container 06.0542

专用数据网 private data network 07.0754

专用铁路 special purpose railway 01.0036

专用通信网 private communication network 07.0062

专用线 private siding 06.0417

专运列车无线电通信 radio communication for special train 07.0646

转臂式轴箱定位装置 rocker type journal box positioning device 05.0152

转动车钩煤车 rotary dumping coal car 05.0804

转动式车钩 rotary dump coupler 05.0529

转动座椅 rotating seat 05.0772

转换塞门 cut-out cock 05.0597

转换时间 transfer time 08.0562

转换锁闭器 switch-and-lock mechanism 08.0302

转换柱 transition mast 04.1408

转极时间 pole-changing time 08.0563

转极值 pole-changing value 08.0570

转盘 turntable 04.0274

转速表传动装置 transmission gear of tachometer 04.0964

转速波动率 rate of speed fluctuation 04.0831

转速继电器 tachometric relay 04.0599

*转体施工 erection by swing method 02.0544

转向杆 deflecting bar 08.0297

转向架 truck, bogie 05.0057

转向架侧架 truck side frame 05.0080

转向架独立供电 bogie individual power supply 04.0366

转向架对角线 truck frame diagonal 05.0043

转向架构架 truck frame 05.0074

转向架换装所 truck changing point 05.0984

转向架基础制动 truck brake rigging 05.0184

转向架架承式牵引电动机 bogie mounted traction motor 04.0402

转向架扭曲刚度 truck rigidity against distorsion 05.0242

转向架式车 bogie car 05.0005

转向架摇枕 truck bolster 05.0090

转向架制动组件 truck-mounted brake assembly 05.0325

转向架中心 bogie pivot center 04.0035

转向架中心距离 distance between bogie pivot centers, bogie pivot pitch 04.0036

转向架组全轴距 wheelbase of combination truck 05.0047

转向角 deflection angle 02.0122

转向设备 turning facilities 04.0273

转向线 turn-around wye, Y-track 05.0940

转辙机 switch machine 08.0602

转辙机安装装置 switch machine installation 08.0629

转辙角 switch angle 03.0341

转辙器 switch 03.0344

转辙锁闭器 plunger lock 08.0301

转轴 shaft 04.0443

转子 rotor 04.0445

桩基础 pile foundation 02.0593

*装车功率 maximum service output power 04.0084

装车数 number of car loadings 06.0313

装车调整 adjustment of car loading 06.0339

装货口 hatch 05.0462

装配式衬砌 precast lining, prefabricated lining 02.0682

装卸搬运 handling 06.0579

装卸定额 handling quota 06.0585

装卸费率 rate of handling charge 06.0593

装卸工作单 handling sheet 06.0586

装卸换算吨 converted tons of handling 06.0588

装卸机械利用率 utilization ratio of machine handling 06.0591

装卸机械完好率 percentage of machine handling in good condition 06.0590

装卸机械作业量 handling volume by machine 06.0589

装卸检修所 inspection and service point for car before loading or after unloading 05.0911

装卸能力 handling capacity 06.0582

装卸事故 loading and unloading accident 06.0648

装卸线 loading and unloading track 06.0086

装卸自然吨 actual tons of handling 06.0587

装卸作业机械化 handling mechanization 06.0580

装卸作业量 handling volume 06.0584

装卸作业自动化 handling automation 06.0581

装卸作业组织 organization of handling 06.0583

装岩机 rock loader 02.0970

装载机 loader 02.0884

装载系数 loading coefficient 01.0073

装载限界 loading clearance limit, loading gauge 01.0062

状态检修 repair based on condition of component 04.0252

状态修 repair according to condition 05.0919

状态转移图 state transition diagram 07.0748

锥体护坡 quadrant revetment, truncated cone banking 02.0557

锥形反转出料混凝土搅拌机 tapered reverse tilting concrete mixer 02.0927

锥形倾翻出料混凝土搅拌机 tapered tilting concrete mixer 02.0928

追钩 catch up 08.0370

追踪列车间隔时间 time interval between trains spaced by automatic block signals 06.0302

追踪运行图 train diagram for automatic block signals 06.0295

坠铊 balance weight 04.1421

坠铊补偿器 balance weight tensioner 04.1422

缀板 stay plate, tie plate 02.0457

缀条 lacing bar 02.0456

准备进路 preparation of the route 06.0244

准峰值检波器 quasi-peak detector 07.0571

准轨铁路 standard-gage railway 01.0011

子车 car without axle generator 05.0751

*自保电路 self-stick circuit 08.0217

自闭电路 self-stick circuit 08.0217

自导向径向转向架 self-steering radial truck 05.0069

自动闭塞 automatic block system 08.0328

自动闭塞联系电路 liaison circuit with automatic blocks 08.0238

自动操纵作业 automatic operation 08.0377

自动抄车号 automatic car identification 08.0390

自动点灯 automatic lighting 08.0125

自动电话机 automatic telephone set 07.0093

自动电连接器 electric automatic coupler 04.0392

自动电平调节系统 automatic level regulating system 07.0327

自动高度调整装置 automatic leveling device 05.0180

自动呼叫 automatic calling 07.0813

自动化编组站 automatic marshalling station 06.0381

自动化驼峰 automatic hump 06.0458

自动化驼峰系统　automatic hump yard system
08.0364

自动缓解　false release by itself　08.0683

自动检错重发设备　automatic retransmission on
request equipment, ARQ equipment　07.0697

[自动]开闭器　switch circuit controller　08.0610

自动空气制动装置　automatic air brake equipment
05.0270

自动栏木　automatic operated barrier　08.0440

自动力矩检定　autostress rating　03.0520

自动喷调装置　injection timer　04.0946

*自动切换　automatic switching over　08.0649

自动撒砂　automatic sanding　04.0182

自动调压　automatic voltage regulation　08.0656

自动调整楔铁装置　automatic compensator
04.1226

自动停车　automatic train stop　08.0413

自动停车装置　automatic train stop equipment
08.0414

自动通过按钮电路　automatically through button
circuit　08.0252

自动限时解锁　automatic time release　08.0149

*自动楔铁调整器　automatic compensator
04.1226

自动液压大型捣固车　auto-leveling-lifting-lining-
tamping machine　03.0418

自动应答　automatic answering　07.0812

自动增益控制　automatic gain control　07.0036

*自动整平－起道－拨道－捣固车　auto-leveling-
lifting-lining-tamping machine　03.0418

自动制动阀　automatic brake valve　04.0167

自动转接　automatic switching over　08.0649

自复式按键开关　self-reset push-key switch
04.0577

自复式按钮　nonstick button　08.0270

自环检测　self-loop test　07.0598

自理装卸　handling by shipper-self　06.0592

自耦变压器　autotransformer, AT　04.1319

自耦变压器供电方式　autotransformer feeding system　04.1278

自耦变压器供电线　AT-feeder　04.1311

自耦变压器所　auto-transformer post, ATP
04.1348

自耦牵引变压器　traction autotransformer　04.0484

自然缓解　unintended release, undesired release
04.0138

自然通风　natural ventilation　02.0803

自然通风装置　natural ventilation equipment
05.0617

自然循环　natural circulation　04.0502

自然制动　unintended braking, undesired braking
04.0132

自适应控制　adaptive control　04.1594

自通风式电动机　self-ventilated motor　04.0413

自卸车　dumping car　05.0805

自卸汽车　dumping truck　02.0985

自由车轮　independent wheel　05.0882

自由内接　free inscribing·　03.0137

自整角机　synchro　04.0479

自重　light weight, tare weight　05.0861

自重系数　ratio of light weight to loading capacity
05.0859

字符　character　07.0819

字符检验　character check　07.0820

综合电缆　composite cable　07.0177

综合概算　comprehensive approximate estimate
02.0021

综合光缆　combined optical fiber cable　07.0481

综合架　composite rack　08.0262

综合屏蔽系数　combination shielding factor
04.1478

综合性货运站　general freight station, general goods
station　06.0077

综合运输　comprehensive transport, multi-mode
tranintermode transport　01.0120

综合折旧率　composite depreciation rate　06.0706

*总出发信号机　advance starting signal　08.0067

总出站信号机　advance starting signal　08.0067

总风缸　main air reservoir　04.0175

总概算　sum of approximate estimate, total estimate,
summary estimate　02.0022

总排量　total displacement　04.0719

总配线架　main distribution frame　07.0130

总容积　total volume　05.0039

总线网[络]　bus network　07.0772

总预算　total budget　02.0851

总重 gross weight 05.0863

总重吨公里 gross ton-kilometers 06.0691

纵电动势 longitudinal electro-motive force 04.1471

纵横奇偶检验 vertical horizontal parity 07.0799

纵横制电话交换机 crossbar telephone switching system 07.0394

纵梁 stringer 02.0462

纵列式区段站 longitudinal type district station 06.0368

纵向冲击 longitudinal impact 01.0107

纵向扼流线圈 longitudinal choke coil 07.0209

纵向钢筋 longitudinal reinforcement 02.0475

纵向轨枕 longitudinal tie 03.0110

纵向力 longitudinal force 05.0887

纵向振动 longitudinal vibration 01.0096

走板 running board 05.0417

走廊 corridor 05.0787

走线架 chute, chamfer 07.0128

*走行部 running gear 05.0056

走行部灯 bogie lamp 04.0644

走行装置 running gear 05.0056

租地 rented land 02.0834

租用电路 leased circuit 07.0111

阻车器 stop device 02.1029

阻抗电桥 impedance bridge 07.0221

[阻抗]匹配变压器 impedance matching transformer 07.0198

阻尼系数 damping coefficient 05.0244

阻尼因数 damping factor 03.0537

阻塞 congestion 07.0870

阻塞干扰 block interference 07.0531

组合 unit block 08.0259

组合衬砌 composite lining 02.0686

组合传动机车 coupled axle drive locomotive 04.0361

组合端子 terminals of a unit block 08.0279

组合钢模板 combined steel formwork 02.0957

组合柜 modular block rack 08.0263

组合活塞 composite piston 04.0901

组合继电器 combination relay 08.0594

组合架 unit block assembly rack 08.0261

组合接触器 grouping contactor 04.0549

组合列车 combined train 06.0220

组合驱动 coupled axle drive 04.0359

组合式电气集中联锁 unit-block type relay interlocking 08.0117

组合式信号机构 modular type signal mechanism 08.0087

组合式转向架 combination truck 05.0071

组合站 combined station 07.0822

组呼 group calling 07.0423

组拼式架桥机 assembly type girder-erecting machine 02.0950

组匣 modular block 08.0260

组匣端子 terminals of a modular block 08.0280

组匣柜 modular block rack 08.0264

组匣式电气集中联锁 modular type relay interlocking 08.0118

*钻孔台车 rock drilling jumbo 02.0962

钻孔桩 bored pile 02.0599

钻探 boring [prospecting], exploration drilling 02.0052

钻[眼]爆[破]法 drilling and blasting method 02.0721

最大常用减压 full service reduction 04.0151

最大常用制动 full service braking, full service application 04.0130

最大磁场 maximum field 04.0434

最大功率 maximum power 04.0816

最大坡度 maximum grade 02.0184

最大倾斜位置 maximum inclining position 05.0848

最大输出功率 maximum output power 04.0425

最大外移位置 maximum outward position 05.0847

最大误差 maximum error, limiting error 02.0111

最大限制信号 most restrictive signal 08.0041

最大允许信号 most favorable signal 08.0043

最大运用功率 maximum service output power 04.0084

最大轴重 maximum allowable axle load 01.0051

最低工作稳定转速试验 minimum steady speed test 04.0858

最低空载稳定转速 minimum idling stabilized speed 04.0830

最低空载转速试验　minimum no-load speed test　04.0859

最高爆发压力　maximum explosive pressure, peak pressure, maximum firing pressure　04.0754

最高燃烧温度　maximum combustion temperature　04.0753

最高水位　highest water level, HWL　02.0079

最高速度　maximum speed　01.0079

最佳含水量　optimum moisture content, best moisture content　02.0260

最佳密度　optimum density, best density　02.0261

最小磁场　minimum field　04.0435

最小接收场强　minimum receiving field strength　07.0519

最小可用接收电平　minimum available receiving level　07.0518

最小曲线半径　minimum radius of curve　02.0171

最小填筑高度　minimum fill height of subgrade, minimum height of fill　02.0245

最易行车　easiest rolling car　06.0492

左开道岔　left hand turnout　03.0288

＊左右水平　track cross level　03.0149

作业标志　working signal　03.0245

作用部　service portion　05.0292

作用阀　application valve　05.0516

坐标方位角　plane-coordinate azimuth　02.0124

坐式便器　western type toilet, seat-type water closet　05.0783

座式继电器　shelf-type relay　08.0585

座席电路　operator's circuit　07.0260

＊座席间中继线　interposition trunk　07.0405